Introduction to ENHANCED RECOVERY METHODS FOR HEAVY OIL AND TAR SANDS

Introduction to ENHANCED RECOVERY METHODS FOR HEAVY OIL AND TAR SANDS

Second Edition

DR. JAMES G. SPEIGHT
Overland Road,
Laramie, WY, USA

ELSEVIER

Amsterdam • Boston • Heidelberg • London
New York • Oxford • Paris • San Diego
San Francisco • Singapore • Sydney • Tokyo
Gulf Professional Publishing is an imprint of Elsevier

G|P
P|

Gulf Professional Publishing is an imprint of Elsevier
50 Hampshire Street, 5th Floor, Cambridge, MA 02139, USA
The Boulevard, Langford Lane, Kidlington, Oxford, OX5 1GB, UK

Library of Congress Cataloging-in-Publication Data

A catalog record for this book is available from the Library of Congress

British Library Cataloguing-in-Publication Data

A catalogue record for this book is available from the British Library

ISBN: 978-0-12-849906-1

For information on all Gulf Professional Publishing visit our website at http://www.elsevier.com/

CONTENTS

Preface to the Second Edition ix
Preface to the First Edition xi

Part I Reservoirs and Reservoir Fluids

1. Heavy Oil and Tar Sand Bitumen **3**
1.1 Introduction 3
1.2 History 4
1.3 Petroleum 12
1.4 Conventional Petroleum 13
1.5 Other Members of the Petroleum Family 22
1.6 Heavy Oil 28
1.7 Tar Sand Bitumen 33
1.8 Rationalization of the Definitions 39
References 45

2. Occurrence and Reservoir Evaluation **49**
2.1 Introduction 49
2.2 Origin of Petroleum and Heavy Oil 51
2.3 Occurrence and Distribution 54
2.4 Exploration 57
2.5 Reserves 68
2.6 Oil Pricing 77
References 82

3. Sediments, Reservoirs, and Deposits **85**
3.1 Introduction 85
3.2 Sediments 87
3.3 Occurrence 87
3.4 Mineralogy 89
3.5 Clastic Sediments and Sedimentary Rocks 91
3.6 Chemical Sediments and Sedimentary Rocks 92
3.7 Biochemical and Organic Sediments and Sedimentary Rocks 93
3.8 Diagenesis 94
3.9 Lithification 94
3.10 Texture 95

3.11 Packing 96
3.12 The Structure of Sediments 97
3.13 Transportation 97
3.14 Stratification and Bedding 98
3.15 Layering and Bedding 98
3.16 Sedimentary Facies 99
3.17 Other Sediment Markers 99
3.18 Classification 101
3.19 Reservoirs and Deposits 102
3.20 Reservoirs 104
3.21 Deposits 116
3.22 Reservoir Productivity 117
References 120

4. Reservoir Fluids 123
4.1 Introduction 123
4.2 Classes of Fluids 124
4.3 Water 131
4.4 Solids Deposition During Recovery 135
References 168

5. Evaluation of Reservoir Fluids 177
5.1 Introduction 177
5.2 Evaluation of Reservoir Fluids and Fluid Assay 181
5.3 Chemical and Physical Properties 189
5.4 Thermal Properties 209
5.5 Metals Content 224
5.6 Bulk Composition and Molecular Weight 228
References 242

Part II Oil Recovery

6. General Methods of Oil Recovery 253
6.1 Introduction 253
6.2 Primary Recovery Methods 256
6.3 Secondary Recovery Methods 264
6.4 Enhanced Oil Recovery Methods 275
6.5 Fluids Production 309
6.6 Microscopic Displacement Efficiency 310
6.7 Hydraulic Fracturing 313
References 319

7. Microbial Enhanced Oil Recovery **323**
7.1 Introduction 323
7.2 Microbial Alteration of Petroleum Constituents 325
7.3 Processes Types 341
References 345

8. Nonthermal Methods of Recovery **353**
8.1 Introduction 353
8.2 Primary Recovery Methods 356
8.3 Secondary Recovery Methods 360
8.4 Enhanced Recovery Methods 369
8.5 Tar Sand Mining 386
8.6 Nonmining Methods 396
References 400

9. Thermal Methods of Recovery **405**
9.1 Introduction 405
9.2 Thermal Stimulation 411
9.3 Steam-Based Processes 418
9.4 In Situ Combustion Processes 431
9.5 Comparison of Recovery Methods 446
References 449

10. Upgrading During Recovery **453**
10.1 Introduction 453
10.2 Upgrading In Situ 458
10.3 Partial Upgrading at the Surface 478
10.4 The Future 486
References 487

Appendix 493
Glossary 495
Index 539

PREFACE TO THE SECOND EDITION

The continued declining of the reserves of conventional (light) crude oil has resulted in an increasing need to develop options to upgrade the abundant supply of known heavy oil and tar sand bitumen reserves. In addition, there is considerable focus and renewed efforts on adapting recovery techniques to the production of heavy oil and tar sand bitumen.

Over the past decade, the demand for crude oil worldwide has substantially increased, straining the supply of conventional (light) oil. Recent price increases and price volatility have emphasized the need for consideration of alternative or insufficiently utilized energy sources, especially heavy crude oil and bitumen, to supplement short- and long-term needs. Heavy oil has been used as refinery feedstock for a considerable time, usually blending with more conventional feedstocks, but has commanded lower prices because of its lower quality relative to conventional oil. Tar sand bitumen has been the subject of much development in Alberta, Canada. It also has commanded lower prices because of its lower quality relative to conventional oil. But both of these heavy feedstocks are sources of liquid fuels and the other products that are used in large amounts and both of these feedstocks are used to supplement supplies of conventional, crude oil to refineries.

Obviously, differences exist between heavy oil and conventional oil, according to the volatility of the constituents. When the lower boiling constituents are lost through natural processes after evolution from organic source materials, the oil becomes heavy, with a high proportion of asphaltic constituents and with substitution in the carbon network of hetero atoms, such as nitrogen, sulfur, and oxygen. Therefore, heavy oil and tar sand bitumen, regardless of the source, always contains the high molecular weight complex asphaltic constituents. Removal or reduction of these constituents, through refining options or, preferably, leaving these constituents in the reservoir during recovery, improves the refinability of heavy oil and bitumen.

The success of the first edition of this book has prompted the need for an update and the creation of this second edition. The objective of this Edition is to present to the reader the current methods of recovery for heavy oil and tar sand bitumen technology by nonthermal and thermal methods. The book also contains a new stand-alone chapter describing the various techniques for biodegradation of petroleum and heavy feedstocks. The techniques could well be the recovery methods of the future, which will also be suitable for (partial or full) upgrading crude oil and heavy feedstocks in situ.

The book is designed to be suitable for undergraduate students, graduate students, and professionals who are working with heavy oil and tar sand bitumen. Each chapter includes a list of references that will guide the reader to more detailed information. In addition, a detailed Glossary is included to assist the reader with any unknown or difficult terminology.

Dr. James G. Speight
CD&W Inc., Laramie, WY, United States
October 2015

PREFACE TO THE FIRST EDITION

The declining reserves of light crude oil have resulted in an increasing need to develop options to upgrade the abundant supply of known heavy oil reserves. In addition, there is considerable focus and renewed efforts on adapting recovery techniques to the production of heavy oil.

Over the past decade, the demand for crude oil worldwide has substantially increased, straining the supply of conventional oil. Recent price increases have emphasized the need for consideration of alternative or insufficiently utilized energy sources, especially heavy crude oil, to supplement short- and long-term needs. Heavy oil has been used as refinery feedstock for considerable time, usually blending with more conventional feedstocks, but has commanded lower prices because of its lower quality relative to conventional oil.

Obviously, differences exist between heavy oil and conventional (light) oil, according to the volatilities of the constituents. When the lower boiling constituents are lost through natural processes after evolution from organic source materials, the oil becomes heavy, with a high proportion of asphaltic molecules, and with substitution in the carbon network of heteroatoms, such as nitrogen, sulfur, and oxygen. Therefore, heavy oil, regardless of source, always contains the heavy fractions, the asphaltic materials, which consist of resins and asphaltenes. Removal or reduction of the asphaltene fraction, through deasphalting or leaving these constituents in the reservoir during recovery, improves the refinability of heavy oil.

The objectives of this book are to present to the reader the current methods of recovery for heavy oil and tar sand bitumen technology by nonthermal and thermal methods. The book is designed to be suitable for undergraduate students, graduate students, and professionals who are working with heavy oil and tar sand bitumen. Each chapter will include a list of references that will guide the reader to more detailed information.

James G. Speight
The University of Trinidad and Tobago, O'Meara Campus,
Arima, Trinidad
July 2008

PART I

Reservoirs and Reservoir Fluids

CHAPTER 1

Heavy Oil and Tar Sand Bitumen

1.1 INTRODUCTION

Petroleum (crude oil; conventional petroleum) is found in the microscopic pores of sedimentary rocks such as sandstone and limestone. Not all of the pores in a rock contain petroleum and some pores will be filled with water or brine that is saturated with minerals. However, not all of the oilfields that are discovered are exploited since the reservoir may be (1) too deep for economic recovery under the prevalent economic conditions, (2) contain crude oil of insufficient volume, (3) so remote that transportation costs would be high, or (4) in a tight formation where additional techniques such as hydraulic fracturing are required to recover the crude oil (see chapter: General Methods of Oil Recovery) (Speight, 2014a, 2015a).

In the typical porous rock reservoir, conventional petroleum is a free-flowing liquid while, on the other hand, heavy oil is a viscous type of petroleum that contains higher levels of sulfur and nonvolatile constituents than conventional petroleum but occurs in similar locations to petroleum (IEA, 2005; Ancheyta and Speight, 2007; Speight, 2014a). The viscous nature of heavy oil is a problem for recovery operations and for refining—the viscosity of the oil may be too high thereby rendering recovery expensive and/or difficult and the sulfur content may be high, which increases the expense of refining the oil.

In any text related to the properties and behavior (recovery or refining) of a natural resource (ie, heavy oil), it is necessary to understand the resource first through the name or terminology or definition. *Terminology* is the means by which various subjects are named so that reference can be made in conversations and in writings and so that the meaning is passed on. *Definitions* are the means by which scientists and engineers communicate the nature of a material to each other and to the world, through either the spoken or the written word. Thus the definition of a material can be extremely important and have a profound influence on how the technical community and the public perceive that material.

Because of the need for a thorough understanding of petroleum and the associated technologies, it is essential that the definitions and the

Introduction to Enhanced Recovery Methods for Heavy Oil and Tar Sands.
DOI: http://dx.doi.org/10.1016/B978-0-12-849906-1.00001-1

terminology of petroleum science and technology be given prime consideration. This will aid in a better understanding of petroleum, its constituents, and its various fractions. Of the many forms of terminology that have been used not all have survived, but the more commonly used are illustrated here. Particularly troublesome, and more confusing, are those terms that are applied to the more viscous materials, for example, the use of the terms *bitumen* and *asphalt*. This part of the text attempts to alleviate much of the confusion that exists, but it must be remembered that the terminology of petroleum is still open to personal choice and historical usage.

The name *heavy oil* can often be misleading as it has also been used in reference to (1) fuel oil that contains residuum left over from distillation, that is, residual fuel oil, (2) coal tar creosote, or (3) viscous crude oil. For the purposes of this text the term is used to mean *viscous crude oil*.

Heavy oil typically has relatively low proportions of volatile compounds with low molecular weights and quite high proportions of high molecular weight compounds of lower volatility. The high molecular weight fraction of heavy oils are comprised of a complex assortment of different molecular and chemical compounds (not necessarily just paraffins or asphaltenes) with high melting points and high pour points that greatly contribute to the poor fluid properties of the heavy oil, thereby contributing to low mobility (compared to conventional crude oil).

More generally, heavy oil typically has low levels (if any at all) of paraffins (straight-chain alkanes) with moderate-to-high levels of asphaltene constituents. The asphaltene constituents are not necessarily the primary cause for the high specific gravity (low API gravity) of the oil nor are they always the prime cause for production problems. It is essential to consider the content of the resin constituents and the aromatic constituents, both of which are capable of hindering the asphaltene constituents from separation during recovery. It is only when the asphaltene constituents separate from the oil as separate phase that they deposit in the formation or in the production train.

1.2 HISTORY

Petroleum, in various forms, is not a recent discovery (Abraham, 1945; Forbes, 1958a, 1958b, 1959, 1964; Speight, 2014a). More than four thousand years ago, bitumen from natural seepages was employed in the construction of the walls and towers of Babylon. Ancient writings indicate the medicinal and lighting uses of petroleum in various societies. In terms of petroleum recovery, the earliest known wells were drilled in China in 347 BC to depths of 800 feet (240 m) and were drilled using bits attached to bamboo poles. The oil was burned to evaporate brine and produce salt. By the tenth century, extensive bamboo pipelines connected oil wells with salt springs.

The use of petroleum in the Middle East was established by the eighth century, when the streets of the newly constructed Baghdad were paved with the nonvolatile residue derived from accessible petroleum and seepages (particularly Hit) in the region. In the ninth century, petroleum was distilled at Baku, Azerbaijan, to produce naphtha which formed the basis of the incendiary *Greek fire* (Cobb and Goldwhite, 1995). These Baku experiences were reported by the geographer Masudi in the tenth century, and by Marco Polo in the thirteenth century, who described the output of those wells as hundreds of shiploads.

The earliest mention of American petroleum occurs in Sir Walter Raleigh's documentation of the Trinidad Asphalt Lake (also called the Trinidad Pitch Lake) in 1595. In 1632, the journal of a Franciscan, Joseph de la Roche d'Allion, describing his visit to the oil springs of New York was published in Sagard's *Histoire du Canada*. A Russian traveler, Peter Kalm, in his work on America published in 1748 showed on a map the oil springs of Pennsylvania. In 1854, Benjamin Silliman, a science professor at Yale University in New Haven, Connecticut followed the work by Arabic alchemists and fractionated petroleum by distillation. These discoveries rapidly spread around the world, and Meerzoeff built the first Russian refinery in the then-mature oil fields at Baku in 1861, where at that time about 90% of the world's oil was produced.

The first commercial oil well drilled in North America was in Oil Springs, Ontario, Canada in 1858 by James Miller Williams. The US petroleum industry began with Edwin Drake's drilling of a 69-foot (21 m) oil well in 1859 at Oil Creek near Titusville, Pennsylvania for the Seneca Oil Company. The well originally yielded 25 barrels per day and by the end of the year output was at the rate of 15 barrels per day. The industry grew through the 1800s, driven by the demand for kerosene and for oil lamps. Petroleum refining became even more popular, perhaps essential, in the early part of the twentieth century with the introduction of the internal combustion engine, which provided a demand that has largely sustained the industry during the past one hundred years. Early finds like those in Pennsylvania and Ontario were quickly outpaced by demand leading to oil booms in Texas, Oklahoma, and California.

By 1910, significant oil fields had been discovered in Canada, the Dutch East Indies (1885, in Sumatra), Iran, (1908, in Masjed Soleiman), Venezuela, and Mexico, which were being developed at an industrial level. Even until the mid-1950s, coal was still the world's foremost fuel, but oil quickly took over. The 1973 energy crisis and the 1979 energy crisis brought to light the concern that oil is a limited resource that will diminish, at least as an economically viable energy source. At the time, the most common and popular predictions were spectacularly dire.

The value of petroleum as a transportable, high-energy source that can be used to power the vast majority of vehicles (gasoline and diesel) and as the base of many industrial chemicals makes it one of the world's most important commodities (Hsu and Robinson, 2006; Gary et al., 2007; Speight, 2014a). Access to it was a major factor in several military conflicts, including World War II and the more recent wars in the Persian Gulf of the twentieth and early twenty-first centuries. Approximately 80% of the world's readily accessible reserves are located in the Middle East with the majority coming from Saudi Arabia. However, when the reserves of heavy oil and tar sand bitumen are taken into account, the balance shifts. Venezuela and Canada have substantial reserves of heavy oil and tar sand bitumen, sufficient to shift the balance of petroleum reserves from the Middle East to the Americas. And it is to this subject that this book is devoted.

The increasing demand for crude oil continues to grow every year, fueled in part by the rapidly growing economies of China and India. The declining availability of conventional oil combined with rising demand has continued to put more pressure on the search for alternative energy sources. Thus, with the demand for petroleum and petroleum-based products reaching new highs, investments in these more challenging reservoirs are rapidly accelerating. In fact, the worldwide importance of heavy oils will continue to emerge as the price of oil remains high and the demand for it remains strong. Although prices are expected to fluctuate, it is worth moving ahead with heavy oil resources on the basis of obtaining a measure (as yet undefined and country-dependent) of oil independence.

However, demand and market pull aside, it is essential to recognize that there are significant differences between conventional crude oil, heavy oil, and tar sand bitumen and, moreover, there are significant variations within the conventional petroleum types as there are also differences between the heavy oil types. While tar sand bitumen is, by definition, different to conventional crude oil and heavy oil, there are also differences between the various types of materials that fall within the definition of tar sand bitumen. This is reflected in the differences observed in various recovery processes as well as in the price variations between conventional crude oil types, the heavy oils, and tar sand bitumen (Tables 1.1 and 1.2) (Speight, 2011, 2013a, b).

Typically, crude oil quality has an effect on the recovery process insofar as crude oil that contains a higher proportion of the more volatile constituents

Table 1.1 API Gravity and Sulfur Content of Selected Crude Oils

Country	Crude oil	API	Sulfur% w/w
Abu Dhabi (UAE)	Abu Al Bu Khoosh	31.6	2.00
Abu Dhabi (UAE)	Abu Mubarras	38.1	0.93
Abu Dhabi (UAE)	El Bunduq	38.5	1.12
Abu Dhabi (UAE)	Murban	40.5	0.78
Abu Dhabi (UAE)	Umm Shaif	37.4	1.51
Abu Dhabi (UAE)	Zakum (Lower)	40.6	1.05
Abu Dhabi (UAE)	Zakum (Upper)	33.1	2.00
Algeria	Zarzaitine	43.0	0.07
Angola	Cabinda	31.7	0.17
Angola	Palanca	40.1	0.11
Angola	Takula	32.4	0.09
Australia	Airlie	44.7	0.01
Australia	Barrow Island	37.3	0.05
Australia	Challis	39.5	0.070
Australia	Cooper Basin	45.2	0.02
Australia	Gippsland	47.0	0.09
Australia	Griffin	55.0	0.03
Australia	Harriet	37.9	0.05
Australia	Jabiru	42.3	0.05
Australia	Jackson	43.8	0.03
Australia	Saladin	48.2	0.02
Australia	Skua	41.9	0.06
Brazil	Garoupa	30.0	0.68
Brazil	Sergipano Platforma	38.4	0.19
Brazil	Sergipano Terra	24.1	0.41
Brunei	Champion Export	23.9	0.12
Brunei	Seria	40.5	0.06
Cameroon	Kole Marine	32.6	0.33
Cameroon	Lokele	20.7	0.46
Canada (Alberta)	Bow River Heavy	26.7	2.10
Canada (Alberta)	Pembina	38.8	0.20
Canada (Alberta)	Rainbow	40.7	0.50
Canada (Alberta)	Rangeland South	39.5	0.75
Canada (Alberta)	Wainwright-Kinsella	23.1	2.58
China	Daqing (Taching)	32.6	0.09

(Continued)

Table 1.1 (Continued)

Country	Crude oil	API	Sulfur% w/w
China	Nanhai Light	40.6	0.06
China	Shengli	24.2	1.00
China	Weizhou	39.7	0.08
Colombia	Cano Limon	29.3	0.51
Congo (Brazzaville)	Emeraude	23.6	0.600
Dubai (UAE)	Fateh	31.1	2.000
Dubai (UAE)	Margham Light	50.3	0.040
Ecuador	Oriente	29.2	0.880
Egypt	Belayim	27.5	2.200
Egypt	Gulf of Suez	31.9	1.520
Egypt	Ras Gharib	21.5	3.640
Gabon	Gamba	31.4	0.090
Gabon	Rabi-Kounga	33.5	0.070
Ghana	Salt Pond	37.4	0.097
India	Bombay High	39.2	0.150
Indonesia	Anoa	45.2	0.040
Indonesia	Ardjuna	35.2	0.105
Indonesia	Attaka	43.3	0.040
Indonesia	Badak	49.5	0.032
Indonesia	Bekapai	41.2	0.080
Indonesia	Belida	45.1	0.020
Indonesia	Bima	21.1	0.250
Indonesia	Cinta	33.4	0.080
Indonesia	Duri (Sumatran Heavy)	21.3	0.180
Indonesia	Ikan Pari	48.0	0.020
Indonesia	Kakap	51.5	0.050
Indonesia	Katapa	50.8	0.060
Indonesia	Lalang (Malacca Straits)	39.7	0.050
Indonesia	Minas (Sumatran Light)	34.5	0.081
Indonesia	Udang	38.0	0.050
Iran	Aboozar (Ardeshir)	26.9	2.480
Iran	Bahrgansar/Nowruz	27.1	2.450
Iran	Dorrood (Darius)	33.6	2.350
Iran	Foroozan (Fereidoon)	31.3	2.500
Iran	Iranian Heavy	30.9	1.730
Iran	Iranian Light	33.8	1.350
Iran	Rostam	35.9	1.550
Iran	Salmon (Sassan)	33.9	1.910
Iraq	Basrah Heavy	24.7	3.500
Iraq	Basrah Light	33.7	1.950
Iraq	Basrah Medium	31.1	2.580
Iraq	North Rumaila	33.7	1.980
Ivory Coast	Espoir	32.3	0.340
Kazakhstan	Kumkol	42.5	0.07
Kuwait	Kuwait Export	31.4	2.52
Libya	Amna	36.0	0.15
Libya	Brega	40.4	0.21
Libya	Bu Attifel	43.3	0.04

(Continued)

Table 1.1 (Continued)

Country	Crude oil	API	Sulfur% w/w
Libya	Buri	26.2	1.76
Libya	Es Sider	37.0	0.45
Libya	Sarir	38.4	0.16
Libya	Sirtica	41.3	0.45
Libya	Zueitina	41.3	0.28
Malaysia	Bintulu	28.1	0.08
Malaysia	Dulang	39.0	0.12
Malaysia	Labuan	32.2	0.07
Malaysia	Miri Light	32.6	0.04
Malaysia	Tembungo	37.4	0.04
Mexico	Isthmus	33.3	1.49
Mexico	Maya	22.2	3.30
Mexico	Olmeca	39.8	0.80
Neutral Zone	Burgan	23.3	3.37
Neutral Zone	Eocene	18.6	4.55
Neutral Zone	Hout	32.8	1.91
Neutral Zone	Khafji	28.5	2.85
Neutral Zone	Ratawi	23.5	4.07
Nigeria	Antan	32.1	0.32
Nigeria	Bonny Light	33.9	0.14
Nigeria	Bonny Medium	25.2	0.23
Nigeria	Brass River	42.8	0.06
Nigeria	Escravos	36.4	0.12
Nigeria	Forcados	29.6	0.18
Nigeria	Pennington	36.6	0.07
Nigeria	Qua Iboe	35.8	0.12
North Sea (Denmark)	Danish North Sea	34.5	0.260
North Sea (Norway)	Ekofisk	39.2	0.169
North Sea (Norway)	Emerald	22.0	0.750
North Sea (Norway)	Oseberg	33.7	0.310
North Sea (UK)	Alba	20.0	1.330
North Sea (UK)	Duncan	38.5	0.180
North Sea (UK)	Forties Blend	40.5	0.350
North Sea (UK)	Innes	45.7	0.130
North Sea (UK)	Kittiwake	37.0	0.65
North Yemen	Alif	40.3	0.10
Oman	Oman Export	34.7	0.94
Papua New Guinea	Kubutu	44.0	0.04
Peru	Loreto Peruvian	33.1	0.23
Qatar	Dukhan (Qatar Land)	40.9	1.27
Qatar	Qatar Marine	36.0	1.42
Ras Al Khaiman (UAE)	Ras Al Khaiman	44.3	0.15
Russia	Siberian Light	37.8	0.42
Saudi Arabia	Arab Extra Light (Berri)	37.2	1.15
Saudi Arabia	Arab Heavy (Safaniya)	27.4	2.80
Saudi Arabia	Arab Light	33.4	1.77
Saudi Arabia	Arab Medium (Zuluf)	28.8	2.49
Sharjah (UAE)	Mubarek	37.0	0.62

(Continued)

Table 1.1 (Continued)

Country	Crude oil	API	Sulfur% w/w
Sumatra	Duri	20.3	0.21
Syria	Souedie	24.9	3.82
Timor Sea (Indonesia)	Hydra	37.5	0.08
Trinidad Tobago	Galeota Mix	32.8	0.27
Tunisia	Ashtart	30.0	0.99
USA (Alaska)	Alaskan North Slope	27.5	1.11
USA (Alaska)	Cook Inlet	35.0	0.10
USA (Alaska)	Drift River	35.3	0.09
USA (Alaska)	Nikiski Terminal	34.6	0.10
USA (California)	Hondo Sandstone	35.2	0.21
USA (California)	Huntington Beach	20.7	1.38
USA (Florida)	Sunniland	24.9	3.25
USA (Louisiana)	Grand Isle	34.2	0.35
USA (Louisiana)	Lake Arthur	41.9	0.06
USA (Louisiana)	Louisiana Light Sweet	36.1	0.45
USA (Louisiana)	Ostrica	32.0	0.30
USA (Louisiana)	South Louisiana	32.8	0.28
USA (Michigan)	Lakehead Sweet	47.0	0.31
USA (New Mexico)	New Mexico Intermediate	37.6	0.17
USA (New Mexico)	New Mexico Light	43.3	0.07
USA (Oklahoma)	Basin-Cushing Composite	34.0	1.95
USA (Texas)	Coastal B-2	32.2	0.22
USA (Texas)	East Texas	37.0	0.21
USA (Texas)	Sea Breeze	37.9	0.10
USA (Texas)	West Texas Intermediate	40.8	0.34
USA (Texas)	West Texas Semi-Sweet	39.0	0.27
USA (Texas)	West Texas Sour	34.1	1.640
USA (Wyoming)	Tom Brown	38.2	0.100
USA (Wyoming)	Wyoming Sweet	37.2	0.330
Venezeula	Lago Medio	32.2	1.010
Venezeula	Leona	24.4	1.510
Venezeula	Mesa	29.8	1.010
Venezuela	Ceuta Export	27.8	1.370
Venezuela	Guanipa	30.3	0.850
Venezuela	La Rosa Medium	25.3	1.730
Venezuela	Lago Treco	26.7	1.500
Venezuela	Oficina	33.3	0.780
Venezuela	Temblador	21.0	0.830
Venezuela	Tia Juana	25.8	1.630
Venezuela	Tia Juana Light	31.8	1.160
Venezuela	Tia Juana Medium 24	24.8	1.610
Venezuela	Tia Juana Medium 26	26.9	1.540
Viet Nam	Bach Ho (White Tiger)	38.6	0.030
Viet Nam	Dai Hung (Big Bear)	36.9	0.080
Yemen	Masila	30.5	0.670
Zaire	Zaire	31.7	0.130

Table 1.2 Simplified Differentiation Between Conventional Crude Oil, Heavy Oil, Extra Heavy Oil, Tar Sand Bitumen, Oil Shale Kerogen, Tight Oil, and Coal

- Conventional crude oil
    - Mobile in the reservoir; API gravity: >25°
    - High-permeability reservoir
    - Primary recovery
    - Secondary recovery
- Tight oil
    - Similar properties to the properties of conventional crude oil; API gravity: >25°
    - Immobile in the reservoir
    - Low-permeability reservoir
    - Horizontal drilling into reservoir
    - Fracturing (typically multifracturing) to release fluids/gases
- Medium crude oil
    - Similar properties to the properties of conventional crude oil; API gravity: 20–25°
    - High-permeability reservoir
    - Primary recovery
    - Secondary recovery
- Heavy crude oil
    - More viscous than conventional crude oil; API gravity: 10–20°
    - Mobile in the reservoir
    - High-permeability reservoir
    - Secondary recovery
    - Tertiary recovery (enhanced oil recovery – EOR; e.g., steam stimulation)
- Extra heavy oil
    - Similar properties to the properties of tar sand bitumen; API gravity: <10°
    - Mobile in the reservoir
    - High-permeability reservoir
    - Secondary recovery
    - Tertiary recovery (enhanced oil recovery—EOR; eg, steam stimulation)
- Tar sand bitumen
    - Immobile in the deposit; API gravity: <10°
    - High-permeability reservoir
    - Mining (often preceded by explosive fracturing)
    - Steam assisted gravity draining (SAGD)
    - Solvent methods (VAPEX)
    - Extreme heating methods
    - Innovative methods[a]

[a]Innovative methods exclude tertiary recovery methods and methods, such as SAGD, VAPEX, but does include variants or hybrids thereof (see chapters: General Methods of Oil Recovery; Microbial Enhanced Oil Recovery).

is somewhat easier to recover from the reservoir but differences in crude oil quality are not the only determinant of recovery differentials, which are not constant over time. In addition, differences in the quality of heavy crude oil also influence the relative ease of recovery, while tar sand bitumen requires different lines of thinking in terms of recovery (Speight, 2013a, b, 2014a). However, factors such as the properties of the reservoir or the deposit, which are outside of the quality of the crude oil, heavy oil, or bitumen, also have an influence on the recovery process.

At this point, the concept of hydraulic fracturing is worthy of a mention—in addition to the recognized recovery processes, hydraulic fracturing is included as an example of technologies that contribute to improvements in production of crude oil (see chapter: General Methods of Oil Recovery). Hydraulic fracturing is used to create additional passageways in the oil reservoir that can facilitate the flow of oil to a producing well. *Tight reservoirs* are oil-containing reservoirs with restricted pore volume and connectivity in the reservoir rock that impede the flow of fluids through the reservoir. Such reservoirs are commonly fractured by injecting a fluid containing sand or other proppant under sufficient pressure to create fractures in the rock through which the oil can more easily flow.

In the process, a viscous fluid containing a proppant, such as sand, is injected under high pressure until the desired fracturing is achieved. The pressure is then released allowing the fluid to flow to the well while the proppant remains in the fractures to prevent them from closing. Care is also taken to contain the fracturing within the oil reservoir to avoid intersecting adjoining aquifers that would introduce excess water into the oil-producing zone. Tailored pulse fracturing is employed to control the extent and direction of the produced fractures. Precise quantities of solid rocket fuel-like propellants are ignited in the wellbore to create a controlled pressure pulse which creates fractures in a more predictable pattern. Foam fracturing, using foam under high pressure in gas reservoirs, has the advantage over high-pressure water injection because it does not create as much damage to the formation, and well cleanup operations are less costly. Carbon dioxide–sand fracturing increases production by eliminating much of the inhibiting effects of pumped fluids, such as plugging by solids, water retention, and chemical interactions.

1.3 PETROLEUM

Petroleum (crude oil) comprises a family of naturally-occurring materials that are predominantly hydrocarbon in nature and mobile within the reservoir and recoverable by application of various methods.

1.4 CONVENTIONAL PETROLEUM

The *definition* of petroleum (also called crude oil or conventional petroleum) has been varied, unsystematic, diverse, and often archaic. In fact, there has been a tendency to define petroleum and heavy oil on the basis of a single property. While this may be suitable for a general understanding it is by no means accurate and does not reflect the true nature of petroleum or heavy oil or the characterization of the material. Unfortunately, this form of *identification* or differentiation is a product of many years of growth and its long established use, however general or inadequate it may be, is altered with difficulty, and a new term, however precise, is adopted only slowly.

Petroleum is a naturally occurring mixture of hydrocarbons, generally in a liquid state, which may also include compounds of sulfur nitrogen oxygen metals and other elements (A0STM D4175; Speight, 2012a; US EIA, 2014). Thus, *petroleum* and the equivalent term *crude oil*, cover a wide assortment of materials consisting of mixtures of hydrocarbons and other compounds containing variable amounts of sulfur, nitrogen, and oxygen, which may vary widely in specific gravity, API gravity, the amount of residuum, as well as the variation in properties (Tables 1.1 and 1.2). Metal-containing constituents, notably those compounds that contain vanadium and nickel, usually occur in the more viscous crude oils in amounts up to several thousand parts per million and can have serious consequences during processing of these feedstocks (Speight, 1984, 2014a). Because petroleum is a mixture of widely varying constituents and proportions, its physical properties also vary widely and the color varies from near colorless to black.

The molecular boundaries of petroleum cover a wide range of boiling points and carbon numbers of hydrocarbon compounds and other compounds containing nitrogen, oxygen, and sulfur, as well as metal-containing (porphyrin) constituents. However, the actual boundaries of such a *petroleum map* can only be arbitrarily defined in terms of boiling point and carbon number. In fact, petroleum is so diverse that materials from different sources exhibit different boundary limits, and for this reason alone it is not surprising that petroleum has been difficult to *map* in a precise manner.

Since there is a wide variation in the properties of crude petroleum, the proportions in which the different constituents occur vary with the origin and the relative amounts of the source materials that form the initial *protopetroleum*, as well as the maturation conditions. Thus, some crude oils have higher proportions of the lower boiling components and others (such as heavy oil and bitumen) have higher proportions of higher boiling components (asphaltic components and residuum).

The fuels derived from petroleum contribute approximately one-third to one-half of the total world energy supply and are used not only for transportation fuels (ie, gasoline, diesel fuel, and aviation fuel, among others) but also to heat buildings. Petroleum products have a wide variety of uses that vary from gaseous and liquid fuels to near-solid machinery lubricants. In addition, the residue of many refinery processes, asphalt—a once-maligned by-product—is now a premium value product for highway surfaces, roofing materials, and miscellaneous waterproofing uses.

Petroleum occurs underground, at various pressures depending on the depth. Because of the pressure, it contains considerable natural gas in solution. Petroleum underground is much more fluid than it is on the surface and is generally mobile under reservoir conditions because the elevated temperatures (the *geothermal gradient*) in subterranean formations decrease the viscosity. Although the geothermal gradient varies from place to place, it is generally on the order of 25°C/km to 30°C/km (15°F/1000 feet or 120°C/1000 feet, ie, 0.015°C per foot of depth or 0.012°C per foot of depth).

Petroleum is derived from aquatic plants and animals that lived and died hundreds of millions of years ago. Their remains mixed with mud and sand in layered deposits that, over the millennia, were geologically transformed into sedimentary rock. Gradually the organic matter decomposed and eventually formed petroleum (or a related precursor), which migrated from the original source beds to more porous and permeable rocks, such as *sandstone* and *siltstone*, where it finally became entrapped. Such entrapped accumulations of petroleum are called *reservoirs*. A series of reservoirs within a common rock structure or a series of reservoirs in separate but neighboring formations is commonly referred to as an *oil field*. A group of fields is often found in a single geologic environment known as a *sedimentary basin* or *province*.

The major components of petroleum are *hydrocarbons*, compounds of hydrogen and carbon that display great variation in their molecular structure. The simplest hydrocarbons are a large group of chain-shaped molecules known as the *paraffins*. This broad series extends from methane, which forms natural gas, through liquids that are refined into gasoline, to crystalline waxes. A series of ring-shaped hydrocarbons, known as the *naphthenes*, range from volatile liquids, such as *naphtha*, to high molecular weight substances, isolated as the *asphaltene* fraction. Another group of ring-shaped hydrocarbons is known as the *aromatics*; the chief compound in this series is benzene, a popular raw material for making petrochemicals. The *nonhydrocarbon constituents* of petroleum include organic derivatives of nitrogen, oxygen, sulfur, and the metals nickel and vanadium. Most of these impurities are removed during refining.

Geologic techniques can determine only the existence of rock formations that are favorable for oil deposits, not whether oil is actually there. Drilling is the only sure way to ascertain the presence of oil. With modern rotary equipment, wells can be drilled to depths of more than 30,000 feet (9000 m). Once oil is found, it may be recovered (brought to the surface) by the pressure created by natural gas or water within the reservoir. Crude oil can also be brought to the surface by injecting water or steam into the reservoir to raise the pressure artificially, or by injecting such substances as carbon dioxide, polymers, and solvents to reduce crude oil viscosity. Thermal recovery methods are frequently used to enhance the production of heavy crude oils, whose extraction is impeded by viscous resistance to flow at reservoir temperatures.

Petroleum is typically recovered from the reservoir by the application of primary and secondary recovery techniques. Although covered elsewhere (see chapter: General Methods of Oil Recovery), mention of (1) primary recovery, (2) secondary recovery, and (3) tertiary recovery is warranted here in terms of the general description of definitions of these techniques.

Primary recovery refers to the process in which the petroleum in the reservoir trap is forced to the surface by the natural pressure contained in the trap. This pressure may result from several forces: (1) when the reservoir is penetrated the pressure release allows the water layer to expand and pushes the oil upwards and replaces it in the rock pores—this is the most effective technique and is known as a water drive system (Fig. 1.1); (2) if the drill penetrates into a layer of oil which has a gas cap above it, the release of pressure allows the gas layer to expand rapidly causing a downward pressure on the oil forcing it to move up through the well (gas cap drive) (Fig. 1.2); and (3) gas dissolved in the oil may be released as bubbles when the trap is pierced; as the oil moves up, the gas in the oil expands and the growing bubbles push the oil to the surface (solution gas drive) (Fig. 1.3). In most reservoir traps, these pressures are sufficient to initially force the petroleum to the surface.

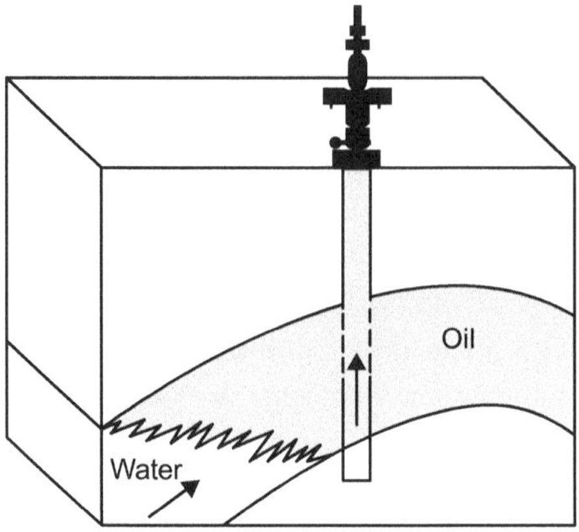

Figure 1.1 Water drive.

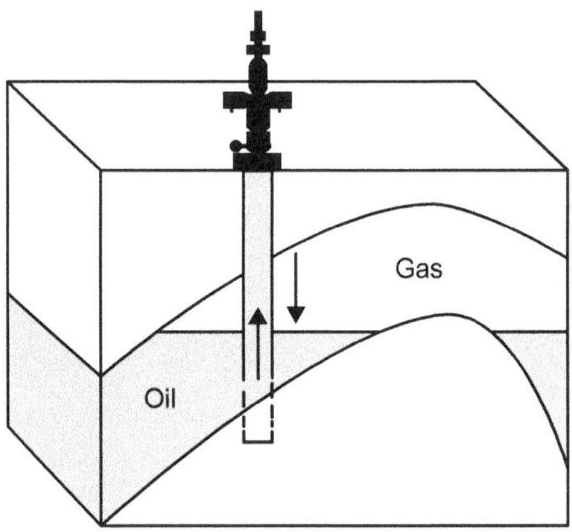

Figure 1.2 Gas cap drive.

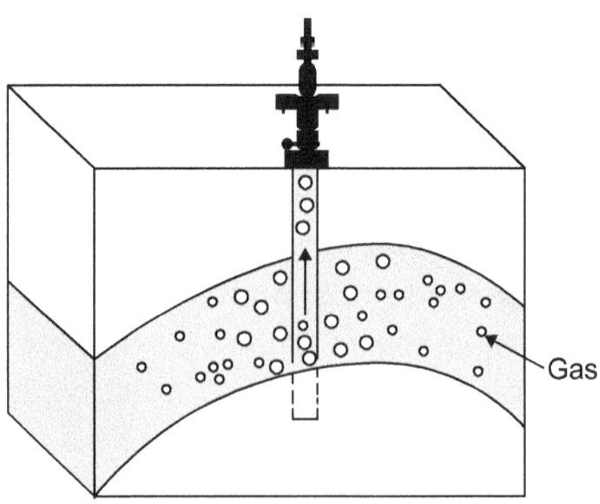

Figure 1.3 Solution gas drive.

At some point in time, this pressure will diminish and the production of petroleum decreases because: (1) there is less force—often referred to as reservoir energy—driving the oil towards the well; and/or (2) the gas that moves into the emptied pore spaces reduces the permeability of the rock, making it more difficult for crude oil to flow through the reservoir to the production well; and/or (3) the fall in pressure and the loss of dissolved gas increases the surface tension and viscosity of the oil. In general terms, primary recovery techniques usually account for less than 30% of the total volume of petroleum recovered. At this stage, depending upon the reservoir condition, additional techniques, such as hydraulic fracturing, may be used to recover the residual crude oil (Speight, 2015a).

Secondary recovery involves trying to maintain reservoir pressure. One technique is to inject natural gas into the reservoir above the oil, forcing the oil downwards, and then injecting water below the oil so forcing it upwards. Sometimes the gas that is used is that which has just been released during *primary recovery*. The disadvantage of using the released gas is that this gas is a marketable product in its own right. However, this is a good method to use if transporting the gas would be costly and, in any case, the re-injected gas can always be collected again if necessary. Alternative secondary techniques involve injecting carbon dioxide or nitrogen into the oil. This makes the oil more fluid and the gas pushes the oil upwards. Again, and depending upon the reservoir condition, additional techniques, such as hydraulic fracturing, may be used to recover the residual crude oil (Speight, 2015a). *Tertiary recovery* is the most expensive approach and involves injecting steam, detergents, solvents, bacteria, or bacterial nutrient solutions into the remaining oil. When high-pressure steam is injected it heats the oil, decreasing its density and viscosity and increasing its flow rate (Fig. 1.4). Detergents which can be injected reduce the viscosity of the oil and act as surfactants reducing the ability of the oil to stick to the rock surface and thus making it easier for it to be flushed up to the surface (Fig. 1.5).

Another *tertiary recovery* technique involves injecting bacteria into the oilfield (microbial enhanced oil recovery, MEOR). Some bacteria produce polysaccharides which reduce the permeability of the water-filled pores of the reservoir rock and this effectively forces injected water into the oil-filled pores, pushing the oil out. Other bacteria produce carbon dioxide which helps to increase pressure within the rock pores, forcing out the oil. Other bacteria produce surfactants and/or chemicals that reduce the viscosity of the oil.

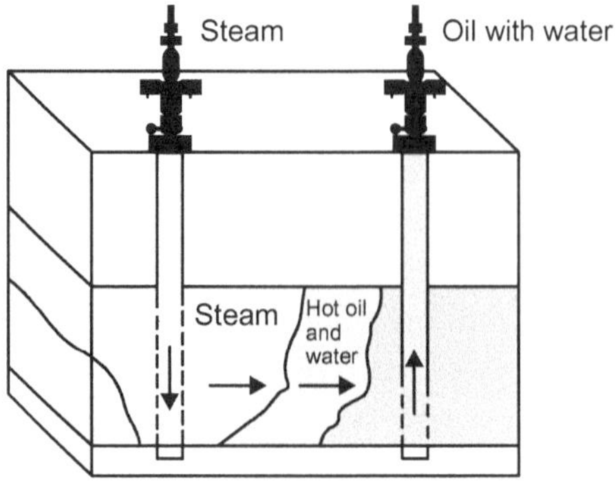

Figure 1.4 Steam flooding.

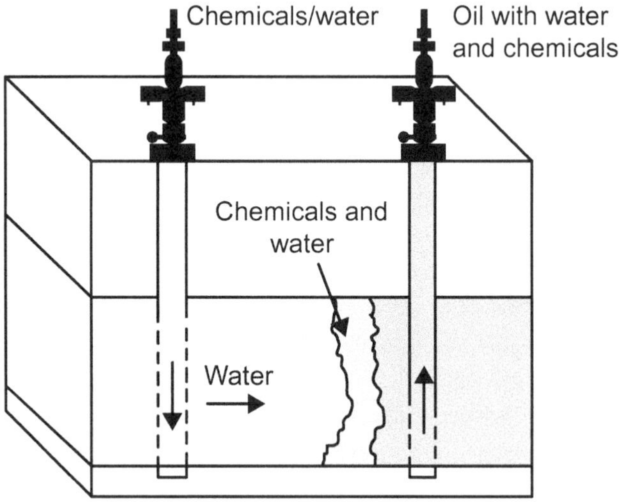

Figure 1.5 Recovery using chemicals or detergents.

MEOR processes involve the use of reservoir microorganisms or specially selected natural bacteria to produce specific metabolic events that lead to enhanced oil recovery (EOR) and the technology is exploited in oil reservoirs to improve recovery (Banat, 1995; Clark et al., 1981; Stosur, 1991). MEOR processes (see chapter: Microbial Enhanced Oil Recovery) are, to a point, similar to in situ bioremediation processes (Speight and Arjoon, 2012). Injected nutrients, together with indigenous or added microbes, promote in situ microbial growth and/or generation of products which mobilize additional oil and move it to producing wells through reservoir repressurization, interfacial tension/oil viscosity reduction, and selective plugging of the most permeable zones (Bryant et al., 1989; Bryant and Lindsey, 1996). Alternatively, the oil-mobilizing microbial products may be produced by fermentation and injected into the reservoir. In the process, nutrients such as sugars, nitrates, or phosphates are regularly injected into the formation to stimulate the growth and improve the performance of the microbes, some of which may be indigenous to the reservoir. The microbes then generate surfactants and carbon dioxide that help to displace the oil in a similar way to other displacement methods. However, the organisms that participate in oil recovery produce a variety of fermentation products, for example, carbon dioxide, methane, hydrogen, bio-surfactants, and polysaccharides from crude oil, pure hydrocarbons, and a variety of nonhydrocarbon substrates. The organic acids produced through fermentation readily dissolve carbonates and can greatly enhance permeability in limestone reservoirs.

More generally, the microbial enhanced recovery technology requires consideration of the physicochemical properties of the reservoir in terms of salinity, pH, temperature, pressure, and nutrient availability (Khire and Khan, 1994a, b). If the reservoir contains high concentrations of sodium chloride, the use of bacteria which can tolerate these conditions is required (Shennan and Levi, 1987; Jenneman, 1989). Bacteria producing biosurfactants and polymers can grow at sodium concentrations up to 8% and selectively plug sandstone to create a biowall that effectively has an adverse effect on the recovery of additional oil (Raiders et al., 1989). Thus the process, like all recovery processes, requires modification to match the reservoir conditions, which vary from reservoir to reservoir and which calls for reservoir-specific customization of the MEOR process, and this alone has the potential to undermine microbial process economic viability. In addition, the process may modify the immediate reservoir environment in a number of ways that could also damage the production hardware or the formation itself. Certain sulfate reducers can produce H_2S, which can corrode pipeline and other components of the recovery equipment. Ensuring success requires an ability to manipulate environmental conditions to promote growth and/or product formation by the participating microorganisms. Exerting such control over the microbial system in the subsurface is itself a serious challenge.

MEOR differs from chemical EOR in the method by which the enhancing products are introduced into the reservoir. Thus, in oil recovery by the *cyclic microbial method*, a solution of nutrients and microorganisms is introduced into the reservoir during injection. The injection well is then shut for an incubation period allowing the microorganisms to produce carbon dioxide gas and surfactants that assist in mobilization of the oil. The well is then opened and oil and oil products resulting from the treatment are produced. The process is repeated as often as oil can be produced from the well. Oil recovery by *microbial flooding* also involves the use of microorganisms but in this case, the reservoir is usually conditioned by a water flush after which a solution of microorganisms and nutrients is injected into the formation. As this solution is pushed through the reservoir by water drive, gases and surfactants are formed, the oil is mobilized, and pumped through the well. However, even though microbes produce the necessary chemical reactions in situ (Table 1.3), whereas surface injected chemicals may tend to follow areas of higher permeability, resulting in decreased sweep efficiency, there is need for caution and astute observation of the effects of the microorganisms on the reservoir chemistry.

The mechanism by which MEOR processes work can be quite complex and may involve multiple biochemical processes. In selective plugging approaches, microbial cell mass or biopolymers plug high permeability

Table 1.3 Microbial Products and Their Contribution to Enhanced Oil Recovery

Microbial	Effect
Acids	Modification of reservoir rock, especially carbonate rock
	Improvement of porosity and permeability
	Reaction of carbonate rocks to produce carbon dioxide
Biomass	Selective or nonselective plugging
	Emulsification through adherence to hydrocarbons
	Modification of solid surfaces, eg, wetting
	Degradation and alteration of oil
	Reduction of oil viscosity and oil pour point
	Desulfurization of oil
Gases (CO_2, CH_4. H_2)	Reservoir repressurization
	Oil swelling
	Viscosity reduction
	Solubilization of carbonate minerals
	Increase of permeability
Solvents	Dissolution of the crude oil
Surface active agent	Lowering interfacial tension
Polymers	Mobility control
	Selective or nonselective plugging

zones and lead to a redirection of the water flood. In other processes, biosurfactants are produced in situ, which leads to increased mobilization of residual oil. In still other processes, microbial production of carbon dioxide and organic solvents reduces the oil viscosity as the primary mechanism for EOR.

Briefly, the term residual oil refers to the oil that remains in the pores of the rock after secondary recovery. Numerous techniques exist for enhanced recovery of residual oil (see chapter: General Methods of Oil Recovery) and many techniques were developed in the early 1980s: (1) polymer flood, (2) surfactant flood, (3) carbon dioxide injection or natural gas injection, and (4) microbial treatment.

The stimulation of oil production by in situ bacterial fermentation is thought to proceed by one or a combination of the following mechanisms: (1) improvement of the relative mobility of oil to water by biosurfactants and biopolymers; (2) partial repressurization of the reservoir by methane and carbon dioxide; (3) reduction of oil viscosity through the dissolution of organic solvents in the oil phase; (4) increase of reservoir permeability and widening of the fissures and channels through the etching of carbonaceous rocks in limestone reservoirs by organic acids produced by anaerobic bacteria; (5) cleaning the wellbore region through the acids and gas from in situ fermentation in which the gas pushes oil from dead space and dislodges debris that plugs the pores. The average pore size is increased and, as a result, the capillary pressure near the wellbore is made more favorable for the flow of oil; and (6) selective plugging of highly permeable zones by injecting slime-forming bacteria followed by a nutrient (such as sucrose solution) which initiates the production of extracellular slimes and aerial sweep efficiency is improved.

The target for EOR processes is the quantity of unrecoverable oil in known reservoirs, highly viscous heavy oil, and bitumen from tar sand deposits. However, there must be the recognition that MEOR is a single process and, in addition to any potential (or perceived) deleterious effects of the microorganisms, scientific knowledge of the fundamentals of microbiology must be coupled with an understanding of the geological and engineering aspects of oil production in order to develop acceptable MEOR technology (see chapter: General Methods of Oil Recovery).

After recovery, petroleum is transported to refineries by pipelines, which can often carry more than 500,000 barrels per day, or by ocean-going tankers. The basic refinery process is distillation, which separates the crude oil into fractions of differing volatility. After the distillation, other physical methods are employed to separate the mixtures, including absorption, adsorption, solvent extraction, and crystallization. After physical separation into such constituents as light and heavy naphtha, kerosene, and light and heavy gas oils, selected petroleum fractions may be subjected to conversion processes, such as thermal cracking (ie, coking) and catalytic cracking. In the most general terms, cracking breaks the large molecules of heavier gas oils into the smaller molecules that form the lighter, more valuable naphtha fractions (Speight and Ozum, 2002; Parkash, 2003; Hsu and Robinson, 2006; Gary et al., 2007; Speight, 2014a).

1.5 OTHER MEMBERS OF THE PETROLEUM FAMILY

Other members of the petroleum family that may invoke the need for enhanced recovery methods, including the application of hydraulic fracturing techniques (Speight, 2015a) and which are worthy of mention here include: (1) crude oil from tight formations, (2) opportunity crudes, (3) high acid crude oils, and (4) foamy oil.

1.5.1 Crude Oil From Tight Formations

Generally, unconventional tight oil resources are found at considerable depths in sedimentary rock formations that are characterized by very low permeability. While some of the tight oil plays produce oil directly from shales, tight oil resources are also produced from low-permeability siltstone formations, sandstone formations, and carbonate formations that occur in close association with a shale source rock. Tight formations scattered throughout North America have the potential to produce crude oil (tight oil) (Fig. 1.5) (US EIA, 2011; Speight, 2013c; US EIA, 2013). Such formations might be composed of shale sediments or sandstone sediments. In a conventional sandstone reservoir the pores are interconnected so gas and oil can flow easily from the rock to a wellbore. In tight sandstones, the pores are smaller and are poorly connected by very narrow capillaries which results in low permeability. Tight oil occurs in sandstone sediments that have an effective permeability of less than 1 mD. A shale play is a defined geographic area containing an organic-rich fine-grained sedimentary rock that underwent physical and chemical compaction during diagenesis to produce the following characteristics: (1) clay to silt sized particles; (2) high % of silica, and sometimes carbonate minerals; (3) thermally mature; (4) hydrocarbon-filled porosity—on the order of 6 to 14%; (5) low permeability—on the order of <0.1 mD; (6) large areal distribution; and (7) fracture stimulation required for economic production.

The most notable tight oil plays in North America include the Bakken shale, the Niobrara formation, the Barnett shale, the Eagle Ford shale, and the Miocene Monterey play of California's San Joaquin Basin (California) and the Cardium play (Alberta, Canada). In many of these tight formations, the existence of large quantities of crude oil has been known for decades and efforts to commercially produce those resources have occurred sporadically with typically disappointing results. However, starting in the mid-2000s, advancements in well drilling and stimulation technologies combined with high oil prices turned tight oil resources into one of the most actively explored and produced targets in North America.

Other known tight formations (on a worldwide basis) include the R'Mah Formation (Syria), the Sargelu Formation (northern Persian Gulf region), the Athel Formation (Oman), the Bazhenov formation and Achimov Formation (West Siberia, Russia), the Coober Pedy formation (Australia), the Chicontepex formation (Mexico), and the Vaca Muerta field (Argentina) (US EIA, 2011, 2013). However, tight oil formations are heterogeneous and vary widely over relatively short distances. Thus, even in a single horizontal production well, the amount of oil recovered may vary, as may recovery within a field or even between adjacent wells. This makes evaluation of *shale plays* and decisions regarding the profitability of wells on a particular lease difficult and tight reservoirs which contain only crude oil (without natural gas as the pressurizing agent) cannot be economically produced (US EIA, 2011, 2013).

Typical of the crude oil from tight formations (*tight oil—tight light oil, and tight shale oil* have been suggested as alternate terms) is the Bakken crude oil which is a light highly volatile crude oil. Briefly, Bakken crude oil is a light sweet (low-sulfur) crude oil that has a relatively high proportion of volatile constituents. The production of the oil also yields a significant amount of volatile gases (including propane and butane) and low-boiling liquids (such as pentane and natural gasoline), which are often referred to collectively as (low-boiling or light) naphtha. By definition, natural gasoline (sometimes also referred to as *gas condensate*) is a mixture of low-boiling liquid hydrocarbon isolates from petroleum and natural gas wells suitable for blending with light naphtha and/or refinery gasoline (Mokhatab et al., 2006; Speight, 2007, 2014a). Because of the presence of low-boiling hydrocarbons, low-boiling naphtha (*light naphtha*) can become extremely explosive, even at relatively low ambient temperatures. Some of these gases may be burned off (flared) at the field wellhead, but others remain in the liquid products extracted from the well (Speight, 2014a).

Oil from tight shale formation is characterized by low-asphaltene content, low-sulfur content, and a significant molecular weight distribution of the paraffinic wax content (Speight, 2014a, 2015b). Paraffin carbon chains of C_{10} to C_{60} have been found, with some shale oils containing carbon chains up to C_{72}. To control deposition and plugging in formations due to paraffins, dispersants are commonly used. In upstream applications, these paraffin dispersants are applied as part of multifunctional additive packages where asphaltene stability and corrosion control are also addressed simultaneously (Speight, 2014a, b, c, 2015b, c). In addition, scale deposits of calcite ($CaCO_3$), other carbonate minerals (minerals containing the carbonate ion, CO_3^{2-}), and silicate minerals (minerals classified on the basis of the structure of the silicate group, which contains different ratios of silicon and oxygen) must be controlled during production or plugging problems arise. A wide range of scale additives is available which can be highly effective when selected appropriately. Depending on the nature of the well and the operational conditions, a specific chemistry is recommended or blends of products are used to address scale deposition.

While the basic approach toward developing a tight oil play are expected to be similar from area to area, the application of specific strategies, especially with respect to well completion and stimulation techniques, will almost certainly differ from play to play, and often even within a given play. The differences depend on the geology (which can be very heterogeneous, even within a play) and reflect the evolution of technologies over time with increased experience and availability.

Finally, the properties of crude oils from tight formations are highly variable. Density and other properties can show wide variation, even within the same field. The Bakken crude is light and sweet with an API of 42° and a sulfur content of 0.19% w/w. Similarly, Eagle Ford is a light sweet feed, with a sulfur content of approximately 0.1% w/w and with published API gravity between 40° API and 62° API.

1.5.2 Opportunity Crudes, High Acid Crudes, and Foamy Oil

There is also the need to understand and accommodate *opportunity crude oils* and/or *high acid crude oils,* which, for many purposes, are often included with heavy oil feedstocks after recovery (Speight, 2014a, b; Yeung, 2014). *Opportunity crude oils* are either new crude oils with unknown or poorly understood properties relating to processing issues or are existing crude oils with well-known properties and processing concerns (Ohmes, 2014). Opportunity crude oils are often, but not always, heavy crude oils but in either case are more difficult to process due to high levels of solids (and other contaminants) produced with the oil, high levels of acidity, and high viscosity.

High acid crude oils are crude oils that contain considerable proportions of naphthenic acids which, as commonly used in the petroleum industry, refers collectively to all of the organic acids present in the crude oil (Shalaby, 2005; Speight, 2014b). By the original definition, a naphthenic acid is a monobasic carboxyl group attached to a saturated cycloaliphatic structure. However, it has been a convention accepted in the oil industry that all organic acids in crude oil are called naphthenic acids. Naphthenic acids in crude oils are now known to be mixtures of low to high molecular weight acids and the naphthenic acid fraction also contains other acidic species.

Naphthenic acids can be very water-soluble to oil-soluble depending on their molecular weight, process temperatures, salinity of waters, and fluid pressures. In the water phase, naphthenic acids can cause stable reverse emulsions (oil droplets in a continuous water phase). In the oil phase with residual water, these acids have the potential to react with a host of minerals, which are capable of neutralizing the acids. The main reaction product found in practice is the calcium naphthenate soap (the calcium salt of naphthenic acids). The total acid matrix is therefore complex and it is unlikely that a simple titration, such as the traditional methods for measurement of the total acid number, can give meaningful results to use in predictions of problems. An alternative way of defining the relative organic acid fraction of crude oils is therefore a real need in the oil industry, both upstream and downstream.

High acid crude oils cause corrosion in the production equipment—corrosion is predominant at temperatures in excess of 180°C (355°F) (Kane and Cayard, 2002; Ghoshal and Sainik, 2013)—and occurs particularly in the atmospheric distillation unit (the first point of entry of the high-acid crude oil) and also in the vacuum distillation units. In addition, overhead corrosion is caused by the mineral salts, magnesium, calcium, and sodium chloride, which are hydrolyzed to produce volatile hydrochloric acid, causing a highly corrosive condition in the overhead exchangers. Therefore these salts present a significant contamination in opportunity crude oils. Other contaminants in opportunity crude oils which are shown to accelerate the hydrolysis reactions are inorganic clays and organic acids.

During primary production of heavy oil from solution gas drive reservoirs, the oil is pushed into the production wells by energy supplied by the dissolved gas. As fluid is withdrawn from the production wells, the pressure in the reservoir declines and the gas that was dissolved in the oil at high pressure starts to come out of solution (*foamy oil*). As pressure declines further with continued removal of fluids from the production wells, more gas is released from solution and the gas already released expands in volume. The expanding gas, which at this point is in the form of isolated bubbles, pushes the oil out of the pores and provides energy for the flow of oil into the production well. This process is very efficient until the isolated gas bubbles link up and the gas itself starts flowing into the production well. Once the gas flow starts, the oil has to compete with the gas for available flow energy. Thus, in some heavy oil reservoirs, due to the properties of the oil and the sand and also due to the production methods, the released gas forms foam with the oil and remains subdivided, in the form of dispersed bubbles, much longer.

Foamy oil is oil-continuous foam that contains dispersed gas bubbles produced at the well head from heavy oil reservoirs under solution gas drive. The nature of the gas dispersions in oil distinguishes foamy oil behavior from conventional heavy oil. The gas that comes out of solution in the reservoir does not coalesce into large gas bubbles nor into a continuous flowing gas phase. Instead it remains as small bubbles entrained in the crude oil, keeping the effective oil viscosity low while providing expansive energy that helps drive the oil toward the production well. Foamy oil accounts for unusually high production in heavy oil reservoirs under solution-gas drive.

However, it has been observed that many heavy oil reservoirs in Alberta and Saskatchewan exhibit foamy oil behavior which is accompanied by sand production, leading to anomalously high oil recovery and lower gas oil ratio (Chugh et al., 2000). These observations suggest that the foamy oil flow might be physically linked to sand production. It is apparent that some additional factors, which remain to be discovered, are involved in making the foamy solution gas possible at field rates of decline. One possible mechanism is the synergistic influence of sand influx into the production wells. Allowing 1–3% sand to enter the wellbore with the fluids can result in propagation of a front of sharp pressure gradients away from the wellbore. These sharp pressure gradients occur at the advancing edge of solution gas drive. It is still unknown how far from the wellbore the dilated zone can propagate.

Thus, foamy oil is formed in solution gas drive reservoirs when gas is released from solution as reservoir pressure declines. It has been noted that the oil at the wellhead of these heavy-oil reservoirs resembles the form of foam, hence the term *foamy oil*. The gas initially exists in the form of small bubbles within individual pores in the rock. As time passes and pressure continues to decline, the bubbles grow to fill the pores. With further declines in pressure, the bubbles created in different locations become large enough to coalesce into a continuous gas phase. Once the gas phase becomes continuous (ie, when gas saturation exceeds the critical level—the minimum saturation at which a continuous gas phase exists in porous media—traditional two-phase (oil and gas) flow with classical relative permeability occurs) (Chen et al., 2006). As a result, the production gas–oil ratio (GOR) increases rapidly after the critical gas saturation has been exceeded.

Reservoirs that exhibit foamy oil behavior are typically characterized by the appearance of an oil-continuous foam at the wellhead. When oil is produced as this nonequilibrium mixture, reservoirs can perform with higher than expected rates of production: up to 30 times that predicted by Darcy's Law, and lower than expected production GORs (Poon and Kisman, 1992). Moreover, foamy oil flow is often accompanied by sand production along with the oil and gas—the presence of sand at the wellhead leads to sand dilation and the presence of high porosity, high permeability zones (wormholes) in the reservoir (Maini, 1999, 2001). It is generally believed that in the field, the high rates and recoveries observed are the combination of the foamy oil mechanism and the presence of these wormholes.

However, the actual structure of foamy oil flow and its mathematical description are still not well understood. Much of the earlier discussion of such flows was based on the concept of microbubbles, that is, bubbles that are much smaller than the average pore-throat size and are thus free to move with the oil during flow (Sheng et al., 1999). Dispersion of this type can be produced only by nucleation of a very large number of bubbles (explosive nucleation) and by the availability of a mechanism that prevents these bubbles from growing into larger bubbles with a decline in pressure. Another hypothesis for the structure of foamy oil flow is that much larger bubbles migrating with the oil, with the dispersion created by breakup of bubbles during migration. The major difference between conventional solution gas drive and foamy solution gas drive is that the pressure gradient in the latter is strong enough to mobilize gas clusters once they have grown to a certain size (Maini, 1999).

1.6 HEAVY OIL

Heavy oil is a *type* of petroleum that is different from conventional petroleum insofar as they are much more difficult to recover from the subsurface reservoir. Heavy oil, particularly heavy oil formed by biodegradation of organic deposits, are found in shallow reservoirs, formed by unconsolidated sands. This characteristic, which brings about difficulties during well drilling and completion operations, may become a production advantage due to higher permeability.

Heavy oil has a much higher viscosity (and lower API gravity) than conventional petroleum, and recovery of these petroleum types usually requires thermal stimulation of the reservoir. When petroleum occurs in a reservoir that allows the crude material to be recovered by pumping operations as a free-flowing dark to light colored liquid, it is often referred to as *conventional petroleum*. During the past two-to-three decades the term *black oil* has been introduced into the petroleum lexicon. This has only served to confuse the somewhat already confusing terminology applied to petroleum, heavy oil, and tar sand bitumen since the term refers to color rather than to any meaningful properties or to recovery behavior (Speight, 2014b, 2015b).

Very simply, heavy oil is a type of crude oil which is very viscous and does not flow easily. The common characteristic properties (relative to conventional crude oil) are high specific gravity, low hydrogen to carbon ratios, high carbon residues, and high contents of asphaltenes, heavy metals, sulfur, and nitrogen. Specialized recovery and refining processes are required to produce more useful fractions, such as naphtha, kerosene, and gas oil.

Heavy oil is an oil resource that is characterized by high viscosities (ie, resistance to flow) and high densities compared to conventional oil. Most heavy oil reservoirs originated as conventional oil that formed in deep formations, but migrated to the surface region where they were degraded by bacteria and by weathering, and where the lightest hydrocarbons escaped. Heavy oil is deficient in hydrogen and has high carbon, sulfur, and heavy metal content. Hence, heavy oil requires additional processing (upgrading) to become a suitable refinery feedstock for a normal refinery.

There are large resources of heavy oil in Canada, Venezuela, Russia, the United States, and many other countries. The resources in North America alone provide a small percentage of current oil production (approximately 2%), and existing commercial technologies could allow for significantly increased production. Under current economic conditions, heavy oil can be profitably produced, but at a smaller profit margin than for conventional oil, due to higher production costs and upgrading costs in conjunction with the lower market price for heavier crude oils. In fact, heavy oil accounts for more than double the resources of conventional oil in the world and heavy oil offers the potential to satisfy current and future oil demand. Not surprisingly, heavy oil has become an important theme in the petroleum industry with an increasing number of operators getting involved or expanding their plans in this market around the world.

However, heavy oil is more difficult to recover from the subsurface reservoir than conventional or light oil. A very general definition of heavy oils has been and remains based on the API gravity or viscosity, and the definition is quite arbitrary, although there have been attempts to rationalize the definition based upon viscosity, API gravity, and density. For example, heavy oils were considered to be those crude oils that had gravity somewhat less than 20° API with the heavy oils falling into the API gravity range 10° to 15°. For example, Cold Lake heavy crude oil has an API gravity equal to 12° and tar sand bitumen usually have an API gravity in the range 5° to 10° (Athabasca bitumen= 8° API). Residua would vary depending upon the temperature at which distillation was terminated but usually vacuum residua are in the range 2° to 8° API (Speight, 2000, 2014a; Speight and Ozum, 2002; Parkash, 2003; Hsu and Robinson, 2006; Gary et al., 2007).

Heavy oil has a much higher viscosity (and lower API gravity) than conventional petroleum and recovery of heavy oil usually requires thermal stimulation of the reservoir. The generic term heavy oil is often applied to a crude oil that has less than 20° API and usually, but not always, sulfur content higher than 2% by weight (Speight, 2000). Furthermore, in contrast to conventional crude oils, heavy oils are darker in color and may even be black.

The term heavy oil has also been arbitrarily (incorrectly) used to describe both the heavy oils that require thermal stimulation of recovery from the reservoir and the bitumen in bituminous sand (tar sand) formations from which the heavy bituminous material is recovered by a mining operation. *Extra heavy oil* is a nondescript term (related to viscosity) of little scientific meaning which is usually applied to tar sand bitumen, which is generally incapable of free flow under reservoir conditions. The general difference is that extra heavy oil may have properties similar to tar sand bitumen but, unlike tar sand bitumen, has some degree of mobility in the reservoir or deposit (Table 1.2) (Delbianco and Montanari, 2009; Speight, 2014a). Extra heavy oils can flow at reservoir temperature and can be produced economically, without additional viscosity-reduction techniques, through variants of conventional processes such as long horizontal wells, or multilaterals. This is the case, for instance, in the Orinoco (Venezuela) or in offshore reservoirs off the coast of Brazil, but once outside of the influence of the high reservoir temperature, these oils are too viscous at the surface to be transported through conventional pipelines and require heated pipelines for transportation. Alternatively, the oil must be partially upgraded or fully upgraded (see chapter: Thermal Methods of Recovery) or diluted with a light hydrocarbon (such as aromatic naphtha) to create a mix that is suitable for transportation.

In the context of the current section, the methods outlined in this book for heavy oil recovery focus on heavy oil with an API gravity of less than 20° with a variable sulfur content that may bear a general relationship to the API gravity (eg, see Table 1.4) (Speight, 2000). However, it must be recognized that some of the heavy oil is sufficiently liquid to be recovered by pumping operations and some is already being recovered by this method. Recovery depends not only on the characteristics of the oil but also on the characteristics of the reservoir—including the temperature

Table 1.4 API Gravity and Sulfur Content of Selected Heavy Oils and Bitumen

Country	Crude Oil	API	Sulfur % w/w
Brazil	Albacor Leste	18.9	0.66
Canada (Alberta)	Athabasca	8.0	4.8
Canada (Alberta)	Cold Lake	13.2	4.11
Canada (Alberta)	Lloydminster	16.0	2.60
Canada (Alberta)	Wabasca	19.6	3.90
Chad	Bolobo	16.8	0.14
Chad	Kome	18.5	0.20
China	Bozhong	16.7	0.30
China	Qinhuangdao	16.00	0.26
China	Zhao Dong	18.4	0.25
Colombia	Castilla	13.3	0.22
Colombia	Chichimene	19.8	1.12
Congo	Yombo	17.7	0.33
Ecuador	Ecuador Heavy	18.2	2.23
Ecuador	Napo	19.2	1.98
Guatemala	Xan-Coban	18.7	6.00
Indonesia	Kulin (South)	19.8	0.30
Iran	Soroosh (Cyrus)	18.1	3.30
Kuwait	Eocene	18.4	4.00

(Continued)

Table 1.4 (Continued)

Country	Crude Oil	API	Sulfur % w/w
USA (California)	Arroyo Grande/Edna	14.9	2.03
USA (California)	Belridge (South)	13.7	1.00
USA (California)	Beta Offshore	15.9	3.60
USA (California)	Beta Offshore	16.9	3.30
USA (California)	Hondo Monterey	19.4	4.70
USA (California)	Hondo Monterey	17.2	4.70
USA (California)	Huntington Beach	19.4	2.00
USA (California)	Huntington Beach	14.4	0.90
USA (California)	Kern River	13.3	1.10
USA (California)	Kern River	14.4	1.02
USA (California)	Lost Hills	18.4	1.00
USA (California)	Midway Sunset	12.6	1.60
USA (California)	Midway Sunset	11.0	1.55
USA (California)	Monterey	12.2	2.30
USA (California)	Mount Poso	16.0	0.70
USA (California)	Newport Beach	15.1	2.00
USA (California)	Point Arguello	19.5	3.50
USA (California)	Point Pedernales	15.9	5.10
USA (California)	San Ardo	12.2	2.30
USA (California)	San Joaquin Valley	15.7	1.20
USA (California)	Santa Maria	13.7	5.20
USA (California)	Sockeye	19.6	5.30
USA (California)	Sockeye	15.9	5.40
USA (California)	Torrance	18.2	1.80
USA (California)	Wilmington	18.6	1.59
USA (California)	Wilmington	16.9	1.70
USA (Mississippi)	Baxterville	16.3	3.02
Venezuela	Bachaquero	16.3	2.35
Venezuela	Bachaquero	12.2	2.80
Venezuela	Bachaquero	14.4	2.52
Venezuela	Bachaquero Heavy	10.7	2.78
Venezuela	Boscan	10.1	5.50
Venezuela	Hamaca	8.4	3.82
Venezuela	Jobo	9.2	4.10
Venezuela	Laguna	10.9	2.66
Venezuela	Lagunillas Heavy	17.0	2.19
Venezuela	Merey	18.0	2.28
Venezuela	Morichal	12.2	2.78
Venezuela	Pilon	14.1	1.91
Venezuela	Tia Juana Pesado	12.1	2.70
Venezuela	Tia Juana Heavy	18.2	2.24
Venezuela	Tia Juana Heavy	11.6	2.68
Venezuela	Tremblador	19.0	0.80
Venezuela	Zuata	15.7	2.69

For reference, Athabasca tar sand bitumen has API = 8° and sulfur content = 4.8–5.0% w/w.

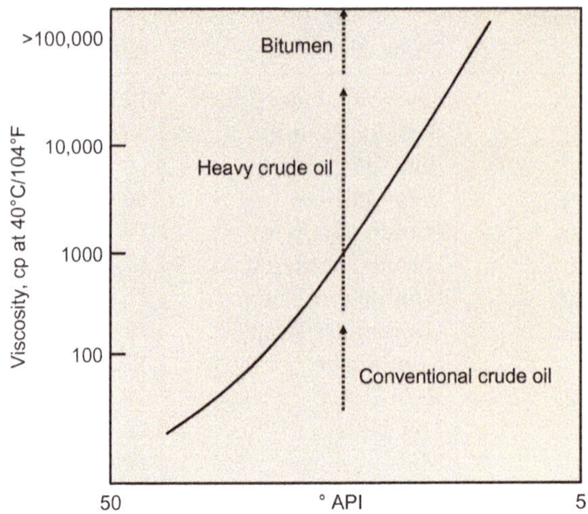

Figure 1.6 General relationship of viscosity to API gravity.

of the reservoir and the pour point of the oil (see also chapter: Evaluation of Reservoir Fluids). These heavy oils fall into a range of high viscosity (Fig. 1.6) and the viscosity is subject to temperature effects (Fig. 1.7), which is the reason for the application of thermal methods to heavy oil recovery.

Finally, the formation of stable emulsions of water and oil is a phenomenon frequently found in the production of heavy oils. Emulsions are formed during simultaneous flow of oil and water, although it is also supposed to occur while still in the reservoir. The flow of the mixture of liquids through devices and equipment that impose a high shear rate, such as in pumps and valves, in pipe singularities and even along lines, will induce emulsion formation. Gas and solid particulates are additional factors that increase the shear rate, intensifying emulsification. The shear rate is a strong factor in emulsion formation, but rheology and fluid properties also play important roles. In addition, the tendency of heavy oil to form foam (foamy oil) must also be taken into account during the release of dissolved gas, in a type of gas and liquid limiting compressible emulsion. All such characteristics will influence the flow in the reservoir, wellbore, and flow lines as well as having an impact on other processes, such as separation.

Finally, some heavy oils found in offshore fields have a high content of organic acids. The acidity rate, measured by TAN (*total acid number*), is particularly important to refining processes. Currently refineries need to take precautions to prevent acidic corrosion of the equipment as well as

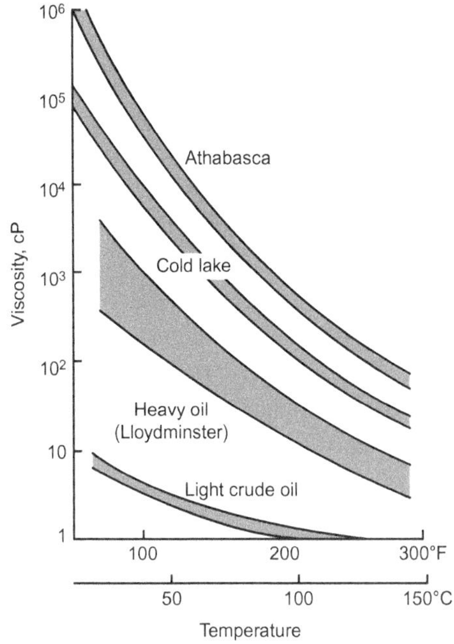

Figure 1.7 Variation of viscosity with temperature.

contributing to the precipitation of organo-metallic salts, causing buildups in processing equipment (Speight, 2014b, c).

1.7 TAR SAND BITUMEN

The term tar sand, also known as oil sand (in Canada), or (more correctly) bituminous sands, commonly describes sandstones or friable sand (quartz) impregnated with viscous bitumen (a hydrocarbonaceous material soluble in carbon disulfide). Significant amounts of fine material, usually largely or completely clay, are also present. The degree of porosity varies from deposit to deposit and is an important characteristic in terms of recovery processes. The bitumen makes up the desirable fraction of the tar sand from which liquid fuels can be derived (Speight, 2013d, 2014a). However, the bitumen is usually not recoverable by conventional petroleum production techniques (Speight, 2013b, 2014a). Furthermore, the properties and composition of the tar sands and the bitumen significantly influence the selection of recovery and treatment processes and vary among deposits. In the tar sands that are recognized as being *water-wet* (rather than *bitumen-wet*) in the Athabasca deposit, a layer of water surrounds the sand grain, and the bitumen partially fills the voids between the wet grains. On the other hand, the Utah tar sands lack the water layer and the bitumen is directly in contact with the sand grains without any intervening water. Typically, more than 99% w/w of the mineral matter is composed of quartz and clay minerals. The Utah deposits range from largely consolidated sands with low porosity and permeability to, in some cases, unconsolidated sands. High concentrations of heteroatoms tend to increase viscosity, increase the bitumen–sand reactions and bitumen–minerals reactions making processing more difficult.

Tar sands are sedimentary rocks containing bitumen, a viscous hydrocarbonaceous mixture. Tar sand deposits may be divided into two major types: (1) a breached reservoir where erosion has removed the capping layers from a reservoir of relatively viscous material, allowing the more volatile petroleum hydrocarbons to escape; and (2) deposits that formed when liquid petroleum seeps into a near-surface reservoir from which the more volatile constituents escaped. In either type of deposit, the lighter, more volatile constituents have escaped to the environment, leaving the less volatile constituents in place which are altered by contact with air, bacteria, and groundwater. Because of the very viscous nature of the bitumen in tar sands, tar sands cannot be processed by the typical petroleum production techniques (Speight, 2013d, 2014a).

Tar sand deposits occur throughout the world and the largest deposits occur in Alberta, Canada (the Athabasca, Wabasca, Cold Lake, and Peace River areas), and in Venezuela. Smaller deposits occur in the United States, with the larger individual deposits in Utah, California, New Mexico, and Kentucky.

The term bitumen (also, on occasion, referred to as native asphalt, and extra heavy oil), which is the organic component of tar sand, includes a wide variety of reddish brown to black materials of semisolid, viscous to brittle character that can exist in nature with no mineral impurity or with mineral matter contents that exceed 50% by weight. Bitumen is frequently found filling pores and crevices of sandstone, limestone, or argillaceous sediments, in which case the organic and associated mineral matrix is known as *rock asphalt* (Abraham, 1945). However, bitumen from different deposits (eg, deposits in the United States and Canada, exhibit a variety of properties) (Tables 1.5 and 1.6).

Bitumen is a naturally-occurring material that is found in deposits where the permeability is low and passage of fluids through the deposit can only be achieved by prior application of fracturing techniques

Table 1.5 Properties of Bitumen From Different California Tar Sand Deposits

Deposit/Field	Reservoir Temperature (F)	Reservoir Permeability (mD)	Reservoir Porosity	API	Viscosity, cP reservoir conditions	Viscosity, cP (80°F)	Viscosity, cP (87°F)	Viscosity, cP (100°F)	Viscosity, cP (200°F)	Sulfur %, w/w
Arroyo Grande		700	38	8				15,000		3–5
Basal Foxen		300	25	9–17 (9.5)		47,000				4–5
Cat Canyon				0–12 (6)						
Brooks Sand	135	1400–5000	32	0–12	15,000					
S Sand	110	3450	37		12,000–1,000,000					
Casmalia		<1	48							
Zaca-Sisquoc			35	4–6						
Oxnard (Vaca)		6000	35	5				500,000	2000	6–7
Paris Valley		3700	32							1.5
Upper Lobe							227,000			
Lower Lobe							23,000			
Midway-Sunset										
Webster Sands	100	1300	28	14				1650		

Table 1.6 Specific Gravity, API Gravity, and Viscosity of Various Bitumen Samples

Source	Specific Gravity	API Gravity	Viscosity, cP	F
Athabasca (Canada)				
Mildred-Ruth Lakes	1.025	6.5	35,000	100
Abasand	1.027	6.3	500,000	100
	1.034	5.4	570,000	100
Ells River	1.008	8.9	25,000	
Utah (United States)				
Asphalt Ridge			610,000	140
Tar Sand Triangle			760,000	140
Sunnyside			1,650,000	100
California				
Arroyo Grande	1.055	2.6	1,300,000	220

(Demaison, 1977). Tar sand bitumen is a high-boiling material with little, if any, material boiling below 350°C (660°F) and the boiling range is approximately the same as the boiling range of an atmospheric residuum.

Tar sands have been defined in the United States (FE-76-4) as:

> *...the several rock types that contain an extremely viscous hydrocarbon which is not recoverable in its natural state by conventional oil well production methods including currently used enhanced recovery techniques. The hydrocarbon-bearing rocks are variously known as bitumen-rocks oil, impregnated rocks, oil sands, and rock asphalt.*

The recovery of the bitumen depends to a large degree on the composition and construction of the deposit. Generally, the bitumen found in tar sand deposits is an extremely viscous material that is *immobile under reservoir conditions* and cannot be recovered through a well by the application of secondary or enhanced recovery techniques. Extra heavy oil, which is often likened to tar sand bitumen because of similarities in the properties of the two, has a degree of mobility under reservoir or deposit conditions.

The expression tar sand is commonly used in the petroleum industry to describe sandstone reservoirs that are impregnated with a heavy, viscous black crude oil that cannot be retrieved through a well by conventional production techniques (FE-76-4, above). However, the term tar sand is actually a misnomer; more correctly, the name *tar* is usually applied to the heavy product remaining after the destructive distillation of coal or other organic matter. Thus, it is incorrect to refer to native bituminous materials as *tar* or *pitch*. Although the word tar is descriptive of the black, heavy bituminous material, it is best to avoid its use with respect to natural materials and to restrict the meaning to the volatile or near-volatile products produced in the destructive distillation of such organic substances as coal. In the simplest sense, pitch is the distillation residue of the various types of tar. In fact, alternative names, such as *bituminous sand* or *oil sand*, are gradually finding usage, with the former name (bituminous sands) more technically correct. The term oil sand is also used in the same way as the term tar sand, and these terms are used interchangeably throughout this text.

Bituminous rock and bituminous sand are those formations in which the bituminous material is found as a filling in veins and fissures in fractured rocks or impregnating relatively shallow sand, sandstone, and limestone strata. These terms are, in fact, correctly geological descriptions of tar sand. The deposits contain as much as 20% bituminous material, and if the organic material in the rock matrix is bitumen, it is usual (although chemically incorrect) to refer to the deposit as rock asphalt to distinguish it from bitumen that is relatively mineral free. A standard test (ASTM D4) is available for determining the bitumen content of various mixtures with inorganic materials, although the use of word bitumen as applied in this test might be questioned and it might be more appropriate to use the term *organic residues* to include tar and pitch. If the material is of the asphaltite-type or asphaltoid-type, the corresponding terms should be used: rock asphaltite or rock asphaltoid.

Bituminous rocks generally have a coarse, porous structure, with the bituminous material in the voids. A much more common situation is that in which the organic material is present as an inherent part of the rock composition insofar as it is a diagenetic residue of the organic material detritus that was deposited with the sediment. The organic components of such rocks are usually refractory and are only slightly affected by most organic solvents. A special class of bituminous rocks that has achieved some importance is the so-called *oil shale* (Scouten, 1990; Speight, 2012b). These are argillaceous, laminated sediments of generally high organic content that can be thermally decomposed to yield appreciable amounts of oil, commonly referred to as *shale oil*. Oil shale does not yield shale oil without the application of high temperatures and the ensuing thermal decomposition that is necessary to decompose the organic material (*kerogen*) in the shale. The kerogen produces a liquid product (shale oil) by thermal decomposition at high temperature (>500°C, >930°F). The raw oil shale can even be used directly as a fuel akin to a low-quality coal. Indeed, oil shale deposits have been exploited as such for several centuries and shale oil has been produced from oil shale since the nineteenth century.

Sapropel is an unconsolidated sedimentary deposit rich in bituminous substances. It is distinguished from peat in being rich in fatty and waxy substances and poor in cellulosic material. When consolidated into rock, sapropel becomes oil shale, bituminous shale, or boghead coal. The principal components are certain types of algae that are rich in fats and waxes. Minor constituents are mineral grains and decomposed fragments of spores, fungi, and bacteria. The organic materials accumulate in water under reducing conditions.

The properties and composition of the tar sands and the bitumen significantly influence the selection of recovery and treatment processes and vary among deposits. The general composition of typical deposits at the P.R. Spring Special Tar Sand Area showed a porosity of 8.4 volume percent with the solid/liquid fraction being 90.5% w/w sand, 1.5% w/w fines, 7.5% w/w bitumen, and 0.5% w/w water. Utah deposits range from largely consolidated sands with low porosity and permeability to, in some cases, unconsolidated sand. High concentrations of heteroatoms (nitrogen, oxygen, sulfur, and metals) tend to increase viscosity, increase the bonding of bitumen with minerals, reduce yields, and make processing more difficult.

Finally, and briefly, *extra heavy oil is* a material that occurs in the solid or near-solid state and is generally has mobility under reservoir conditions. However, extra heavy oil is a recently-evolved term (related to viscosity) of little scientific meaning. While this type of oil may resemble tar sand bitumen and does not flow easily, extra heavy oil is generally recognized as having mobility in the reservoir compared to tar sand bitumen, which is typically incapable of mobility (free flow) under reservoir conditions. For example, the tar sand bitumen located in Alberta Canada is not mobile in the deposit and requires extreme methods of recovery to recover the bitumen. On the other hand, much of the extra heavy oil located in the Orinoco belt of Venezuela requires recovery methods that are less extreme because of the mobility of the material in the reservoir. Whether the mobility of extra heavy oil is due to a high reservoir temperature (that is higher than the pour point of the extra heavy oil) or due to other factors is variable and subject to localized conditions in the reservoir.

In order to utilize the extra heavy oil produced in Venezuela, the government-run oil company (PDVSA) has developed Orimulsion fuel, which is a dispersion of extra heavy oil and approximately 30% water which is targeted for use as a boiler fuel in applications such as power generation and industrial use (Shah et al., 2010). However, concerns about the environmental impact of the use of Orimulsion have been raised owing to the relatively high levels of sulfur, nickel, and vanadium compared with other fuel oils (Miller and Srivastava, 2000). Technologies such as *selective catalytic reduction*, *flue gas desulfurization*, and *electrostatic precipitation* are suitable for cleanup of the exhaust emissions (Mokhatab et al., 2006; Speight, 2007).

1.8 RATIONALIZATION OF THE DEFINITIONS

Generally, heavy oil and tar bitumen are characterized by high viscosities (ie, resistance to flow measured in centipoises or cP) and high density compared to conventional oil. Attempts have been made to define heavy oil as an oil with a gas-free viscosity between 100 cP and 10,000 Cp at reservoir temperature (tar sand bitumen is incorrectly defined to have a viscosity greater than 10,000 cP, and may be as high as 10,000,000 cP). Attempting to draw a finite line in order to define heavy oil on the one side of the line and bitumen on the other side of the line is meaningless and incorrect and offers very little in the way of scientific integrity since the standard test methods that are used to determine viscosity suffer from reproducibility and accuracy (Speight, 2015b).

On the other hand, heavy oil is slightly less dense than water with an API gravity between 10° and 20°. In addition, heavy oil can flow in some reservoirs at downhole temperatures and/or with in situ solution gas, but at the surface, it is a thick, black, viscous fluid, which typically has an API gravity <10°. It should also be noted that the term oil sands as used in Canada was created for incentive tax purposes for those heavy crude oils found above a certain latitude where the infrastructure was almost nonexistent. For correctness, the terms heavy oil, extra-heavy oil, and bitumen should not be used interchangeably.

As already noted, the generic term heavy oil is often applied to petroleum that has an API gravity of less than 20° and those materials having less than 10° API than have been referred to as bitumen (Speight, 2014a, 2015b). Following from this convenient generalization, there has also been an attempt to classify petroleum, heavy oil, and tar sand bitumen using the viscosity scale, and 10,000 centipoise being the fine line of demarcation between heavy oil and tar sand bitumen. Use of such a system leads to confusion when having to differentiate between a material having a viscosity of 9950 cP and one having a viscosity of 10,050 cP as well as taking into account the limits of accuracy of the method of viscosity determination (Speight, 2014a, 2015b). Such definitions based on a single property should be discounted as being of little scientific value. Whether the limits are the usual laboratory experimental difference (±3%) or more likely the limits of accuracy of the method (±5% to ±10%) there is the question of viability and reliability of the definition. In fact, the, inaccuracies (ie, the limits of *experimental difference*) of the method of measuring viscosity also increase the potential for misclassification using this (or any) single property for classification purposes.

Thus, in order to classify petroleum, heavy oil, and bitumen the use of a single parameter such as viscosity is not enough and any attempt to classify petroleum, heavy oil, and bitumen on the basis of a single property is no longer sufficient to define the nature and properties of petroleum and petroleum-related materials. The general classification method used for conventional petroleum, heavy oil, extra heavy oil, and tar sand bitumen involves not only an inspection of several properties but also some acknowledgment of the method of recovery.

Petroleum is referred to generically as a *fossil energy resource* and is further classified as a *hydrocarbon resource* and, for illustrative (or comparative) purposes in this text, coal and oil shale kerogen have also been included in this classification. However, the inclusion of coal and oil shale under the broad classification of hydrocarbon resources has required (incorrectly) that the term hydrocarbon be expanded to include the high molecular weight (macromolecular) nonhydrocarbon heteroatomic species that constitute coal and oil shale kerogen. Heteroatomic species are those organic constituents that contain atoms other than carbon and hydrogen, for example, nitrogen, oxygen, sulfur, and metals (nickel and vanadium) as an integral part of the molecular matrix.

Use of the term organic sediments is more correct and to be preferred (Fig. 1.8). The inclusion of coal and oil shale kerogen in the category hydrocarbon resources is due to the fact that these two natural resources (coal and oil shale kerogen) will produce hydrocarbons on high-temperature processing (Fig. 1.9). Therefore, if either coal and/or oil shale kerogen is to be included in the term hydrocarbon resources, it is more appropriate that they be classed as *hydrocarbon-producing resources* under the general classification of organic sediments. Thus, fossil energy resources divide into two classes: (1) naturally occurring hydrocarbons (petroleum, natural gas, and natural waxes); and (2) hydrocarbon sources (oil shale and coal), which may be made to generate hydrocarbons by the application of conversion processes. Both classes may very aptly be described as organic sediments.

Whenever attempting to define or classify tar sand bitumen, it is always necessary to return to the definition as given by the United States Federal Energy Administration (FE-76-4), thus:

> *...the several rock types that contain an extremely viscous hydrocarbon which is not recoverable in its natural state by conventional oil well production methods including currently used enhanced recovery techniques. The hydrocarbon-bearing rocks are variously known as bitumen-rocks oil, impregnated rocks, oil sands, and rock asphalt.*

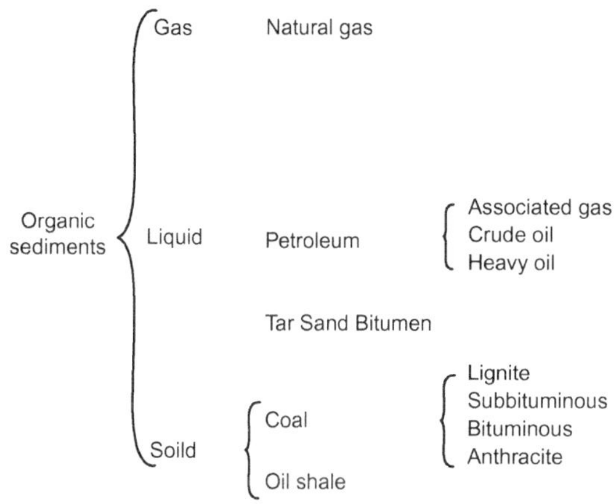

Figure 1.8 Classification of fossil fuel as organic sediments.

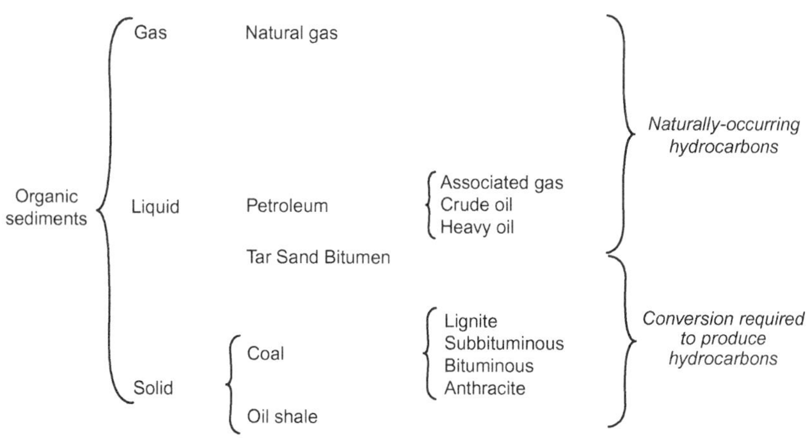

Figure 1.9 Classification of fossil fuels as hydrocarbon resources and hydrocarbon producing resources.

By inference, petroleum and heavy oil are recoverable by well production methods and currently used enhanced recovery techniques. For convenience, it is assumed that before depletion of the reservoir energy, conventional crude oil is produced by primary and secondary techniques while heavy oil requires tertiary (enhanced) oil recovery (EOR) techniques. While this is an oversimplification, it may be used as a general guide. The term *natural state* cannot be defined out of context and in the context of FEA Ruling 1976-4 and the term is defined in terms of the composition of the heavy oil or bitumen. The final determinant of whether or not a reservoir is a tar sand deposit is the character of the viscous phase (bitumen) and the method that is required for recovery. Generally, bitumen in the natural state is solid or near solid at room temperature and is solid or near solid at reservoir temperature. In other words, tar sand bitumen in the natural state is immobile in the deposit and requires conversion or extreme stimulation for recovery as might be afforded by an in situ combustion process. In the in situ combustion process, the goal can be either (1) conversion of a low portion of the bitumen, or (2) conversion of as much as possible of the bitumen to low-boiling products (see chapters: Nonthermal Methods of Recovery; Thermal Methods of Recovery). In the former case (ie, production of the controlled amount of a low-boiling product), the product can act as a solvent to increase the flow rate of the oil ahead of the combustion front. In the latter case (ie, production of as much as possible of a low-boiling product), the goal is to convert as much as possible of the bitumen to low-boiling products thereby reducing the amount of upgrading in a surface refinery.

Thus, by this definition (FE-76-4), tar sand bitumen is not crude oil and it is set apart from conventional crude oil and heavy crude oil insofar as it cannot be recovered from a deposit by the use of conventional (including enhanced) oil recovery techniques as set forth in the June 1979 Federal Energy Regulations. To emphasize this point, bitumen has been recovered commercially by mining and the hot water process, and is currently upgraded (converted to synthetic crude oil) by a combination of a thermal or hydrothermal process followed by product hydrotreating to produce a low-sulfur hydrocarbon product known as *synthetic crude oil*.

Tar sand bitumen is a naturally occurring material that is immobile in the deposit and cannot be recovered by the application of EOR technologies, including steam-based technologies. On the other hand, heavy oil is mobile in the reservoir and can be recovered by the application of EOR technologies, including steam-based technologies. Furthermore, since the most significant property of tar sand bitumen is its immobility under the conditions of temperatures and pressure in the deposit, the interrelated properties of API gravity (ASTM D287) and viscosity (ASTM D445) may present an indication (but only an indication) of the mobility of oil or immobility of bitumen, but in reality these properties only offer subjective descriptions of the oil in the reservoir. The most pertinent and objective representation of this oil or bitumen mobility is the *pour point* (ASTM D97) (see also chapter: Evaluation of Reservoir Fluids).

By definition, the pour point is the lowest temperature at which oil will move, pour, or flow when it is chilled without disturbance under definite conditions (ASTM D97). In fact, the pour point of an oil when used in conjunction with the reservoir temperature gives a better indication of the condition of the oil in the reservoir than the viscosity. Thus, the pour point and reservoir temperature present a more accurate assessment of the condition of the oil in the reservoir, being an indicator of the mobility of the oil in the reservoir. Indeed, when used in conjunction with reservoir temperature, the pour point gives an indication of the liquidity of the heavy oil or bitumen and, therefore, the ability of the heavy oil or bitumen to flow under reservoir conditions. In summary, the pour point is an important consideration because, for efficient production, additional energy must be supplied to the reservoir by a thermal process to increase the reservoir temperature beyond the pour point.

For example, Athabasca bitumen with a pour point in the range 50°C to 100°C (122°F to 212°F) and a deposit temperature of 4°C to 10°C (39°F to 50°F) is a solid or near solid in the deposit and will exhibit little or no mobility under deposit conditions. Pour points on the order of 35°C to 60°C (95°F to 140°F) have been recorded for the bitumen in Utah deposits with formation (deposit) temperatures on the order of 10°C (50°F). This indicates that the bitumen is solid (and immobile) within the deposit. The injection of steam to raise and maintain the reservoir temperature above the pour point of the bitumen and to enhance bitumen mobility is difficult, in some cases almost impossible. Conversely, when the reservoir temperature exceeds the pour point the oil is fluid in the reservoir and therefore mobile. The injection of steam to raise and maintain the reservoir temperature above the pour point of the bitumen and to enhance bitumen mobility is possible and oil recovery can be achieved.

A method that uses the pour point of the oil and the reservoir temperature (Fig. 1.10) adds a specific qualification to the term extremely

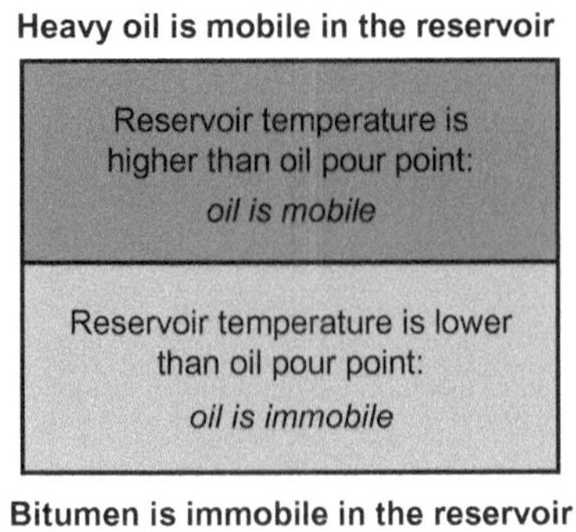

Figure 1.10 Use of pour point to define heavy oil and bitumen.

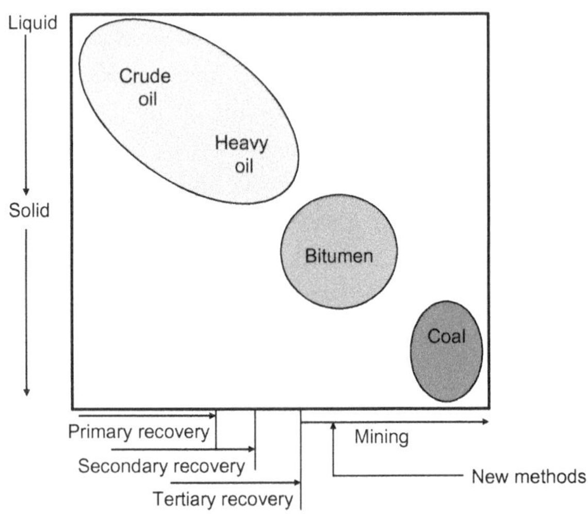

Figure 1.11 Schematic representation of the properties and recovery methods for crude oil, heavy oil, bitumen, and coal.

viscous as it occurs in the definition of tar sand and coupled with the measurement of the viscosity provides a more complete indication of the properties of the bitumen under the conditions prevalent in the deposit. In fact, when used in conjunction with the recovery method (Fig. 1.11), pour point offers more general applicability to the conditions of the oil in the reservoir or the bitumen in the deposit and comparison of the two temperatures (pour point and reservoir temperatures) shows promise and warrants further consideration. The measurement of the viscosity adds an extra dimension in terms of the degree of mobility of the bitumen.

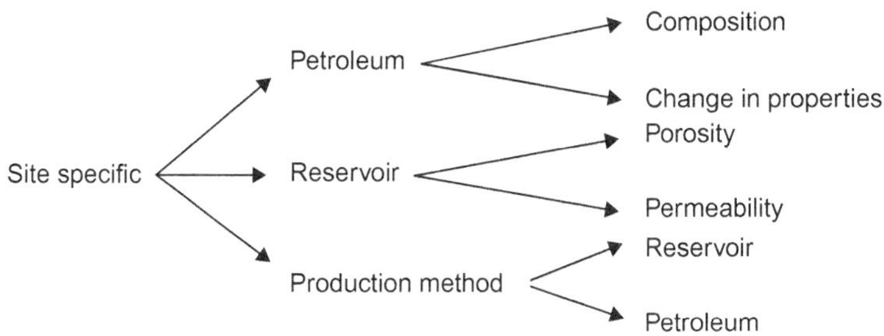

Figure 1.12 Representation of the changing parameters for crude oil and/or heavy oil.

In summary, supplies of conventional petroleum are depleted and heavy oil with tar sand bitumen resources are becoming recognized as important contributors to liquid fuels production. However, heavy oil cannot be defined or classified using a single property and, likewise, tar sand bitumen cannot be defined using a single property but bitumen can be defined by the recovery method. Furthermore, heavy oil is usually mobile in the reservoir whereas tar sand bitumen is immobile in the deposit.

In conclusion, heavy oil and tar sand bitumen are extremely viscous and production of these materials is difficult. Although the worldwide reserves of heavy oil and bitumen are enormous, difficult technical and economic hurdles are involved in their recovery methods. There are numerous alternative techniques for heavy oil and bitumen production (such as steam-based processes and cold production) but a prime factor limiting the efficiency of the recovery methods is the heterogeneity of the reservoir system and the deposit. The rocks in these reservoirs and deposits containing the heavy oils have porosity, permeability, connectivity, as well as mineral content that vary over short (vertical and horizontal) distances.

Finally, it is essentials to realize that (in the current context) of conventional petroleum and heavy oil, there are several parameters that can influence properties and recovery. These properties are usually site specific to the particular reservoir in which the crude oil or heavy oil is located (Fig. 1.12).

REFERENCES

Abraham, H., 1945. Asphalts and Allied Substances. Van Nostrand, New York.

Ancheyta, J., Speight, J.G., 2007. Hydroprocessing of Heavy Oils and Residua. CRC-Taylor and Francis Group, Boca Raton, Florida.

ASTM D4. Standard Test Method for Bitumen Content. Annual Book of Standards. American Society for Testing and Materials, West Conshohocken, Pennsylvania.

ASTM D97. Standard Test Method for Pour Point of Petroleum Products. Annual Book of Standards. American Society for Testing and Materials, West Conshohocken, Pennsylvania.
ASTM D287. Standard Test Method for API Gravity of Crude Petroleum and Petroleum Products (Hydrometer Method). Annual Book of Standards. American Society for Testing and Materials, West Conshohocken, Pennsylvania.
ASTM D445. Standard Test Method for Kinematic Viscosity of Transparent and Opaque Liquids (and Calculation of Dynamic Viscosity). Annual Book of Standards. American Society for Testing and Materials, West Conshohocken, Pennsylvania.
ASTM D4175. Standard Terminology Relating to Petroleum, Petroleum Products, and Lubricants. Annual Book of Standards. American Society for Testing and Materials, West Conshohocken, Pennsylvania.
Banat, I.M., 1995. Biosurfactant production and possible uses in microbial enhanced oil recovery and oil pollution remediation. Biores. Technol. 51, 1–12.
Bryant, R.S., Donaldson, E.C., Yen, T.F., Chilingarian, G.V., 1989. Microbial enhanced oil recovery. In: Donaldson, E.C., Chilingarian, G.V., Yen, T.F. (Eds.), Enhanced Oil Recovery II: Processes and Operations Elsevier, Amsterdam, Netherlands, pp. 423–450.
Bryant, R.S., Lindsey, R.P., 1996. World-wide applications of microbial technology for improving oil recovery Proceedings. SPE Symposium on Improved Oil Recovery. Society of Petroleum Engineers, Richardson, Texas, pp. 27–134.
Chen, Z., Huan, G., Ma, Y., 2006. Computational Methods for Multiphase Flows in Porous Media, in the Computational Science and Engineering Series, vol. 2. Society for Industrial and Applied Mathematics (SIAM), Philadelphia, Pennsylvania.
Chugh, S., Baker, R., Telesford, A., Zhang, E., 2000. Mainstream options for heavy oil: Part I—cold production. J. Can. Petrol. Technol. 39 (4), 31–39.
Clark, J.B., Munnecke, D.M., Jenneman, G.E., 1981. In situ microbial enhancement of oil production. Dev. Ind. Microbiol. 15, 695–701.
Cobb, C., Goldwhite, H., 1995. Creations of Fire: Chemistry's Lively History From Alchemy to the Atomic Age. Plenum Press, New York.
Delbianco, A., Montanari, R., 2009. Encyclopedia of Hydrocarbons, Volume III/New Developments: Energy, Transport, Sustainability. Eni S.p.A, Rome, Italy.
Demaison, G.J., 1977. Tar sands and supergiant oil fields. Am. Assoc. Petrol. Geol. Bull. 61, 1950–1961.
Forbes, R.J., 1958a. A History of Technology. Oxford University Press, Oxford, United Kingdom.
Forbes, R.J., 1958b. Studies in Early Petroleum Chemistry. E. J. Brill, Leiden, Netherlands.
Forbes, R.J., 1959. More Studies in Early Petroleum Chemistry. E.J. Brill, Leiden, Netherlands.
Forbes, R.J., 1964. Studies in Ancient Technology. E. J. Brill, Leiden, Netherlands.
Gary, J.G., Handwerk, G.E., Kaiser, M.J., 2007. Petroleum Refining: Technology and Economics, fifth ed. CRC Press, Taylor & Francis Group, Boca Raton, Florida.
Ghoshal, S., Sainik, V., 2013. Monitor and minimize corrosion in High-TAN Crude processing. Hydrocarbon Process 92 (3), 35–38.
Hsu, C.S., Robinson, P.R., 2006. Practical Advances in Petroleum Processing, vols. 1 and 2. Springer, New York.
IEA, 2005. Resources to Reserves: Oil & Gas Technologies for the Energy Markets of the Future. International Energy Agency, Paris, France.
Jenneman, G.E., 1989. The potential for in-situ microbial applications. Dev. Petrol. Sci. 22, 37–74.
Kane, R.D., Cayard, M.S., 2002. A Comprehensive Study on Naphthenic Acid Corrosion. NACE International, Houston, Texas, Corrosion 2002.

Khire, J.M., Khan, M.I., 1994a. Microbially enhanced oil recovery (MEOR). Part 1. Importance and mechanisms of microbial enhanced oil recovery. Enzyme Microb. Technol. 16, 170–172.

Khire, J.M., Khan, M.I., 1994b. Microbially enhanced oil recovery (MEOR). Part 2. Microbes and the subsurface environment for microbial enhanced oil recovery. Enzyme Microb. Technol. 16, 258–259.

Maini, B.B., 1999. Foamy oil flow in primary production of heavy oil under solution gas drive. Paper No. SPE 56541. Proceedings. Annual Technical Conference and Exhibition, Houston, Texas USA, October 3–6. Society of Petroleum Engineers, Richardson, Texas.

Maini, B.B., 2001. Foamy oil flow. Paper No. SPE 68885. SPE J. Pet. Tech., Distinguished Authors Series. Page 54–64, October. Society of Petroleum Engineers, Richardson, Texas.

Miller, C.A., Srivastava, R.K., 2000. The combustion of orimulsion and its generation of air pollutants. Energy Combust. Sci. 26, 131–160.

Mokhatab, S., Poe, W.A., Speight, J.G., 2006. Handbook of Natural Gas Transmission and Processing. Elsevier, Amsterdam, Netherlands.

Ohmes, R., 2014. Characterizing and tracking contaminants in opportunity crudes. Digital refining. <http://www.digitalrefining.com/article/1000893, Characterising_and_tracking_contaminants_in_opportunity_crudes_.html#.VJhFjV4AA> (accessed 1.11.14.).

Parkash, S., 2003. Refining Processes Handbook. Gulf Professional Publishing, Elsevier, Amsterdam, Netherlands.

Poon, D., Kisman, K., 1992. Non-Newtonian effects on the primary production of heavy oil reservoirs. J. Can. Petrol. Technol. 31 (7), 1–6.

Raiders, R.A., Knapp, R.M., McInerney, M.J., 1989. Microbial selective plugging and enhanced oil recovery. J. Ind. Microbiol. 4, 215–230.

Scouten, C., 1990. In: Speight, J.G. (Ed.), In Fuel Science and Technology Handbook Marcel Dekker Inc., New York, pp. 795–1053. (Chapters 25–31).

Shah, A., Fishwick, R., Wood, J., Leeke, G., Rigby, S., Malcolm Greaves, M., 2010. A review of novel techniques for heavy oil and bitumen extraction and upgrading. Energy Environ. Sci. 3, 700–714.

Shalaby, H.M., 2005. Refining of Kuwait's heavy crude oil: Materials challenges. Proceedings. Workshop on Corrosion and Protection of Metals. Arab School for Science and Technology. December 3–7, Kuwait.

Sheng, J.J., Maini, B.B., Hayes, R.E., Tortike, W.S., 1999. Critical review of foamy oil flow. Trans. Porous Media 35, 157–187.

Shennan, J.L., Levi, J.D., 1987. In situ microbial enhanced oil recovery. In: Kosaric, N., Cairns, W.L., Gray, N.C.C. (Eds.), Biosurfactants and Biotechnology Marcel Dekker, New York, pp. 163–180.

Speight, J.G., 1984. In: Kaliaguine, S., Mahay, A. (Eds.), Characterization of Heavy Crude Oils and Petroleum Residues Elsevier, Amsterdam, pp. 515.

Speight, J.G., 2000. The Desulfurization of Heavy Oils and Residua, second ed. Marcel Dekker, New York.

Speight, J.G., 2007. Natural Gas: A Basic Handbook. GPC Books, Gulf Publishing Company, Houston, Texas.

Speight, J.G., 2011. An Introduction to Petroleum Technology, Economics, and Politics. Scrivener Publishing, Salem, Massachusetts.

Speight, J.G., 2012a. Crude Oil Assay Database. Knovel, Elsevier, New York, Online version available at: <http://www.knovel.com/web/portal/browse/display?_EXT_KNOVEL_DISPLAY_bookid=5485&VerticalID=0>.

Speight, J.G., 2012b. Shale Oil Production Processes. Gulf Professional Publishing, Elsevier, Oxford, United Kingdom.

Speight, J.G., 2013a. Heavy Oil Production Processes. Gulf Professional Publishing, Elsevier, Oxford, United Kingdom.
Speight, J.G., 2013b. Oil Sand Production Processes. Gulf Professional Publishing, Elsevier, Oxford, United Kingdom.
Speight, J.G., 2013c. The Chemistry and Technology of Coal, third ed. CRC Press, Taylor & Francis Group, Boca Raton, Florida.
Speight, J.G., 2013d. Heavy and Extra Heavy Oil Upgrading Technologies. Gulf Professional Publishing, Elsevier, Oxford, United Kingdom.
Speight, J.G., 2014a. The Chemistry and Technology of Petroleum, fifth ed. CRC-Taylor and Francis Group, Boca Raton, Florida.
Speight, J.G., 2014b. High Acid Crudes. Gulf Professional Publishing, Elsevier, Oxford, United Kingdom.
Speight, J.G., 2014c. Oil and Gas Corrosion Prevention. Gulf Professional Publishing, Elsevier, Oxford, United Kingdom.
Speight, J.G., 2015a. Handbook of Hydraulic Fracturing. John Wiley & Sons Inc., Hoboken, New Jersey.
Speight, J.G., 2015b. Handbook of Petroleum Product Analysis, second ed. John Wiley & Sons Inc., Hoboken, New Jersey.
Speight, J.G., 2015c. Fouling in Refineries. Gulf Professional Publishing, Elsevier, Oxford, United Kingdom.
Speight, J.G., Arjoon, K.K., 2012. Bioremediation of Petroleum and Petroleum Products. Scrivener Publishing, Salem, Massachusetts.
Speight, J.G., Ozum, B., 2002. Petroleum Refining Processes. Marcel Dekker Inc., New York.
Stosur, G.J., 1991. Unconventional EOR concepts. Crit. Rep. Appl. Chem. 33, 341–373.
US EIA, 2011. Review of Emerging Resources. US Shale Gas and Shale Oil Plays. Energy Information Administration, United States Department of Energy, Washington, DC.
US EIA, 2013. Technically Recoverable *Shale Oil and Shale Gas R*esources: An Assessment of 137 Shale Formations in 41 Countries outside the United States. Energy Information Administration, United States Department of Energy, Washington, DC.
US EIA, 2014. Crude Oils and Different Quality Characteristics. Energy Information Administration, United States Department of Energy, Washington, DC, http://www.eia.gov/todayinenergy/detail.cfm?id=7110.
Yeung, T.W., 2014. Evaluating opportunity crude processing. Digital refining. <http://www.digitalrefining.com/article/1000644> (Accessed October 25, 2014).

CHAPTER 2

Occurrence and Reservoir Evaluation

2.1 INTRODUCTION

The Earth contains a finite supply of fossil fuels—the major fossil fuels are natural gas, petroleum, and coal—although there are questions about the real amounts of these fossil fuels remaining. The best current estimates for the longevity of each fossil fuel is estimated from the reserves–production ratio (BP, 2014), which gives an indication (in years) of how long each fossil fuel will last at the current rates of production. Thus, estimates vary from at least 50 years of crude oil at current rates of consumption to 300 years of coal at current rates of consumption with natural gas varying between the two extremes (Speight, 2011a, 2013). In addition, the amounts of natural gas and crude oil located in tight sandstone formations and in shale formations has added a recent but exciting twist to the amount of these fossil fuels remaining.

Thus the declining reserves of light crude oil have resulted in an increasing need to develop options to upgrade the abundant supply of known heavy oil reserves (IEA, 2005; Meyer and Attanasi, 2003; Meyer et al., 2007). In addition, there is considerable focus and renewed efforts on adapting recovery techniques to the production of heavy oil. Over the past decade, the demand for crude oil worldwide has substantially increased, straining the supply of conventional oil. This has led to consideration of alternative or insufficiently utilized energy sources, especially heavy crude oil to supplement short- and long-term needs. Heavy oil has been used as refinery feedstock for a considerable time, usually blending with more conventional feedstocks, but has commanded lower prices because of its lower quality relative to conventional oil.

Even though heavy oil is a member of the petroleum family, differences exist between heavy oil and conventional (light) oil, according to the relative amounts of volatile constituents. When the lower-boiling constituents are lost through natural processes after evolution from organic source materials, the oil becomes heavy, with a high proportion of

Introduction to Enhanced Recovery Methods for Heavy Oil and Tar Sands.
DOI: http://dx.doi.org/10.1016/B978-0-12-849906-1.00002-3

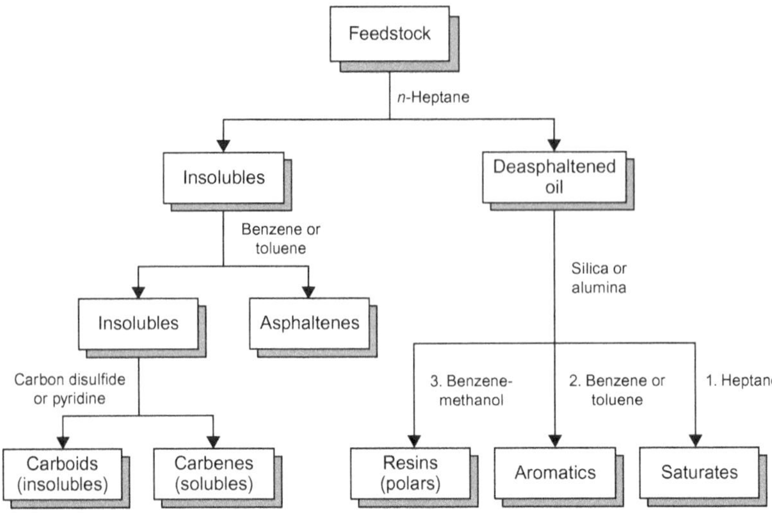

Figure 2.1 Separation scheme and nomenclature of different fractions of petroleum and heavy oil. Note: The nomenclature is an artifact of the separation process and does not guarantee chemical types or chemical homogeneity.

asphaltic molecules, and with substitution in the carbon network of heteroatoms, such as nitrogen, sulfur, and oxygen. Therefore, heavy oil, regardless of source, always contains the higher boiling constituents, the asphaltic materials (ie, the asphaltene and resin constituents) (Fig. 2.1). Typically, the asphaltene constituents produce coke (50%, w/w yield) as do the resins (35%, w/w coke) during thermal processes. Removal or reduction of the asphaltene constituents as well as the resin constituents, through deasphalting or, better still, leaving these constituents in the reservoir or deposit during recovery, improves the refinability of heavy oil and bitumen.

Detection of hydrocarbons in the subsurface during exploration takes a number of forms: direct identification of hydrocarbons at the surface, direct hydrocarbon indicators (DHI) in the subsurface and indirect indicators both at the surface and in the subsurface. Traditionally, oil exploration was primarily conducted by recognizing seeps of hydrocarbons at the surface. The Chinese, for example, used oil (mostly bitumen) obtained from seeps for use in medication, waterproofing, and warfare several thousand years ago. The ancient Chinese frequently dug shallow pits or horizontal tunnels at the seep locations in order to recover the oil. In Baku, Azerbaijan, there are still gas and oil seeps that are permanently alight and have been used to light caravanserai since the times of Marco Polo and the Silk Road. With the dawning of the modern era in Oil Creek, Pennsylvania, Colonel Edwin Drake drilled the first well to intentionally look for oil in the subsurface in 1859. Again, this was based on direct identification of seeped hydrocarbons at the surface. Initially the oil was used to provide kerosene for lamps but the later invention of automobiles drove up demand and ushered in the modern methods of oil exploration.

Around the turn of the 20th century and up until the 1950s, the main exploration tool used for finding oil was the use of intensive and detailed geological mapping. This was frequently in terrain that was remote and inhospitable. The early pioneers working their way through the jungles of Burma, the deserts of Iraq, or the mountains of Iran, would conduct detailed evaluations of the nature and distribution of rock units that could represent potential reservoirs, seals, and source units as well as frequency, orientation, and geological history of folds or faults that could act as traps for the migrating hydrocarbons. If all four of the features required for oil or gas to be created and trapped can be recognized in a region, then a variety of play concepts can be generated. Detailed local study might identify a suitable target (prospect) and then a shallow well would be drilled to test the features.

2.2 ORIGIN OF PETROLEUM AND HEAVY OIL

There are two theories on the origin of carbon fuels: (1) the abiogenic theory and (2) the biogenic theory (Kenney et al., 2001). The two theories have been intensely debated since the 1860s, shortly after the discovery of widespread occurrence of petroleum. It is not the intent of this section to sway the reader in his or her views of the origin of petroleum and natural gas. The intent is to place before the reader both points of view from which the reader can do further research and decide.

Generally, heavy oil was originally conventional oil that migrated from deep source rocks or deep reservoirs to the near surface, where the oil was biologically degraded and weathered by water. Bacteria feeding on the migrated conventional oil removed hydrogen and produced the denser, more viscous heavy oil. Lighter hydrocarbons may also have evaporated from the shallow, uncapped formations.

Therefore the origins of heavy oil are the same as the origins of conventional oil and a brief discussion of the means by which oil is formed is warranted here as a point of reference for heavy oil properties and behavior.

2.2.1 Abiogenic Origin

There have been several attempts at formulating theories that describe the detail of the origin of petroleum, of which the early postulates started with inorganic substances as a source material. For example, in 1866, Berthelot considered acetylene the basic material and crude oil constituents were produced from the acetylene.

$$CaCO_3 + \text{alkali metal} \rightarrow CaC_2\,(\text{calcium carbide})$$

$$CaC_2 + H_2O \rightarrow HC \equiv CH\,(\text{acetylene}) \rightarrow \text{petroleum}$$

There are also several more recent theories related to the formation of petroleum from nonbiogenic sources in the earth (Speight, 2014 and references cited therein).

Thus, the idea of abiogenic petroleum origin proposes that large amounts of carbon exist naturally, some in the form of hydrocarbons. Hydrocarbons are less dense than aqueous pore fluids, and migrate upward through deep fracture networks. Thermophilic, rock-dwelling microbial life-forms are in part responsible for the biomarkers found in petroleum. However, their role in the formation, alteration, or contamination of the various hydrocarbon deposits is not yet understood. Thermodynamic calculations and experimental studies confirm that *n*-alkanes (common petroleum components) do not spontaneously evolve from methane at pressures typically found in sedimentary basins, and so the theory of an abiogenic origin of hydrocarbons suggests deep generation (below 120 miles).

From the chemical point of view the inorganic theories are interesting because of their historical importance, but these theories have not received much attention. Geological and chemical methods have demonstrated the optical activity of petroleum constituents, the presence of thermally labile organic compounds, and the almost exclusive occurrence of oil in sedimentary rocks.

2.2.2 Biogenic Origin

It is now generally accepted, but not conclusively proven beyond a reasonable doubt, that the sole method of petroleum formation arises from the decay of organic matter in the earth. It is therefore from this scientific aspect that petroleum formation is referenced in this text. Nevertheless, alternative theories should not be dismissed until it can be conclusively established that petroleum formation is due to one particular aspect of geochemistry. In this process, the formation of petroleum constituents is believed to occur through the progressive chemical change of materials provided by microscopic aquatic organisms that were incorporated over eons in marine or near-marine sedimentary rocks. In fact, the details of petroleum genesis (*diagenesis*, *catagenesis*, and *metagenesis*) have long been a topic of interest. However, the details of this *transformation* and the mechanism by which petroleum is expelled from the source sediment and accumulates in the reservoir rock are still uncertain.

Transformation of some of this sedimentary material to petroleum probably began soon after deposition, with bacteria playing a role in the initial stages and clay particles serving as catalysts. Heat within the strata may have provided energy for the reaction, temperatures increasing more or less directly with depth. Some evidence indicates that most petroleum has formed at temperatures not exceeding about 100– 120°C (210–250°F), with the generation of petroleum hydrocarbons (and heteroatom constituents) commencing at temperatures as low as 65°C (150°F).

For this type of beginning, it is possible to formulate the production of heavy oil by any one or more of several processes. First, the petroleum-type product (often referred to as protopetroleum) may be expelled from its source rock as immature oil (Speight, 2014). There is general agreement that immature oils account for a small percentage of the heavy oil (Larter et al., 2006). Most heavy oil and what became tar sand bitumen is thought to begin by being expelled from source rocks as light or medium oil and migrating into a geologic trap. If the trap is later elevated into an oxidizing zone, several processes (such as water washing, bacterial degradation, and evaporation) can convert the oil to heavy oil and/or bitumen—the biodegradation may be aerobic and occur near the surface or anaerobic and occur in deep (air-deficient) subsurface reservoirs (Head et al., 2003; Larter et al., 2003, 2006). This explanation permits biodegradation to occur in any reservoir that has a water leg and has not been heated to more than 80°C (176°F). The controls on the biodegradation depend on local factors rather than basin-wide factors.

In addition, and like its conventional or light oil counterpart, the composition of heavy oil is greatly influenced not only by the nature of the precursors that eventually form the heavy oil but also by the relative amounts of these precursors (that are dependent upon the local flora and fauna as well as the prevalent conditions) that occur in the source material. Hence, it is not surprising that heavy oil, light conventional petroleum, can vary in composition with the location and age of the reservoir. The lower mobility of heavy oil also makes it extremely likely that two wells in the same reservoir will produce heavy oil with different characteristics. In fact, the immobility of bitumen within a tar sand deposit can show horizontal and vertical variations in composition and properties (Speight, 2014).

In a little more detail, the origin of the bitumen in the tar sand in Alberta (Canada) which have been investigated more than any other tar sand resources offers an explanation for the origin of the bitumen. However, the origin of the Alberta tar sand is a controversial subject. Two primary theories for the origin of these sands have been proposed: (1) the resource is the remnants of a once vast reserve of crude oil that, over extremely long periods of time, has escaped or been destroyed microbiologically, thus leaving behind some bitumen and also converting the lighter crude oil into bitumen through bacterial processes; and/or (2) the bitumen evolved from highly organic cretaceous shale formations and underground pressure forced the bitumen out of the organic-rich shale where it soaked into existing silt grade sediments and sand bodies.

In the former case, petroleum would be formed in the traditional manner, and then converted to bitumen by additional processes. As stated above, as organic matter is buried, the process of diagenesis compacts it into organic-rich sedimentary rocks. Further degrees of diagenesis then lead to the generation of chemically difficult-to-define protopetroleum which is then, through deeper burial and catagenesis, transformed through various means into crude oil, heavy oil, or bitumen, depending on the maturation conditions.

As temperatures and pressures increase (deeper burial) the process of catagenesis begins, which is the thermal degradation of kerogen to form hydrocarbon chains. The conditions of catagenesis determine the product, such that higher temperature and pressure lead to more complete transformation of the organic precursors and progressively lower-boiling lower-molecular weight products. On the other hand, if the catagenesis process did not proceed to completion (or near completion), bitumen was formed, which then (under the conditions of temperature and pressure) seeped into the surrounding sands.

Whatever the process, bitumen formation is complex and it is available for development to produce liquid fuels and other products.

2.3 OCCURRENCE AND DISTRIBUTION

Petroleum is found in sedimentary rocks throughout the world. In many places the oil has been degraded and the result is heavy oil or tar sand bitumen, depending upon the degree of degradation (Meyer and Attanasi, 2004).

Four main issues control the occurrence and distribution of conventional oil, heavy oil, and tar sand bitumen: (1) source; (2) reservoir; (3) seal; and (4) trap (Meyer et al., 2007; Speight, 2014). A source is a fine-grained rock unit containing sufficient organic matter so that when it is heated and/or placed under pressure (maturation), hydrocarbons are generated. If the organic matter is of marine algal origin, then the source rock is most likely to generate oil under optimum maturation conditions, whereas rocks dominated by land plant matter will tend to create gaseous hydrocarbons. The hydrocarbons are of lower gravity than the surrounding groundwater and, therefore, move away from and generally upwards (migration) from the source rock until they are trapped in a reservoir.

Geologists generally agree that petroleum deposits were formed from the remains of enormous quantities of aquatic plants and organisms which became mixed with sand and mud at the bottom of bodies of water. If a certain regime of temperatures and pressures existed, this biological material was converted into petroleum during geologic time. The pressure in the earth's crust forced the petroleum into the tiny open spaces between the grains of imbedded sandstone and other coarse textured strata (and not into vast underground pools). For petroleum to be present in commercial quantities some sort of trapping mechanism must have been present to prevent its escape. In addition, these traps must have not been breached by natural means after accumulation occurred. There are two major kinds of trapping mechanisms. Structural traps are created by a deformation of the earth's crust; the folding or faulting of rocks results in the entrapment of petroleum. Stratigraphic traps result from the relative difference in the porosity and permeability of the oil-bearing rocks compared to less porous and permeable adjacent rocks, which then serve combinations to prevent having both the further movement of oil. Many traps are structural and stratigraphic features.

A reservoir is a rock unit that acts as a storage device for the hydrocarbons that migrate from the source rock. Hydrocarbons are retained within the reservoir because these rocks contain numerous pores (essentially microscopic-sized holes) between the mineral grains making up the fabric of the reservoir. In good quality reservoirs the porosity is frequently over 20% of the rock volume. However, the pores need to be interlinked in such a manner that the fluids can move into (and out of, if we are to exploit the oil and gas) the reservoirs over geological time. This is known as permeability. There are two main rock types that make up the giant reservoirs around the world—sandstones that are made up of sand grains (quartz and feldspar in the majority) and carbonates that are made up of organically created calcium carbonate grains (corals, algae, and shells) or mud. In order to stop the upward movement of hydrocarbons and constrain them to one zone of the subsurface (trap), there must be a barrier to prevent fluid migration. Generally this mechanism or seal consists of rocks that are impermeable to fluid flow. The most effective of these seals are mudstone or shale, very fine-grained rocks containing abundant clay minerals. Occasionally the impermeable layers are dense igneous rocks and in rare situations, there may be significant rock and fluid pressure differences in a region that prevents fluid flow and acts as a seal.

Equally important is the presence of a trapping mechanism otherwise the geological formation does not form a reservoir. The trapping mechanism is either of a structural form where the reservoir rock unit is contorted to produce a zone where fluids naturally accumulate against a seal, or of a stratigraphic nature where a reservoir rock unit changes laterally into an impermeable unit reflecting changes in depositional environment along one bed. Warping of the strata in an upward direction to form anticlinal folds are suitable formations for trapping hydrocarbons along with faults where tilted (but permeable) strata are moved against impermeable strata. In addition, since the oil reservoir (a rock formation or trap holding an accumulation of petroleum) was formed from sea sediments, most reservoirs also contain salt water. In addition, natural gas is almost always present, either dissolved in the oil or as free gas separate from the oil. The associated water and natural gas are important in maintaining pressure during petroleum production.

The biogenic theory of petroleum formation and occurrence has led to the search for containing thick beds of sedimentary rocks. Geophysical techniques have led to the discovery of basins (a tract of land in which the rock strata are tilted toward a common center), both onshore and offshore, which may contain crude oil or natural gas. Sufficient seismic work has been done to give an indication of their prospective petroleum area and their general structural aspects. In many of these basins exploratory drilling with varying amounts and degrees of success have taken place while others (such as tight sands and shale formations) are only now the subject of intensive drilling and development programs.

Many experts expect the importance of large fields and tight formations for new discoveries to continue, for both geologic and economic reasons. In new offshore areas, only large fields are economical to develop, although in places where the associated infrastructure (pipelines, etc.)

exists, smaller finds may be sought and developed (Speight, 2015). The earliest discoveries of oil and gas made over a century ago were based on surface seepages and analyses of surface geology. After a century of exploration, petroleum exploration has become extremely sophisticated. New techniques and instrumentation, such as geophysical surveys, geochemical analysis, and subsurface logging, combined with the use of computers, provide information on the structural characteristics of the subsurface.

These techniques provide only raw data which must be evaluated by geologists and geophysicists; the room for judgment and, therefore, differences in interpretation remains great. Drilling remains the only way to determine the actual existence of petroleum resources, with the other exploration techniques providing guidance as to the most promising locations within a basin.

2.4 EXPLORATION

Exploration for hydrocarbons (oil, gas, and condensate) is commonly acknowledged to have begun with the discovery at Oil Creek, Pennsylvania, by (self-styled Colonel) Edwin Drake in 1859. However, this was only the start of the modern global era of technology-driven advances in exploration. Traditionally, oil exploration was conducted by recognizing seeps of hydrocarbons at the surface. The Chinese, for example, used oil (mostly bitumen) obtained from seeps in medication, waterproofing, and warfare several thousand years ago. They frequently dug shallow pits or horizontal tunnels at seep locations but also, as early as 200 BC, drilled down as far as 3500 feet (1067 m) using rudimentary bamboo poles (making Drake's 69.5 feet more than 2000 years later seem puny by comparison). In Baku, Azerbaijan, there are still gas and oil seeps that are permanently on fire and have been used to light caravanserai since the times of Marco Polo and the Silk Route. Similarly, seeps were recognized and exploited in the Caucasus (Groznyy region of Chechnya), Ploesti in Romania, Digboi in Assam, Sanga Sanga in eastern Borneo, and Talara in Peru.

Even Drake's well, the first to intentionally look for oil in the subsurface, was based on direct identification of seeped hydrocarbons at the surface. Initially, the oil produced was used to provide kerosene for lamps, but the later invention of automobiles drove up demand and ushered in the modern methods of oil exploration. In fact, most oil until the turn of the 20th century was in one form or another related to seep identification.

However, one theory developed during this time was to have a profound impact on exploration. In the mid-1800s, William Logan, first Director of the Geological Survey of Canada, recognized oil seeps associated with the crests of convex-upward folded rocks and employed a geologist, Thomas Hunt, to formalize an anticlinal theory. This idea, however, was only recognized as a viable tool for exploration when Spindletop was discovered on the Gulf Coast of Texas in 1901. For the next 30 years, the anticlinal theory dominated exploration, to the extent that many believed that there were no other types of hydrocarbon accumulation. As a result, geologists became critical to understand the structural configurations of rock sequences which, when combined with seep occurrences, proved to be the keys to discovering the main oil-producing provinces of the United States, Mexico, and Venezuela. For a period of time before World War I, Oklahoma, Texas, and California were the World's leading production areas.

Heavy oil and bitumen occurrence have been found (historically) by the presence of seepages from the near-surface formation, whereas locating crude oil reservoirs required a more different scientific approach. In the case of conventional crude oil, from the beginning of the 20th century until the 1950s, the main exploration tool used for finding crude oil was the use of intensive and detailed geological mapping. This was frequently in terrain that was remote and inhospitable. The early pioneers working their way through the jungles of Burma, India (Burma Oil Company), and Borneo (Shell Petroleum), the deserts of Iraq or the mountains of Iran (the Anglo-Persian Oil Company that became British Petroleum, now BP), would conduct detailed evaluations of the nature and distribution of rock units. These rock units represented potential reservoirs, seals, and source units, as well as frequency, orientation, and geological history of folds or faults that could act as traps for the migrating hydrocarbons.

The typical trap was the anticlinal formation that was created by folds in the strata and it took until the 1920s for explorers to realize that crude oil could occur in situations where no anticline was preserved. For example, it was noted as far back as 1880 that oil was trapped in the Venango Sands of Pennsylvania, not in the form of an anticlinal structure, but by the lithology occurring in a moving paleo-shoreline. After the 1920s, mapping of surface features was complimented by the development of seismic refraction, gravity, and magnetic geophysical methods. In particular, gravity and seismic methods proved effective in locating oil trapped against buried salt domes in the onshore Gulf of Mexico. At this time, another significant advance in exploration of the subsurface took place with the application of geophysical techniques to measuring properties of rocks and fluids encountered whilst drilling for hydrocarbons. In France in 1927, they initially measured the resistivity of the rocks in shallow wells (drilled primarily for water distribution), but later went on to add other electric, sonic, and radioactive logging tools. It is now even possible to log porosity, permeability, mineralogy, and fluids and image the structures and downhole rock types.

Aerial remote sensing for features favored for hydrocarbon accumulation became an important and effective technique, particularly in areas of sparse vegetation cover following World War II when low-cost, rapid reconnaissance of large areas became feasible. Large-scale features such as faults and folds could be identified and targeted for detailed seismic acquisition. In the 1970s, this capability was improved dramatically by the use of satellite remote sensing technologies (LANDSAT). From the 1940s to the 1960s, there were important developments in the understanding of the controls on lateral and vertical variations within reservoir sequences. In particular, the new discipline of sedimentology used modern depositional analogs from around the world to understand the nature, distribution, and controls over ancient reservoir sequences. There was also much interest generated over the discovery of carbonate oil-bearing reservoirs in West Texas and Alberta, Canada (the Leduc Reef that lies just south of Edmonton), and recognition that modern intertidal carbonate-evaporate sequences in the United Arab Emirates had equivalents in ancient reservoirs. These developments led to the discovery of many supergiant carbonate oil fields in the United States (Yates Field), Mexico (Posa Rica), Iraq (Kirkuk), and Russia (a number of Siberian oil fields).

Other tools such as geochemistry, developed during this period, have helped to quantify the level of maturity and the nature and distribution of source potential in a region. Micropaleontology was developed in Tertiary Basins such as in the Republic of Trinidad and Tobago and in the Caucasus for horizon identification and correlation using planktonic foraminifera, but spread rapidly to the United States Gulf Coast. Now, geochemistry and biostratigraphy, including palynology (the study of spores and other organic matter), have become standard tools that are used in exploration.

Also beginning in the 1970s, there was a significant advance in the power and reduction in size and cost of computers that has led directly to a dramatic increase in the ability of geophysicists to acquire, process, and interpret large quantities of seismic data. Initially, this was in the form of 2D reflection seismic onshore, but this trend has continued to the present day and now oil companies regularly undertake, mostly offshore, 3D seismic surveys and even 4D field surveys. Three-dimensional surveys are repeated over the same area every few years to monitor fluid movement within reservoirs and thereby optimally manage hydrocarbon recovery.

Exploration for oil and gas has progressed dramatically in the last 30 years, driven forward by the ever increasing power and capabilities of the computer. As a result, it now takes only a fraction of the time required 40 years ago to find and develop oil fields. However, technology in itself does not find oil or gas fields; it frequently requires the personal knowledge and inspiration that is the mark of a true explorer to discover some of the major new exploration plays in such areas as Equatorial Guinea, Angola, Nigeria, Trinidad, the Gulf of Mexico, and the northern Canadian Rockies. After an exploration effort has successfully discovered petroleum within an acceptable range of reserve potential, the challenge becomes how to best optimize extraction of recoverable reserves in a manner yielding an acceptable economic return on total cash expenditures required over the life of the project. Surface and subsurface conditions of a discovery have considerable impact on the extraction process, its related costs, and ultimate project success or failure.

The process of evaluating how to best optimize extraction of recoverable reserves begins with a development plan which must include consideration of all of the available geologic and engineering data to make an initial estimate of reserves in-place, to project recovery efficiency and optimal recoverable reserve levels under various producing scenarios, and to evaluate development plan alternatives. Development alternatives include (1) the number of wells to be drilled and completed for production or injection, (2) well spacing and pattern, (3) surface processing requirements, (4) transportation options, (5) cost projections, (6) project schedules, (7) depletion plans, (8) operational programs, and (9) logistics and economic studies.

The ultimate recovery of original in-place volumes may be as high as one-third for oil and 80% or more for gas. There are three phases of recovering reserves (1) *primary recovery*, (2) *secondary recovery*, and (3) *tertiary recovery*, also called *enhanced oil recovery* (EOR) (Fig. 6.1). Primary recovery occurs as wells produce because of natural energy from expansion of gas and water within the producing formation, pushing fluids into the wellbore and lifting them to the surface. During primary recovery, the natural pressure of the reservoir or gravity drive oil into the wellbore, combined with artificial lift techniques (such as pumps) which bring the oil to the surface. But only about 10% of a reservoir's original oil in-place is typically produced during primary recovery. Secondary recovery occurs as artificial energy is applied to lift fluids to the surface. This may be accomplished by injecting gas down a hole to lift fluids to the surface, installation of a subsurface pump, or injecting gas or water into the formation itself. Secondary recovery is done when well, reservoir, facility, and economic conditions permit. Secondary recovery techniques add to the field's productive life generally by injecting water or gas to displace oil and drive it to a production wellbore, resulting in the recovery of 20–40% of the original oil in-place.

Tertiary recovery processes were developed for application in situations in which secondary processes had become ineffective. However, the same tertiary processes were also considered for reservoir applications for which secondary recovery techniques were not used because of low recovery potential—in the latter case, the name tertiary is a misnomer. For most reservoirs, it is advantageous to begin a secondary or a tertiary process concurrent with primary production. For these applications, the term EOR was introduced and has become popular in referring to, in general, any recovery process that enhances the recovery of oil beyond what primary and secondary production would normally be expected to yield.

EOR processes can be classified into five categories: (1) chemical processes, (2) miscible processes, (3) miscible fluid displacement processes, (4) thermal processes, and (5) microbial flooding processes. Chemical processes are polymer, micellar–polymer, and alkaline flooding. The category of miscible displacement includes single contact and multiple-contact miscible processes. Thermal processes include hot water, steam cycling, steam drive, and in situ combustion. In general, thermal processes are applicable in reservoirs containing heavy crude oils, whereas chemical and miscible displacement processes are used in reservoirs containing light crude oils. Microbial processes use microorganisms to assist in oil recovery.

Tertiary recovery occurs when means of increasing fluid mobility in oil reservoirs within the reservoir are introduced in addition to secondary techniques. This may be accomplished by introducing additional heat into the formation to lower the viscosity (thin the oil) and improve its ability to flow to the wellbore. Heat may be introduced by either injecting steam in a *steam flood* or injecting oxygen to enable the ignition and combustion of oil within the reservoir in a *fire flood*. Tertiary oil recovery (EOR) methods take oil recovery one step further and rely on methods that reduce the viscosity of the oil to increase recovery, compared to the natural or induced energy methods of primary recovery and secondary recovery. Tertiary recovery is started *before* secondary oil recovery techniques are no longer enough to sustain production. For example, thermally EOR methods are recovery methods in which the oil is heated making it easier to extract; usually steam is used for heating the oil.

Generally the first stage in the extraction of crude oil is to drill a well into the underground reservoir. Often many wells (*multilateral wells*) will be drilled into the same reservoir, to ensure that the extraction rate will be economically viable. Also, some wells (*secondary wells*) may be used to pump water, steam, acids, or various gas mixtures into the reservoir to raise or maintain the reservoir pressure, and so maintain an economic extraction rate. However, over the lifetime of the well the pressure will fall, and at some point there will be insufficient underground pressure to force the oil to the surface. *Secondary oil recovery* uses various techniques to aid in recovering oil from depleted or low-pressure reservoirs. Sometimes pumps, such as beam pumps (*horsehead pumps*) and electrical submersible pumps are used to bring the oil to the surface. Other secondary techniques for oil recovery increase the reservoir's pressure by water injection, natural gas reinjection, and gas lift, which injects air, carbon dioxide, or some other gas into the reservoir.

Conventional oil production methods may be unsuccessful because the management of the reservoir was poor or because reservoir heterogeneity has prevented the recovery of crude oil in an economical manner. Reservoir heterogeneity, such as fractures and faults, can cause reservoirs to drain inefficiently by conventional methods. Also, highly cemented or shale zones can produce barriers to the flow of fluids in reservoirs and lead to high residual oil saturation. Reservoirs containing crude oils with low API gravity often cannot be produced efficiently without application of EOR methods because of the high viscosity of the crude oil. In some cases, the reservoir pressure was depleted prematurely by poor reservoir management practices to create reservoirs with low energy and high oil saturation.

As might be expected, the type of exploration technique employed depends upon the nature of the site and the properties of the crude oil, heavy oil, or bitumen contained therein. In other words, and as for many environmental operations, the recovery techniques applied to a specific site are dictated by the nature of the site and are, in fact, site specific (Fig. 6.2) For example, in areas where little is known about the subsurface, preliminary reconnaissance techniques are necessary to identify potential reservoir systems that warrant further investigation. Techniques for reconnaissance that have been employed to make inferences about the subsurface structure include *satellite and high-altitude imagery* and *magnetic and gravity surveys.*

Once an area has been selected for further investigation, more detailed methods (such as the *seismic reflection* method) are brought into play. Drilling is the final stage of the exploratory program and is in fact the only method by which a petroleum reservoir can be conclusively identified. However, in keeping with the concept of site specificity, drilling may be the only option in some areas for commencement of the exploration program. The risk involved in the drilling operation depends upon previous knowledge of the site subsurface. Thus there is the need to relate the character of the exploratory wells at a given site to the characteristics of the reservoir.

In addition, drilling technology has been improved to such an extent that in addition to the recovery processes featured in this chapter, improved drilling technology is included as an example of technologies that contribute to improvements in oil production. Some of the improved drilling technology in current use includes (1) mud pulse telemetry, (2) pressure coring systems, (3) downhole drill motors, and (4) polycrystalline drill bits.

Mud pulse telemetry is a drilling system that transmits the location of a drill bit by sending pressure pulses through the drilling mud that circulates from the drill bit to the drilling rig on the surface. The technology, which was developed during the 1970s, eliminates the need for costly drill pipe removal and lowering an instrument in the hole to obtain the data. This *measurement-while-drilling* capability enables the continuous monitoring of drilling progress, improves drilling accuracy, and saves the industry millions of dollars in operational costs. The *pressure coring system* provides the key capability to obtain accurate measurements of underground reservoir conditions—a core barrel for retrieving reservoir rock samples while maintaining the ambient pressure of the reservoir. This technology is now included in the range of tools and services provided in specialized coring operations.

The *downhole drill motor* concept has been largely responsible for advances now routinely employed by industry in the varieties of slant-hole and horizontal drilling. Improvements in accuracy, equipment, and technique, combined with slim-hole drilling technology, have revolutionized access to oil reservoirs and minimized the environmental impact of drilling by enabling multiple drill hole emplacement from drill pads to have a greatly reduced *footprint*. Early use of the superior capabilities of *polycrystalline diamond drill bits* was impeded by severe problems of the diamond cutting bits breaking away from the drill bit substrata, requiring costly, time-consuming drill bit retrieval. The development of a diffusion technique that permanently bonded the cutters to the bit improved the technology and in one test, only three such bits were needed instead of the 13 conventional bits that would have been required to drill a 5000-foot well.

Recovery, as applied in the petroleum industry, is the production of oil from a reservoir. There are several methods by which this can be achieved that range from recovery due to reservoir energy (ie, the oil flows from the well hole without assistance) to enhanced recovery methods in which considerable energy must be added to the reservoir to produce the oil. However, the effect of the method on the oil and on the reservoir must be considered before application.

Thus, once the well is completed, the flow of oil into the well is commenced. For limestone reservoir rock, acid is pumped down the well and out through the perforations. The acid dissolves channels in the limestone that lead oil into the well. For sandstone reservoir rock, a specially blended fluid containing *proppants* (sand, walnut shells, aluminum pellets) is pumped down the well and out through the perforations. The pressure from this fluid makes small fractures in the sandstone that allow oil to flow into the well, while the proppants hold these fractures open. Once the oil is flowing, the oil rig is removed from the site and production equipment is set up to extract the oil from the well.

A well is always carefully controlled in its flush stage of production to prevent the potentially dangerous and wasteful *gusher*. This is actually a dangerous condition, and is (hopefully) prevented by the blowout preventer and the pressure of the drilling mud. In most wells, acidizing or fracturing the well starts the oil flow.

Generally, crude oil reservoirs sometimes exist with an overlying *gas cap*, in communication with aquifers, or both. The presence of a gas cap is a somewhat lesser occurrence in heavy oil reservoirs and tar sand deposits. The oil resides together with water and free gas in very small holes (pore spaces) and fractures. The size, shape, and degree of interconnection of the pores vary considerably from place to place in an individual reservoir. Below the oil layer the sandstone is usually saturated with salt water. The oil is released from this formation by drilling a well and puncturing the limestone layer on either side of the limestone dome or fold. If the peak of the formation is tapped, only the gas is obtained. If the penetration is made too far from the center, only salt water is obtained. Oil wells may be either on land or under water. In North America many wells are offshore in the shallow parts of the oceans. The crude oil or unrefined oil is typically collected from individual wells by small pipelines.

Conventional crude oil is a brownish green to black liquid with a specific gravity in a range from about 0.810 to 0.985 and having a boiling range from about 20°C (68°F) to above 350°C (660°F), above which active decomposition ensues when distillation is attempted. The oils contain from 0% to 35% or more of gasoline, as well as varying proportions of kerosene hydrocarbons and higher boiling constituents up to the viscous and nonvolatile compounds present in lubricant oil and in asphalt. The composition of the crude oil obtained from the well is variable and depends not only on the original composition of the oil in situ but also on the manner of production and the stage reached in the life of the well or reservoir.

New methods to drill for oil are continually being sought, including directional or horizontal drilling techniques, to reach oil under ecologically sensitive areas, and using lasers to drill oil wells. For example, *directional drilling* is also used to reach formations and targets not directly below the penetration point or drilling from shore to locations under water (Fig. 6.3). A controlled deviation may also be used from a selected depth in an existing hole to attain economy in drilling costs. Various types of tools are used in directional drilling along with instruments to help orient their position and measure the degree and direction of deviation; two such tools are the *whipstock* and the *knuckle joint*. The whipstock is a gradually tapered wedge with a chisel shaped base that prevents rotation after it has been forced into the bottom of an open hole. As the bit moves down, it is deflected by the taper about 5° from the alignment of the existing hole.

Approximately one-third of the world's crude oil is produced from offshore fields, usually from steel drilling platforms set on the ocean floor (Speight, 2015). In shallow, calm waters, these may be little more than a wellhead and workspace. The larger ocean rigs, however, include not only the well equipment but also processing equipment and extensive crew quarters. Recent developments in ocean drilling include the use of floating tension leg platforms that are tied to the sea floor by giant cables and drill ships, which can hold a steady position above a sea floor well using constant, computer-controlled adjustments. Subsea satellite platforms, where all of the necessary equipment is located on the ocean bed at the well site, have been used for small fields located in producing areas. In Arctic areas islands are built from dredged gravel and sand to provide platforms capable of resisting drifting ice fields.

When tight formations (Chapter 1) are encountered and it becomes necessary to encourage flow, several methods are used, one of which involves hydraulic fracturing to fracture the rock. If the formation is mainly limestone, hydrochloric acid is sent down the hole to dissolve channels in the rock. The acid is inhibited to protect the steel casing. In sandstone, the preferred method is hydraulic fracturing. Finally, the drilling job is complete when the drill bit penetrates the reservoir and the reservoir is evaluated to see whether the well represents the discovery of a *prospect* or whether it is a dry hole.

Evaluation of the reservoir is usually initiated by examining the cuttings from the wellbore for evidence of hydrocarbons while the drill bit passes through a reservoir trap. The evaluation of these cuttings helps pinpoint the possible producing intervals in the wellbore. At this time, a wire-line, lowered into the hole and an electric log is run to help define possible producing intervals, the presence of hydrocarbons, and detailed information about the different formations throughout the wellbore. Further tests can also be run on individual formations within the wellbore, such as pressure tests, formation fluid recovery and sidewall core analysis. On the other hand, if the hole is dry, it is plugged and abandoned.

One of the most important recent discoveries in petroleum studies has been plate tectonics. Not only has this revolutionized the earth sciences, but also it has provided a conceptual setting for oil exploration. The movement of plates around the surface of Earth creates large-scale depressions into which substantial quantities of sediments eroded from the surrounding high ground may accumulate. These accumulations can exceed thicknesses of several thousand kilometers and are referred to as sedimentary basins. By the comparison of basins around the world and by analogy to existing producing hydrocarbon regions, an exploration team can say which basins are worth looking at in more detail. Then an exploration team will spend time ensuring that within the basin there are present all the key elements that control the presence of hydrocarbons. Assuming that all the needed features are present, the team would agree that the basin contained a viable petroleum system and prospect generation can proceed.

In modern exploration programs, the mapping of gravity and magnetic anomalies would normally be the first two methods to be applied to a new basin or region being evaluated. These techniques would be used to identify large-scale changes in the structure of the basement and sedimentary basins together with major differences in rock density, such as the influx of dense igneous rocks or light salt into a sedimentary sequence. These techniques are large-scale, can be applied over both land and water and can even be collected remotely from plane or satellite.

At the same time, remote sensing of onshore areas initially based on large-scale photogeological surveys and, since the 1970s, by satellite imaging, can identify areas with the anticlinal and faulted structural features, seeps, or salt domes frequently associated with oil occurrences. Offshore remote sensing of the sea surface can lead to the identification of slicks associated with the seepage of oil (both natural and man-made) into the water column. A coarse two-dimensional grid of seismic data is then collected to obtain a picture of the subsurface in the area to be targeted. Seismic data collection involves the generation of a seismic wave using an energy source such as an air-gun in water, dynamite in drill holes inland, or a truck with a plate that is thumped down onto the road/soil surface (vibroseis). The wave travels through the earth's rock layers and reflects back off key surfaces. The time taken for the waves to be received back at the surface along with their strength is recorded via geophones and displayed on a seismic section. Processing the two-dimensional seismic sections using highly sophisticated software reveals the detailed structure of the subsurface and in certain circumstances shows the presence of DHI such as bright spots associated with gas/water differences. Primarily, though, seismic is used to indicate the nature of folded and faulted structures that could prove to be suitable hydrocarbon traps.

The objective of seismic acquisition and processing is to acoustically image the subsurface in a geologically accurate manner with as high resolution as possible. For a detailed analysis of a small area representing a field or prospect, a high density and calibrated three-dimensional seismic is collected. Modern technology also allows scientists to accurately map changes in fluid movements through time (repeat multiple 3D seismic surveys, known as 4D seismic) and this technique is now particularly important in monitoring production performance of the reservoir.

Ultimately, however, the only way of confirming the presence or absence of hydrocarbons at depth is by drilling the prospect. In certain areas of the world where drilling is cheap and the subsurface has been explored extensively, such as certain onshore basins of the United States, drilling is commonly preferred to extensive and expensive seismic acquisition. Wells are then analyzed using electric, sonic, and radioactive logging techniques that measure characteristics of the rocks and fluids. These methods can identify the presence of oil and gas, which can then be tested to see if they occur at commercially viable production levels.

On the other hand, at a cost of over 10 million dollars per offshore exploration well, the oil companies are also likely to employ the sophisticated battery of direct and indirect detection techniques first before resorting to drilling in these areas.

The analysis of a basin is usually part of the first study a company conducts prior to moving into an area for future exploration; it is also sometimes conducted during the exploration phase. Exploration geology comprises all the activities and studies necessary for finding where oil exists in subterranean formations. Usually seismic (or 3D seismic) studies are performed and older exploration data (seismic lines, well logs, reports) are used to guide and expand upon the new studies. On occasion, gravity and magnetic studies are conducted, and oil seepages are mapped to find potential areas for the occurrence of crude oil. As soon as a significant crude oil occurrence is identified by an exploration- or wildcat-well the *appraisal stage* commences and reserve estimates are made.

2.5 RESERVES

The *appraisal stage* is used to delineate the extent of the discovery and to estimate the reserves in terms of barrels of crude oil in-place. Reservoir properties and connectivity (Chapter 3), crude oil properties (Chapter 4) and gas–oil and oil–water contacts are determined to calculate potential recoverable volumes of the crude oil. This is usually achieved by drilling appraisal wells within the initial exploration area or initial well and production tests add valuable information related to reservoir pressure and connectivity. Geochemical and petrophysical analysis gives information on the type (viscosity, chemistry, API, carbon content, etc.) of the hydrocarbon and the nature of the reservoir (porosity, permeability, etc.). Once this information has been collected and assessed, reserve estimates can be made.

Over the past several decades, the assessment of petroleum, heavy oil, and tars and bitumen resources and reserves has gone through several iterations, partly because of the evolution of the definitions of resources and reserves that are key factors in the continued development of resources. It is essential for future planning that governments and industry have a clear assessment of the quantities of petroleum, heavy oil, and tar sand bitumen that are available for production and the quantities that are anticipated to become available within a practical time frame through additional field development, technological advances, or exploration. To achieve such an assessment, it is imperative that the industry adopt a consistent nomenclature for assessing the current and future quantities of petroleum expected to be recovered from naturally occurring underground accumulations. Such quantities are defined as reserves, and their assessment is of considerable importance to governments, international agencies, economists, bankers, and the international energy industry.

A conscious effort has been made to keep the recommended terminology as close to current common usage as possible in order to minimize the impact of previously reported quantities and changes required to bring about wide acceptance. Due to the many forms of occurrence of petroleum, heavy oil, and tar sand bitumen, the wide range of characteristics of these materials, the uncertainty associated with the geological environment, and the constant evolution of evaluation technologies, a precise classification and definition system is not always practical and the complexity required for a precise system would detract from its understanding by those involved in the development of crude oil, heavy crude oil, and tar sand bitumen resources.

In fact, the terminology used in classifying petroleum, heavy oil, and tar sand bitumen is still in an evolutionary stage with new terminologies being added to the petroleum lexicon on a regular basis, sometimes with justification but in many cases without any form of scientific justification. An example is the use of the term *black oil* which, besides the color of the oil, offers nothing in the way of explanation of the material and certainly adds nothing to science and/or engineering. Thus, it is not surprising that the various categories of reserves have been the subject of much study and discussion for many years. In fact, attempts to standardize reserves terminology began in the mid-1930s when the American Petroleum Institute considered classification for petroleum and definitions of various reserves categories. Since then, the evolution of technology has yielded more precise engineering methods to determine reserves and has intensified the need for an improved nomenclature to achieve consistency among professionals working with reserves terminology.

Oil reserves are the estimated quantities of conventional petroleum and/or heavy crude oil that are claimed to be recoverable under existing economic and operating conditions. Many oil-producing nations do not reveal their reservoir engineering field data, and instead provide unsubstantiated claims for their oil reserves. In most cases, *oil* refers to conventional oil but, depending on the source, tar sand bitumen has not always been included but now is recognized as a resource (BP, 2014). However, the exact definitions and data vary from country to country which means that national statistics are not always comparable and may even be manipulated for political reasons (Speight, 2011b).

The total amount of conventional oil or heavy oil in a reservoir is known as oil–in-place. However, because of reservoir characteristics and limitations in production technology, only a fraction of this oil can be brought to the surface, and it is only this producible fraction that is considered to be *reserves*. The ratio of reserves to oil–in-place for a given field is often referred to as the *recovery factor*. The recovery factor of a field may change over time based on operating history and in response to changes in technology and economics. The recovery factor may also rise over time for conventional oil if additional investment is made in EOR techniques, such as gas injection or water-flooding. For heavy oil, the evolution of older methods and the use of newer and innovative methods may increase the recovery factor.

Because the geology of the subsurface cannot be examined directly, indirect techniques must be used to estimate the size and recoverability of the resource. While new technologies have increased the accuracy of these techniques, significant uncertainties still remain. In general, most early estimates of the reserves of an oil field are conservative and tend to grow with time; this phenomenon is referred to as *reserves growth* (Morehouse, 1997).

2.5.1 Definitions

The definitions that are used to describe petroleum and heavy oil reserves are often misunderstood because they are not adequately defined at the time of use. Therefore, as a means of alleviating this problem, it is pertinent at this point to consider the definitions used to describe the amount of petroleum that remains in subterranean reservoirs.

Petroleum and heavy oil are a *resource*; in particular, a *fossil fuel resource* (Chapter 1). A *resource* is the entire commodity that exists in the sediments and strata whereas the *reserves* represent that fraction of a commodity that can be recovered economically. However, the use of the term reserves as being descriptive of the resource is subject to much speculation. In fact, it is subject to word variations! For example, reserves are classed as *proved*, *unproved*, *probable*, *possible*, and *undiscovered*. By the same rationale, tar sand bitumen, although not considered to be a member of the petroleum family by virtue of the methods used for recovery (Chapter 1) is also a fossil fuel resource and the reserves can also be classed as *proved*, *unproved*, *probable*, *possible*, and *undiscovered*.

Proven reserves are those reserves of petroleum that are actually found by drilling operations and are recoverable by means of current technology. They have a high degree of accuracy and are frequently updated as the recovery operation proceeds. They may be updated by means of reservoir characteristics, such as production data, pressure transient analysis, and reservoir modeling.

Probable reserves are those reserves of petroleum that are nearly certain but about which a slight doubt exists. *Possible reserves* are those reserves of petroleum with an even greater degree of uncertainty about recovery but about which there is some information. An additional term *potential reserves* is also used on occasion; these reserves are based upon geological information about the types of sediments where such resources are likely to occur and they are considered to represent an educated guess. Then, there are the so-called *undiscovered reserves*, which are little more than figments of the imagination! The terms undiscovered reserves or undiscovered resources should be used with caution, especially when applied as a means of estimating reserves of petroleum reserves. The data are very speculative and are regarded by many energy scientists as having little value other than unbridled optimism.

The term *inferred reserves* is also commonly used in addition to, or in place of, potential reserves. Inferred reserves are regarded as of a higher degree of accuracy than potential reserves, and the term is applied to those reserves that are estimated using an improved understanding of reservoir frameworks. The term also usually includes those reserves that can be recovered by further development of recovery technologies.

The differences between the data obtained from these various estimates can be considerable, but it must be remembered that any data about the reserves of petroleum (and, for that matter, about any other fuel or mineral resource) will always be open to questions about the degree of certainty. Thus, in reality, and in spite of the use of self-righteous word manipulation, proven reserves may be a very small part of the total hypothetical and/or speculative amounts of a resource.

At some time in the future, certain resources may become reserves for righteous or unrighteous reasons (Speight, 2011b). Such a reclassification can arise as a result of improvements in recovery techniques which may either make the resource accessible or bring about a lowering of the recovery costs and render winning of the resource an economical proposition. In addition, other uses may also be found for a commodity, and the increased demand may result in an increase in price. Alternatively, a large deposit may become exhausted and unable to produce any more of the resource thus forcing production to focus on a resource that is lower grade but has a higher recovery cost.

More recently, the Society for Petroleum Engineers, with the American Association of Petroleum Geologists (AAPG), the World Petroleum Council (WPC), and the Society of Petroleum Evaluation Engineers (SPEE) (SPE, 2007) has developed a resource classification system (Fig. 2.2) that moves away from systems in which all quantities of petroleum that are estimated to be initially-in-place are used. In these definitions, the quantities estimated to be initially-in-place are (1) total petroleum-initially-in-place, (2) discovered petroleum-initially-in-place,

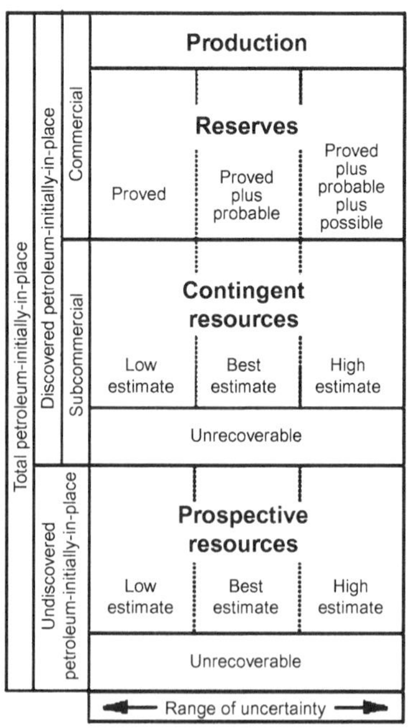

Figure 2.2 Representation of resource estimation. Note: The horizontal axis represents the range of uncertainty in the estimated potentially recoverable volume for an accumulation whereas the vertical axis represents the level of status/maturity of the accumulation. The vertical axis can be further subdivided to classify accumulations on the basis of the commercial decisions required to move an accumulation towards production.

and (3) undiscovered petroleum-initially-in-place. The recoverable portions of petroleum are defined separately as (1) reserves, (2) contingent resources, and (3) prospective resources. In any case and whatever the definition, reserves are a subset of resources and are those quantities of petroleum that are discovered (ie, in known accumulations), recoverable, commercial, and remaining.

The *total petroleum-initially-in-place* is that quantity of petroleum (or heavy oil) that is estimated to exist originally in naturally occurring accumulations. The total petroleum-initially-in-place is, therefore, that quantity of petroleum that is estimated, on a given date, to be contained in known accumulations, plus those quantities already produced therefrom, plus those estimated quantities in accumulations yet to be discovered. The total petroleum-initially-in-place may be subdivided into *discovered petroleum-initially-in-place* and *undiscovered petroleum-initially-in-place*, with discovered petroleum-initially-in-place being limited to known accumulations.

It is recognized that the quantity of petroleum-initially-in-place may constitute *potentially recoverable resources* since the estimation of the proportion that may be recoverable can be subject to significant uncertainty and will change with variations in commercial circumstances, technological developments, and data availability. A portion of those quantities classified as unrecoverable may become recoverable resources in the future as commercial circumstances change, technological developments occur, or additional data are acquired.

Discovered petroleum-initially-in-place is that quantity of petroleum that is estimated, on a given date, to be contained in known accumulations, plus those quantities already produced therefrom. Discovered petroleum-initially-in-place may be subdivided into *commercial* and *subcommercial* categories, with the estimated potentially recoverable portion being classified as *reserves* and *contingent resources*, respectively (as defined below).

Reserves are those quantities of petroleum which are anticipated to be commercially recovered from known accumulations from a given date forward (BP, 2014). Estimated recoverable quantities from known accumulations that do not fulfill the requirement of commerciality should be classified as *contingent resources* (as defined below). The definition of commerciality for an accumulation will vary according to local conditions and circumstances and is left to the discretion of the country or company concerned. However, reserves must still be categorized according to specific criteria and, therefore, proved reserves will be limited to those quantities that are commercial under current economic conditions, while probable and possible reserves may be based on future economic conditions. In general, quantities should not be classified as reserves unless there is an expectation that the accumulation will be developed and placed on production within a reasonable time frame.

Contingent resources are those quantities of petroleum that are estimated, on a given date, to be potentially recoverable from known accumulations, but which are not currently considered as commercially recoverable. Some ambiguity may exist between the definitions of contingent resources and unproved reserves. This is a reflection of variations in current industry practice but if the degree of commitment is not such that the accumulation is expected to be developed and placed on production within a reasonable time frame, the estimated recoverable volumes for the accumulation be classified as contingent resources. Contingent resources may include, for example, accumulations for which there is currently no viable market, or where commercial recovery is dependent on the development of new technology, or where evaluation of the accumulation is still at an early stage.

Undiscovered petroleum-initially-in-place is that quantity of petroleum that is estimated, on a given date, to be contained in accumulations yet to be discovered. The estimated potentially recoverable portion *of undiscovered petroleum-initially-in-place* is classified as prospective resources, which are those quantities of petroleum that are estimated, on a given date, to be potentially recoverable from undiscovered accumulations. In addition, the *estimated ultimate recovery* (EUR) is the quantity of petroleum which is estimated, on a given date, to be potentially recoverable from an accumulation, plus those quantities already produced therefrom. EUR not a resource category but it is a term that may be applied to an individual accumulation of any status/maturity (discovered or undiscovered).

Petroleum quantities classified as reserves, contingent resources or prospective resources should not be aggregated with each other without due consideration of the significant differences in the criteria associated with their classification. In particular, there may be a significant risk that accumulations containing contingent resources or prospective resources will not achieve commercial production.

The *range of uncertainty* (Fig. 2.2) reflects a reasonable range of estimated potentially recoverable volumes for an individual accumulation. Any estimation of resource quantities for an accumulation is subject to both technical and commercial uncertainties, and should, in general, be quoted as a range. In the case of reserves, and where appropriate, this range of uncertainty can be reflected in estimates for *proved reserves* (1P), *proved plus probable reserves* (2P), and *proved plus probable plus possible reserves* (3P) scenarios. For other resource categories, the terms *low estimate, best estimate,* and *high estimate* are recommended.

The term best estimate is used as a generic expression for the estimate considered to be the closest to the quantity that will actually be recovered from the accumulation between the date of the estimate and the time of abandonment. If probabilistic methods are used, this term would generally be a measure of central tendency of the uncertainty distribution. The terms low estimate and high estimate should provide a reasonable assessment of the range of uncertainty in the best estimate.

For undiscovered accumulations (prospective resources) the range will, in general, be substantially greater than the ranges for discovered accumulations. In all cases, however, the actual range will be dependent on the amount and quality of data (both technical and commercial) that is available for that accumulation. As more data become available for a specific accumulation (eg, additional wells, reservoir performance data) the range of uncertainty in the EUR for that accumulation should be reduced.

The low estimate, best estimate, and high estimate of potentially recoverable volumes should reflect some comparability with the reserves categories of proved reserves, proved plus probable reserves, and proved plus probable plus possible reserves, respectively. While there may be a significant risk that subcommercial or undiscovered accumulations will not achieve commercial production, it is useful to consider the range of potentially recoverable volumes independently of such a risk.

In terms of actual numbers, on a worldwide basis the produced conventional crude oil is estimated to be approximately 784 billion (784×10^9) bbl with approximately 836 billion bbl remaining as reserves. It is also estimated that there are 180 billion bbl which remain to be discovered with approximately 1 trillion (1×10^{12}) bbl yet-to-be-produced. The annual depletion rate is estimated to be 2.6%.

2.5.2 The Real Numbers

One of the issues that arises when delineating an placing a value on heavy oil reserves is that because of the confusing and often unclear nomenclature, many estimates of heavy oil reserves also contain numbers related to the tar sand bitumen reserves. Until a unified system of nomenclature is accepted, there will always be differences between the various estimates leading to publication of these differences in the relevant literature (Kovarik, 2003; Meyer et al., 2007; Speight, 2011b).

Generally, the total resources of heavy oil in known accumulations are 3396 billion barrels (3396×10^9 bbls) of original oil in-place, of which 30 billion barrels (30×10^9 bbls) are included as prospective additional oil. The total natural bitumen resource in known accumulations amounts to 5505 billion barrels (5505×10^9 bbls) of oil originally in place, which includes 993 billion barrels (993×10^9 bbls) as prospective additional oil. This resource is distributed in 192 basins containing heavy oil and 89 basins with natural bitumen. Of these basins, some with subdivisions, the most prolific by far for known heavy oil and natural bitumen volumes are continental multicyclic basins, either basins on the craton margin or closed basins along convergent plate margins. The former includes 47% of the natural bitumen, the latter 47% of the heavy oil and 46% of the natural bitumen. Little if any heavy oil occurs in fore-arc basins, and natural bitumen does not occur in either fore-arc or delta basins (Meyer et al., 2007).

Venezuela has 47–76 billion barrels ($47–76 \times 10^9$ bbls) of proven reserves, according to oil industry/DOE estimates. The United States Geological Survey (USGS) puts Venezuelan reserves around the same level, at 48 identified and 110 ultimately recoverable. Venezuela claims 1.2 trillion barrels (1.2×10^{12} bbls) of unconventional oil reserves in the supergiant heavy oil field stretching from the mouth of the Orinoco River near Trinidad down the east side of the Andes mountains (Arcaya, 2001). The oil is located in a geosynclinal trough that is theorized to be continuous through the Falkland Islands off the coast of Argentina. Only parts of the heavy oil field have been fully explored, but those parts have been estimated at some three to four trillion barrels of heavy oil in-place, with perhaps one-third recoverable using current technology.

Because conventional light oil can typically be produced at a high rate and a low cost, it has been used before other types of oil. Bitumen-impregnated sand (tar sand, oil sand) and heavy oil differ from conventional light oil by the high viscosity (resistance to flow) at reservoir temperatures, high density (low API gravity), and significant contents of nitrogen, oxygen, and sulfur compounds and heavy-metal contaminants. In fact, heavy oil and bitumen resemble the residuum from the refining of conventional crude oil (Speight and Ozum, 2002; Parkash, 2003; Hsu and Robinson, 2006; Gary et al., 2007; Speight, 2014). Most heavy oil is found at the margins of geologic basins and is thought to be the residue of formerly light oil that has lost its light-molecular-weight components through degradation by bacteria, water-washing, and evaporation.

Table 2.1 Regional Distribution of Estimated Technically Recoverable Heavy Oil and Natural Bitumen in Billions of Barrels (bbls × 10^9)

	Heavy Oil		Bitumen	
Region	**Recovery Factor**	**Technically Recoverable bbl × 10^9**	**Recovery Factor**	**Technically Recoverable bbl × 10^9**
North America	0.19	35.3	0.32	530.9
South America	0.13	265.7	0.09	0.1
W. Hemisphere	0.13	301.0	0.32	531.0
Africa	0.18	7.2	0.10	43.0
Europe	0.15	4.9	0.14	0.2
Middle East	0.12	78.2	0.10	0.0
Asia	0.14	29.6	0.16	42.8
Russia	0.13	13.4	0.13	33.7
E. Hemisphere	0.13	133.3	0.13	119.7
World		434.3		

The estimated volume of technically recoverable heavy oil (>400 billion barrels, >400 × 10^9 bbls) and tar sand bitumen (>600 billion barrels, >600 × 10^9 bbls) in known accumulations is at least equal to the remaining conventional (light) oil reserves (Table 2.1) (Speight, 2011b).

2.6 OIL PRICING

Currently, politics and other factors aside, crude oil remains the primary energy source in the world (Speight, 2011b; BP, 2014). For a century the world has depended on low cost oil to stimulate and maintain economic growth (Yergin, 1991). However, sustaining the rate of economic growth is open to question because the volume of oil that can ultimately be recovered is subject to much speculation because of the uncertainties of reserve estimation and this, in turn, affects the price of oil.

At this stage, dealing with petroleum reserves and resources, it is appropriate to deal with a related topic and that is *crude oil prices*. However, it is not the intent here to move into predictions of the future. There is nothing difficult about making predictions. But predictions of future events are always difficult! It is difficult to be correct! Everyone can justify with amazingly accurate 20-20 hindsight why his/her predictions were correct, even though the predictions have been shown and proven to be incorrect! These erstwhile mediums will use statistics to show that, after several rounds of mathematical manipulation, that their predictions were very close. Even though the outcome bears no relationship to what really happened. It is easy to make a statement that oil prices will continue to rise (after all, the pessimist is never disappointed) but the predictability is determining when and by how much. In fact, since the release of the First Edition of this book, oil prices (although still volatile) have decreased markedly! Therefore, it is the purpose of this section to forgo any predictions. It is, however, the purpose of this section, to present a brief history of oil prices from which the reader can make his/her own predictions.

2.6.1 Oil Price History

Crude oil prices have seen wide price swings over the past decade, whether it is due to apparent shortage or oversupply. At the time of writing, prices are above $120 per barrel. Even when adjusted for inflation to current dollars, an average price per barrel bears little relationship to reality, especially when reality is a much higher price per barrel.

Historically (or, some might say, hysterically), crude oil prices varied from $2.50 and $3.00 from 1948 through the end of the 1960s. The price rose from $2.50 in 1948 to about $3.00 in 1957. From 1958 to 1970 prices were stable at about $3.00 per barrel. Organization of Petroleum Exporting Countries (OPEC) was formed in 1960 with five founding members Iran, Iraq, Kuwait, Saudi Arabia, and Venezuela. By the end of 1971 six other nations (namely Qatar, Indonesia, Libya, United Arab Emirates, Algeria, and Nigeria) had swelled the membership ranks of OPEC.

Throughout this period, the petroleum exporting countries found increasing demand for their crude oil. In 1972 the price of crude oil was about $3.00 per barrel and by the end of 1974 the price of oil had quadrupled to over $12.00. The Yom Kippur War started with an attack on Israel by Syria and Egypt on October 5, 1973. Many countries in the western world showed strong support for Israel and, as a result, several of the Middle Eastern oil exporting nations imposed an embargo on those countries by decreasing oil.

From 1974 to 1978 the price of crude oil was relatively flat ranging from $12.21 per barrel to $13.55 per barrel. When adjusted for inflation the price over that period of time exhibited a moderate decline. Then events in Iran and Iraq (overthrow of the Shah of Iran and the Iran-Iraq war) led to another round of crude oil price increases and crude oil prices rose to $35 per barrel in 1981.

The higher prices resulted in, amongst other actions usually involving energy conservation, increased exploration and production in the non-OPEC world. In mid-1985, oil prices were linked to the spot market for crude and by early 1986 (with increased production by some OPEC members), crude oil prices moved downward to $8–10 per barrel. The price of crude oil rose again in 1990 with the Iraqi invasion of Kuwait and the ensuing Gulf War, but following the war crude oil prices entered a steady decline. The price cycle then turned up and from 1990 to 1997 world oil consumption increased by in excess of six million barrels per day. The price increases came to a rapid end when, due to a downward trend in several Asian economies, higher OPEC production sent prices downward.

A low point was reached in January 1999 after increased oil production from Iraq coincided with the Asian financial crisis, which reduced demand. The prices then rapidly increased, more than doubling by September 2000, then fell until the end of 2001 before steadily increasing, reaching $40–50 per barrel by September 2004. In October 2004, the price of crude oil exceeded $53 per barrel and for December delivery exceeded $55 per barrel. Crude oil prices surged to a record high above $60 a barrel in June 2005, sustaining a rally built on strong demand for gasoline and diesel and on concerns about refiners' ability to keep up. This trend has continued and, at the time of writing the First Edition of this book, crude oil prices had surged past $120 per barrel. At the time of writing of this edition, West Texas Intermediate (WTI) crude oil is priced at $45 per barrel and Brent crude oil is priced at $48 per barrel. While these prices will change by the time this edition is printed, the price of crude oil has dropped considerably from 2007 (the time of release of the First Edition of this book) thereby upsetting the prediction of the pessimistic naysayers who foresaw a continuous rise in crude oil prices and predicted doomsday scenarios where the world ran out of crude oil, if not yesterday then certainly tomorrow!

2.6.2 Pricing Strategies

Crude oil is of little use before refining and is traded for the final petroleum products that consumers demand. The intrinsic properties of crude oils determine the mix of final petroleum products. The two most important qualities of crude oil are viscosity (thickness or density) and sulfur content.

Light conventional crude oil is contrasted with heavy crude oil that has a low share of light hydrocarbons and requires much more complex refining process than distillation (such as coking and cracking) to produce similar proportions of the more valuable petroleum products. Sulfur is an undesirable property of crude oil and refiners make heavy investments in order to remove it. Tar sand bitumen also offers a contrast to both conventional crude oil and heavy crude oil insofar as recovery of the bitumen is more complex and the refining scenarios are not readily amenable to high yields of liquid products and result in substantial yields of coke (Speight, 2014).

Since the type of crude oil has a bearing on refining yields, different types of crude oil fetch different prices. Crude oils that yield a higher proportion of the more valuable final petroleum products and require simple refining processes (the light/sweet crude variety) usually command a premium over those that yield a lower percentage of the more valuable petroleum products and require more complex refining processes (the heavy/sour crude variety).

However, differences in quality are not the only determinant of oil price differentials and hence differentials are not constant over time. For example, changes in the prices of different petroleum products (or the gross product worth) lead to changes in crude oil differentials. Another factor that may influence the differential is the heating season and whether the season turns out to be colder than expected, which suggests that movements in crude oil price differentials are likely to exhibit a seasonal behavior. Factors outside the oil sector can also have an influence.

Briefly, the price of a barrel of oil is highly dependent on both its grade (which is determined by factors such as its specific gravity or API and its sulfur content) and location. The vast majority of oil will not be traded on an exchange but on an over-the-counter basis, typically with reference to a marker crude oil grade that is typically quoted via the pricing agency IPE claim that 65% of traded oil is priced off their Brent benchmarks. The Energy Information Administration (EIA) uses the Imported Refiner Acquisition Cost, the weighted average cost of all oil imported into the United States as the *world oil price*.

Heavy crude oil provides an interesting situation for the economics of petroleum development. The resources of heavy oil in the world are more than twice the resources of conventional crude oil. On one hand, due to increased refining costs and high sulfur content, heavy crude oil is often priced at a discount to lighter ones and the increased viscosity and density also makes production more difficult. On the other hand, large quantities of heavy crude oil exist in the Americas including Canada (particularly in Alberta and Saskatchewan), the United States (particularly in Northern California), and Venezuela (particularly in Orinoco). The relatively shallow depth of these reservoirs (often less than 3000 feet) contributes to lower drilling costs but recovery costs (steam generation) are relatively high.

Since the type of crude oil has a bearing on refining yields, different types of crude oil fetch different prices, that is, there is a price differential between conventional (light) crude oil and heavy crude oil. Crude oils that yield a higher proportion of the more valuable final petroleum products and require simple refining processes (the light/sweet crude variety) usually command a premium over those that yield a lower percentage of the more valuable petroleum products and require more complex refining processes (the heavy/sour crude variety). However, crude oil price differentials are influenced by a wide array of factors and are highly volatile.

Large discounts increased the attractiveness of heavy crude oils to refineries which responded by increasing their imports of heavy crude and increasing the production of refined petroleum products to meet the rise in demand. Thus, the incremental demand in growth for light petroleum products, mainly gasoline, can be met by imports of cheaper heavier crude oils. However, the use of more heavy crude oil in-place of conventional light crude oil may be detrimental not only to the ability of many refineries to cope with the increased amount of high-sulfur feedstocks but it may also be a burden on refining capacity when the specter of more-feedstock-for-the-same-amount-of-gasoline raises its head.

2.6.3 The Role of Heavy Oil and Tar Sand Bitumen in the Future

The Hubbert theory (Hubbert, 1956, 1962) assumes that oil reserves will not be replenished (ie, that abiogenic replenishment is negligible) and predicts that future world oil production must inevitably reach a peak and then decline as these reserves are exhausted. Controversy surrounds the theory since predictions for the time of the global peak are dependent on the past production and discovery data used in the calculation. However, there has been a general misinterpretation of the Hubbert theory insofar as the proponents of the Hubbert theory (a reasonable postulate that a natural resource is finite) seemed to omit that accessing greater resource bases is not a matter of more difficult technology but rather a matter of producing crude oil, heavy oil, and tar sand bitumen with available and sustainable techniques. For the United States, the prediction that a natural-occurring resource is finite turned out to be correct and, after crude oil production in the United States peaked in 1971 and thus lost its excess production capacity, OPEC was able to manipulate oil prices

(Speight, 2011b). However, for a variety of reasons, it is difficult to predict the oil peak in any given region. Based on available production data, proponents have previously (and incorrectly) predicted the peak for the world to be in years 1989, 1995, or in the 1995–2000 period. Other predictions have chosen 2007 and beyond for the peak of oil production. While this may have happened, discoveries of new fields (not necessarily giant oil fields) have continued to provide sustenance to the refining industry.

In summary, the petroleum industry is indeed on the verge of a major decision period with the onset of the processing of high volumes of heavy crude oil and tar sand bitumen. Several technology breakthroughs have made this possible but many technical challenges still remain. More important, several trends that should have been established in the wake of the decreasing crude prices have never been put into practice. Some would argue that the periods of oil price decline were the impetus to start development of better technology and expertise. They would also argue that politicians at various levels of government have failed to recognize the need for a measure of energy independence through the development of alternate resources (Speight, 2008) as well through the development of new and innovative technologies that would assist in maximizing domestic oil recovery (Speight, 2011b).

REFERENCES

Arcaya, I., 2001. Venezuela and the United States: a four-pillar strategy for energy security. Remarks to the Business Council for International Understanding Petroleum Club of Houston, Houston, TX, July 24.

BP, 2014. Statistical Review of World Energy. British Petroleum Company, London, June.

Dusseault, M.D., 2001. Comparing Venezuelan and Canadian Heavy Oil and Tar Sands. Paper No. 2001-061. Proceedings. International Petroleum Conference, Calgary Alberta Canada, June 12–14. Canadian Institute of Mining, Westmount, Quebec, Canada.

Gary, J.G., Handwerk, G.E., Kaiser, M.J., 2007. Petroleum Refining: Technology and Economics, fifth ed. CRC Press, Taylor & Francis Group, Boca Raton, FL.

Head, I.M., Jones, D.M., Larter, S.R., 2003. Biological activity in the deep subsurface and the origin of heavy oil. Nature 426 (20), 344–352.

Hsu, C.S., Robinson, P.R., 2006. Practical Advances in Petroleum Processing, vols 1 and 2. Springer, New York.

Hubbert, M.K., 1956. Nuclear Energy and the Fossil Fuels—Drilling and Production Practice. American Petroleum Institute, Washington, DC.

Hubbert, M.K., 1962. Energy resources. Report to the Committee on Natural Resources. National Academy of Sciences, Washington, DC.

IEA, 2005. Resources to Reserves: Oil and Gas Technologies for the Energy Markets of the Future. International Energy Agency, Paris.

Kenney, J.F., Shnyukov, A.Y.F., Krayushkin, V.A., Karpov, I.K., Kutcherov, V.G., Plotnikova, I.N., 2001. Dismissal of the claims of a biological connection for natural petroleum. Energia 22 (3), 26–34.

Kovarik, W., 2003. The oil reserve fallacy: proven reserves are not a measure of future supply. <http://www.runet.edu/~wkovarik/oil/3unconventional.html> (accessed 15.01.15.).

Larter, S., Huang, H., Adams, J., Bennett, B., Jokanola, O., Oldenburg, T., et al., 2006. The controls in the composition of biodegraded oils in the deep subsurface: part II. Geological controls on subsurface biodegradation fluxes and constraints on reservoir-fluid property prediction. Am. Assoc. Petroleum. Geol. Bull. 90 (6), 921–938.

Larter, S., Whilhelms, A., Head, I., Koopmans, M., Aplin, A., Di Primio, R., et al., 2003. The controls on the composition of biodegraded oils in the deep subsurface—part 1: biodegradation rates in petroleum reservoirs. Org. Geochem. 34 (3), 601–613.

Meyer, R.F., and Attanasi, E.D., 2003. Heavy oil and natural bitumen—strategic petroleum resources. Fact Sheet 70-03. U.S. Geological Survey, Washington, DC, August. <http://pubs.usgs.gov/fs/fs070-03/fs070-03.html>.

Meyer, R.F., Attanasi, E.D., 2004. Natural Bitumen and Extra Heavy Oil. World Energy Council 2004 Survey of Energy Resources. Elsevier, Amsterdam, Chapter 4.

Meyer, R.F., Attanasi, E.D., and Freeman, P.A. 2007. Heavy oil and natural bitumen resources in geological basins of the world: open-file report 2007-1084, U.S. Geological Survey, Washington, DC. <http://pubs.usgs.gov/of/2007/1084/>.

Morehouse, D.F., 1997. The intricate puzzle of oil and gas reserves growth Natural Gas Monthly. U.S. Energy Information Administration, Washington, DC, July.

Parkash, S., 2003. Refining Processes Handbook. Gulf Professional Publishing, Elsevier, Amsterdam.

SPE, 2007. Petroleum Resources Management System. Society for Petroleum Engineers, Richardson, TX.

Speight, J.G., 2008. Synthetic Fuels Handbook: Properties, Processes, and Performance. McGraw-Hill, New York.

Speight, J.G., 2011a. The Refinery of the Future. Gulf Professional Publishing, Elsevier, Oxford.

Speight, J.G., 2011b. An Introduction to Petroleum Technology, Economics, and Politics. Scrivener Publishing, Salem, MA.

Speight, J.G., 2013. The Chemistry and Technology of Coal, third ed. CRC-Taylor and Francis Group, Boca Raton, FL.

Speight, J.G., 2014. The Chemistry and Technology of Petroleum, fifth ed. CRC-Taylor and Francis Group, Boca Raton, FL.

Speight, J.G., 2015. Handbook of Offshore Oil and Gas Operations. Gulf Professional Publishing, Elsevier, Oxford.

Speight, J.G., Ozum, B., 2002. Petroleum Refining Processes. Marcel Dekker, Inc., New York.

Yergin, D., 1991. The Prize: The Epic Quest for Oil, Money, and Power. Simon & Schuster, New York.

CHAPTER 3

Sediments, Reservoirs, and Deposits

3.1 INTRODUCTION

Sediments and sedimentary rocks are an important, but unfortunately an often-ignored, aspect of petroleum geology and hence petroleum recovery operations. Knowledge of the geology and mineralogy of a reservoir or deposit leads to deriving the means by which to find and penetrate the reservoir (or deposit), which would be sorely lacking and, most likely, cause problems for recovery. In addition, such knowledge is advantageous when in situ upgrading is considered as a process option during recovery. The minerals might well (hopefully) exhibit beneficial catalytic activity on the in situ upgrading process or, on the other hand, the minerals might have an adverse effect on the process chemistry and physics.

Typically, the majority (in excess of 90%) of all sedimentary rocks consists of (1) sandstone rocks, which consist of sand sized fragments and are important when dealing with conventional petroleum reservoirs, (2) mudrocks, which consist of silt and clay sized fragments and are important when dealing with crude oil and natural gas from tight formations, such as shale formations, and (3) carbonate rocks, which consist of calcite ($CaCO_3$), aragonite ($CaCO_3$), or dolomite ($CaCO_3{\cdot}MgCO_3$). The carbonate minerals are typically formed as a result of biological and physical processes, including precipitation from marine and fresh water environments. Of these minerals, the mudrocks are most abundant, making up about 65% of all sedimentary rocks. Sandstones make up 20% to 25% of all sedimentary rocks, and carbonate rocks account for about 10% to 15% of all sedimentary rocks.

Sedimentology involves a study of sand, mud (or silt), and clay, and the various ways they are deposited (Prothero and Schwab, 1996; Nichols, 1999). Using these studies, sedimentologists can apply their understanding of modern processes to ancient rock, to try to understand how it formed. Most of the rocks of interest to the petroleum scientist and petroleum engineer are sedimentary rocks, and it is in these kinds of rocks that crude

Introduction to Enhanced Recovery Methods for Heavy Oil and Tar Sands.
DOI: http://dx.doi.org/10.1016/B978-0-12-849906-1.00003-5

oil and natural gas are found. In addition, sedimentology is tied to stratigraphy, which studies the relationships between rock layers and how they can shift and move. This also affects where crude oil reservoirs, heavy oil reservoirs, and tar sand deposits can be found, as well as how the extraction of petroleum affects the sediment around the deposit. Within the sediments, events that are of importance to the petroleum chemist, geochemist, geologist, and engineer occur.

The importance of knowledge of the properties and mode of formations is essential to the geologist (who should be part of any multidisciplinary team for a drilling project) because of the need to drill into or drill through these formations in the search for crude oil. A major factor in drill bit selection is the type of formation to be drilled and the effectiveness of a drill bit varies by formation type. There are three types of formations: soft formations, medium formations, and hard formations. A soft formation includes unconsolidated sand, clay minerals, and shale formations. Medium formations include formations originating from calcite (a polymorph of calcium carbonate, $CaCO_3$) and dolomite sediments ($CaCO_3$ $MgCO_3$). Hard formations include hard shale, calcite formations, mudstone formations, cherty limestones, and hard and abrasive formations.

Briefly, chert is a fine-grained silica (SiO_2)-rich microcrystalline or cryptocrystalline, or microfibrous, sedimentary rock that may contain small fossils. This mineral varies in color (from white to black), but most often occurs as gray, brown, grayish brown, and light green to rusty red. The color is an expression of the presence of the various trace elements in the sediment rock—the red and green colors are most often related to traces of iron (in the oxidized form, Fe_2O_3, and in the reduced form, FeO, respectively).

Thus, multiple factors affect the selection of the drill bit and if a large number of wells have been drilled in the same or into a similar formation, then information from an adjacent well is most often used to make the appropriate selection of the drill bit. For example, a fixed cutter bit is one where there are no moving parts and drilling occurs due to shearing, scraping or abrasion of the rock. These bits can either be (1) polycrystalline diamond compact drill bits—PDC drill bits; (2) grit-hot-pressed insert drill bits—GHI drill bits; or (3) natural diamond drill bits. Roller cone bits can either be (1) tungsten carbide insert bits—TCI bits for harder formations; or (2) milled tooth bits—MT bits—for softer rock. The manufacturing process and composites used in each type of drill bit allow the bits to be enhanced and tailored to the specific drilling situations.

3.2 SEDIMENTS

Sediments are formed by various processes and, therefore, each sediment should not be expected to be precisely the same as any other sediment, unless the process of formation and the rocks from which the sediment was formed are identical and the conditions of formations were also identical. In short, for any sediment to be the same as another sediment, the conditions prevalent during the initiation period (ie, source of the sediment such as the rocks and minerals) and the external influence (ie, the conditions in the surrounding environment) must also be identical.

In terms of sediment formation, water courses, such as rivers and oceans, as well as rain runoff have the ability to carry the particles (detritus) washed from eroding rocks and these particles consist of fragments of rocks and minerals; to some extent wind is also an agent in sediment formation. When the energy of the transporting medium is insufficient to carry the detritus, the particles separate from the carrier (*sedimentation*). When the particles are broken pieces of rock the deposition process is usually referred to as *clastic sedimentation* but if sedimentary deposition is due to the soluble-in-water material chemically precipitating from the water this type of sedimentation is referred to as *chemical sedimentation*. A third process, wherein living organisms extract ions dissolved in water to create protective shells and/or skeletal tissue is known as *biochemical sedimentation*. The accumulation of plant matter or any other matter from a biological origin, such as occurs at the bottom of a swamp, is referred to as *organic sedimentation*. Thus, the origin of the sediment gives rise to the categorization of sedimentary rocks as: (1) *clastic sediments and sedimentary rocks*; (2) *chemical sedimentary rocks*; (3) *biochemical sedimentary rocks*; and (4) organic *sedimentary rocks*.

3.3 OCCURRENCE

Sedimentary rocks occur on all continents and, because sediment is buried and if exposed is eroded, older sedimentary rocks show less exposed outcrop area than younger sedimentary rocks—more than 40% of the exposed sedimentary rocks are younger than Cretaceous in age. But to the petroleum scientist or engineer it is the buried sedimentary rocks that tend to be the most important, because of their potential use as a reservoir or deposit for crude oil with natural gas, heavy oil, and tar sand deposits.

In order for sediment to be deposited, a low lying area (a basin), is required and the largest basins are the ocean basins, which currently cover

Table 3.1 The Geologic Timescale

Era	Period	Epoch	Duration (Millions of Years)	Years Ago (Millions of Years)
Cenozoic	Quaternary	Holocene	10,000 years ago to the present	
		Pleistocene	2	0.01
	Tertiary	Pliocene	11	2
		Miocene	12	13
		Oligocene	11	25
		Eocene	22	36
		Paleocene	71	58
Mesozoic	Cretaceous		71	65
	Jurassic		54	136
	Triassic		35	190
Paleozoic	Permian		55	225
	Carboniferous		65	280
	Devonian		60	345
	Silurian		20	405
	Ordovician		75	425
	Cambrian		100	500
Precambrian			3380	600

about 70% of the surface of the Earth. However, with the passage of time, sea levels have often changed giving rise to events where the continents were covered by shallow seas (epeiric sea or inland sea and the epicontinental sea, which is that portion of the sea lying on the continental shelf). Major transgressions occurred during the Cretaceous, and from the Early Cambrian through Mississippian (Table 3.1)—a transgression occurs when the sea level rises and the sea invades the continent. Thus, it is not surprising that most of the sedimentary rocks preserved in the geologic record are marine sediments. Basins result from plate tectonics, and even the large transgressions appear to be related to tectonic factors, as increased spreading rates in the ocean basins can result in changes in the configuration of ocean basins that result in flooding of the continents. A detailed basin analysis is usually carried out prior to defining leads and prospects for future drilling. This involves a full study of the petroleum system and studies of source rock (presence and quality); burial history; maturation (timing and volumes); migration and focus; and potential regional seals and major reservoir units (that define carrier beds). These parameters are then used to investigate where potential hydrocarbons might migrate towards

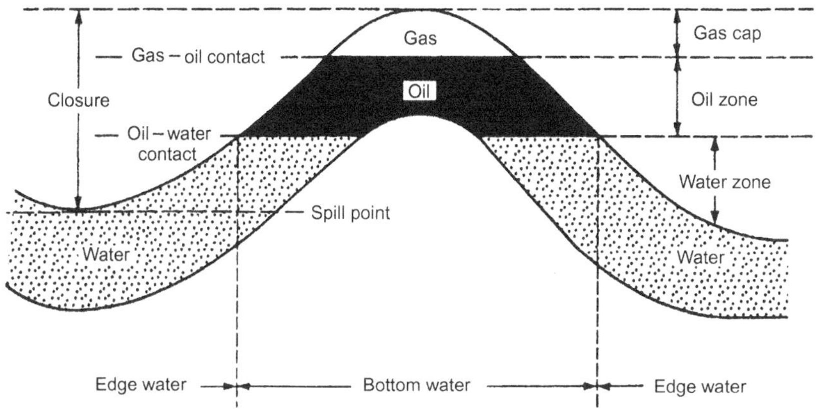

Figure 3.1 Schematic representation of a syncline-anticline petroleum-natural gas reservoir.

after which traps and potential leads and prospects are then defined in the area that is likely to have received hydrocarbons.

A *structural basin* is a large-scale structural formation of rock strata formed by tectonic warping of previously flat-lying strata. Structural basins are geological depressions, and are the inverse of domes—some elongated structural basins are also known as synclines, which are folds with younger layers closer to the center of the structure. Structural basins are often important sources of coal and petroleum, especially the syncline-anticline system.

Synclines are typically a downward fold (a synformal syncline or trough) but synclines that point upwards, or perched, can be found when strata have been overturned and folded (an antiformal syncline). On the other hand, anticlines are folds in which each half of the fold dips away from the crest. Synclines and anticlines usually occur together and are ideal locations for a petroleum reservoir (Fig. 3.1) (Speight, 2014).

Finally, on the issue of basins, structural basins may also be sedimentary basins, which are aggregations of sediment that filled up a depression or accumulated in an area; however, many structural basins were formed by tectonic events long after the sedimentary layers were deposited.

3.4 MINERALOGY

The detrital nature of sediments leads to the distinct possibility that a variety of minerals can occur in a sediment or sedimentary rock. Clay minerals, the dominant mineral produced by chemical weathering of rocks, is the most abundant mineral in mudrocks. Quartz (SiO_2), because it is stable under conditions present at the surface of the Earth, and because it is also a product of chemical weathering, is the most abundant mineral in sandstone sediments and rocks and the second most abundant mineral in mudrock sediment. Feldspar minerals (a group of rock-forming tectosilicate minerals, $KAlSi_3O_8$–$NaAlSi_3O_8$–$CaAl_2Si_2O_8$) are the most common minerals in igneous and metamorphic rocks. Although feldspar eventually breaks down to clay minerals and quartz, it is still the third most abundant mineral in sedimentary rocks. Carbonate minerals, either precipitated directly or by organisms, make up most biochemical and chemical sedimentary rocks, but carbonates are also common in mudrocks and sandstones.

Minerals that are found in sedimentary rocks can be divided into 2 classes: (1) allogenic minerals, which are formed elsewhere and transported into the area of deposition, and (2) authigenic minerals, which are formed at the site of deposition, either by direct chemical precipitation or by later diagenetic processes. Any mineral can be an allogenic mineral, but some are more stable under the conditions present at the surface of the Earth than others. Conditions that are present at the surface of the Earth and which differ from those where most minerals form are: (1) low temperature, (2) low pressure, (3) high concentration of free oxygen, and (4) high amounts of free liquid water. Furthermore, because these conditions differ from those under which most rocks form, allogenic minerals can be classified based on their stability under near-surface conditions. Minerals that are very stable at the surface of the Earth are minerals that either form as a result of chemical weathering, or crystallize at the lowest temperatures.

Authigenic minerals can also be allogenic minerals, but some are formed during diagenesis but are not very stable in the transportation cycle either because they dissolve readily in water or because they are easily abraded during transportation. Thus, authigenic minerals can be subdivided into (1) minerals that are stable during diagenesis and transportation, and (2) minerals that are unstable during transportation. Moreover, the longer that a mineral is in the weathering and transportation cycles of sedimentary rock-forming processes, the more likely the mineral will suffer breakdown to a more stable mineral or even cease to be recognized as a mineral and the chemical components become part of a completely different mineralogical entity. Thus, knowing the chemical mineralogy of the various minerals systems, sediments can be classified on the basis to which the sediment(s) has (have) achieved *mineralogical maturity*. Mineralogically mature sediments and sedimentary rocks—such as sediments and rocks that have been subject to the weathering and transportation cycle for a considerable time—consist entirely of minerals that are stable near the surface.

3.5 CLASTIC SEDIMENTS AND SEDIMENTARY ROCKS

Siliclastic sedimentary rocks form by the accumulation of mostly silicate mineral fragments and include most sandstones, mudrocks, conglomerates, and breccia. The formation of a clastic sediment and sedimentary rocks involves processes such as: (1) *weathering*, which is the first stage in the transformation of solid rock into smaller fragments or into dissolved ions; (2) erosion, which is the cumulative effects of many processes which act together and which begins by moving the weathered products from their original location.

Conglomerate and breccia are rocks that contain substantial amounts of coarse-grained clasts (pebbles, cobbles, or boulders). In a *conglomerate*, the coarse-grained clasts are well rounded, indicating that they have spent considerable time in the transportation process and were ultimately deposited in a high energy environment capable of carrying the large clasts. In a *breccia*, the coarse grained clasts are very angular, indicating that the clasts spent little time in the transportation cycle.

Of importance to the petroleum chemist, geochemist, geologist, and engineer are the sandstone sediments and the mudrock sediments. As the name implies, a sandstone sediment (one of the most common types of sedimentary rocks) is composed of silica particles and can be found in many different depositional settings. The texture and composition permit assessment and interpretation of the transport and depositional cycle and may even allow identification of the source. Quartz (SiO_2) is, by far, the dominant mineral in sandstones but there are other varieties: (1) quartz arenite is approximately 100% quartz grains; (2) arkose contains abundant feldspar and is commonly coarse-grained and usually either pink of gray, depending on the color of feldspar—feldspars ($KAlSi_3O_8$–$NaAlSi_3O_8$–$CaAl_2Si_2O_8$) are a group of rock-forming tectosilicate minerals; (3) in a lithic sandstone, the grains are mostly small rock fragments; and, finally, (4) a Wacke sandstone is a poorly sorted sandstone containing fragments of rock and minerals in a clay matrix that contains more than 15% mud (silt-sized and clay-sized grains).

Mudrocks are made of fine-grained clasts (silt-sized and clay-sized)—*siltstone* is one variety that consists of silt-sized fragments. A *shale* is composed of clay sized particles and is a rock that tends to break into thin flat fragments. A *mudstone* is similar to a shale, but does not break into thin flat fragments. Organic-rich shales are the source of petroleum and lately have been shown to be productive reservoirs for production of crude oil and natural gas. Fine-grained clastic materials are deposited in nonagitated water, calm water, where there is little energy to continue to transport the small grains and, thus, mudrocks form in deep water ocean basins and lakes.

3.6 CHEMICAL SEDIMENTS AND SEDIMENTARY ROCKS

Chemical sedimentary rocks are formed by direct chemical precipitation from water. However, while some limestones and cherts may form in this manner, evaporite deposits consisting of halite, gypsum, and other salts are the most common. Chemical sediments are typically formed from dissolved ions released into water by the weathering process, are carried in streams or in groundwater, and eventually these dissolved ions end in up in the lakes and oceans. With the passage of time, these ions recombine with other ions and, depending upon the nature of the combined chemical, will precipitate as deposited minerals that accumulate as chemical sediments and chemical sedimentary rocks. Among these are such minerals as: (1) evaporites; (2) travertine; (3) dolostone; and (4) chemical chert.

As the name implies *evaporites* are formed by the evaporation of lacustrinbe (lake) water or sea water which produces halite (the mineral form of sodium chloride, NaCl) and gypsum ($CaSO_4\ 2H_2O$) deposits by chemical precipitation. This type of process occurs in lakes that have no outlets (such as the Great Salt Lake, Utah) or in restricted ocean basins, such as past events in the Mediterranean Sea or in the Gulf of Mexico. On the other hand, *travertine* (a form of calcite, $CaCO_3$) is formed from the precipitation of calcium (Ca^{2+}) and bicarbonate (HCO_3^-) ions in groundwater and typically occurs in lakes, hot springs, and caves:

$$Ca^{2+} + 2HCO_3^- \rightarrow CaCO_3 + H_2O + CO_2$$

On the other hand, dolostone is limestone that has been chemically modified by magnesium-rich (Mg-rich) fluids flowing through the rock and, in the process, a portion of the calcite is recrystallized to dolomite ($CaCO_3$ $MgCO_3$).

The formation of chemical chert occurs when groundwater flowing through sedimentary rock precipitates silica (SiO_2) and replaces some of the minerals that were originally present; the material is a *nonbiogenic chert* (abiogenic chert). There are several varieties of the nonbiogenic chert that are named according to the mode of formation, and/or for the most part, properties. For example, *flint* is black or gray as a result of the content of varying amounts of entrapped organic matter, while *jasper* is red or yellow due to the presence of iron oxides (Fe_nO_m) and another chert—petrified wood—is a rock in which the wood grain is preserved by silica and agate and consists of concentrically layered rings.

3.7 BIOCHEMICAL AND ORGANIC SEDIMENTS AND SEDIMENTARY ROCKS

Biochemical sediments and organic sediments and sedimentary rocks are those derived from living organisms and consist of fragments of particles produced by precipitation from once living organisms—most of these rocks are limestones and chert. Under the appropriate conditions, which vary depending upon the surrounding environment, the remains of dead organisms accumulate to become a sediment and eventually sedimentary rock. Among the types of rock produced by this process are: (1) biochemical limestone; (2) biochemical chert; (3) diatomite; and (4) coal, which matures into various forms.

Biochemical limestone (calcite; $CaCO_3$) is precipitated by organisms usually to form a shell or other skeletal structure and the accumulation of such remains results in a limestone sediment and, eventually, a limestone formation. Typically, the fossilized remains of the organism are preserved in the rock or they go through a recrystallization process during lithification which destroys the remains. The formation of *biochemical chert* is the silica-product result of silica-secreting planktonic organisms (such as Radiolaria and Diatoms) which can accumulate on the sea floor and recrystallize during lithification to form the chert. The result is a hard rock that is usually accumulates in the form of thin beds. *Diatomite*, also known as *diatomaceous earth*, is the naturally occurring fossilized remains of diatoms. Diatoms are single-celled aquatic algae (Bacillariophyceae). Diatomite is a near pure sedimentary deposit consisting almost entirely of silica. The properties which make diatomite valuable include low density, high porosity, high surface area, abrasiveness, insulating properties, inertness, absorptive capacity, brightness, and high silica content.

Finally, *coal* is an organic rock that is the result of the chemical changes that occur during the maturation process. The original proto-coal is formed from the organic carbon, which is, in turn, the remains of fossilized plant matter. The carbon-bearing species typically accumulates in tropical wetland settings and under anaerobic (oxygen-deficient) conditions the complete conversion of the carbon to carbon dioxide does not occur and carbonaceous residue is formed that eventually becomes coal.

3.8 DIAGENESIS

The lithification of sediment into sedimentary rocks takes place after the sediment has been deposited and buried. The processes by which the sediment becomes lithified into a hard sedimentary rock (diagenesis) includes all physical, chemical, and biological processes that act on the sediment.

The first step in diagenesis is the compaction of the sediment and loss of water as a result of the weight of the overlying sediment. Compaction and burial may cause recrystallization of the minerals to make the rock even harder. Fluids flowing through the rock and organisms may precipitate new minerals in the pore spaces between grains to form a cement that holds the sediment together—common cementitious materials include quartz (SiO_2), calcite ($CaCO_3$), and hematite, the mineral form of ferric oxide (Fe_2O_3).

Other conditions present during diagenesis, such as the presence or absence of free oxygen may cause other alterations to the original sediment. In an environment where there is excess oxygen (aerobic conditions), organic remains will be converted to carbon dioxide and water. Iron will change from ferrous iron (Fe^{2+}) to ferric iron (Fe^{3+}), and will change the color of the sediment to a deep red (rust) color. In an environment where there is a depletion of oxygen (anaerobic conditions), organic material may be transformed to solid carbon in the form of coal, or may be converted to hydrocarbon derivatives or to hydrocarbonaceous compounds (also called protopetroleum), the source of crude oil, heavy oil, and, most likely, tar sand bitumen also (Speight, 2014). Diagenesis is also a response to increasing the temperature and pressure as sediment gets buried deeper. As temperature increases beyond about 200°C (390°F), metamorphosis commences.

3.9 LITHIFICATION

Lithification (also a form of diagenesis) is the process or processes by which unconsolidated materials (noncemented materls or noncoherent materials) are converted into coherent solid rock by compaction or by cementation. In the initial stage of the process, compaction occurs as the weight of the overlying material (overburden) increases and forces the grains closer together, reducing pore space and eliminating some of the contained water. Some of this water may carry mineral components in solution, and these constituents may later precipitate as new minerals in the pore spaces. This causes cementation, which will then start to bind the individual particles together.

Cementation is the most common method of lithifying coarse grained sediments and, in the process, sedimentary particles are deposited in contact with each other, but there is also a certain amount of empty space in a pile of sediment. As water carrying dissolved ions fills in that empty space, the ions may crystallize new minerals between the grains. In the process of growth these newly forming crystals become enmeshed with each other, as we saw in recrystallization, and the result is once again a more cohesive, harder mass than the original loose sediment.

A second natural method of lithifying sediment is by *recrystallization* of some or all of the constituent minerals. Limestone rocks and other chemical sedimentary rocks are most susceptible to this process because their minerals are fairly easy to modify in surface environments. Much of the so-called natural calcite ($CaCO_3$) in the oceans is not calcite but a different mineral—*aragonite*. Aragonite is a carbonate mineral and one of the two common, naturally occurring, crystal forms of calcium carbonate (the other form being the mineral calcite) and is formed by biological and physical processes, including precipitation from marine and fresh water environments. Aragonite takes up less volume per molecule than calcite and is not truly stable in fresh water, so once rainwater comes into contact with it, it changes into calcite. The crystals grow larger because of this transformation and tend to mesh themselves together, which locks the particles making the entire structure stronger and harder.

Quartz cement is quite common in nature, and tends to make the hardest rocks. Calcite and hematite are also rather common, but tend not to lithify the rock as tightly. Halite and gypsum cements occur in some special environments, but they are poor cementing agents. Many other minerals are known to act as interparticle cements—all but those mentioned above are rare and, at best, only play a very minor role as cementitious materials.

3.10 TEXTURE

Since most sedimentary rocks are derived by processes of weathering, transportation, deposition, and diagenesis, the texture of the sediment or rock is dependent on the processes that occur during each stage and include: (1) the nature of the source rock, which determines the original shape of the grains and the mineralogical composition of the original sediment; (2) the distance of transportation, which is determined by the strength of the wind or water currents that carry and deposit the sediment and determines whether or not grains are transported or deposited; (3) the presence of any biological activity; and (4) the chemical environment under which diagenesis occurs. Typically, the longer a sediment is involved in the transportation cycle, the more time it has to become well-sorted and the more time is available for grains to lose their rough edges and corners by abrasion. Thus, a texturally mature sediment is sediment that is well-sorted and well-rounded and is generally mineralogically mature.

The deposition process also controls structures that could be preserved in the sediment and thus give clues to the environment of deposition. The longer that grains are in the transportation process the more likely they are to change shape and become sorted on the basis of size and mineralogy, which also controls the extent to which they break down to stable minerals during the transportation process. Biological activity with the sediment prior to diagenesis is an important aspect since burrowing organisms can redistribute sediment after it has been deposited, thus erasing some of the clues to the original environment of deposition. Also, during diagenesis grains are compacted, new minerals precipitate in the pore spaces, some minerals continue to react to produce new minerals, and some minerals recrystallize. The true course of events depends on (1) the composition of fluids moving through the rock, (2) the composition of the mineral grains, as well as (3) the pressure and temperature conditions attained during diagenesis.

3.11 PACKING

Packing refers to the arrangement of clastic grains entirely apart from any authigenic cement that may have later crystallized between them. If the clastic grains touch each other throughout, the rock is said to be *grain supported* but if the rock is poorly sorted and the grains are separated by a mud or silt matrix, the rock is *matrix supported*. If the rock is poorly or moderately sorted, the percentage of matrix and texture of the matrix should also be described.

Porosity is the percentage of the volume of the rock that is open space (pore space) and is an important aspect of reservoir rock character and dependent upon the packing of the grains. The porosity determines the amount of water or other fluids, such as petroleum, that a rock can contain. In sediments or sedimentary rocks the porosity depends on grain packing which, in turn, depends on grain size, the shapes of the grains, the degree of sorting, and the degree of cementation. Well-rounded coarse-grained sediments usually have higher porosity than fine-grained sediments—the grains do not fit together closely—but angular grains of fine-grained sediment can be compacted to a close fit and, thus, porosity is reduced in angular grain sediments. Mudrocks, because of their fine grain size and close packing, contribute to shale formations and typically have very low porosity resulting in the need for hydraulic fracturing of such formations to release any trapped crude oil.

Finally, induration refers to the hardness of the rock or how easily it breaks apart. Well indurated rocks, which have undergone cementation, are difficult to break whereas moderately indurated rocks can be easily broken and poorly indurated or friable rocks can be hand-broken. A nonindurated rock has not undergone any cementation

3.12 THE STRUCTURE OF SEDIMENTS

The deposition process generally causes variations in layering, bed forms, or other structures that present indications of the environment in which deposition occurs. Data such as water depth, flow velocity, and current direction can often be determined from sedimentary structures. Furthermore, the various structures of sediments and formation are the result of the depositional process. Each process is sufficiently different from another process that the structure of each sediment must also be assumed to be different from sediment to sediment. The most important aspect of sediment structure is the transportation of the material to the place where the sediment forms.

3.13 TRANSPORTATION

Sediment can be transported by sliding down slopes, being picked up by the wind, or by being carried by running water in streams, rivers, or ocean currents. The distance the sediment is transported and the energy of the transporting medium affects the *deposition*. Sediment is deposited when the energy of the transporting medium becomes too low to continue the transport process. In other words, if the velocity of the transporting medium becomes too low to transport sediment, the sediment will fall out and become deposited. The final sediment thus reflects the energy of the transporting medium.

3.14 STRATIFICATION AND BEDDING

Typically, and often without exception, sediments are deposited in low lying areas that often extend over wide areas. Successive depositional events produce layers (bedding or stratification) that are often an evident feature of sedimentary rocks. Furthermore, in some circumstances, the strata can exhibit different colors (due to the presence of different minerals) as well as differences in grain size. All of these differences can be related to differences in the environment present during the depositional events that led to the formation of the different beds or strata. In terms of nomenclature, this is varied but in some instance a series of beds is referred to as a stratum (pl.: strata) while a sequence of strata that is sufficiently different to be recognized as such, is termed a *formation*, which is a basic geologic mapping unit. Also, because sedimentary rocks can be deformed by folding and faulting long after the deposition process has ended, it is important to be able to determine the original orientation of the rocks after deposition.

3.15 LAYERING AND BEDDING

Sedimentary rocks typically exhibit a layered structure, which is one of the most obvious features of these rocks. The layers may be formed because of differences in mineralogy, clast size, degree of sorting, or color of the different layers. In rocks, these differences may be made more prominent by the differences in resistance to weathering or color changes brought out by weathering. The layering is usually described on the basis of layer thickness or by the structure of the layer. For example, there are (1) rhythmic layering, (2) cross-bedding, and (3) graded bedding.

Rhythmic layering is evident when there is the deposition of alternating parallel layers of sediment having different properties. This may be caused by seasonal changes during deposition (*varves*, which are annual sediment layers), such as is seen in lacustrine (lake) deposits when coarse sediment is deposited in summer months and fine sediment is deposited in the winter at a time when the surface of the lake is frozen.

Cross-bedding refers to sets of beds that are inclined relative to one another. This phenomenon occurs when the beds are inclined in the direction of the wind or water flow that was prevalent at the time of deposition. Boundaries between sets of cross-beds usually represent an erosional surface and are often noticeable in beach deposits, sand dunes, and river deposited sediment. Individual beds within cross-bedded strata are useful indicators of current direction and tops and bottoms.

Graded bedding occurs as a result of a decrease in the velocity of the water flow. In such a case, the larger or denser particles are deposited first followed by smaller particles. This results in bedding showing a decrease in grain size from the bottom of the bed to the top of the bed. Sediment can also be added as a pulse of turbid water and as the pulse decreases the water loses velocity and sedimentation occurs. *Reverse graded bedding* cannot occur as current velocity increases mean that each layer will simply be removed as the current achieves a velocity high enough to carry sediment of a particular size. Multiple graded-bed sequences turbidites are common. *Nonsorted sediment* is sediment in which a mixture of grain sizes occurs and is a result of events such as rock falls, debris flows, mudflows, and deposition from melting ice.

Imbricate bedding occurs when elongated grains pile up on each other, which can be an indicator of current direction if some other means is present to provide top/bottom directions

3.16 SEDIMENTARY FACIES

Facies are bodies of rock with specified characteristics and are typically a distinctive rock unit that forms under certain conditions of sedimentation, reflecting a particular process or environment (Reading, 1996). Thus, *sedimentary facies* are groups of characteristics that reflect a sedimentary environment different from those elsewhere in the same deposit. Thus, facies may change vertically through a sequence as a result of changing environment through time and, in addition, facies may change laterally through a deposit as a result of changing environments with distance at the same time. Individual facies are generally described in terms of the environment is which deposition occurred (Table 3.2).

3.17 OTHER SEDIMENT MARKERS

Ripple marks are caused by water flowing over loose sediment which creates bed forms by moving sediment with the flow. *Bed forms* are linked to flow velocity and sediment size, whereas ripples are characteristic of

Table 3.2 Types of Facies

I. Terrigenous or Nonmarine Facies	**II. Near Shore Facies**
Alluvial fan	Deltaic
Floodplain	Estuarine
Lacustrine (lakes)	Bay or Lagoon
Fluvial (streams)	Marsh
Eolian (wind deposited, sand dunes)	Intertidal
Swamp	Beach or Barrier Island
Glacial	Glacial Marine
III. Shallow Marine Facies (Depth <650 feet)	**IV. Deep Marine Facies (Depth >650 feet)**
Continental shelf	Submarine canyon
Carbonate bank and shelf	Submarine fan
Forearc basin	Back arc basin
Evaporite basin	Deep ocean basin

shallow water deposition and can also be caused by wind blowing over the surface. Ripples are commonly preserved in sedimentary rocks and asymmetric ripples indicate flow direction, with the steep slope on the down-current direction. Ripples preserved in ancient rocks can also be indicators of up/down direction in the original sediment and symmetric ripples form as a result of constant water (wave) energy oscillating back and forth. Symmetrical ripple marks occur in environments where there is a steady back and forth movement of the water. Such ripple marks can still be used as top and bottom indicators.

There are other indicators of the depositional environment that offer information to the geologist about the depositional environment. For example, *mud cracks* are the result of the drying out of wet sediment at the surface of the Earth. The cracks form due to shrinkage of the sediment as it dries. When present in rock, they indicate that the surface was exposed at the Earth's surface and then rapidly buried. *Sole marks are* flutes eroded in soft sediment that can become filled with mud. Both the flutes and the resulting casts (*flute casts*) can be preserved in rock. *Raindrop marks* are pits (or tiny craters) created by falling rain. If present, these suggest that the sediment was exposed to the surface of the Earth just prior to burial. *Fossils* are the remains of once living organisms and offer an important and reasonably accurate indicator of the environment of deposition.

Rock Color also offers an overview of the type and time of the depositional environment. For example, sulfide minerals, along with buried

Table 3.3 Characteristics of Sediments Based on Grain Size and Shape

Particle-Type	Size Range, mm	Sediment-Type	Rock-Type
Boulder	>256	Gravel	Conglomerate or Breccia
Cobble	64–256		
Pebble	4–64		
Granule	2–4		
Very coarse sand	1–2	Sand	Sandstone
Coarse sand	0.5–1		
Medium sand	0.25–0.5		
Fine sand	0.125–0.25		
Very fine sand	0.0625–0.125		
Coarse silt	0.031–0.625	Silt	Siltstone
Medium silt	0.016–0.031		
Fine silt	0.008–0.016		
Very fine silt	0.004–0.008		
Clay	<0.004	Clay	Claystone, mudstone, shale

organic matter give rocks a dark color and indicate deposition in a reducing environment. Deposition in oxidizing environment produces red colored iron oxides and is often indicative of deposition in a nonmarine environment and such red colored rocks are often referred to as red beds.

3.18 CLASSIFICATION

Sediments have characteristics that are unique to the environment in which the sediments were formed. Each environment has its own energy regime as well as sediment delivery, transport, and depositional conditions that are reflected in the sediment deposited.

Clastic sedimentary particles and sedimentary rocks which are classified in terms of grain size and shape (Table 3.3), among other factors, leave indications about the mode of transportation. In general, the coarser sediment gets left behind by the transportation process. Thus, coarse sediment is usually found closer to its source and fine-grained sediment is found farther from the source.

When sediment is transported and deposited, it leaves indications of the mode of transportation and deposition. For example, if the mode of transport is by sliding/tumbling down a slope, the deposits show a wide variety of particle sizes—the grain size and the interrelationship between grains gives the resulting sediment texture and sediment can be used as an indicator of

Figure 3.2 Representation of poorly-sorted sediment and well-sorted sediment.

the mode of transport and deposition. In addition, the existence of *sorting*—the degree of uniformity of grain size—where particles become sorted on the basis of density, is an indicator of the energy of the transporting medium since high energy transportation enables the carrying of larger fragments and, as the energy decreases, heavier particles are deposited and lighter fragments continue to be transported. This results in sorting due to density and, if the particles have the same density, the heavier particles will also be larger, so the sorting will take place on the basis of size. Thus, size sorting can be classified on a relative basis—well-sorted to poorly-sorted (Fig. 3.2).

Rounding, which also occurs during the transportation process, causes the grains to be reduced in size due to abrasion. Random abrasion results in the eventual rounding of the sharp corners and edges of grains and indicates the amount of time a sediment has been in the transportation cycle.

Sediment maturity refers to the length of time that the sediment has been in the sedimentary cycle. Texturally mature sediment is sediment that is well rounded, (as rounding increases with transport distance and time) and well sorted and sorting gets better as larger clasts are left behind and smaller clasts are carried away. Furthermore, since the weathering processes continues during sediment transport, mineral grains that are unstable near the surface become less common as the distance of transport or time in the cycle increases and compositionally mature sediment is typically composed of the most stable minerals.

3.19 RESERVOIRS AND DEPOSITS

For the purposes of this text, a reservoir is a geological structure where the contents (conventional crude oil, heavy oil) have a degree of mobility under the prevalent conditions that is dependent upon the properties of the crude oil and heavy oil as well as the permeability of the reservoir.

The reservoir is a porous and permeable lithological unit or set of units that holds the crude oil reserves. Analysis of the suitability of a reservoir requires an assessment of the porosity (in order to estimate the volume of crude oil present) and the permeability (in order to estimate the ease of flow of the crude oil within the reservoir). Each reservoir must have a seal in the form of an impermeable (or low permeability) basement rock *and* an impermeable cap rock. Both the basement rock and the cap rock impede the escape of the crude oil from the reservoir rock. Common impermeable seals include evaporite minerals (any of a variety of individual minerals found in the sedimentary deposit of soluble salts that results from the evaporation of water) and shale minerals (a fine-grained, clastic sedimentary rock composed of mud that is a mix of flakes of clay minerals and silt-sized particles of other minerals, especially quartz, SiO_2, and calcite, $CaCO_3$). The analysis of the seals involves assessment of the thickness and extent, such that the effectiveness of the seals can be estimated and can be quantified.

On the other hand, a deposit is (essentially) a reservoir in which the contents (in this case, bitumen) are immobile under the prevalent conditions and which is dependent upon the properties of bitumen as well as the permeability of the reservoir. In either case, (reservoir or deposit) there are three stages of development: (1) primary recovery; (2) secondary recovery; and (3) tertiary recovery.

The existence of a rock formation (typically, sandstones and fractured limestones) that constitutes a reservoir is determined through a combination of regional studies (ie, analysis of other wells in the area), stratigraphy and sedimentology (to quantify the pattern and extent of sedimentation), and seismic interpretation (Frehner, 2011). Once a possible crude oil reservoir has been identified, key physical characteristics of a reservoir that are of interest are determined and include: (1) bulk rock volume; (2) net-to-gross ratio; (3) porosity; and (4) permeability.

Bulk rock volume, or the gross rock volume of rock above any hydrocarbon-water contact, is determined by mapping and correlating sedimentary data. The net-to-gross ratio, typically estimated from analogs and wireline logs, is used to calculate the proportion of the sedimentary packages that contains reservoir rocks. The bulk rock volume multiplied by the net-to-gross ratio gives the net rock volume of the reservoir. The net rock volume multiplied by porosity gives the total hydrocarbon pore volume, that is, the volume within the sedimentary package that fluids (importantly, hydrocarbons and water) can occupy. The summation of these volumes for a given prospect will allow determination of whether or not the prospect is commercially viable.

Traditionally, porosity and permeability are determined through the study of drilling samples, analysis of cores obtained from the wellbore, examination of contiguous parts of the reservoir that outcrop at the surface (Guerriero et al., 2011), and by the technique of formation evaluation using wireline tools passed down the well itself. Modern advances in seismic data acquisition and processing have meant that the seismic attributes of subsurface rocks are readily available and can be used to infer physical/sedimentary properties of the rocks themselves.

In *primary recovery*, the crude oil or (in rare cases) the heavy oil is forced out of the reservoir by pressure generated from gas present in the liquid or reservoir. In *secondary recovery*, the reservoir is subjected to water flooding or gas injection to maintain a pressure that continues to move oil to the surface. In *tertiary recovery*, also known as *enhanced oil recovery*, a fluid (steam) or heat (in situ combustion) is introduced into the reservoir or deposit to reduce the viscosity of the crude oil, heavy oil, or bitumen and to improve the flow. These fluids could consist of gases that are miscible with oil (typically carbon dioxide), steam, air or oxygen, polymer solutions, gels, surfactant-polymer formulations, alkaline-surfactant-polymer formulations, or microorganism formulations (see chapters: General Methods of Oil Recovery; Microbial Enhanced Oil Recovery; Nonthermal Methods of Recovery).

However, prior to reservoir development it is necessary to assess the morphology of the reservoir in terms of location, shape, and mineralogy. This will assist in determining the optimal choice of the recovery method and formulation of the injected fluid. From the data obtained from an examination of cores and core flooding, the best method for oil recovery can be devised.

3.20 RESERVOIRS

A reservoir is a subsurface porous permeable rock body or formation that has the capability to store and transmit fluids, such as crude oil, water, and natural gas. A reservoir is characteristically large and extensive in volume and capacity and is created by the sequential steps of deposition, conversion, migration, and entrapment.

Deposition occurs when organic rich sediment occurring in areas of high primary productivity is deposited and settles due to low energy currents. Such environments include rivers deltas, estuaries, and continental shelves. Further deposition of sediment results in the build-up of layers of sediment, for example, silts or clays atop the previously organic rich sediment reduce oxygen from effectively getting into the system effectively preserving organic matter. Further burial of the organic matter occurs due to more deposition and pressures and temperatures of the layers increase due to increasing depth and the geothermal gradient, which is generally on the order of 25°C/km to 30°C/km (15°F/1000 feet or 120°C/1000 feet, ie, 0.015°C per foot of depth or 0.012°C per foot of depth) (see chapter: Heavy Oil and Tar Sand Bitumen) (Speight, 2014).

Crude oil (conventional oil or heavy oil) cannot be retained as an accumulation unless there is a trap, and this requires that the boundary between the cap rock or other sealing agent and the reservoir rock generally be convex upward, but the exact form of the boundary varies widely. The simplest forms are the flat-lying convex lens, the anticline, and the dome, each of which has a convex upper surface (see chapter: Heavy Oil and Tar Sand Bitumen). Many oil and gas accumulations are trapped in anticlines or domes, structures that are generally more easily detected than some other types of traps.

The most common reservoir rocks are sandstone (SiO_2), limestone ($CaCO_3$), and dolomite ($CaCO_3$ $MgCO_3$). The five basic elements of a reservoir system include: (1) source rock—the rock containing the organic material that converted into petroleum reservoir fluid; (2) migratory pathway; (3) reservoir rock—rock that has sufficient porosity and permeability to store and yield the fluid; (4) a seal—an impermeable cap rock to prevent the upward escape of petroleum to the Earth's surface; and (5) a trap—a physical arrangement or space that prevents migration of reservoir fluid whereby source, reservoir rock, basement rock, and seal are arranged to trap petroleum. The reservoir rock is a porous medium made up of pores conventionally envisaged as irregularly shaped holes in rock some 1 to 100 μm in length and diameter connected to may be six other pores—there can be some 10^6 pores in 1 cc of rock and possibly over 10^{22} pores/1 cc in a typical reservoir.

However, for the purposes of reservoir definition and understanding reservoir behavior as it relates to heavy oil recovery, the process involves using data from a variety of length scales (pore-core-reservoir) and sources (laboratory-field) to improve understanding of reservoirs and their petrophysical properties (Kovscek, 2002). In addition, there is a wide variety of mineralogical composition in the various heavy oil reservoirs and tar sand deposits (Dusseault, 2001).

3.20.1 Structural Types

Reservoirs are created by structural deformation of the geological strata whereby these traps are formed by tectonic processes after the deposition of the beds involved. There are three basic forms of a structural trap in petroleum geology: (1) anticline trap; (2) fault trap; and (3) salt dome trap. The anticline is a typical *structural trap* that is produced by compressional folding, by uplift and by drape over older tectonically created features. An anticline is an example of rocks which were previously flat, but have been bent into an arch. The rocks have been folded or bucked into the form of a dome and hydrocarbons accumulate in the hinge area of an anticline. Another form of trap is the *stratigraphic trap* that is formed by a change in lithology when another stratum seals a reservoir bed or when the permeability changes (through a change in facies, ie, a change due to the presence of rock with characteristics different to those of the reservoir rock or a change in lithology) within the reservoir bed itself. Salt domes, formed by flow of salt at substantial depths, also have created numerous traps that are both a structural trap and a stratigraphic trap.

In order to become a reservoir, the rock must possess *fluid-holding capacity* (*porosity*) and also *fluid-transmitting capacity* (*permeability*); a variety of different types of openings in rocks are responsible for these properties in reservoir rocks. The most common occurrences when reservoir fluids are held are (1) the pores between the grains of which the rocks are made or the cavities inside fossils, (2) openings formed by solution or fractures, and (3) joints that have been created in various ways. The relative proportion of the different kinds of openings varies with the rock type, but pores usually account for the bulk of the storage space. The *effective porosity* for oil storage results from continuously connected openings which also provide the permeability but, although a rock must be porous to be permeable, there is no simple quantitative relationship between the porosity and permeability.

Reservoir rocks tend to show far greater variations in permeability than in porosity, and in addition, these two properties, as measured on core samples from reservoir rocks, are not always identical with the values indicated for the rock in bulk underground. The differences arise from the nonrepresentative nature of cores, especially when there are wide variations in the sizes of the openings in the rocks and irregularities in their distribution. Porosity is generally on the order of 5% to 30%; while permeability is commonly between 0.005 Darcy (5 mD) and several Darcys (several thousand milliDarcys) as measured on small samples.

3.20.2 Reservoir Heterogeneity

In addition to understanding the petrophysics of the reservoir, recovery of crude oil and heavy oil requires an understanding of displacement and flow through porous media (Dawe, 2004). However, flow through porous media is complicated. Inside a reservoir there can be displacements and miscible and/or immiscible flow, with one, two, or sometimes three mobile phases (oil, gas, and water) (Grattoni and Dawe, 2003). Understanding the physics of displacement is important for the correct interpretation of laboratory core data, for the assessment of quantities and position of residual oil, and for reservoir simulation. Thus, heterogeneity in the form of layers, lenses, cross-beds, and quadrants can have a profound effect on reservoir properties, especially fluid displacement patterns. Any changes (in fact, even modest changes in rock permeability) can give rise to distortions in displacement profiles and disperse the streamlines and lower sweep efficiencies and recovery.

Physically, crude oil and heavy oil reservoirs themselves are complicated geological heterogeneous bodies. They are not the homogeneous porous media that are often envisaged on paper and used in calculations. Heterogeneity means that a specific property of interest will vary vertically and longitudinally within the reservoir (Dawe, 2004) much like the coal in a seam that varies in composition from one part of the seam to another.

For example, well log and core analysis reports show that all reservoirs are heterogeneous, with rock properties (such as porosity and saturation) varying within the reservoir. As we shall show, permeability heterogeneities cause variations in the fluid movements compared to the equivalent homogeneous system. Often, the effects of heterogeneities are generally not well accounted for at the planning stage of an operation and only become evident when it may be too late and water has started to be produced before the predicted time (Dawe, 2004). An understanding of the movement of fluids within such heterogeneous porous media is therefore fundamental to petroleum production and its efficient management.

The reservoir rock is a porous medium made up of pores conventionally envisaged as irregularly shaped holes in rock some 1 to 100 μm in length and diameter connected to may be six other pores—there can be some 10^6 pores in 1 cc of rock and possibly over 10^{22} pores in a typical reservoir. This can cause heterogeneity within the reservoir. An understanding of the movement of fluids within such heterogeneous porous media is therefore fundamental to petroleum production and its efficient management. Therefore, it is necessary to understand the means by which heterogeneity affects reservoir behavior and predictions of oil recovery.

Wettability is the tendency of one fluid to spread on or adhere to a solid surface in the presence of other immiscible fluids and is determined by the interface boundary conditions acting within the pore space of sedimentary rocks. These conditions have a dominant effect on interface movement and associated oil displacement. In fact, the phenomenon of wettability has a major effect in multiphase flow systems and affect such processes as oil migration from source rocks to enhanced recovery processes, such as alkaline flooding.

In general, at least one of the two immiscible fluids in a porous medium will be the wetting phase. When the system is in equilibrium, the wetting fluid will completely occupy the smallest pores and be in contact with the majority of the rock surface. The nonwetting fluid will occupy the center of the large pores and form globules that extend over several pores. When the reservoir rock is water wet, there is a tendency for water to occupy the small pores and to contact the majority of the rock surface. The term *oil wet* is referred to the rock, which is preferentially in contact with the oil, and oil will occupy the small pores and contact the majority of the rock surface.

The minerals present in reservoir rocks are generally known as being intrinsically *hydrophilic*, that is, preferentially water-wet and it has been thought that all reservoirs were hydrophilic. However, problems connected with reservoir production sparked laboratory investigations that revealed the existence of reservoirs of intermediate wettability, or even clearly oil-wet (Chilingar and Yen, 1992; Robin, 2001).

From reservoir studies, the heterogeneities that are needed to be studied are those that interfere with the flow of fluids. Geological variations themselves that do not create flow pattern changes are not important for flow. The heterogeneities can be permeability or wettability variations (Caruana and Dawe, 1996a, b; Dawe, 2004). The effects of layer thickness, permeability contrast, angle of layer to flow direction, mobility ratio, wettability, and flood rate have been examined. Each of these parameters influences the displacement profiles, and disperses the flood front. It is more than likely that wettability of the reservoir rock by heavy oil (particularly adsorption of the polar constituents) can have major effects on heavy oil recovery.

Briefly, *wettability* of reservoirs rocks is the tendency of one fluid to spread on or adhere to a solid surface in the presence of other immiscible fluids and is determined by complex *interface boundary conditions* acting within pore space of sedimentary rocks.

Wettability is generally considered to be one of the most important parameters influencing saturation, distribution and flow of fluids in porous media. Knowledge of the wettability of reservoir rock is important to petroleum engineers and geologists. For example, a waterflood on a strongly oil-wet rock is much less efficient than one in water-wet rock.

Heavy oil typically occurs in geologically young reservoirs (from the Cretaceous) (Table 3.1) and, because these reservoirs are shallow, they have less effective seals and are thus exposed to conditions conducive to the formation of heavy oils. The fact that most heavy oil reservoirs are shallow is an indication that many of them were discovered as soon as human beings settled nearby. Collecting oil from seeps and digging by hand were the earliest and most primitive means of recovery, followed by mining and tunneling. However, heavy oil resources, along with tar sand bitumen resources can be subdivided into a number of different categories based on their location, environment, and characteristics.

Most of the heavy oil currently recovered is produced from underground reservoirs. However, surface seepage of crude oil and natural gas are common in many regions. In fact, it is the surface seepage of oil that led to the first use of the high-boiling material (heavy oil or bitumen) in the Fertile Crescent (see chapter: Heavy Oil and Tar Sand Bitumen) (Abraham, 1945; Forbes, 1958a, b, 1959; Speight, 1978, 2014). It may also be concluded that the presence of active seeps in an area is evidence that oil and gas are still migrating.

The following categorization is not all-encompassing but does illustrate the wide variety among heavy oil resources: (1) shallowest resources, <150 feet depth; (2) shallow resources, 150 to 300 feet deep with no cap rock seal; (3) medium-depth resources, 300 to 1000 feet deep, cap rock seal, pressure <200 psi; (4) intermediate-depth resources, 1000 to 3000 feet, pressure >200 psi; (5) deep resources, >3000 feet deep; (6) carbonate resources—tight rock formations, variable porosity; (7) thinly bedded resources, <30 feet thick; and (8) highly laminated resources (low vertical permeability, often shale layering). Furthermore, the properties of the heavy oil such as composition and viscosity are equally important but have not been included in the above list. If the oil properties were to be included, the complexity of the heavy oil resources can increase by at least an order of magnitude. In addition, many heavy oil reservoirs have been found in Arctic regions and offshore beneath the continental shelves of

Africa and North and South America. Heavy oil has also been discovered beneath the Caspian, Mediterranean, Adriatic, Red, Black, North, Beaufort, and Caribbean seas, as well as beneath the Persian Gulf and the Gulf of Mexico.

Heavy oil reservoirs are usually shallow (up to 1000 m below the surface line) and therefore present low reservoir temperatures (between 40°C and 60°C). Low sedimentary overburden tends to ease the biodegradation of the oil mainly when associated with bottom aquifers and also contributes to creating fragile geological structures where faults occur, thereby creating geological compartments and heterogeneities. Such kinds of reservoir may also have low seal pressure which may cause the dissolved gas to leave the oil, increasing its viscosity. The reservoir lithology is usually sandstones deposited as turbidites; high permeability (on the order of Darcys) and porosity are quite common. High permeability may compensate the elevated oil viscosity resulting in high well productivities.

Although one of the characteristics of heavy oil is that it is mobile in the reservoir, that is, the pour point of the oil is lower that the reservoir temperature (Chapter 1), some reservoirs have a sufficiently high temperature that heavy oil can be produced with essentially conventional methods. Once the oil is produced at the surface, the temperature differential between the reservoir (higher temperature) and the surface (lower temperature) might be such that the oil resorts to the more familiar extremely viscous fluid.

Just as the properties of the oil can dictate the relative ease or difficulty of production of heavy oil, the same rationale can be applied to the reservoir properties (see chapter: Heavy Oil and Tar Sand Bitumen). Heavy oil usually requires specialized (tertiary, enhanced) production methods. Very shallow heavy oil reservoirs can be mined and the oil allowed to drain into a mine tunnel. Slightly deeper deposits can be produced by increasing reservoir contact with horizontal wells and multilateral wells, producing the oil with large amounts of sand, or by injecting steam, which lowers the viscosity and reduces the residual oil saturation, thus improving recovery efficiency.

In terms of the reservoir itself, heavy oil reservoirs are low-energy systems in the sense of a typical conventional crude oil reservoir. The low molecular weight gases and highly-volatile liquids are missing from heavy oil with the result that the self-contained energy of the reservoir is mainly due to the amount of gas it holds, either in the form of free gas or as gas dissolved in oil. In other words, the small content of light fractions in heavy oils implies a reservoir with low energy density, which causes low recovery factors and low productivity rates.

Finally, a word of caution. In the typical laboratory core tests, reference to the internal structure of the core samples is not always addressed and the cores are usually assumed to be uniform and homogeneous. Therefore, the input and output data measurements taken in core test experiments are used to calculate the permeability properties by standard methods which neglect capillary pressure effects if immiscible displacements are being used and only generalized total effects are observed. If the core material is homogeneous, the observed data could be truely representative of the core but, if the core contains heterogeneity (which must be assumed to some degree), the chemistry and physics (ie, the mechanisms) of displacement and fluid flow is strongly affected, even to the point of being misleading. Thus, failure to acknowledge that the core to some (or great) degree has internal heterogeneity (that may or may not be representative of the in-place reservoir) will give an erroneous interpretation of the reservoir and, hence, an incorrect characterization of the porous medium. This will lead to an inaccurate assessment of the chosen recovery process since the interpretation of the data may be grossly inaccurate and any effects or heterogeneity within the core will (unknowingly) render the parameters derived from these interpretations unreliable (Caruana and Dawe, 1996a; Dawe, 2004).

3.20.3 Reservoir Evaluation

It has been customary to attempt to define conditions—usually a limited list of reservoir and fluid properties and operating conditions—under which thermal recovery operations are likely to be commercially attractive. These conditions have been called by a variety of names, including screening criteria, screening guidelines, preferred criteria, and selection criteria. Any such set of criteria reflects its authors' opinions at a particular time and usually are affected by the current and local economic climate.

Although most available guidelines attempt to include the effect of economic considerations in the technical variables, there are a few technical considerations that in themselves limit the operability of thermal recovery processes. Important technical considerations are those that affect (1) the ability to generate heat within or inject heat into an oil-containing reservoir at efficient rates, (2) the ability to displace the oil, and (3) the ability to recover the oil, all in a controlled manner. It is difficult if not impossible to translate these broad considerations into specific values of properties or groups of properties that would limit the range of applicability of a thermal process. Some examples will help clarify this point.

In steam drives, the gravity of the crude plays no role in the technical considerations just listed except as it might affect plugging and, thus, the ability to maintain adequate communication between wells in reservoirs containing relatively heavy crudes. With regard to loss in injectivity or productivity, gravity override (or bypassing) of steam reduces the tendency of the formation to plug. Also, hydraulic fracturing, control of the injection temperature, and cyclic steam injection have been used successfully in thermal operations to avoid or minimize plugging. Thus, it seems impractical to place a limit on the range of API gravity of crudes to be considered for steam drive processes.

Injectivity is another parameter to consider. Reduced ability to inject (or produce) fluids is a factor affecting the economics of all enhanced recovery processes, especially those requiring the injection of heated fluids. When the injection rate is reduced, the project life must be longer and more of the heat entering the formation is lost to the adjacent zones. This tends to reduce the oil displacement and production rates (ie, the rate at which income is generated), which in turn reduces the economic attractiveness. But what may be an acceptable level of injectivity in one project could be economically disastrous in another.

Although guidelines giving parameter limits are useful in helping government bodies and oil companies decide how to allocate funds to support their general research and development efforts, they are of little use to an operator having a specific reservoir requiring either improved performance or special development. Each reservoir should be examined individually (at least briefly) as though there were no guidelines, especially where the reserves are great enough to support some engineering studies. A screening economic analysis for each likely process is the preferred method for forming initial decisions.

The approach is to indicate some of the elements necessary to screen a potential venture, including an understanding of the possible effects of uncertainties on a parameter or group of parameters. This approach, which requires only a few hours to perform, should allow the engineer to make preliminary decisions or recommendations as to the merits of a prospective thermal project.

3.20.3.1 Reservoir Thickness

Reservoir thickness is particularly important because it has a pronounced effect on the fraction of the introduced heat that remains in the reservoir—a quantity known as the heat efficiency. For a given reservoir (which has a fixed thickness), the only way to increase the heat efficiency is to reduce the time t required to produce the oil from the zone between an injector and a producer. This generally can be done by increasing the injection rate (one also must be able to recover the oil at an increased rate) or by reducing the spacing. Each of these alternatives would cost money to implement but may be attractive compared with a slowly expanding steam front that results in low oil production rates and a long operating life.

3.20.3.2 Well Spacing

The main consideration regarding well spacing is sand continuity, which is vital in the selection of well spacing for processes involving interwell fluid displacement. Partial communication in sands where only a fraction of the interval correlates between well locations invariably leads to poor recoveries and poor effective transmissivity. Knowledge of the depositional setting of the sands may be crucial in establishing whether open intervals in nearby wells are part of the same blanket sand or are likely to be in unconnected and meandering river channels. Regional trends in the direction of river channels may result in an apparent anisotropy in reservoir permeability, with relatively good communication between wells in the same channels and poor or no communication in a transverse direction. Displacement processes are not indicated where communication is poor over relatively short distances. Instead, cyclic steam injection or some other stimulation treatment that does not depend on interwell communication is preferred.

Another factor related to well spacing is the well pattern. Because of the extensive use of repeated well patterns (such as five-spots) in water flooding, they were considered necessary in the early years of thermal recovery. Now, however, there is considerable evidence that the use of repeated regular well patterns is not always essential for the success of field projects. The choice of well locations is important in reservoirs having significant dip or a strong lateral aquifer influx.

Although fractures and zones of relatively high permeability to the injected fluid may be useful for initially introducing heat into the reservoir at adequate rates, such zones may be a detriment to the efficiency of later operations. For example, steam and unreacted air might circulate through these high-permeability zones and be produced at unacceptably high rates.

The application of multilateral well technology could reduce the number of wells required in very low productivity fields, increasing the economic viability of these developments. However, a key requirement for the successful application of multilateral well technology is the capability to control the offtake rate from individual wells. Without this, the lower drawdowns required to achieve a given flow rate as water breaks through, would mean that an increasing volume of fluid would be produced from the first lateral cutting water, at the expense of the other laterals. The option of completely shutting off the offending lateral would not be acceptable, since a significant fraction of the recoverable reserves are produced after water breakthrough, with wells typically needing to produce for extended periods at high water cuts.

In relation to water production, the *downhole separation* with the disposal of water downhole can improve oil recovery in circumstances where production facilities are constrained. Hydrocyclones are used to separate off a purified water stream and downhole pumps to provide the power to produce fluids to the surface and to pump excess water into a disposal zone. In applications where heavy oil is the oil to be produced, difficulties in separation may arise because of the small difference in density (specific gravity or API gravity) between the heavy oil and water and because high bulk fluid viscosities can arise through the formation of emulsions at intermediate volume fractions of oil and water. High viscosity is a constraint to the practical range over which heavy oil recovery is possible. However, removing as little as 50% v/v of the water downhole can allow increased production of heavy oil. The reduction in fluid production to surface provides an alternative to expanding the existing free water knockout capacity, however, this is at the cost of installing more complex equipment downhole and providing an appropriate water disposal route in each well.

3.20.3.3 Other Considerations

In addition to the factors already discussed, there are a few other important considerations in selecting a reservoir for thermal projects. Foremost are any factors that may affect the control of the operation, especially where safety is involved. Control means that the fluids are injected only into the target reservoir and that both displaced and injected fluids are collected efficiently and safely. The leakage of produced or injected fluids to the surface is an extreme example of loss of control.

Conditions that should be considered and evaluated carefully are injection pressures close to the fracture initiation pressure or close to the estimated pressure limits of well equipment (especially in old wells) and reservoirs where communication with nearby outcrops is likely. These could lead to (1) fluid contamination of overlying formations, (2) loss of production, (3) loss of pressure, (4) loss of control, and (5) early termination of the project.

Where displaced oil can bypass a producing well, it is generally desirable on technical grounds (but not necessarily justifiable economically) to fracture the well and to maintain it in unimpaired condition, especially where low-flow-resistance zones near the well compete for the flow of oil. Stimulation of producers by cyclic steam injection has long been recognized as an excellent method for reducing the resistance to flow into these wells. Special attention should be paid to methods for avoiding or minimizing the resaturation of previously depleted zones or the saturation of previously clean sands (such as zones below a water/oil contact).

Finally, data obtained from core-related laboratory experiments may not be totally reliable for further planning. For example, flow in heterogeneous porous media is important for predicting breakthrough times for reservoir engineers but laboratory tests can use realistic porous media (often specially selected "homogeneous" samples) as well as reservoir fluids, but they are essentially *blind* techniques that produce data based on overall effects and any interpretation of the data can be grossly distorted, often unknowingly, by artificial mixing, end effects, or heterogeneities within the core (Dawe, 2004). Furthermore, heterogeneity in the form of geological features such as layers, lenses, cross-beds, and quadrants will have a significant effect on oil displacement. In fact, even slight changes in the permeability of the reservoir rock can cause to distortions in the displacement profile, giving rise to lower sweep efficiencies and recovery.

The use of simple assumptions that may ignore the true condition (structure, especially any heterogeneity) of the reservoir rock will, more often than not, present a false indication of the reservoir even though the mathematics is tractable (Dawe, 2004). This may lead to statements that the data were tweaked or fine-tuned to accommodate the model (whatever that means). Thus, a thorough and realistic characterization of the reservoir is necessary, plus a good understanding of how fluids can flow through such a heterogeneous system leading to more accurate predictions of recovery efficiency.

In many heavy oil reservoirs, especially those with a thick sediment layer or high inclination angle and high vertical permeability, oil compositional segregation can and will occur. This causes the higher molecular weight constituents (or more dense components) to remain in (or, through density segregation, to migrate to) the lower domains of the reservoir rock. Therefore, the difference in the specific gravity of the heavy oil may vary within a range of several degrees API along the depth of the reservoir and there may even by horizontal variations in the API gravity depending upon the mobility of the heavy oil and the permeability of the reservoir. Any such differences will be reflected in variations in the recovery factor for wells at different points of the reservoir.

3.21 DEPOSITS

Conventionally, the bitumen in tar sands cannot be pumped from the ground in its natural state; instead tar sand deposits are mined, usually using strip mining or open pit techniques, or the oil is extracted by underground heating with additional upgrading. In accordance with geological terminology, the immobile bitumen with the sand is often referred to as an ore body and, hence the use of the term *deposit* to describe a tar sand (oil sand) formation.

More correctly, the tar sand deposits are sedimentary rocks containing bitumen, a high-boiling hydrocarbonaceous material, and the deposits may be divided into two major types: (1) a breached petroleum reservoir where erosion has removed the capping layers from a reservoir of relatively heavy petroleum, allowing the more volatile petroleum hydrocarbons to escape; and (2) a tar sand deposit which forms when liquid petroleum seeps into a near-surface reservoir from which the more volatile constituents escape. In either type of deposit, the more volatile constituents have escaped to the environment, leaving the less volatile constituents in place. The material left in place is altered by contact with air, bacteria, and groundwater.

Tar sand deposits are not uniform. Differences in the permeability and porosity of the reservoir rock and varying degrees of alteration by contact with air, bacteria, and groundwater leave a high degree of uncertainty in the estimates of the bitumen content of a specific deposit. The bitumen in tar sand deposits cannot be pumped to the surface in its natural state (see chapter: Heavy Oil and Tar Sand Bitumen)—instead tar sand deposits are mined, usually using strip mining or open pit techniques, or the oil is extracted by underground heating with additional upgrading.

In addition to understanding the petrophysics of the deposit, recovery of tar sand bitumen requires an understanding of displacement and flow through porous media when the bitumen is heated or particularly decomposed to produce a solvent that will move the nondecomposed bitumen. Therefore, understanding the physics of displacement and the flow properties of heated bitumen or bitumen-solvent mixtures is important for (1) the correct interpretation of any core data, and (2) the assessment of quantities and the properties of the bitumen or bitumen-solvent mixture for deposit simulation. Thus, as for the reservoir, heterogeneity in the form of layers, lenses, cross-beds, and quadrants can have a profound effect on deposit properties and, by inference, fluid flow patterns and fluid displacement patterns. Any changes (in fact, even modest changes in deposit permeability) can give rise to distortions in displacement profiles and disperse the streamlines and lower sweep efficiencies and recovery.

3.22 RESERVOIR PRODUCTIVITY

After a crude oil occurrence has been identified and appraised with the outcome that the reservoir can be produced commercially, the production stage is initiated. This stage focuses on extracting the crude oil (and natural gas) hydrocarbons in a controlled manner so that reservoir development management allows commercially favorable volumes to be produced.

The productivity of a reservoir is controlled by the reservoir geometry (primarily formation thickness and reservoir continuity) and the ratio of the permeability of the formation to the viscosity of the oil. As a first approximation, combinations of permeability-thickness and viscosity need to be carried out, before sufficient information is available to proceed to a development.

3.22.1 Production

Thus, once oil is found in commercial quantities, there is a substantial delay until production begins, and such delays can last several years in many offshore areas. After the initial discovery, further drilling and planning is necessary to determine the optimum way to exploit the find. In addition, the necessary infrastructure (pipelines, other supporting equipment, and people) must be located and brought. In many Arctic areas the technology to allow acceptable environmental recovery of oil is still being developed.

Production begins initially with the oil allowed to flow naturally from the high pressure area underground to the surface. This natural recovery, called primary recovery, depends on reservoir pressure and the natural drive mechanism. The drive mechanism refers to the sources of energy within the reservoir that will assist production, and the properties of the oil, gas, and water found in relative proportions and locations.

Secondary recovery is another procedure used to boost recovery from a reservoir. This procedure consists of reinjection of either associated gas or water near the well to maintain underground pressure. In the past, secondary recovery techniques followed the use of primary recovery techniques; today these techniques may be used simultaneously to increase total production.

The recovery percentage of the oil in place recovered varies greatly from field to field. Thirty three percent (33% v/v) is sometimes used as an average figure based upon experience in the United States but recovery in the East Texas oil field which has a natural water drive may exceed 75% v/v. Production rates are limited by the drive mechanism and other geologic factors. In addition, economic and resource maximization considerations (production at too fast a rate would lead to a reduction in total recovery) also influence production rates. A general planning guide for a country or region is that 1/15 of the recoverable oil remaining can be produced each year; this is referred to as the 15 to 1 reserves to production ratio. This planning guide is only a rough estimate and is, to some extent, a function of well density. Production from United States, Canadian, and Australian fields, which are in a mature state of development, exceed the 1/15 ratio. In the United States the figure is approximately 1/10.

With the increased price of oil since the 1973 embargo, other procedures have become economical for recovering additional oil from existing fields where primary and secondary techniques have already been used. This is especially true for the United States, since many early US fields were exploited ineffectively causing a large portion of oil to be left in the ground. These additional techniques, called enhanced recovery techniques, consist of injecting heat or chemicals into the reservoir. Except for heat injection, most of these techniques are at the developmental stage. The success of these techniques varies greatly from field to field. A range of approximately 2 to 10% v/v has been given as the additional percentage of oil in place that can be recovered in the United States using enhanced recovery techniques. The recovery of crude oil from tight formations and from shale formations and the application of hydraulic fracturing techniques (Speight, 2015) has modified this figure somewhat and modifications to the data will continue as long as these formations are discovered and developed.

Since enhanced recovery techniques are expensive and require energy for their utilization, the extent of the use of these techniques will depend upon the future price of oil and the price of the chemicals and steam required for their implementation. The production of heavy oil requiring steam for its primary recovery is not included in enhanced recovery estimates. However, heavy oil fields for which some production is possible without steam injection are included in enhanced recovery estimates. Thus most heavy oil fields in California are included in enhanced recovery estimates.

3.22.2 Production Techniques

Although the recovery techniques are presented in more detail elsewhere, mention of the various techniques is necessary here to place the perspective of the reservoir (Jayasekera and Goodyear, 1999). These techniques are

(1) polymer flooding, (2) gas injection, (3) hot water flooding, (4) steam injection, and (5) in situ combustion.

Polymer flooding is used in heavy crude oil reservoirs to reduce the local effective residual oil saturation by changing the fractional flow curve and to improve sweep in less permeable zones. However, polymer flooding can also be effective in gravity-dominated heavy oil reservoirs. In the absence of underlying water, injected water channels along the base of the reservoir before coning into the production well. In polymer flooding, the ratio of viscous to gravity forces is increased, reducing the slumping and improving vertical sweep in the reservoir. Even in the presence of an underlying aquifer (provided the thickness is no greater than the thickness of the oil column), polymer flooding can still be effective, by restricting water channeling through the aquifer. In addition, heavy oil reservoirs are often sufficiently shallow in depth (below the surface) for the temperature to be low enough for biopolymers to have sufficient thermal stability to be applicable.

Gas injection into a reservoir gas cap, where this exists, may provide an alternative strategy to water flooding, giving lower effective residual oil saturations (because of the much greater density contrast between oil and gas compared to that between water and oil). Recent applications to heavy oil reservoirs have shown that the relative permeability of the reservoir is independent of the oil viscosity.

The displacement mechanisms associated with *hot water flooding* are complex, with even the density difference between reservoir brine and the injected hot water being important in some circumstances. A potential problem with hot water flooding is the unstable nature of the displacement, triggered by the unstable miscible displacement between cold and hot water. For moderate temperature reservoirs with oil viscosity on the order of 10 cP, there is little difference to recovery in the absence of any underlying water, but recovery is reduced in the presence of limited thicknesses of underlying water. On the other hand, in low-temperature reservoirs with oil viscosities on the order of 100 cP there may be no significant benefit from hot water flooding.

Steam injection is a highly successful recovery method for heavy oil reservoirs but may not be successful (economically) for heavy oil reservoirs at depths of 3000 to 5000 feet. Steam flooding is only likely to be economically feasible if the reservoir pressure is significantly reduced prior to steam injection, which may disqualify reservoirs with significant aquifer pressure support from this process. In addition, it may be difficult to implement a steam flood, since the well positions required for a basic water flood may be quite different from those needed for a steam flood.

In situ combustion using air injection provides an alternative to hydrocarbon gas or nitrogen injection, with potential benefits from reduced gas costs and lowered effective residual oil saturation. However, certain reservoirs, such as carbonate reservoirs, may produce undue amounts of carbon dioxide that could adversely affect the process. In addition, the same overall considerations relevant to the reservoir and gas injection apply, with additional concerns including the effect of the breakthrough of corrosive combustion products on the production facilities and well casing integrity in the combustion zone.

REFERENCES

Abraham, H., 1945. Asphalts and Allied Substances. Van Nostrand Publishers, New York.

Caruana, A., Dawe, R.A., 1996a. Effect of heterogeneities on miscible and immiscible flow processes in porous media. Trends Chem. Eng. 3, 185–203.

Caruana, A., Dawe, R.A., 1996b. Flow behavior in the presence of wettability heterogeneities. Trans. Porous Media 25, 217–233.

Chilingar, G.V., Yen, T.F., 1992. Some notes on the wettability and relative permeability of carbonate reservoir rock. Energy Sources 7 (1), 21–27.

Dawe, R.A., 2004. Miscible displacement. In: Heterogeneous Porous Media. Proceedings. 6th Caribbean Congress of Fluid Dynamics, University of the West Indies, St. Augustine, Trinidad, Republic of Trinidad and Tobago. January 22–23.

Dusseault, M.D., 2001. Comparing Venezuelan and Canadian Heavy Oil and Tar Sands. Paper No. 2001-061. Proceedings International Petroleum Conference, Calgary Alberta Canada. June 12–14. Canadian Institute of Mining, Westmount, Quebec, Canada.

Forbes, R.J., 1958a. A History of Technology. Oxford University Press, Oxford, United Kingdom.

Forbes, R.J., 1958b. Studies in Early Petroleum Chemistry. E.J. Brill, Leiden, Netherlands.

Forbes, R.J., 1959. More Studies in Early Petroleum Chemistry. E.J. Brill, Leiden, Netherlands.

Frehner, B., 2011. Finding Oil: The Nature of Petroleum Geology, 1859–1920. University of Nebraska Press, University of Nebraska, Lincoln, Nebraska.

Grattoni, C.A., Dawe, R.A., 2003. Consideration of wetting and spreading in three-phase flow in porous media. In: Lakatos, I. (Ed.), Progress in Mining and Oilfield Chemistry, Volume 5. Recent Advances in Enhanced Oil and Gas Recovery Akad. Kiado, Budapest.

Guerriero, V., Vitale, S., Ciarcia, S., Mazzoli, S., 2011. Improved statistical multi-scale analysis of fractured reservoir analogues. Tectonophysics 504 (1), 14–24.

Jayasekera, A.J., Goodyear, S.G., 1999. The development of heavy oil fields in the UK continental shelf: past, present, and future. Paper No. SPE 54623. Proceedings. SPE Western Regional Meeting, Anchorage, Alaska. May 26–28. Society of Petroleum Engineers, Richardson, Texas.

Kovscek, A.R., 2002. Heavy and thermal oil recovery production mechanisms. Quarterly Technical Progress Report. Reporting Period: April 1 through June 30, 2002. DOE Contract Number: DE-FC26-00BC15311.

Nichols, G., 1999. Sedimentology & Stratigraphy. Wiley-Blackwell, Malden, Massachusetts.

Prothero, D.R., Schwab, F., 1996. Sedimentary Geology: An Introduction to Sedimentary Rocks and Stratigraphy. W.H. Freeman Publishers, Macmillan Group, New York.
Reading, H.G., 1996. Sedimentary Environments and Facies. Blackwell Scientific Publications, Wiley-Blackwell Publications, John Wiley & Sons Inc., Hoboken, New Jersey.
Robin, M., 2001. Interfacial phenomena: reservoir wettability in oil recovery. Oil Gas Sci. Technol. Revue de l'Institut Français de Pétrole 56 (1), 55–62.
Speight, J.G., 1978. The use of Bitumen as an adhesive and building material at the ancient cities of Babylon, Calah, Nineveh, and Ur. Report to the Dean, College of Science, University of Mosul, Iraq. 1978.
Speight, J.G., 2011. An Introduction to Petroleum Technology, Economics, and Politics. Scrivener Publishing, Salem, Massachusetts.
Speight, J.G., 2014. The Chemistry and Technology of Petroleum, fifth ed. CRC-Taylor and Francis Group, Boca Raton, Florida.
Speight, J.G., 2015. Handbook of Hydraulic Fracturing. John Wiley & Sons Inc., Hoboken, New Jersey.

CHAPTER 4

Reservoir Fluids

4.1 INTRODUCTION

Reservoir fluids are the fluids (including gases and solids) that exist in a reservoir. The fluid type must be determined very early in the life of a reservoir (often before sampling or initial production) because fluid type is the critical factor in many of the decisions that must be made about producing the fluid from the reservoir.

Reservoir-fluid properties play a key role in the design and optimization of injection/production strategies and surface facilities for efficient reservoir management. Inaccurate fluid characterization often leads to high uncertainties in in-place-volume estimates and recovery predictions, and hence affects asset value. Indeed, evaluating reservoir fluids is an important aspect of prospect evaluation, development planning, and reservoir management. Although many different types of fluid exist, the composition of the reservoir fluids provides important information that can influence the recovery process. Prior to production, these measurements will represent static data but once production commences, dynamic data will become available and variations in fluid composition, which are effective on a production timescale, will become apparent.

The reservoir fluid data are an important compliment to the pressure data and have the advantage of providing a direct measure of reservoir fluid continuity. Many different measurements of the reservoir fluid are available, both at reservoir and atmospheric conditions, but it has become increasingly common to apply some type of detailed oil fingerprinting for this purpose. Typically the analysis of reservoir pressures and the oil fingerprinting data will give similar results. In some situations, however, this is not the case and can be a source of confusion in determining the true nature of the reservoir architecture. When these situations occur, it is often because the processes that control the reservoir pressures and the timing of their development are different from those controlling the reservoir fluid filling history. In this paper we will show several case studies illustrating the complimentary nature of reservoir pressure and fluid composition data. We will also discuss geologic settings where these two types of data give

Introduction to Enhanced Recovery Methods for Heavy Oil and Tar Sands.
DOI: http://dx.doi.org/10.1016/B978-0-12-849906-1.00004-7

divergent views of reservoir continuity and offer an explanation for this behavior.

Reservoir fluids, including heavy oil, vary greatly in composition. In some fields, the fluid is in the gaseous state and in others it is in the liquid state but gas and liquid frequently coexist in a reservoir. The rocks which contain these reservoir fluids also vary considerably in composition and in physical and flow properties and this can serve to complicate the sampling procedure. Other factors, such as producing area, height of the column of hydrocarbon fluid, fracturing or faulting, and water production also serve to distinguish one reservoir from another. The combination of all these factors affects the choice of sampling methods and preparations for sampling.

4.2 CLASSES OF FLUIDS

The pore spaces in the rock formations that form crude oil reservoirs (and deposits) are always usually completely filled (saturated) with a combination of the following fluids: (1) crude oil in the liquid phase with any associated impurities; (2) natural gas in the vapor phase with any associated impurities; and (3) water. The arrangement for a heavy oil reservoir, an extra heavy oil reservoir/deposit, or a tar sand deposit may be somewhat different because of the variable phase nature of these viscous materials and the way in which they exist in the reservoir or deposit. Nevertheless, in most cases there are very few (if any) locations where the pores are completely empty. There is no true *void space*—nature not only abhors a vacuum but also seems to abhor an empty space!

During deposition, when sediments were being deposited (usually in an aqueous environment), the pores were completely saturated with water (ie, water saturation was 100% of the pore space). During the process of deep burial, when compaction and partial cementation was occurring, the water may have changed in composition, but the saturation remained 100% unless other invaders (such as hydrocarbons) entered the pores and displaced some or all of the water.

If the water-saturated pores happen to be near an active petroleum-generating source rock (such as an organic-rich shale formation) and the pores are in pressure communication with the source rock, hydrocarbons can enter the pores and occupy space. Since hydrocarbons are typically less dense than water, and the resulting buoyant force causes the oil or gas to migrate through the porous, permeable rock until it escapes at the surface or is stopped by an impermeable layer that forms a seal (basement rock, cap rock, or side-seal). Thus, in the pores of crude oil reservoirs, there always remains some water that was there before the hydrocarbon entrapment. At any time during the life of a crude oil reservoir, the following relationship must hold true.

$$S_o + S_w + S_g = 1.0$$

In this equation, S_o is the volume of oil divided by the pore volume, S_w is the volume of water divided by the pore volume, and S_g is the volume of gas divided by the pore volume. It is common for oil or gas saturation to be zero, but water saturation is always greater than zero. Saturation is a direct measure of the fluid content of the porous rock. It therefore directly influences the hydrocarbon storage capacity of the reservoir or deposit. Other uses are the identification of gas/oil or oil/water contacts by changes of residual saturation with depth, and indirectly it is used as a correlation variable to estimate the productivity of reservoir rocks.

In addition, reservoir fluids vary greatly in composition and chemical properties. The term *fluid* can mean gaseous state or in the liquid state and the fluids exist either as solid, liquid or gaseous state. Gas and liquids coexist in a given reservoir. Heavy oil has an American Petroleum Institute (API) gravity between 10° and 20° (see chapter: "Heavy Oil and Tar Sand Bitumen"), is more viscous than conventional petroleum, and has the commonality of being, or ability to be, in the liquid state and, therefore, has mobility in the reservoir.

For the purpose of this text, the focus is on the heavy oil and bitumen to the exclusion of any reservoir water and gas. Some possibilities for the inclusion of water and gas include: (1) the presence or absence of bottom water; (2) the presence of gas without any bottom water; and (3) the absence of gas and water. In addition, petrophysical analyses of the composition of the reservoir rock and deposit are not included. For details of the analysis of gas and water as well as analysis of the reservoir/deposit rock, it is recommended that the reader consult the relevant texts.

In order to produce a wellhead product from a reservoir there has to be flow of the fluids to the wellbores through the heterogeneous porous media of the reservoir. Fluid movements within the reservoir are governed by the local fluid potential gradients and reservoir effective permeability, the injection and production points, and the fluid viscosities. However, fluid flow is also governed by the character of the fluids in the reservoir. In the current context, reservoirs usually contain three main *fluids*: (1) natural gas;

(2) crude oil or heavy oil; and (3) water, with minor constituents being acid gases (carbon dioxide and hydrogen sulfide). These components will vary greatly in combination and proportion within each reservoir and in the case of heavy oil; the amount of gas will be substantially less than would be found in a conventional oil reservoir.

The distribution of the fluids in a reservoir rock is dependent on the densities of the fluids as well as on the properties of the rock. If the pores are of uniform size and evenly distributed, there is: (1) an upper zone where the pores are filled mainly by gas (the gas cap), (2) a middle zone in which the pores are occupied principally by oil with gas in solution, and (3) a lower zone with its pores filled by water. A certain amount of water (approximately 10–30%) occurs along with the oil in the middle zone. There is a transition zone from the pores occupied entirely by water to pores occupied mainly by oil in the reservoir rock, and the thickness of this zone depends on the densities and interfacial tension of the oil and water as well as on the sizes of the pores. Similarly, there is some water in the pores in the upper gas zone that has at its base a transition zone from pores occupied largely by gas to pores filled mainly by oil.

The water found in the crude oil and natural gas zones (Fig. 4.1) is known generally as *interstitial water* and usually occurs as collars around grain contacts, as a filling of pores with unusually small throats connecting with adjacent pores, or, to a much smaller extent, as wetting films on the surface of the mineral grains when the rock is preferentially wet by water. However, in actual reservoirs and deposits composed of porous rock, the fluid interfaces are not so distinct. Not only does water exist throughout the oil and gas zones at a saturation of at least irreducible water saturation, but the fluid contacts are generally spread over a distance of a few feet to

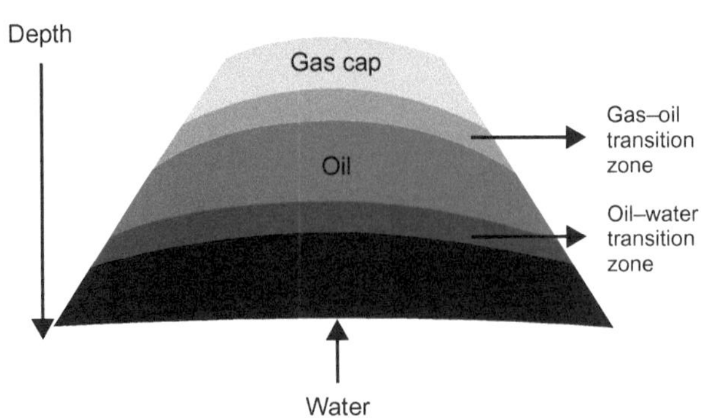

Figure 4.1 Representation of the zones in a reservoir.

tens of feet, depending on the density difference between the fluids and the permeability of the rock. This is, of course subject to the mobility of the fluid, especially the organic fluid (heavy oil, extra heavy oil, bitumen) and the temperature of the reservoir (deposit) that may or may not promote fluid mobility.

The water may occur as wetting films, or collars, around the sand grains as well as in some completely filled pores. The three-dimensional network allows continuity to exist for the hydrocarbons by means of connections on every side of the sand grains. The so-called gas–oil and oil–water contacts are generally horizontal but have been known to exist as a very gentle incline. On occasion, part of an accumulation of the oil or gas has its lower boundary marked not by the water-bearing zone of the reservoir rock but by an adjacent sealing rock that has characteristics similar to those of the cap rock. When the pressure and temperature conditions are suitable in relation to the proportions and the nature of the gas and oil, there may be no gas cap but only oil, with dissolved gas overlying the water.

The water (brine) produced with oil has been trapped with the oil and is brought to the surface along with oil. Because the water has been in contact with the oil, it contains some of the chemical characteristics of the formation and the oil itself. Oil and gas wells produce more water than oil (7 bbl/1 bbl oil in some fields). The composition (salt content) of coproduced water determines the need for antiscaling additives. There are strict regulations to limit disposal and beneficial use options as well as environmental impacts that pertain to oil field waters.

Reservoirs contain complex fluid mixtures whose behavior is strongly dependent on chemical make-up. Heavy oil is a fluid that is also a multicomponent mixture composed of nonhydrocarbons and a variety of hydrocarbons—especially of the alkane series. Typical hydrocarbons encountered in heavy oil are the higher boiling hydrocarbons—the amount depending on the original source materials and the maturation pathways; volatile hydrocarbons boiling lower than C_{12} are not present in a ready abundance.

Reservoir temperatures may vary up to 90°C or more, while surface conditions are around 20°C. Pressure can vary from its atmospheric value (or lower in the case of vacuum distillation) to a number in the hundred million Pascal (Pa). Within such an ample range of conditions, hydrocarbon fluids undergo severe transformations and exist as a single phase (gas, liquid, or solid) or coexist in several forms (liquid plus gas, solid plus liquid, vapor plus solid, or even in liquid plus liquid combinations). Understanding how hydrocarbon fluids interact with and react to their thermodynamic surroundings is essential to adequately analyze systems of interest and one of the most useful phase behavior visualizations is the pressure–temperature (p–T) diagram or p–T envelope. Each envelope represents a thermodynamic boundary separating the two-phase conditions (inside the envelope) from the single-phase region (outside).

The correct identification of the type of hydrocarbon fluid is critical for the proper design and development of the correct production strategy for the field under consideration. This identification is critical for proper hydrocarbon reservoir modeling as well. Numerical modeling is a useful technique that allows engineers to predict the oil recovery performance by solving a set of flow equations governing the recovery process mechanisms. For nonthermal oil recovery, there are *black oil models* and *compositional models*—unfortunately the definition of black oil is ill-advised because of the poor definition of this material (see chapter: "Heavy Oil and Tar Sand Bitumen") and compositional models for, say, asphaltene deposition (Section 4.4.2), do not take not consideration the variations of the properties of the constituents of the bulk asphaltene fraction. For thermal oil recovery, there are the same limitations as presented above but the models have had a modicum of success in solving the flow equations and the energy equations that take into account the heat transfer between the reservoir fluids and the rock. Modeling is designed to take out the guesswork from recovery methods but still needs considerable work. Thus, when a fluid is assumed to behave as heavy oil, it is assumed that its behavior is complex and selecting an appropriate model for behavior becomes a limiting factor.

Reservoir fluids are brought to the surface as a mixture of oil, gas, and water, which is sent to a surface production facility before they can be disposed or sold to an industrial costumer (eg, a refinery). A surface production facility is the system in charge of the separation of the well stream fluids into its three single phase components—oil, gas, and water—and of their transport and processing into marketable products and/or their disposal in an environmentally acceptable manner. Once separated, the oil, natural gas, and water follow different paths. Water is typically reinjected for reservoir pressure maintenance operations. The oil usually goes through a process of dehydration, which removes basic sediments and hydrocarbon fluids are assumed to comprise two components—stock tank oil and surface gas.

When comparing heavy oil and tar sand bitumen with conventional crude oil (see chapter: "Heavy Oil and Tar Sand Bitumen") the differences in properties and behavior immediately become obvious. This is due to the differences in chemical composition of each fluid which ultimately affects the properties (Speight, 2014a), especially the viscosity and the pour point. To the recovery engineer, viscosity influences every other aspect of heavy oil bitumen recovery. Furthermore, the various technologies that have been developed for recovery of conventional crude oil (see chapter: "General Methods of Oil Recovery") do not typically address the issues of producing heavy oil and/or bitumen. Furthermore, mathematical and physical models developed for the recovery of conventional oil are not typically applicable to the recovery of heavy oil, and especially not to tar sand bitumen.

Heavy oil, albeit even with some mobility in the reservoir, requires some form of enhanced oil recovery (EOR)—typically a steam-based process (see chapter: "Nonthermal Methods of Recovery"), a solvent-based process (see chapter: "Microbial Enhanced Oil Recovery"), or a hybrid (combination) of both types of process (see chapters: "Microbial Enhanced Oil Recovery" and "Nonthermal Methods of Recovery"). Without enhanced recovery operations (especially steam-based operations), the recovery factor (oil recovered/total oil-in-place) for a heavy oil reservoir might be as low as 1% (v/v) of the oil-in-place with a higher recovery factory being little more than a dream. With thermal recovery, the recovery factory can be boosted considerably. In some thermal recovery operations, thermal recovery options in some reservoirs include the use of the cyclical steam (huff "n" puff) process, downhole heaters, or the recently developed steam assisted gravity drainage process (see chapter: "Nonthermal Methods of Recovery").

At this point, it is also necessary to consider the effect of foaming oil (foamy oil) (see chapter: "Heavy Oil and Tar Sand Bitumen") on the recovery process. In many cases, the properties of the heavy oil—especially when the heavy oil contains associated gas—in the reservoir cause the oil to foam during the production process. Foaming complicates the process even more and may occur at any point in the reservoir, wellbore, flowlines, or production equipment at the surface. Foaming occurs when the gas reaches the bubble point and comes out of suspension as the pressure and temperature of the fluids change. Gas separates from conventional crude oil much faster than it does from heavy oil. The ability of the oil to foam depends on its viscosity, so higher viscosity oil is more likely to foam, and the foam it makes will be longer lasting. However, the occurrence of foaming is not always an adverse effect. Since any foam that forms in the reservoir increases pressure, it can serve as a temporary gas drive, pushing oil toward the wellbore. It is necessary, if possible, to have a sufficient understanding of fluid properties to know (or predict) at what point and when foaming is likely to occur.

In addition, oil that contains oil-based drilling mud can suffer changes in viscosity over a fluid's bubble point (compared to the unadulterated oil). If surface equipment has been designed based on the properties of the unadulterated oil, gas can come out of suspension much sooner than expected thereby overwhelming the surface facilities. Furthermore, when reservoir fluids interact when mixed with other fluids, such as gas, solvents, or heavy crude oil mixed with conventional crude oil, phase separation (such as aspajltene separation from the heavy oil) can occur thereby adding further confusion to the recovery process. The latter phenomenon (phase separation) can be offset by prior testing of the various fluids and their behavior on mixing (or blending).

In addition to identifying the fluids and the respective properties (see chapter: "Evaluation of Reservoir Fluids") there is also the issue of phase behavior of the fluids. The phase behavior of hydrocarbons, hydrocarbon mixtures, and crude oil are considered an important part of petroleum behavior (Pedersen and Christensen, 2006). The production and treatment of crude oil requires knowledge of phase behavior in which pressure–volume–temperature (PVT) properties play an important role throughout the life of a reservoir. Furthermore, accurate PVT properties are required for the design of well, surface facilities, and refineries. Characterization of the reservoir fluids (in this context the term *reservoir fluids* includes gases, liquids, and solids) and distribution within the reservoir help in defining the continuity, communication, and behavior within the various zones of the reservoir (Dandekar, 2006). Such information allows well operators to reasonably predict, although with some degree of uncertainty, the potential for solids waxes and asphaltene derivatives to deposit anywhere from the near-wellbore region, surface facilities, and during transportation in pipelines.

Experimental phase behavior studies include measuring volumetric properties as functions of PVT properties which are critical for efficient reservoir management throughout the life of the reservoir, from discovery to abandonment (Honarpour et al., 2006; Nagarajan et al., 2007a, 2007b). Constant composition expansion, differential liberation, and separator tests are conducted to analyze the volumetric properties of the heavy oils. The process of phase behavior analysis starts with single-phase sample collection, leading to development of a fine-tuned equation of state (EOS) that accurately predicts the equilibrium phase properties for petroleum fluids starting with fluid sampling to tuning the derived EOS for all types of reservoir fluids.

However, current methods that are used to assess (or estimate) the tendency for solids deposition and the factors affecting such events still leave much to be understood. Until such capabilities are available, continued efforts to define the phase boundaries and production pathway for reservoir fluids are the key to providing an understanding of the potential for hydrocarbon solid deposition and the subsequent impact of these solids on a given fluid/production system. Some account (perhaps even more account) should also be given to the chemistry involved as well as the real properties (not estimated properties) of the various crude oil systems.

The predictions of an EOS cannot be relied upon directly, as an EOS cannot accurately simulate the interactions between numerous hydrocarbon and nonhydrocarbon components present in petroleum crude oil. In order to have meaningful and accurate estimates of fluid properties and phase behavior, an EOS requires some amount of tuning to match with experimental data. And as long as the equation has to be *tuned* to match the real data, questions must be raised about the information used during the derivation of any EOS. For example, while the average properties of the asphaltene fraction are employed in equation derivation, there are very few examples where the spread of properties (that represent the real world of the asphaltene fraction) are often ignored. Thus, the need arises to tune the derived equation as it is applied to asphaltene fraction behavior.

In terms of behavior of the organic fluids, detailed characterization studies (see chapter: "Evaluation of Reservoir Fluids") are beneficial in choosing the correct (or as near-correct as possible) recovery process. The ability to predict the behavior of a reservoir fluid (such as heavy oil, extra heavy oil, and tar sand bitumen) dictates not only the choice of the recovery process but also the efficiency of the chosen process.

4.3 WATER

The water in the reservoir or deposit, often referred to as *connate water* (and also known as fossil water), is a critical component of the reservoir fluids. Sedimentary rocks (see chapter: "Sediments, Reservoirs, and Deposits") are usually deposited in water, with layers of sediment accumulating in rivers, lakes, shallow seas, and on the ocean floor over many millions of years. Consequently, most sedimentary rocks are water-wet and as the sediments are buried and lithified, connate water will undergo major changes in composition. In some cases it may be diluted or displaced by other waters, while in other reservoirs some connate water has been displaced by hydrocarbons. Formations in which crude oil exists generally contain several immiscible fluids. Water that does not flow as reservoir pressure falls is known as irreducible water.

The chemistry of connate water can change in composition throughout the history of the rock and, thus, the connate water can be dense and saline compared with sea water. In contrast, formation water (also called interstitial water) is water that exists in the pore spaces of the reservoir (deposit) rock but might not have been present when the rock was formed.

The water–rock interactions determine whether or not the rock is water-wet or oil-wet with either phenomenon making a different to the recovery process. Consider, for example, the difference between the Canadian (Athabasca) tar sand deposits and the US tar sand deposits. The former (the Canadian tar sand deposits), which have been postulated to be water-wet, lend themselves conveniently to bitumen recovery by the hot water process (see chapter: "Microbial Enhanced Oil Recovery"). On the other hand, the US tar sand deposits, which have been postulated to be bitumen-wet, do not lend themselves to the convenience of bitumen recovery by the hot water process but require more selective processing conditions (see chapter: "Microbial Enhanced Oil Recovery").

Water that can be displaced from the reservoir during production is referred to as free water. The total water content in a hydrocarbon-bearing reservoir rock (free and irreducible) is formation water. In the simplest terms, the water saturation in a formation is the fraction of its pore volume occupied by water. A formation that contains only water has a water saturation of 100. The water saturation of any formation can vary from 100 to quite low values, but it is rarely, if ever, zero. No matter how rich the oil or gas layers in a reservoir, there is always a small amount of capillary water that cannot be displaced by the hydrocarbons. A formation at irreducible water saturation will produce water-free oil.

The nature and quantity of formation waters have a direct influence on oil exploration and exploitation. Typically, sea water contains around 3.5% (w/w) dissolved solids (often expressed as 35,000 ppm) of which approximately 90% of the dissolved material is sodium chloride (NaCl). However, the water in a reservoir (or deposit) can be very different insofar as at depths in excess of a several hundred feet the concentration of dissolved solids formation waters can be extremely high (>100,000 ppm, and even in excess of 300,000 ppm)—formation water with mineral concentrations in excess of 100,000 ppm dissolved solids are referred to as brine. The most concentrated brines are found in nondeformed basins, such as those which overlie geologically stable formations. Brines may also be particularly concentrated in areas where the sedimentary sequence contains evaporite units.

Evaporites are sediments that have been chemically precipitated due to evaporation of water from an aqueous solution and common evaporates can be dominated by halite (salt, NaCl), anhydrite (anhydrous calcium sulfate, $CaSO_4$), and gypsum (hydrated calcium sulfate, $CaSO_4 \cdot 2H_2O$). Evaporites are often interbedded with other sedimentary rocks, such as mudstones, marls, and siltstones.

When water is present, and it is present in almost every oil field (some heavy oil reservoirs and tar sand deposits show little evidence for the presence of water) this can influence a water-based recovery process. Connate formation water, injected water, and produced water must all be dealt with if production rates, and total recovery are to be maximized. Current production technologies aim to identify and assess the overall distribution of reservoir (deposit) water and to monitor and control the movement of formation and injection water. In addition, there has been an increasing awareness of water control issues in the crude oil reservoirs. Furthermore, issues related to the presence of reservoir (deposit) water can be encountered at every stage of development and at every point in the injection/production cycle. Efforts to monitor and control oil field water are now a central part of field development and management. Effective water control can help to reduce the costs associated with production facilities, and to protect the environment by reducing the volumes of produced water for disposal. In shale formations the bound water in the pores is not normally considered part of the fluid flow. However, shale-bound water makes it much harder to estimate water saturation accurately from logs.

To determine the fluid saturations from a core sample, two methods are commonly employed: (1) evaporation of the fluids in the pore space, known as the retort method and (2) the leaching of fluids in the pore space, known as the Dean-Stark extraction method.

In the retort technique the sample is sealed inside an aluminum cell and then heated in stages from 205°C to 595°C (400–1100°F). The advantages to this method are that the time to conduct the test is typically less than 24 hours, and multiple samples can be run simultaneously. However, the disadvantage is the *coking effect*, in which some of the crude oil constituents form coke within the pores or on the surface of the core sample under the thermal conditions employed, which results in a low estimate of the oil content of the core. This (the *coking effect*) is especially true when dealing with core samples from heavy oil reservoirs, extra heavy oil reservoirs, or deposit and core samples from tar sand deposits. A correction factor (based on the thermal properties sample) might be estimated but this is not truly accurate because it may fail to acknowledge the catalytic effect of any minerals in the core sample as well as any other thermal peculiarities of the crude oil, the heavy oil, the extra heavy oil, and the bitumen (Speight, 2014a, 2015). A second disadvantage of the heating process is the removal of both pore water and water of crystallization. The latter is the bound water in clay minerals and other minerals as water of hydration leading to overestimates of the water content of the core. A final disadvantage of this method is it destroys the sample, therefore no further testing can be achieved.

The Dean-Stark extraction method (Speight, 2015) uses the vapor of a solvent to rise through the core and leach out the oil and water. The water condenses and is collected in a graduated cylinder. The solvent and oil continuously cycle through the extraction process. A typical solvent used for this method is toluene, which is miscible with crude oil (heavy oil, extra heavy oil, and bitumen) but immiscible with water. The advantages of the Dean-Stark method are the accuracy, the oil and water measurement are on the same sample, and the core can be used for further analysis. The primary disadvantage of this method is the time necessary to assure complete extraction of the core material and there is always the potential for the solvent (toluene) to bypass crude oil in small pore throats and channels. In addition, application of the method to low-permeable samples (from clay formation and from tight sandstone formations) may also create doubts about the accuracy of the data.

Determining reliable measurement for connate water saturation (S_{wc}) in an oil reservoir can be achieved using the single well chemical tracer test method (Deans and Mut, 1997). The test method is usually carried out on wells that are essentially 100% oil producers and the procedure takes into account that oil is the mobile phase and water is stationary in the pore space. Because oil is the mobile phase, it is used to carry the chemical tracers into and back out of the formation. The chemical (typically an organic alkyl ester), which is dissolved in the crude oil, is injected and then pushed away from the well. The first bank carries the ester plus material-balance tracer while the push bank contains only the material-balance tracer. During the flow into the formation, the chemical partitions between the oil and water phases. This partitioning slows the ester velocity only slightly because the ester is more soluble in the oil phase. After the push step, the ester bank is located 10–20 feet from the wellbore. During a 2-day to 10-day shut-in period, a portion of the ester hydrolyzes (reacts with the connate water) to form the alcohol product tracer, which is much more soluble in the connate water than in the oil phase. The well then is placed on production and samples of the produced oil are taken frequently and analyzed for tracer content. The unreacted ester comes back first because it travels at nearly the same velocity as the carrier oil while the product alcohol is produced later because it travels more slowly than the oil. The difference in arrival volumes between the ester and the delayed alcohol tracer thus is a function of S_{wc} (Deans and Mut, 1997).

4.4 SOLIDS DEPOSITION DURING RECOVERY

Solids deposition can cause (1) reservoir channel plugging, (2) wellbore plugging, (3) deposition in surface equipment, and (4) deposition in pipelines. In fact, the presence and separation of high molecular weight constituents (such as resin and asphaltene constituents) from the liquid phase is valuable (even necessary) in order to know the point (related to the operating conditions) at which the separation will occur and the quantity of material that will separate. Accurate determination of the threshold points for phase separation (Figs. 4.2 and 4.3) from crude oil is required in order to (1) characterize crude oil behavior, (2) predict phase separation, and (3) develop preventive measures for phase separation (Mitchell and Speight, 1973; Kokal et al., 1992; Dandekar et al., 2004; Mousavi-Dehghania et al., 2004; Udourioh et al., 2014).

Under reservoir conditions (for a specific reservoir and a specific crude oil), the asphaltene constituents, resin constituents, maltenes, and the oil phase are in thermodynamic equilibrium. This equilibrium can be disturbed by a number of factors such as pressure reductions, change in temperature, change in crude oil chemical composition, addition of miscible gases and liquids to the oil as applied in various EOR techniques, and

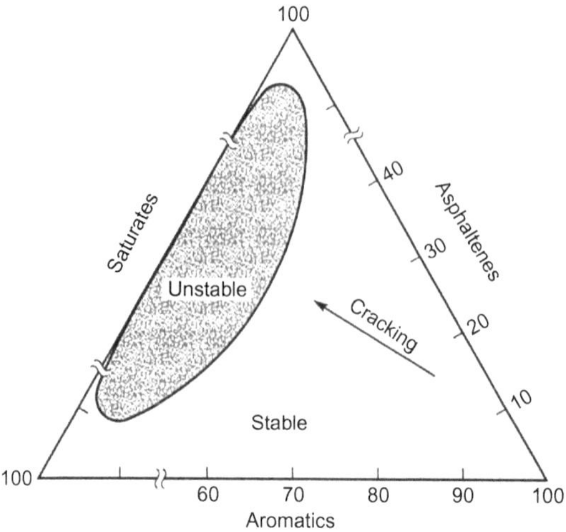

Figure 4.2 Schematic representation of phase relationships during feedstock cracking.

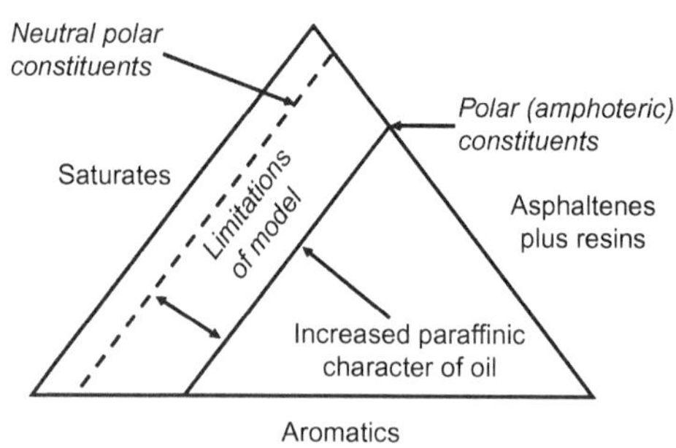

Figure 4.3 Different asphaltene constituents will phase separate at different times depending upon the composition of the feedstock.

during acid simulation (Andersen and Stenby, 1996; Hirschberg et al., 1984; Gawrys et al., 2003).

Phase separation leading to solids deposition and deposition from oil reservoir fluids during production is a serious problem as it can cause plugging of reservoir formation, wellbore, tubing, and production facilities. Asphaltene precipitation during primary depletion of highly undersaturated reservoirs due to the changes in pressure, temperature, and compositions or during application of any of the (improved oil recovery) processes has been described by many authors (Karan et al., 2003; Dandekar et al., 2004).

Thus, one of the key issues that can affect recovery of crude oil, heavy oil, extra heavy oil, and tar sand bitumen is the potential for the deposition of solids during the recovery process. Thus, any observations relating to the properties and behavior of these materials would not be complete without some attention to the phenomena of solids deposition, often referred to as *instability* or, when other fluids are involved, as *incompatibility* (Speight, 2014a, 2015). Each of these phenomena result in formation of a separate phase or, in some cases such as when high temperatures are used for recovery, degradation products or other undesirable changes in the original properties of petroleum products. And it is the analytical methods that provide the data that point to the reason for problems during recovery operations.

Furthermore, it is appropriate here to define some of the terms that are used in respect of instability and incompatibility so that their use later in this text or on any other text will be more apparent and will also alleviate some potential for misunderstanding. The general scientific areas of *instability* and *incompatibility* are complex and have been considered to be nothing better than a black art because not all of the reactions that contribute to instability and incompatibility have been defined (Mushrush and Speight, 1995; Speight, 2014a). Nevertheless, studies over the past three decades have made valuable contributions to an understanding of instability and incompatibility in fuels.

Briefly, the term *incompatibility* refers to the formation of a *precipitate* (or *sediment*) or *separate phase* when two liquids are mixed. The term *instability*, in the context pf recovery operations, is often used in reference to the formation of sediment which is usually due to chemical reactions, such as oxidation, and is chemical rather than physical. This term may be used to contrast the formation of a precipitate in the near term (almost immediately). The phenomenon of *instability* is often referred to as *incompatibility*, and more commonly known as *sludge formation*, and *sediment formation*, or *deposit formation*. In petroleum and its products, *instability* often manifests itself in the form of various properties (Table 4.1) (Stavinoha and Henry, 1981; Hardy and Wechter, 1990; Power and Mathys, 1992). Hence, there are different ways of defining each of these terms but the terms are often used interchangeably.

Thermal stability is the ability of the crude oil, heavy oil, extra heavy oil, and bitumen to withstand relatively high temperatures for short periods of time without the formation of sediment (ie, carbonaceous deposits and/or coke). *Thermal oxidative stability* is the ability of the material to withstand relatively high temperatures for short periods of time in the presence of oxidation and without the formation of sediment or deterioration

Table 4.1 Examples of Properties Related to Instability in Petroleum and Petroleum Products

Property	Comments
Asphaltene constituents	Influence oil–rock interactions
	Separates from oil when gases are dissolved
	Thermal alteration can cause phase separation
Heteroatom constituents	Provide polarity to oil
	Preferential reaction with oxygen
	Preferential thermal alteration
Aromatic constituents	May be incompatible with paraffin medium
	Phase separation of paraffin constituents
Nonasphaltene constituents	Thermal alteration causes changes in polarity
	Phase separation of polar species

of properties (ASTM D3241) and there is standard equipment for various oxidation tests (ASTM D4871). *Stability* is also the ability of the liquid to withstand long periods at temperatures up to 100°C (212°F) without degradation and, in addition, the determination of the *reaction threshold temperature* for various liquid and solid materials is also beneficial (ASTM D2883).

Other terms that are more often applied to petroleum products are (1) dry sludge and (2) existent dry sludge, but these may be used in reference to insoluble material formed during recovery operations. Thus, instability may also be manifested in the appearance of an insoluble material that is often referred to as *dry sludge*, which is the material separated from the crude oil by filtration and which is insoluble in *n*-heptane. *Existent dry sludge* is the dry sludge in the original sample as received and is distinguished from the accelerated dry sludge obtained after aging the sample by chemical addition or heat. The *existent dry sludge* is distinguished from the *potential dry sludge* that is obtained by aging the sample at an elevated temperature. The test for *existent dry sludge* may be used as an indicator of process operability and as a measure of potential downstream solids deposition (eg, solids deposition in separators and/or pipelines).

In the current context of recovery operations, the term *incompatibility* is used when solids deposition occurs in the formation or wellpipe recovery and the product is incompatible with (immiscible with or insoluble in) the original crude oil, heavy oil, extra heavy oil, or bitumen. Such an example is the formation and deposition of wax and other solids during recovery (of conventional crude oil) or the deposition of asphaltene-type materials from the more viscous materials during solvent-based recovery operations or during thermal recovery operations.

There is also the suggestion (often, but not always, real) that the sediments and deposits originate from the inorganic constituents of petroleum, heavy oil, and bitumen. They may be formed from the inherent components of the crude oil (ie, the metalloporphyrin constituents) or from the ingestion of contaminants by the crude oil during the passage into the wellbore and to the surface. The inorganic materials can be salt, sand, rust, and other contaminants that are insoluble in the crude oil. In addition, crude oil is known to assimilate (*pick up*) iron and other metal contaminants from contact with pipelines and pumps.

The chemistry and physics of incompatibility can, to some extent, be elucidated (Por, 1992; Power and Mathys, 1992; Mushrush and Speight, 1995; Speight, 2014a) but many unknowns remain. In addition to the chemical aspects, there are also aspects such as the attractive force differences, such as: (1) specific interactions between like/unlike molecules (eg, hydrogen bonding and electron donor–acceptor phenomena); (2) field interactions, such as dispersion forces and dipole–dipole interactions; and (3) any effects imposed on the system by the size and shape of the interacting molecular species. Such interactions are not always easy to define and, thus, the measurement of instability and incompatibility has involved visual observations, solubility tests of hot filtration sediment, and gum formation. However, such methods are often considered to be *after-the-fact methods* insofar as they do not offer much in the way of predictability. In recovery processes, predictability is not just a luxury, it is a necessity. The same principle must be applied to the measurement of instability and incompatibility. Therefore, methods are continually being sought to aid in achieving this goal.

The focus of incompatibility studies has usually been on the whole crude oil and specifically on the asphaltene fraction. The asphaltene constituents are the highest molecular weight and most polar fractions found in crude oil, heavy oil, and bitumen (Speight, 1994, 2014a). The characteristics of the asphaltene constituents and the amount in crude oil depend to a greater or lesser extent on the crude oil source. The problems with the asphaltene constituents have increased due to the need to extract even the heaviest crude oil and bitumen from reservoirs and deposits. Some examples of problems that occur during recovery operations and which are due to flocculation and/or sedimentation of asphaltene constituents are: (1) *w*ellbore plugging and pipeline deposition; (2) the formation of degraded asphaltene constituents during thermal processes which are more aromatic and less compatible with the crude oil; (3) the formation of emulsions if there is a high degree of water contamination, and the asphaltene constituents, which are highly polar and surface active, are responsible for the undesired stabilization of emulsions; (4) the occurrence of preheating as might occur during in situ combustion operations which can promote reaction of the asphaltene constituents and separation precipitation of the reacted asphaltene constituents as a separate phase; and (5) blending of the heavier materials with lighter crude oil for transportation purposes, whereupon the change of medium during mixing may cause destabilization of asphaltene (and reacted asphaltene) constituents.

Thus, wax deposition combined with the deposition of asphaltene or reacted asphaltene constituents can lead to severe problems during recovery operations. The deposition of wax is less likely in the case of the recovery of heavy crude oil than it is in recovery of conventional (light) crude oil just as the deposition of asphaltene constituents is less likely in the case of conventional (light) oil recovery than it is in recovery of heavy crude oil. Nevertheless with the increased application of thermal process to heavy oil recovery and the increased potential for the application of processes based on in situ combustion, there is the possibility that thermal products and the original heavy material will form incompatible mixtures. In addition, the thermal decomposition of asphaltene constituents can lead to the formation of long-chain paraffin products (Speight, 1970b, 2014a) which, in turn, could lead to wax deposition during the recovery operations. Both wax deposition and the deposition of (unreacted or reacted) asphaltene constituents are worthy of further comment.

4.4.1 Wax Deposition

Wax deposition is the formation of a layer of the separated solid wax and the eventual growth of this layer on a surface in contact with the crude oil. Precipitation does not necessarily lead to deposition since the precipitated wax may not deposit due to the prevailing operating conditions. Thus, precipitation, though an important condition for deposition, is not necessarily sufficient for wax deposition (Singh et al., 2001).

Wax deposits typically contain crude oil, water, gum, resin constituents, and asphaltene constituents which are entrapped during the crystallization and deposition process but the presence and amounts of these extraneous components depend on the properties of the particular crude oil.

If present, the trapped oil causes diffusion of wax molecules into the gel deposit and counter-diffusion of oil out of the gel deposit—a process that depends on the critical carbon number of the oil, which is a function of the crude oil and the prevailing operating conditions (Singh et al., 2000). In the gel deposit, the fraction of molecules with carbon numbers greater than the critical carbon number increases, while that of molecules with carbon numbers lower than the critical carbon number decreases. The process of diffusion and counter-diffusion leading to hardening of the gel deposit, increase in size of deposit, and increase in the amount of wax in gel deposit, is called *aging*, the second stage of wax deposition. Molecular diffusion, therefore, is critical to aging and the hardening of wax gel deposits.

The deposition of wax gel has been suggested to follow a process that can be described by the following five steps (Singh et al., 2000): (1) gelation of the waxy oil or formation of an incipient gel layer on the cold surface; (2) diffusion of waxy hydrocarbons with carbon numbers greater than the critical carbon number towards the gel layer from the bulk oil; (3) internal diffusion of the waxy hydrocarbons through the trapped oil; (4) precipitation of the waxy hydrocarbons through the trapped oil; and (5) counter-diffusion of dewaxed oil—hydrocarbons with carbon numbers lower than the critical carbon number—out of the gel deposit layer. The last three steps—steps 3, 4, and 5—are proposed to be responsible for the increase in solid wax content of the wax gel deposit.

The deposition of paraffin wax has significant effects on crude oil production, transportation, storage, and refining. The buildup of wax deposits decreases flow patterns in the reservoir, decreases the cross-sectional area in pipes and pipelines, restricts operating capacity, and places additional strain on pumping equipment. In addition, wax precipitation and deposition in flowlines and pipelines is an issue impacting the development of deep-water subsea hydrocarbon reservoirs.

Wax deposition in crude oil pipelines at or below the wax appearance temperature (WAT) can lead to gelling that inhibits flow by increasing the viscosity as the temperature approaches the pour point (Pedersen and Rønningsen, 2003). When the temperature of the pipeline wall is below WAT, the deposition of a layer of wax constituents will form near the pipe wall, gradually increasing in thickness, and constricting fluid flow. Typically, and anticipating such an event, the insertion of a wax-removal pig is often conducted periodically for a hot waxy crude oil pipeline to maintain a steady throughput and operation pressure. Solid deposits of wax and asphaltene constituents (along with the associated resin constituents) can pose major problems in reservoirs, pipelines, and separators, hampering recovery efficiency, transport efficiency, and storage. The terms *crystallization* and *precipitation* have been used interchangeably as a means of describing the mechanism of wax deposition and will be used in a similar manner in this chapter.

Crystallization is the process of separation of solid phase from a homogenous solution, the separated solid phase appearing as crystals. Paraffin wax constituents remain in solution as natural components of crude oil until temperature decreases to the point of the wax solubility (Hammami et al., 2003). Wax crystals form through (1) nucleation and (2) growth, in that order, but nucleation and growth occur simultaneously in the oil system, with one or the other predominating at a given time (Domanska and Morawski, 2005). As the solubility limit is approached, the wax molecules form clusters which grow larger and become stable upon reaching a certain critical size. The critical size is dependent upon the prevailing condition but the clusters redissolve when the critical size is not attained and become unstable. Clusters that achieve the critical size will have an increasing number of molecules associated with the clusters, leading to an increase in size of the crystals (*crystal growth stage*).

Generally, wax deposition can be mild, or it can be severe enough that it is unmanageable, and the sooner the issues are diagnosed the easier it is to design a preventive or control management plan that will reduce or eliminate the technical and economic problems. Technical issues associated with wax deposition include: (1) permeability reduction and formation damage when it occurs in the wellbore and in the vicinity of the wellbore; (2) reduction in the interior diameter and eventual plugging of flow channels and production lines; (3) changes in the crude oil composition and behavior due to phase separation of wax; and (4) additional strain on pumping equipment owing to increased pressure drop along flow channels as wax deposition increases (García et al., 2001; García and Urbina, 2003; Zhu et al., 2008).

In addition, asphaltene constituents offer molecular sites for wax crystal buildup, which leads to increased wax deposition. Furthermore, coprecipitation of asphaltene constituents and associated resin constituents with wax could result in excessive solids buildup with significantly different properties (hardness). Pressure, temperature, fluid composition, and (in a pipeline system) flow characteristics—such as turbulent or laminar flow—and geometry of the pipeline are the key factors that affect solid deposition (Burger et al., 1981; Bhat and Mehrotra, 2004, 2008). Solid deposition causes reduction in effective pipe diameter, which means reduction in the line fill of the pipeline. It also causes increased pressure losses in the pipeline, malfunction of pipeline instrumentation, and contamination of the transported fluid. Wax constituents and asphaltene constituents (with any associated resin constituents) are the solid deposition problems associated with transporting blends of many crude oils through pipelines (García and Urbina, 2003).

Recall, asphaltene constituents are a class of compounds in crude oil defined as high molecular weight polar constituents that are insoluble in *n*-heptane but soluble in aromatic solvents such as toluene (Speight, 2014a). In addition, the asphaltene constituents contain the majority of the heteroelements (nitrogen, oxygen, sulfur, and metals) in crude oil (Speight, 2014a, 2015). The resin's constituents stabilize the asphaltene constituents in crude oil by associating with the asphaltene constituents (Andersen and Speight, 2001; Speight, 2014a). If the concentration of resin constituents (measured as the resin–asphaltene ratio) is insufficient to stabilize the asphaltene constituents—under conditions such as a change in temperature, pressure, or pH—the asphaltene constituents will separate out of the crude oil and cause recovery problems (see chapters: "Heavy Oil and Tar Sand Bitumen" and "Reservoir Fluids").

Pressure has a significant effect on the temperature at which wax appears in crude oil (Weingarten and Euchner, 1988; Brown et al., 1994), although it has been proposed (Pan et al., 1996) that a decrease in the WAT is only pronounced at pressures below the bubble point pressure of the crude oil, and that above the bubble point pressure, the WAT generally increases with pressure. Hence, control of wax deposition is essential—wax precipitation is a thermodynamic phenomenon that will lead to deposition of solid wax crystals (Venkatesan and Creek, 2007). To address the issue, three phenomena have to be considered: (1) wax precipitation; (2) dynamic wax deposition; and (3) heat transfer from the wellbore.

Wax deposition is particularly evident when crude oils from tight shale formations are being recovered, transported, processed, and stored. The properties of these from tight shale formations are significantly different to those of typical crude oils because of the high paraffinic content. In addition, the varying content of filterable solids, hydrogen sulfide (H_2S), and mercaptan derivatives (RSH) also offer challenges during recovery, transportation, and storage. Solids content (and the potential for inorganic deposition) of samples from a single producing region can be highly variable and associated with the stage of fracturing and production from which the oil is produced.

The challenges associated with the production of crude oil from tight shale formations are a function of the wax content which leads to wax deposition. To control deposition and plugging in formations due to paraffins, dispersants are commonly used and, in upstream applications, the dispersants are applied as part of multifunctional additive packages where asphaltene stability and deposition control are also addressed simultaneously. Chemical and mechanical solutions are used to mitigate wax deposition and these deposit problems and preventive deposition control programs have been developed to manage the wax deposition occurring in storage tanks. Use of the appropriate chemical treatment to control wax buildup during production and during storage can accommodate and transfer larger quantities of waxy crude oil without significant plugging issues.

The high paraffin content of the oil is a main contributor that leads to the issue of deposition during blending and wax deposition. Mixing oil from tight shale with an asphaltene-containing oil leads to destabilization of the asphaltene constituents, separation, and deposition (asphaltene deposition). In addition, several crude oils from tight shale formations have a high content of hydrogen sulfide (H_2S) that can lead to the need for hydrogen sulfide scavengers prior to distillation. The scavengers are often amine-based products—such as methyl triazine—that are converted into mono-ethanolamine (MEA) in the atmospheric distillation unit and which contribute to problems in the unit. Once MEA forms, it rapidly reacts with chlorine [from any associated formation water (brine)] to form chloride salts which lose solubility in the hydrocarbon phase and become solids at the processing temperatures of the atmospheric unit forming deposits on the trays or overhead system. The deposits are hygroscopic, and, once water is absorbed, the deposits become very corrosive.

4.4.1.1 Factors Leading to Wax Deposition

Wax can form anywhere in the producing system when the conditions due to temperature and pressure changes become favorable for the precipitation of the paraffin constituents. However, wax deposition typically occurs in wells that produce at lower rates and leads to severe well downtime and, although wax deposition is a commonly encountered problem in production operations, there is no universally effective treatment for the problem.

Wax deposition occurs when the temperature of crude oil containing wax (long-chain paraffin) constituents falls below the cloud point and the *n*-paraffin components begin to crystallize into solid wax particles

(Hammami et al., 2003; Leontaritis and Leontaritis, 2003; Coutinho and Daridon, 2005). These particles can adhere to each other and to surfaces, such as when oil comes in contact with any surface that has a temperature below the WAT, thereby acting as a heat sink. The predominant mechanisms that have been proposed to describe paraffin deposition are *shear dispersion* and *molecular dispersion*. The former term, *shear dispersion*, is the relationship between deposition rate and shear rate—shearing of the wax molecules occurs due to the hydrodynamic drag of the flowing fluid which depends on the flow rate and viscosity of the fluid. Higher viscosity and low flow rates result in high wax deposition rates but in conditions of highly turbulent flow, wax deposition rates decrease with increased flow because wax is mechanically sheared off the deposits on the pipe wall (Tiwary and Mehrotra, 2009). The shear rate increases with deposit thickness increases due to a decrease in the flow area and increase in flow velocity which is reflected in a decrease in the wax deposition rate.

On the other hand, *molecular diffusion* is the process by which the radial temperature gradient in the line causes a concentration gradient of dissolved paraffin components in the liquid phase. This concentration gradient causes paraffin to diffuse to the pipe wall, where it is assumed to deposit. The widely recognized transport methods contributing to wax thickness on the pipe wall are molecular diffusion of dissolved wax, particle transport of precipitated wax, and sloughing of previously deposited wax.

Treatment methods are usually highly case dependent, requiring the proper identification of the mechanisms for wax deposition and the development of a predicting technique that is specific for the target field (Coutinho and Ruffier-Meray, 1997; Coutinho et al., 2001, Zhu et al., 2008). Thus, the factors that influence wax deposition are: (1) crude oil composition; (2) temperature; and (3) pressure as well as other factors.

4.4.1.1.1 Crude Oil Composition

Crude oil is variable in composition but is typically composed of fractions that are named according to the method of separation: (1) saturates; (2) aromatics; (3) resin constituents; and (4) asphaltene constituents (see chapters: "Reservoir Fluids" and "Evaluation of Reservoir Fluids") (García et al., 2001; Speight, 2001; García and Urbina, 2003; Speight, 2014a, 2015). These components are in thermodynamic equilibrium at initial reservoir conditions and the aromatic constituents serve as solvents for high molecular weight saturates, which are the sources of paraffin waxes in crude oil, while the polar components, especially asphaltene constituents, induce wax nucleation (Hammami and Raines, 1997, 1999; Singh et al., 2001). However, the solubility of paraffins in aromatic, naphthenic, and other organic solvents is diminished at low temperatures and lower-boiling constituents of crude oil may assist in maintaining the high molecular weight constituents in solution.

The onset of crude oil production results in the loss of the lower-boiling constituents as they are first to leave the reservoir and this alters the original composition of the oil system, resulting in decreased solubility of the wax constituents. This loss of solubility could lead to precipitation and deposition of wax (Pan et al., 1997). Generally, the weight percent of the saturates in the crude oil, the structural distribution of the paraffin components, and the occurrence of other solids, such as formation fines and the presence of asphaltene constituents, which could form nucleating sites, all contribute to wax precipitation and deposition. Therefore, knowledge of the oil composition gives an indication of the propensity of a crude oil for wax deposition and, hence, the oil stability. Oil stability has been reported to depend on its solids content and the balance between aromatics and saturates as well as between asphaltene constituents and alkanes (García and Carbognani, 2001). Saturates, aromatics, resins, and asphaltenes (SARA) analysis can measure the distribution by weight percent of saturates, aromatics, resin constituents, and asphaltene components, for stable and unstable crude oils (Carbognani et al., 1999).

4.4.1.1.2 Temperature

Wax precipitation occurs when the wax molecules contained in the crude oil reach their solubility limit due to one or more changes in the equilibrium conditions (stability or instability) in the crude oil (see chapter: "Reservoir Fluids"), which causes a decrease in solubility of the wax constituents. The solubility limit for the wax constituents is dependent on the temperature and, as such, can be defined by the temperature, given other specified factors. While some of the factors influence wax precipitation by increasing or decreasing the solubility limit (depending on the temperature), other factors may provide a favorable environment for deposition to occur. Such factors include oil composition plus available solution gas, and pressure of the oil which affects the amount of gas in solution. Other factors are flow rate, completion, and pipe or deposition surface roughness. Typically, paraffin solubility increases with increasing temperature and decreases with decreasing temperature. Thus, temperature appears to be the predominant and most critical factor in wax precipitation and deposition due to its direct relationship with the solubility of paraffin hydrocarbons (Sadeghazad and Christiansen, 1998; Singh et al., 2000).

Wax precipitates from crude oil when the operating temperature is at or below the WAT (cloud point temperature) and wax deposition will not occur until the operating temperature falls to or below the WAT (Erickson et al., 1993; Coutinho and Daridon, 2005). All other factors actually lead to wax deposition when the temperature is already at or below the cloud point.

4.4.1.1.3 Pressure

Pressure, as an important parameter in the exploitation of reservoir fluids, plays a significant role in wax precipitation and deposition. The pressure profile during oil production is such that the reservoir pressure declines with production, and the pressure of the flow stream drops all the way from the reservoir to the surface. The lighter components of the reservoir fluid tend to be the first to leave the reservoir as pressure depletes. This causes an increase in the solute solvent ratio, since the light ends serve as solvent to the wax components. Hence, the solubility of wax is reduced with the loss of these light ends.

The WAT increases with increase in pressure above the bubble point at constant composition, which indicates that an increase in pressure in the one-phase liquid region (above the bubble point pressure) will favor wax deposition (Brown et al., 1994). The situation is different below the bubble point where there is two-phase existence and the WAT decreases with increase in pressure up to bubble point pressure (Brown et al., 1994) due to dissolution of light ends back into the liquid phase. The WAT increases with increase in pressure for some oils (Brown et al., 1994; Karan et al., 2000) and increases with increase in pressure for a fixed component liquid mixture (Pan et al., 1997).

4.4.1.1.4 Other Factors

Although crude oil composition, temperature, and pressure have the most significant effects on wax deposition, other factors that have been identified as contributing to wax deposition include (1) flow rate, (2) gas–oil ratio, and (3) pipeline wall roughness. In terms of flow rate, wax deposition is influenced more by laminar flow than when flow is in the turbulent regime, and increasing the flow rate from laminar to turbulent reduces maximum deposition rate and at the same time lowers the temperature at which maximum deposition rate occurs. Low flow rates offer the moving oil an increased residence time in the flow channel which allows more heat loss to the surroundings, leading to a higher chance of the bulk oil temperature falling below the WAT and sufficient time for wax separation and deposition. In addition, there is a difference in texture between wax deposited at high flow rates and wax deposited at low flow rates (Jessen and Howell, 1958; Tronov, 1969). Paraffin wax deposited at high flow rates appears harder, being more compact and more firmly attached to the deposition surface.

The gas–oil ratio influences wax deposition in a manner that depends on the pressure—above the bubble point, where all gases remain in solution, solution gas helps to keep wax in solution (Luo et al., 2001; Singh et al., 2004). A high gas–oil ratio would result in more expansion and subsequent cooling as pressure of the oil system depletes—a situation that can increase wax deposition.

4.4.2 Asphaltene Deposition

Asphaltene constituents are best known for the problems they cause as solid deposits that obstruct flow in the petroleum production systems and asphaltene formation and deposition in oil reservoirs and transmission lines is one of the most important problems of oil industry (Ali and Islam, 1997; Minssieux, 1997; Wang and Civan, 2001; Almehaideb, 2004; Hong and Watkinson, 2004; Soltani Soulgani et al., 2011; Kord and Ayatollahi, 2012). In fact, asphaltene deposition in producing wells, pipes, and surface facilities has been a serious problem with a wide economic import to the oil industry. Deposition of asphaltene on the wall of production tubing reduces available diameter to oil flow and subsequently, oil production rate decreases. Crude oil is considered as a colloidal mixture consisting of four major groups namely (1) saturate constituents, (2) aromatic constituents, (3) resin constituents, and (4) asphaltene constituents. Generally, asphaltene particles are stable in oil phase at reservoir conditions but variations in temperature, pressure, and composition of the oil can induce the asphaltene precipitation process. The existing models for asphaltene stability can be classified as either thermodynamic or scaling models. Proponents of thermodynamic models have used the Flory-Huggins theory that relates to polymer chemistry (Flory, 1953) or equations of state to model asphaltene phase behavior. In the single-component solid model, which is another thermodynamic model for predicting asphaltene phase behavior in oil, the precipitated asphaltene is represented as a pure solid. In the similar form of this model (Nghiem et al., 1993), a new method for characterizing the asphaltene solid phase was proposed in which the heaviest component splits into two components, a nonprecipitating component and a precipitating component. The two components were considered identical in critical properties and acentric factors, but their interaction coefficients with the light components are different.

However, models for predicting the precipitation (separation) of asphaltene constituents, although critical, is only a part of the solution. It is equally important to predict the radial transportation of these precipitated particles and quantify their deposition tendencies (Escobedo and Mansoori, 1995) and it has also been logically proposed that in a complex system where asphaltene deposition and asphaltene removal are occurring, the net rate of asphaltene deposition can be considered as the difference between the rate of deposition and removal of asphaltene (Ramirez-Jaramillo et al., 2006).

In order to resolve such issues the concepts and mechanisms of asphaltene precipitation and deposition including instability and incompatibility of the asphaltene constituents with the surrounding oil medium and other effects must be properly considered. Furthermore, the experimental data for any attempts at the predictability of asphaltene deposition must be gathered from the most reliable sources and be based upon data from recognized standard analytical methods (Speight, 2014a, 2015).

A better understanding of the effect of asphaltene constituents and resin constituents is key to preventing the formation of deposits during production and refining and mitigating the deleterious effects of these petroleum constituents (Speight, 2014a, 2015). In addition, the resin constituents of petroleum will also be considered—for the purposes of this chapter, the resin constituents are regarded as those materials soluble in *n*-pentane or *n*-heptane (ie, whichever hydrocarbon is used for the separation of asphaltene constituents) but insoluble in liquid propane. The *resin constituents* are also those materials soluble in *n*-pentane or *n*-heptane but cannot be extracted from an adsorbent such as clay or alumina by *n*-pentane or *n*-heptane The oils fraction is, therefore, that fraction of crude oil soluble in *n*-pentane or in *n*-heptane that is extractable from an earth by *n*-pentane or by *n*-heptane and is composed of saturates and aromatics fractions (Fig. 4.4).

The resin constituents are closely related to the asphaltene constituents and for the purposes of this text, the term *resin* generally implies material that has been eluted from various solid adsorbents (Koots and Speight,

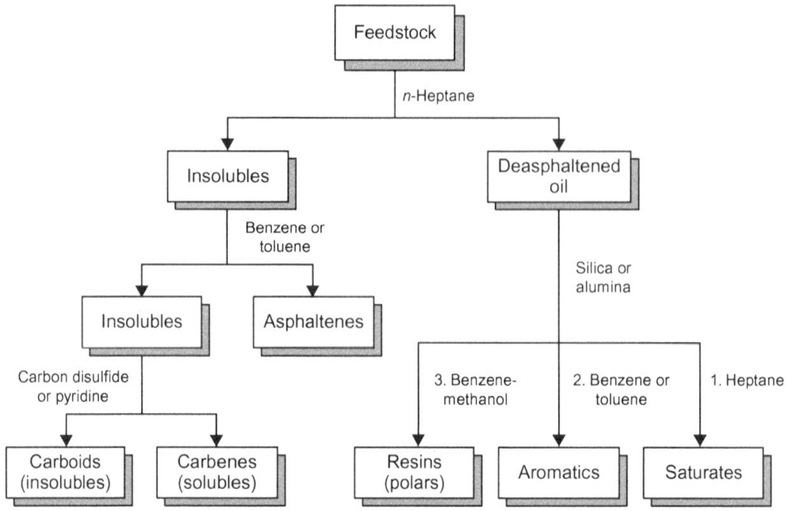

Figure 4.4 Schematic of the separation of heavy oil into various bulk fractions—the carbenes and carboids are generally considered to be the organic product of thermal treatment and are not typically found in untreated crude oil, heavy oil, or bitumen.

1975; Speight, 2001, 2014a, 2015). Thus, after the asphaltene constituents are precipitated, adsorbents are added to the *n*-pentane or *n*-heptane solutions of the resin constituents and oils, by which process the resin constituents are adsorbed and subsequently recovered by use of a more polar solvent and the oils remain in solution. The term *maltenes* (sometime called *petrolenes*) indicates a mixture of the resin constituents and oil constituents obtained in the filtrates from the asphaltene precipitation (Speight, 2014a, 2015).

It is generally recognized and accepted that that the crude oil and heavy oil system (which has been extended to include extra heavy oil and tar sand bitumen) is a colloid-type system comprising fractions of saturates, aromatics, resin constituents, and asphaltene constituents (Speight, 1994, 2014a). The asphaltene constituents are stabilized by the resin constituents but the asphaltene constituents can be naturally or artificially precipitated if the molecular association with the resin constituents is disturbed.

As a result, deposition of the asphaltene constituents can occur in different parts of the recovery system including (1) the reservoir, (2) the well tubing, and (3) the surface flowlines. The deposition depends on the changes in flow conditions, pressure, temperature, and oil composition. Such occurrences will decrease well productivity and reservoir productivity and correction requires frequent chemical treatments for the removal of the asphaltene constituents. Destabilization (ie, flocculation) of the asphaltene constituents depends on breaking up the balance of attraction forces between the associated resin molecule constituents and the asphaltene constituents (Koots and Speight, 1975; Branco et al., 2001; Speight, 2014a).

In attempts to define and understand the nature of deposition by asphaltene constituents, there are many models for asphaltene deposition, but very few if any consider the complexity of the asphaltene fraction, preferring instead to use *average parameters*; therefore, there are frequent references to models needing *fine tuning*. Furthermore, the effect of resins in the feedstock have been largely ignored in such models and there has been a general failure of the model proponents to recognize that the resin constituents also play a major role in crude oil chemistry and physics insofar as the resin constituents stabilize (peptize, disperse) the asphaltene constituents (see chapter: "Reservoir Fluids") (Koots and Speight, 1975; Hammami and Freworn, 1998; Carnahan et al., 1999; Gawrys et al., 2003). Without the presence (or a change in the bonding ability) of resin constituents, the asphaltene constituents might well remain adsorbed on to the reservoir rock—not such a bad idea as it would allow a more acceptable (asphaltene-free) crude oil to be pulped to the surface. For the remainder of this section, although the focus may appear to be on the asphaltene constituents, the resin constituents are also included.

In fact, the properties and behavior of petroleum (heavy oil, extra heavy oil, and tar sand bitumen) as applied to recovery, transportation, storage, and refining are dictated by the chemical composition, which includes each component (bulk fraction) of crude oil. Once the chemical composition has been established, it becomes possible to correlate composition, properties, and behavior in various recovery, transportation, storage, and refining scenarios (Speight and Ozum, 2002; Hsu and Robinson, 2006; Gary et al., 2007; Speight, 2008, 2014a).

In terms of composition, it is generally recognized that crude oil and heavy feedstocks are composed of four major fractions (saturate constituents, aromatic constituents, resin constituents, and asphaltene constituents) (Fig. 4.4) that differ from one another sufficiently in solubility and adsorptive character that the separation can be achieved by application of relevant methods (Andersen and Speight, 2001; Speight, 2014a, 2015). Indeed, although these four fractions are chemically complex, the methods of separation have undergone several modifications to such an extent that the evolution of the separation techniques is a study in itself (Speight, 2001, 2014a, 2015). And because the fractions are in a balanced relationship in crude oil, the chemical and physical character of these constituents needs reference in this section.

4.4.2.1 Composition

Resin and asphaltene fractions isolated from different sources tend to exhibit a relatively constant carbon and hydrogen content, although there can be variations in terms of heteroatom content because of local and regional variations in the plant precursors and mineralogical composition of the nearby geological formations. This is reflected, in part, by changes in the functional group composition of the fractions from different crude oils (Koots and Speight, 1975; Huc et al., 1984; Ancheyta et al., 2002). However, exposing resin constituents and asphaltene constituents to atmospheric oxygen can substantially alter the oxygen content, and exposing a crude oil to elemental sulfur or even to sulfur-containing minerals can result in excessive sulfur uptake. Perhaps oxygen and sulfur contents vary more markedly than do nitrogen contents because of these conditions.

One aspect of asphaltene characterization that has provided strong evidence for the complexity of the fraction arises from composition studies using fractionation techniques. Indeed, the asphaltene fraction can be subdivided by a variety of techniques into a series of subfractions with sufficiently different properties (Bestougeff and Darmois, 1947, 1948; Bestougeff and Mouton, 1977; Long, 1979, 1981; Bestougeff and Byramjee, 1994; Francisco and Speight, 1984; Speight, 1986, 1994; Andersen et al., 1997; Speight, 2014a). Indeed, the separation of the asphaltene fraction into a variety of functional (and polar) types (Fig. 4.5) (Francisco and Speight, 1984) has confirmed the complexity of the asphaltene fraction. Other studies (Schmitter et al., 1984) have brought to light the occurrence of four-ring aromatic nitrogen species in petroleum. These findings are of particular interest since they correspond to the ring systems that have been tentatively identified by application of the high performance liquid chromatography (HPLC) technique to the basic nitrogen fraction of asphaltene constituents (Speight, 1986b, 1994). These studies refute the concept of an average structure in which it is claimed that the behavior of this complex fraction can be explained by use of a single structural entity, which often bears no relationship to the reality of the actual behavior of the fraction and the separated subfractions. Of specific interest is the observation that when the asphaltene fraction is fractionated on the basis of aromaticity *and* polarity, it appears that the more aromatic species contain higher amounts

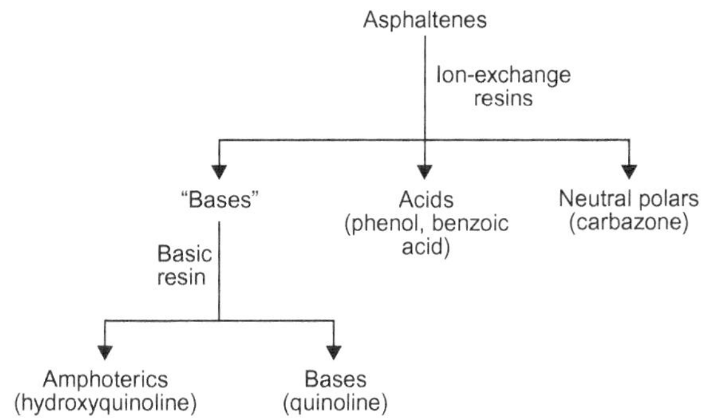

Figure 4.5 Separation of asphaltene constituents based on functionality (polarity).

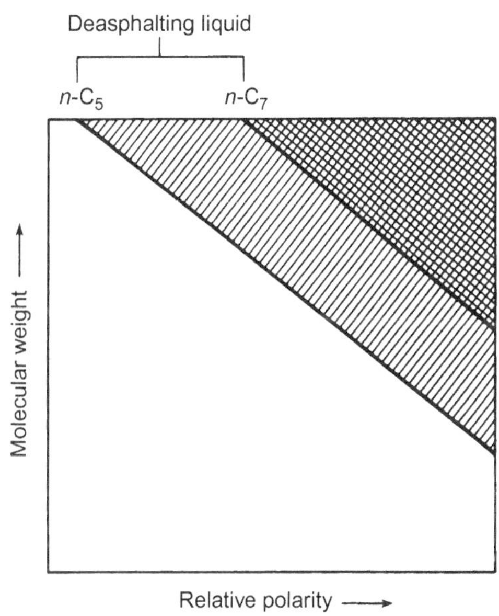

Figure 4.6 Representation of the asphaltene fraction as a collection of species of different molecular weight and polarity.

of nitrogen, suggesting that the nitrogen species are located predominantly in aromatic systems (Speight, 1984, 1994, 2014a).

On the other hand, the molecular weight/polarity concept, in which asphaltene fractions are defined on the basis of molecular weight and polarity (Fig. 4.6), can be used to indicate the relative nature of the arbitrarily named carbenes and the carboids (thermal decomposition products of asphaltene constituents), which are lower molecular weight and more polar than the asphaltene constituents from which they are formed

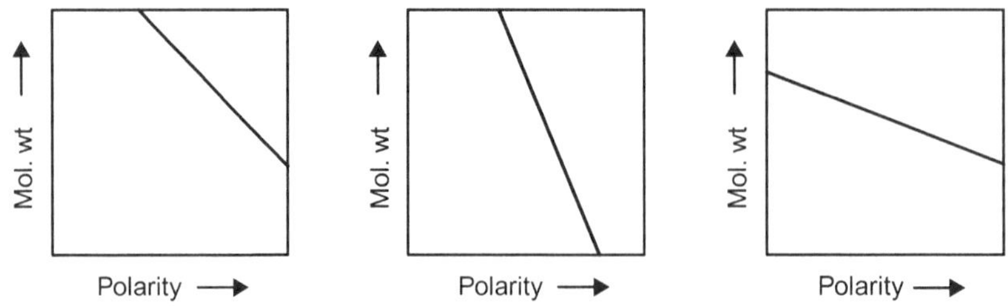

Figure 4.7 Asphaltenes fractions from different crude oils will vary in the relationship between molecular weight and polarity depending on the relative amounts of the precursors and the maturation process parameters (see Fig. 4.6 for comparison).

(Speight, 1994, 2014a). Furthermore, the slope of the line as illustrated is purely illustrative and will vary with the source and composition of different fractions (Fig. 4.7). However, it should not be construed that this two-dimensional diagram is truly representative of the nature of asphaltene constituents and a three-dimensional diagram involving functionality or molecular weight might be more appropriate for showing the chemical and physical differences between the various asphaltene constituents, carbene constituents, and carboid constituents. Nevertheless, the concept does emphasize the lower molecular weight and increased polarity of the various asphaltene constituents.

Studies on the disposition of *nitrogen* in petroleum asphaltene constituents indicated the existence of nitrogen in various heterocyclic locations (Moschopedis and Speight, 1976b, 1979; Jacobson and Gray, 1987). The more conventional (from the organic chemist's viewpoint) nitrogen species, that is, primary, secondary, and tertiary aromatic amines, have not been established as present in petroleum asphaltene constituents. It is reasonable to assume that the organic nitrogen in petroleum asphaltene constituents can be defined in terms of basic and nonbasic types. Spectroscopic investigations (Moschopedis and Speight, 1979) suggest that carbazole derivatives occur in asphaltene constituents. The application of X-ray absorption near-edge structure spectroscopy to the study of asphaltene constituents has led to the conclusion that a large portion of the nitrogen is present in the pyrrole form rather than the pyridine form (Mitra-Kirtley et al., 1993).

Thermal studies have shown that only 1% of the nitrogen is lost during the thermal treatment, but substantially more of the sulfur (23%) and almost all of the oxygen (81%) is lost as a result of this treatment. The tendency for nitrogen and sulfur to remain in the nonvolatile residue produced during thermal decomposition, as opposed to the relatively facile elimination of oxygen, supports the concept that nitrogen and sulfur have stability because of their location in ring systems.

Oxygen has been identified in carboxylic, phenol, and ketone function (Moschopedis and Speight, 1976a, 1976b; Ritchie et al., 1979; Speight and Moschopedis, 1981; Rose and Francisco, 1987) locations but is not usually regarded as located primarily in heteroaromatic ring systems. Some evidence for the location of oxygen within the asphaltene fraction has been obtained by infrared spectroscopy. Examination of dilute solutions of the asphaltene constituents in carbon tetrachloride show that at low concentration (0.01%, w/w) of asphaltene constituents a band occurs at 3585 cm^{-1}, which is within the range anticipated for free nonhydrogen-bonded phenolic hydroxyl groups. In keeping with the concept of hydrogen bonding, this band becomes barely perceptible, and the appearance of the broad absorption in the range 3200–3450 cm^{-1} becomes evident at concentrations above 1% by weight. Other evidence for the presence and nature of oxygen functions in asphaltene constituents has been derived from infrared spectroscopic examination of the products after interaction of the asphaltene constituents with acetic anhydride (Moschopedis and Speight, 1976a, 1976b) as well as from thermal studies (Ritchie et al., 1979).

Sulfur occurs as benzothiophene derivatives, dibenzothiophene derivatives, and naphthene-benzothiophene derivatives (Yen, 1974; Speight and Pancirov, 1984; Rose and Francisco, 1988)—more highly condensed thiophene types may also exist but are precluded from identification by low volatility. Other forms of sulfur that occur in asphaltene constituents include the alkyl–alkyl sulfides, alkyl–aryl sulfides, and aryl–aryl sulfides (Yen, 1974).

Metals (ie, nickel and vanadium) are much more difficult to integrate into the asphaltene system. Nickel and vanadium occur as porphyrins (Baker, 1969; Yen, 1975, 1994), but whether these are an integral part of the asphaltene structure is not known. Some of the porphyrins can be isolated as a separate stream from petroleum (Branthaver, 1990; Reynolds, 1998).

4.4.2.2 Molecular Weight

The molecular weights of asphaltene fractions span a wide range from several hundred to several million Daltons leading to speculation about self-association (Reerink, 1973; Koots and Speight, 1975; Speight et al., 1985; Speight, 1994, 2014a; Galtsev et al., 1995; Yarranton, 2005). The tendency of the asphaltene constituents to form aggregates in hydrocarbon solution is one of their most characteristic features and complicates the determination of asphaltene molecular weight (Speight and Moschopedis, 1977; Speight et al., 1985; Speight, 2014a, 2015).

For any one method, the observed molecular weights suggest that asphaltene constituents form molecular aggregates, even in dilute solution, and this association is influenced by solvent polarity, asphaltene concentration, and the temperature at which the determination is made. The precise mechanism of the association has not been conclusively established, but hydrogen bonding and the formation of charge-transfer complexes have been cited as being responsible for intermolecular association (Yen, 1974, 1994; Moschopedis and Speight, 1976a, 1976b; Moschopedis et al., 1976). In fact, intermolecular hydrogen bonding could be involved in asphaltene association and may have a significant effect on observed molecular weights (Moschopedis and Speight, 1976c; Speight et al., 1985).

Although it is of some importance to know under which conditions that the asphaltene constituents associate and the forces active during the association of the asphaltene constituents associate, caution is advised in using the data from various analytical methods. Once the asphaltene fraction is removed for the oil environment a different set of forces are operative. The aggregation of asphalt constituents in solvents or when adsorbed on minerals is not the same as the association of asphaltene constituents in the oil medium. The conclusion is that asphaltene constituents do associate and aggregate in solution but the conditions will not be the same from one system to another, especially from a solvent system to the oil system, thereby throwing doubt on many of the conclusions related to asphaltene aggregation.

The higher molecular weights recorded when solvents of low polarity are employed are undoubtedly the result of intermolecular association between the asphaltene nuclei, but solvents of high dielectric constant are able to bring about dissociation of these asphaltene agglomerations to what, in fact, appear to be single asphaltene particles. Furthermore, this observation precludes asphaltene structures that invoke the concept of a polymer molecule to account for the high molecular weights observed in nonpolar solvents.

The tendency of the asphaltene constituents to self-associate can have adverse effects on any recovery process. Association of the asphaltene constituents in crude oil and the appearance of a separate phase (often referred to as flocculation) in paraffinic crude oils or when an asphaltic crude oil (asphaltene-containing crude oil) is often irreversible, even when the conditions are returned to preflocculation point (Koots and Speight, 1975; Fuhr et al., 1991; Speight, 1994; Abebi et al., 1998; Acevedo et al., 1995; Hammami et al., 2000; Karan et al., 2003; Gharfeh et al., 2004). This is the major cause of unrepairable arterial blockage damage to the flow of petroleum fluids. Due to the large size and their adsorption affinity to solid surfaces, flocculated asphaltene constituents can cause irreversible deposition which may not be easy to remove.

In summary, the molecular weight of the asphaltene fractions and of subfractions of the asphaltene fraction are variable and method dependent (Yen, 1974; Speight et al., 1985; Speight, 1994, 2001, 2014a, 2015). There is a tendency for asphaltene constituents to associate even in dilute solution in nonpolar solvents.

4.4.2.3 Reactions

Even though asphaltene chemistry might be represented by relatively simple equations, the chemistry of conversion process may be quite complex. In fact, the complexity of the individual reactions occurring in an extremely complex mixture and the *interference* of the products with those from other components of the mixture or from the original unchanged or partially changed feedstock is unpredictable and also can lead to phase separation and deposition. Similarly, the *interference* of secondary and tertiary products with the course of a reaction and, hence, with the formation of primary products may also be cause for concern, also leading to deposition.

The different reactions of asphaltene constituents, although largely ignored, can be major events in recovery operations. In particular, the changes that occur as a result of thermal reactions (in steam based and other thermal processes) and oxidative reactions (in processes which require air to be pumped into the formation) can upset the chemical and physical balance of the heavy feedstock systems thereby causing solids deposition during recovery operations. It is for this reason, that these reactions are considered here.

In fact, with respect to petroleum, since it is a system in which the various constituents are in balance, any change in the physical and chemical properties of crude oil can be a key factor in the stability of the system (Koots and Speight, 1975; Speight, 1994, 2014a; Otzisk and Kempen, 2009). For example, as the degree of conversion *in a thermal process* increases, the solubility power of the medium towards the heavy and polar molecules decreases due to the formation of saturated products (Speight,

1994). This is reflected in a relative change in the solubility parameters of the dispersed and solvent phases leading to a phase separation (Speight, 1992, 1994; Wiehe, 1993, 1994).

4.4.2.3.1 Thermal Reactions

The relative reactivity of petroleum constituents can be assessed on the basis of bond energies and the asphaltene constituents are no exception to this generalization. Whilst the use of bond energy data is a method for predicting the reactivity or the stability of specific bonds under designed conditions, it must be remembered that the reactivity of a particular bond is also subject to its environment. Thus, it is not only the reactivity of the constituents of petroleum that are important in processing behavior but also the stereochemistry of the constituents as they relate to one another that is also of some importance.

The thermal reactions of the constituents of the asphaltene fraction have received considerable attention (Speight, 2014a) and this information is valuable when perturbations to the petroleum system are given consideration vis-à-vis the potential effects that can cause deposition (deposition of sediment) after the delicate balance of the system has been changed or even destroyed (Leontaritis and Mansoori, 1988; Speight, 1994, 2014a; Mansoori, 1996; Ancheyta et al., 2009). The asphaltene constituents undergo a wide range of interactions based not only on their condensed aromatic structure but also on the attending alkyl and naphthenic moieties (Speight, 1984).

Asphaltene constituents can be *thermally decomposed* under conditions similar to those employed for visbreaking [viscosity breaking; approximately 470°C (880°F)] to afford, on the one hand, light oils that contain higher (to > C_{40}) paraffins (leading to the potential for wax deposition) and, on the other hand, coke:

$$\begin{aligned}\text{Asphaltene fraction} \rightarrow\ & H_2, CO, CO_2, H_2S, SO_2, H_2O \\ & + CH_4, CH_3(CH_2)_n CH_3, + \text{olefins}\end{aligned}$$

In fact, the thermal decomposition of asphaltene constituents provides an excellent example of inconsistencies in the derivation of average structural types (for the asphaltene fraction) in which alkyl side chains are deduced to contain approximately four carbon atoms (Speight, 1970a, 1971, 1972). Asphaltene pyrolysis (350–800C, 660–1470F) produces substantial amounts of alkanes (having 22–40 carbon atoms in the molecule) in the distillate that can only be presumed to reflect the presence of such chains in the original asphaltene (Speight, 1970b).

The application of thermal techniques to study the nature of the volatile thermal fragments from petroleum asphaltene constituents has produced some interesting data relating to the polynuclear aromatic systems (Speight, 1971, 1972; Ritchie et al., 1979; Schucker and Keweshan, 1980; Gallegos, 1981). These thermal techniques have produced strong evidence for the presence of small (one-ring to four-ring) polynuclear aromatic systems (Speight and Pancirov, 1984) and now, application of the technique to the various functional fractions confirmed the general but unequal distribution of these systems throughout asphaltene constituents. While these systems may seem innocuous, the potential is high for deposition of the systems from the liquid as the liquid composition changes with reaction time.

Each asphaltene fraction produced the same type of polynuclear aromatic systems (ie, alkyl derivatives of benzene, naphthalene, phenanthrene, chrysene, benzothiophene, and dibenzothiophene) in the volatile matter but the distribution was not constant. It was also possible to compute the hydrocarbon distribution; a noteworthy point here is the overall preponderance of single-ring species (cycloparaffin derivatives and alkylbenzene derivatives) as well as the domination of saturated material over aromatic material. The preponderance of the low-molecular-weight material in the volatile products is anticipated because the more complex systems remain as nonvolatile material and, in fact, are converted to coke (Speight, 1971; Speight and Pancirov, 1984).

Paraffins are not the only hydrocarbon products of the thermal reactions of asphaltene constituents. The reaction paths are extremely complex; spectroscopic investigations indicate an overall dealkylation of the aromatics to methyl (predominantly) or ethyl (minority) groups. This is in keeping with a mass spectroscopic examination of asphaltene fractions (by direct introduction into the ionization chamber), which indicates a progressive increase with increasing temperature (50–350C, 120–660F) of ions attributable to low-molecular-weight hydrocarbons. Higher temperatures (500C, 932F) promote the formation of benzene and naphthalene nuclei as the predominant aromatics in the light oil, but unfortunately an increase in coke production also occurs.

In conclusion, thermal decomposition of asphaltene constituents affords light oil having a similar composition to the crude oil from which the asphaltene fraction was produced, long-chain paraffin derivatives, and a hydrocarbon gas composed of the lower molecular weight paraffin derivatives. Either one or more of these thermal products can influence recovery operations based on thermal processes.

4.4.2.3.2 Oxidation

Oxidation of resin and asphaltene constituents is an important aspect of the contribution of such reactions to deposition. Oxidation of the resin and asphaltene constituents introduces oxygen-containing functional groups into the molecular structures of the constituents when it disturbs the resin–asphaltene relationships leading to phase separation of the oxidized products. The oxidation is exemplified by the use of chemical reagents and aerial oxygen—the latter form of oxidation is directly related to oxidation during storage.

Oxidation of asphaltene constituents with common oxidizing agents, such as acid and alkaline peroxide, acid dichromate, and alkaline permanganate, is a slow process. The occurrence of a broad band centered at $3420\,cm^{-1}$ and a band at $1710\,cm^{-1}$ in the infrared spectra of the products indicates the formation of phenolic and carboxyl groups during the oxidation. Elemental analyses of the products indicate that there are two predominant oxidation routes, notably (1) the oxidation of naphthene moieties to aromatics as well as the oxidation of *active* methylene groups to ketones, and (2) severe oxidation of naphthene and aromatic functions resulting in degradation of these systems to carboxylic acid functions.

Oxidation of asphaltene constituents in solution, by air, and in either the presence or absence of a metal compounds is also possible (Moschopedis and Speight, 1978). There is some oxygen uptake, as can be seen from the increased O–C atomic ratios, but the most obvious effect is the increase in the amount of *n*-heptane-insoluble material. And analysis of the data show that it is the higher heteroatom (more polar constituents) of the asphaltene constituents that are more susceptible to oxidation leaving the suggestion that the polarity of the constituents may be determined by the incorporation of the heteroatoms into ring systems.

Air-blowing of asphaltene constituents at various temperatures brings about significant oxygen uptake. This is accompanied by a marked decrease in the molecular weight (vapor pressure osmometry, benzene solution) of the product. This indicates that intermolecular hydrogen bonding of oxygen functionality may play a part in the observed *high* molecular weights and physical structure of petroleum (Moschopedis and Speight, 1978; see also Taft et al., 1996).

4.4.2.3.3 Other Reactions

Asphaltene constituents may also be hydrogenated to produce resin constituents and oils at elevated temperatures (>250°C, >480°F). Chemical hydrogenation under much milder conditions, for example, with lithium-ethylenediamine or sodium-liquid ammonia, also produces lower molecular weight species together with marked reductions in the sulfur and oxygen contents.

Reactions of asphaltene constituents with sulfur have also received some attention and have yielded interesting results. For example, treatment of the asphaltene constituents with oxygen or with sulfur at 150–250°C (300–480°F) yields a condensed aromatic product [H/C = 0.97; H/C (asphaltene constituents) = 1.20] containing very little additional sulfur. The predominant reaction appears to be condensation between the aromatic and aliphatic moieties of the asphaltene constituents caused by elemental sulfur, which are in turn converted to hydrogen sulfide. Condensation appears to proceed in preference to molecular degradation, and treatment of the condensed products at 200–300C (390–570F) for 1–5 hours again affords good grade cokes (H/C = 0.54–0.56). In all instances the final products contain only very low amounts of elements other than carbon and hydrogen (ΣNOS <5%, w/w), a property that could lead to deposition of the carbonaceous solid during recovery.

4.4.2.4 Solubility Parameter

The solubility parameter (Table 4.2) is a measure of the ability of the compound to act as a solvent or nonsolvent and as well as a measure the intermolecular and intramolecular forces between the solvent molecule and between the solvent and solute molecules (Barton, 1991). The solubility parameter is also a major parameter in consideration of the behavior of certain petroleum products, especially in material such as residua and asphalt (Redelius, 2000, 2004) that are chemically and physically change during a recovery process.

The most prevalent thermodynamic approach to describing asphaltene solubility has been the application of the solubility parameter or the concept of cohesive energy density. The application of solubility parameter data to correlate asphaltene precipitation, and, hence, crude oil–solvent interaction, has been used on prior occasions (Mitchell and Speight, 1973; Wiehe, 1995). The solubility parameters of asphaltene constituents can be estimated from the properties of the solvent used for separation (Long and Speight, 1989, 1998) or be measured by the titration method (Andersen and Speight, 1992) or even from the atomic hydrogen–carbon ratio (Speight, 2014a).

Table 4.2 Hansen Solubility Parameters for Selected Hydrocarbons

Compound	dD	dP	dH
Toluene	18.0	1.4	2.0
Cyclopentane	16.4	0.0	1.8
Cyclohexane	16.8	0.0	0.2
o-Xylene	17.8	1.0	3.1
trans-decalin	18.0	0.0	0.6
p-Diethylbenzene	18.0	0.0	0.6
Mesitylene	18.0	0.0	0.6
Benzene	18.4	0.0	2.0
cis-decalin	18.8	0.0	0.0
Naphthalene	19.2	2.0	5.9
Tetrahydronaphthalene	19.6	2.0	2.9
1-Methylnaphthalene	20.6	0.8	4.7
Biphenyl	21.4	1.0	2.0

Note: dD, the energy from dispersion forces between molecules; dP, the energy from dipolar intermolecular forces between molecules; dH, the energy from bonds between molecules.

The solubility parameter difference that results in a phase separation of two materials, such as asphaltene constituents in a solvent, can be estimated using the Scatchard-Hildebrand equation:

$$\ln a_a = \ln x_a + M_a / \left\{ RT\rho_a \left[\Phi_s^{\,2} (\delta_s - \delta_a)^2 \right] \right\}$$

In this equation, a_a is the activity of the solute *a*, x_a is the mole fraction solubility of *a*, M_a is the molecular weight of *a*, ρ_a is the density of *a*, Φ_s is the volume fraction of solvent, and $(\delta_s - \delta_{aa})$ is the difference between the solubility parameters of the solute *a* and the solvent *s*. Assuming that the activity of the asphaltene constituents a_a is 1 (solid asphaltene constituents in equilibrium with dissolved asphaltene constituents) and the volume fraction of an excess of solvent is essentially 1, the equation can be rearranged into a form that can be used to gain insight into the solubility of asphaltene constituents:

$$\ln x_a = -M_a / RT\Delta_a [(\delta_s - \delta_a)^2]$$

Assuming a density for asphaltene constituents of 1.28 g/cc and a molecular weight of 1000 g/mol, the solubility of asphaltene constituents as a function of the differences between solubility parameters of the asphaltene constituents and precipitating solvent can be calculated. Thus, the solubility of asphaltene constituents can be shown to decrease as the difference between solubility parameters increases, with the limit of solubility attained at a difference of about 3. Thus, if the asphaltene constituents are part of a polarity and molecular weight continuum in a crude oil, their precipitation is not as straightforward as it would be for a single species with a particular solubility parameter.

Additional information can be gleaned by calculating the solubility parameter difference at several molecular weights ranging from 50 to 10,000 g/mol at which the solubility of asphaltenic or other material is a mole fraction of 0.001 (0.1% or 1000 ppm). As a result, a phase diagram that is a function of molecular weight and solubility parameter difference can be defined. Both polarity and molecular weight of asphaltene constituents in a solvent define the solubility boundaries and explains conceptually how asphaltene constituents are precipitated from the mixture in crude oils that can be considered a type of continuum of molecular weights and polarities. Furthermore, as the molecular weight of a particular solute decreases, there is an increased tolerance of polarity difference between solute and solvent under miscible conditions.

Incompatibility phenomena can be explained by the use of the solubility parameter for asphaltene constituents and other petroleum fractions (Speight, 1992, 1994, 2014a). As an extension of this concept, there is sufficient data to draw an approximate correlation between hydrogen-to-carbon atomic ratio and the solubility parameter, δ, for hydrocarbon solvents and petroleum constituents (Speight, 1994). In general, hydrocarbon liquids can dissolve polynuclear aromatic hydrocarbons where there is, usually, less than a three-point difference between the lower solubility parameter of the solvent and the solubility parameter of the hydrocarbon.

By this means, the solubility parameter of asphaltene constituents can be estimated to be a range of values that is also in keeping with the asphaltene constituents being composed of a mixture of different compounds types with the accompanying variation in polarity. Removal of alkyl side chains from the asphaltene constituents will decrease the H–C ratio and increase the solubility parameter thereby bringing about a concurrent decrease of the asphaltene product in the hydrocarbon solvent. In fact, on the molecular weight polarity diagram for asphaltene constituents, carbenes and carboids can be shown as lower molecular weight, highly polar entities in keeping with molecular fragmentation models.

Despite numerous investigations, determination of the actual molecular structures that exist in petroleum has proved difficult. It is no doubt the great complexity of these materials, which are actually isolated by physical phenomena rather than by virtue of their chemical structures, that has hindered the formulation of individual molecular structures. Nevertheless, the various investigations have brought to light some significant facts about the structures that exist in petroleum.

In summary, disturbance of the crude oil system can result in changes to the system balance and interactions of the various constituents leading to irreversible flocculation of asphaltene constituents which can severely reduce the permeability of the reservoir, cause formation damage, and can also plug up the wellbore and tubing. This phenomenon is largely ascribed to the different extents of compressibility of the low-boiling petroleum constituents and the higher boiling constituents (resin and asphaltene constituents) of the undersaturated crude. In fact, the relative volume fraction of the lower-boiling constituents within the liquid phase increases as the pressure of the undersaturated reservoir fluid approaches its bubble point. Thus, any factor such as changes in pressure, temperature, or composition that disrupt this adsorption equilibrium and precipitation can cause asphaltene separation and deposition. Precipitation and deposition may occur during primary production, during the displacement of reservoir oil by carbon dioxide, hydrocarbon gas, or water and gas application (Khanifar et al., 2011).

4.4.3 Remediation Strategies

Deposition produces an unnecessary interruption and expense to refineries and, thus, there are large incentives to mitigate deposition in refining. The occurrence of deposition not only reduces refinery efficiency but also the effectiveness of catalysts. The majority of the deposition in recovery operations is caused by means of four general causes: (1) the presence of inorganic contaminants in the crude oil; (2) formation of separate phases—incompatibility due to changes in the oil composition; (3) the formation of carbonaceous products during thermal recovery processes; and (4) the production of insoluble asphaltene (and resin) constituents during or after recovery. Other causes, such as emulsion formation and polymerization though the aerial oxidation of the constituents also influence recovery. Focus has been mainly on deposition due to the presence of the asphaltene fraction (Andersen and Stenby, 1996).

4.4.3.1 Wax Deposition

Wax deposition can cause a multitude of operational challenges, not the least of which is restriction of flow patterns by reduction of the internal diameter of pipelines. In addition, there is increased surface roughness on the pipe wall which causes increased back-pressure and reduced throughput. Wax deposits that occur in process units and storage tanks lead to system disturbances that give rise to intensive cleanup and disposal operations. These occurrences may result in production shutdowns, hazardous conditions, and damage to equipment and it is essential that wax mitigation and remediation techniques be developed and on-hand.

Deposition mitigation/prevention practices can be either preventive or corrective. Preventive practices are proactive and involve in-operation planned steps to avoid wax deposits and growth. On the other hand, corrective practices are reactive (not in the chemical sense) and require the periodic removal of any wax deposits. Either practice can be selected before any application is applied but requires a thorough cost/benefit analysis before application and before the final selection of a wax management strategy. In many cases, a combination of these approaches may be selected in order to provide an optimized wax control strategy.

When wax deposition occurs, there must be adequate observation of the system to control the production of wax deposits and monitor the wax layer thickness. At this time, deposition mitigation techniques can be applied and involve actions such as: (1) optimum wellbore size and layout, which also refers to pipeline diameter and as few elbows as possible where deposition is likely to occur; (2) prevention of heat loss by insulating the wellbore and pipeline to prevent heat loss and to maintain flowing temperatures above the WAT; (3) use of wax inhibitors or dispersants that should be injected into the system above the WAT; (4) use of solvents to remove existing wax deposits and dispersants when it is not possible to inject above the WAT; (5) use of nonmetallic linings and coatings to reduce the frictional drag and thereby reduce the effects of shear dispersion and molecular diffusion; and (6) in the case of the pipeline, selection and use of a suitable pig design and periodic pigging of the line. The use of such nonmetallic lining and coating has an additional benefit insofar as they will also reduce the potential for wax deposition of the lines and serve to reduce corrosion caused by deposition (Speight, 2014b).

The primary remediation technique to remove wax deposits foulant that is common in general use in the crude oil and natural gas industry typically involves the use of pigging programs to remove accumulated deposits—the program must also assure that the pig will not result in the pig becoming stuck behind the wax accumulation. In addition, there are also programs which apply heat to melt the wax—accomplished by the use of by the injection of hot oil, steam, or hot water or the use of electrical heating to melt the wax deposits. When using this strategy it is necessary to note that the wax disappearance temperature is typically higher than the WAT—there is no equality of the two temperatures.

The development of control techniques to be applied to wax deposition typically require a thermodynamically modeling wax precipitation and does not take into account any effect of flow rate or other dynamic factors. Dynamic wax deposition describes the formation of gel and the amount of wax deposited on pipeline walls, taking into account effects such as shear rate and flow rate (Ahn et al., 2005). As noted, temperature (in this case, the temperature of the wellbore) is one of the most important parameters controlling wax deposition. If the temperature (which is controlled by heat transfer from the wellbore to the surroundings) is above the WAT, wax will not precipitate. Hence, it is necessary to thermodynamically describe wax precipitation (Lira-Galeana et al., 1996; Coutinho et al., 2001; Weispfennig, 2006).

4.4.3.2 Asphaltene Deposition

One of the effective remediation methods of asphaltene deposition, such as the deposition of (unreacted or reacted) asphaltene constituents (sometimes referred to as heavy organics deposition), is the use of strong aromatic/polar solvents that could dissolve the asphaltene deposits (Waxman et al., 1980; Kim et al., 1990; Acevedo et al., 1995; Dubey and Waxman, 1995; Sanchez and Mansoori, 1998). Flocculated asphaltene constituents (aggregates) can form steric colloids in the presence of excessive amounts of resin constituents (Swanson, 1942; Lichaa, 1977; Koots and Speight, 1975; Park and Mansoori, 1988) but, knowing this, there is still the need to predict when and how phase separation will occur. Deposition of phase-separated material (such as asphaltene constituents) on the wall of production strings reduces available diameter to oil flow, and, subsequently, oil production rate decreases. Gradually, the oil flow path is plugged by an increasing thickness of deposited asphaltene (Soltani Soulgani et al., 2010). The conventional way to treat asphaltene deposition was through remediation that included solvent soaks with aromatic solvent and/or aromatic solvent blended with dispersants and physical removal, such as wirelining, hydroblasting, and drilling.

Traditionally the deposition phenomenon is caused by the separation of inorganic and/or organic constituents or the reacted products of these constituents in a thermal recovery operation (Speight, 2014a). Further, with exposure to heat, the stability of the petroleum system changes and any products formed during thermal recovery processes may also appear as a phase-separated sediment which can result in flow restrictions and changes to reservoir permeability and loss of recovery efficiency.

In addition, the complexity and variation in the properties of subfractions of the resin and asphaltene fraction (Figs. 4.8 and 4.9) detracts from the concept of any form of meaningful average structure for these fractions. The complexity and variation of subfractions of the asphaltene fraction also throws doubt on models of asphaltene deposition that are derived using the bulk fraction rather than giving consideration to the differences of the constituents within this fractions and their behavior during recovery and processing.

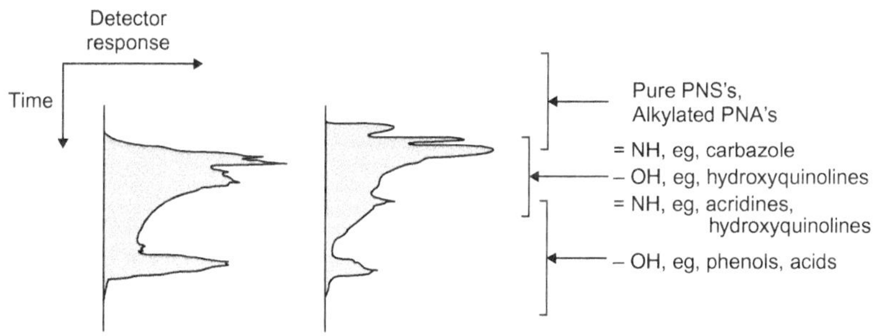

Figure 4.8 Illustration of the make-up of two different asphaltene fractions by HPLC.

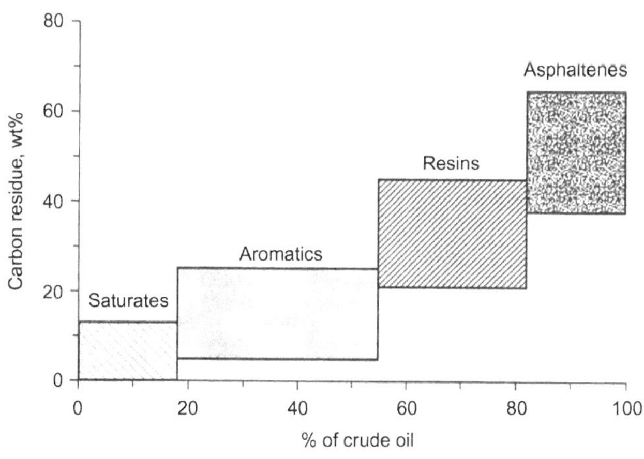

Figure 4.9 Subfractionation of the various fractions of crude oil yield subfractions with different carbon residues.

Thus, the mechanism of deposition of the asphaltene constituents is a multipath approach where each constituent (chemical-type) must be considered on the basis of individual chemistry and chemical properties. The chemistry and physics of deposition can, to some extent, be elucidated but many unknowns remain—particularly unknowns in the areas of the chemistry of deposition by the resin and asphaltene constituents (Speight, 2014a).

REFERENCES

Abebi, S.J., Seifried, S., Shaw, J.M., 1998. Unusual retrograde condensation and asphaltene precipitation in a model heavy oil system. J. Pet. Sci. Technol. 16 (3/4), 209–226.

Acevedo, S., Ranaudo, M.A., Escobar, G., Gutierrez, L., Ortega, P., 1995. Adsorption of asphaltene constituents and resins on organic and inorganic substrates and their correlation with precipitation problems in production well tubing. Fuel 74, 595–598.

Ahn, S., Wang, K.S., Shuler, P.J., Creek, J.L., Tang, Y., 2005. Paraffin crystal and deposition control by emulsification. Paper No. 93357, Proceedings. SPE International Symposium on Oilfield Chemistry, Houston, TX, February 2–4. Society for Petroleum Engineers, Richardson, TX.

Ali, M.A., Islam, M.R., 1997. The effect of asphaltene precipitation on carbonate rock permeability: an experimental and numerical approach. Paper No. SPE 38856, Proceedings. SPE Annual Technical Conference and Exhibition, San Antonio, TX, October 5–8. Society of Petroleum Engineers, Richardson, TX.

Almehaideb, R.A., 2004. Asphaltene precipitation and deposition in the near wellbore region: a modeling approach. J. Petrol. Sci. Eng. 42, 157–170.

Ancheyta, J., Centeno, G., Trejo, F., Marroquin, G., Garcia, J.A., Tenorio, E., et al., 2002. Extraction and characterization of asphaltene constituents from different crude oils and solvents. Energy Fuels 16, 1121–1127.

Ancheyta, J., Trejo, F., Rana, M.S., 2009. Asphaltenes: Chemical Transformation During Hydroprocessing of Heavy Oils. CRC Press, Taylor & Francis Group, Boca Raton, FL, (Chapter 5).

Andersen, S.I., Keul, A., Stenby, E., 1997. Variation in composition of the sub-fractions of petroleum asphaltene constituents. Petrol. Sci. Technol. 15, 611–645.

Andersen, S.I., Speight, J.G., 1992. Asphaltene precipitation and incipient flocculation in mixed solvents. Div. Fuel Chem. Am. Chem. Soc. 37 (3), 1335. (Preprints).

Andersen, S.I., Speight, J.G., 2001. Petroleum reins: separation, character, and role in petroleum. Petrol. Sci. Technol. 19, 1.

Andersen, S.I., Stenby, E.H., 1996. Thermodynamics of asphaltene precipitation and dissolution investigation of temperature and solvent effects. Fuel Sci. Technol. Int. 14 (1, 2), 231–287.

Baker, E.W., 1969. Porphyrins. In: Eglinton, G., Murphy, M.T.J. (Eds.), In Organic Geochemistry Springer-Verlag, New York.

Barton, A.F.M., 1991. CRC Handbook of Solubility Parameters and Other Cohesion Parameters, second ed. CRC Press, Boca Raton, FL.

Bestougeff, M.A., Byramjee, R.J., 1994. Chemical constitution of asphaltene constituents In: Yen, T.F. Chilingarian, G.V. (Eds.), Asphaltene Constituents and Asphalts. Developments in Petroleum Science., vol. 40 Elsevier, Amsterdam. (Chapter 3).

Bestougeff, M.A., Darmois, R., 1947. Détermination des Points de Fusion et des Poids Moléculaires des Asphaltènes. Comptes Rend 224, 1365.

Bestougeff, M.A., Darmois, R., 1948. Études des Asphaltènes. Comptes Rend 227, 129.

Bestougeff, M.A., Mouton, Y., 1977. Bull. Liason Lab. Pont. Chaus, 79. Special Volume.

Bhat, N.V., Mehrotra, A.K., 2004. Measurement and prediction of the phase behavior of wax–solvent mixtures: significance of wax disappearance temperature. Ind. Eng. Chem. Res. 43 (12), 3451–3461.
Bhat, N.V., Mehrotra, A.K., 2008. Modeling the effect of shear stress on the composition and growth of the deposit layer from waxy mixtures under laminar flow in a pipeline. Energy Fuels 22 (5), 3237–3248.
Branco, V.A.M., Mansoori, G.A., De Almeida Xavier, L.C., Park, S.J., Manafi, H., 2001. Asphaltene flocculation and collapse from petroleum fluids. J. Petrol. Sci. Eng. 32, 217–230.
Branthaver, J.F., 1990. Porphyrins. In: Speight, J.G. (Ed.), Fuel Science and Technology Handbook Marcel Dekker Inc., New York.
Brown, T.S., Niesen, V.G., Erickson, D., 1994. The effects of light ends and high pressure on paraffin formation. Paper No. 28505 Proceedings. 69th Technical Conference and Exhibition; New Orleans, Louisiana, September 25–28. Society for Petroleum Engineers, Richardson, TX.
Burger, E.D., Perkins, T.K., Striegler, J.H., 1981. Studies of wax deposition in the Trans-Alaska pipeline. J. Petrol. Technol. 31 (6), 1075–1086.
Carbognani, L., Orea, M., Fonseca, M., 1999. Complex nature of separated solid phases from crude oils. Energy Fuels 13, 351–358.
Carnahan, N.J., Salager, R., Davila, A., 1999. Properties of resins extracted from boscan crude oil and their effect on the stability of asphaltene constituents in Boscan and hamaca crude oils. Energy Fuels 13, 309.
Coutinho, J.A.P., Daridon, J.L., 2005. The limitations of the cloud point measurement techniques and the influence of the oil composition on its detection. Petrol. Sci. Technol. 23, 1113–1128.
Coutinho, J.A.P., Pauly, J., Daridon, J.L., 2001. A thermodynamic model to predict wax formation in petroleum fluids. Braz. J. Chem. Eng. 18 (4), 411–422.
Coutinho, J.A.P., Ruffier-Meray, V., 1997. Experimental measurements and thermodynamic modeling of paraffinic wax formation in undercooled solutions. Ind. Eng. Chem. Res. 36, 4977–4983.
Dandekar, A.Y., 2006. Petroleum Reservoir Rock and Fluid Properties. CRC Press, Taylor and Francis Group, Boca Raton, FL.
Dandekar, A.Y., Andersen, S.I., Stenby, E.H., 2004. Solid organic deposition during gas injection studies. Petrol. Sci. Technol. 22 (7, 8), 1209–1229.
Deans, H.A., Mut, A.D., 1997. Chemical tracer studies to determine water saturation at prudhoe bay. SPE Res. Eng. 12 (1), 52–57. Paper No. SPE 28591, Society of Petroleum Engineers, Richardson, TX.
Domanska, U., Morawski, P., 2005. Influence of size and shape effects on the high-pressure solubility of n-alkanes: experimental data, correlation and prediction. J. Chem. Thermodyn. 37, 1276–1287.
Dubey, S.T., Waxman, M.H., 1995. Asphaltene adsorption and desorption from mineral surfaces. SPE Reserve. Eng. J., 389–395.
Erickson, D.D., Niesen, V.G., Brown, T.S., 1993. Thermodynamic measurement and prediction of paraffin precipitation in crude oil. Paper No. 26604. Proceedings. 68th Annual Technical Conference and Exhibition, Houston, TX, October 3–6. Society for Petroleum Engineers, Richardson, TX.
Escobedo, J., Mansoori, G.A. 1995. Asphaltene and other heavy organic particle deposition during transfer and production operations. Paper No. 30672. Proceedings. SPE Annual Technical Conference and Exhibition, Dallas, TX, October 22–25. Society of Petroleum Engineers, Richardson, TX.
Flory, P.J., 1953. Principles of Polymer Chemistry. Cornell University Press, Ithaca, NY.

Francisco, M.A., Speight, J.G., 1984. Asphaltene characterization by a non-spectroscopic method. Div. Fuel. Chem. Am. Chem. Soc. 29 (1), 36. (Preprints).
Fuhr, B.J., Cathrea, C., Coates, L., Kalia, H., Majeed, A.I., 1991. Properties of asphaltenes in waxy crude. Fuel 70, 1293–1297.
Gallegos, E.J.J., 1981. Alkylbenzenes derived from carotenes in coals by GC/MS. Chromatogr. Sci. 19, 177–182.
Galtsev, V.E., Ametov, I.M., Grinberg, O.Y., 1995. Asphaltene association in crude oil as studied by ENDOR. Fuel 74, 670.
García, M.D., Carbognani, L., 2001. Asphaltene-Paraffin structural interactions. Effect on crude oil stability. Energy Fuels 15 (5), 1021–1027.
García, M.D., Orea, M., Carbognani, L., Urbina, A., 2001. The effect of paraffinic fractions on crude oil wax crystallization. Petrol. Sci. Technol. 19 (12), 189–196.
García, M.D., Urbina, A., 2003. Effect of crude oil composition and blending on flowing properties. Petrol. Sci. Technol. 21 (5–6), 863–878.
Gary, J.G., Handwerk, G.E., Kaiser, M.J., 2007. Petroleum Refining: Technology and Economics, fifth ed. CRC Press, Taylor & Francis Group, Boca Raton, FL.
Gawrys, K., Spiecker, L., Matthew, P., Kilpatrick, P.K., 2003. The role of asphaltene solubility and chemical composition on asphaltene aggregation. Petrol. Sci. Technol. 21 (3, 4), 461–489.
Gharfeh, S., Yen, A., Asomaning, S., Blumer, D., 2004. Asphaltene flocculation onset determinations for heavy crude oil and its implications. Petrol. Sci. Technol. 22 (7, 8), 1055–1075.
Hammami, A., Freworn, K.A., 1998. Asphaltic crude oil characterization: an experimental investigation of the effect of resins on the stability of asphaltene constituents. Petrol. Sci. Technol. 16 (3, 4), 227–249.
Hammami, A., Phelps, H., Little, T.M., 2000. Asphaltene precipitation from live oils: an experimental investigation of the onset conditions and reversibility. Energy Fuels 14, 14–18.
Hammami, A., Raines, M.A., 1997. Paraffin deposition from crude oils: comparison of laboratory results to field data. Paper No. SPE-38776. Proceedings. SPE Annual Technical Conference and Exhibition, San Antonio, TX, USA, October 5–8.
Hammami, A., Raines, M.A., 1999. Paraffin deposition from crude oils: comparison of laboratory results with field data. SPE J. 4 (1), 9–18.
Hammami, A., Ratulowski, J., Coutinho, J.A.P., 2003. Cloud points: can we measure or model them? Petrol. Sci. Technol. 21 (3–4), 345–358.
Hardy, D.R., Wechter, M.A., 1990. Insoluble sediment formation in middle-distillate diesel fuel: the role of soluble macromolecular oxidatively reactive species. Energy Fuels 4, 270–274.
Hirschberg, A., deJong, L., Schipper, B., Maijer, J., 1984. Influence of temperature and pressure on asphaltene flocculation. SPE J. 24, 283–293.
Honarpour, M.M., Nagarajan, N.R., Sampath, K., 2006. Rock/fluid characterization and their integration—implication on reservoir management. J. Pet. Tech. 58 (9), 120.
Hong, E., Watkinson, P., 2004. A study of asphaltene solubility and precipitation. Fuel 83, 1881–1887.
Hsu, C.S., Robinson, P.R., 2006. Practical Advances in Petroleum Processing, vols. 1 and 2. Springer, New York.
Huc, A.Y., Behar, F., Roussel, J.C., 1984. Geochemical variety of asphaltene constituents from crude oils Proceedings. Symposium on Characterization of Heavy Crude Oils and Petroleum Residues. Editions Technip, Paris.99–103.
Jacobson, J.M., Gray, M.R., 1987. The use of infrared spectroscopy and nitrogen titration data in the structural group analysis of bitumen. Fuel 66, 749–752.

Jessen, F.W., Howell, J.N., 1958. Effect of flow rate on paraffin accumulation in plastic, steel and coated pipes. J. Petrol. Technol. Trans. AIME 213, 80–83.
Karan, K., Hammami, A., Flannery, M., Stankiewicz, B.A., 2003. Evaluation of asphaltene instability and a chemical control during production of live oils. Petrol. Sci. Technol. 21 (3, 4), 629–645.
Karan, K., Ratulowski, J., German, P., 2000. Measurement of waxy crude properties using novel laboratory techniques. Paper No. 62945. Proceedings. Annual Technical Conference and Exhibition, Dallas, TX, October 1–4. Society for Petroleum Engineers, Richardson, TX.
Khanifar, A., Demiral, B., Darman, N., 2011. Modeling of asphaltene precipitation and deposition during WAG application. Proceedings. International Petroleum Technology Conference, Bangkok, Thailand, November 15–17. International Petroleum Technology Conference (IPTC), Richardson, TX.
Kim, S.T., Boudh-Hir, M.E., Mansoori, G.A., 1990. The role of asphaltene in wettability reversal. Paper No. 20700. Proceedings. Annual Meeting SPE. Society of Petroleum Engineers, Richardson, TX.
Kokal, S.L., Najman, J., Sayegh, S.G., George, A.E., 1992. Measurement and correlation of asphaltene precipitation from heavy oils by gas injection. J. Can. Petrol. Technol. 31, 24.
Koots, J.A., Speight, J.G., 1975. The relation of petroleum resins to asphaltene constituents. Fuel 54, 179.
Kord, S., Ayatollahi, S., 2012. Asphaltene precipitation in live crude oil during natural depletion: experimental investigation and modeling. Fluid Phase Equilibria 336, 63–70.
Leontaritis, K.J., Leontaritis, J.D., 2003. Cloud point and wax deposition measurement techniques. Paper No. 80267. Proceedings. SPE International Symposium on Oil Field Chemistry held in Houston, TX, February 5–7. Society for Petroleum Engineers, Richardson, TX.
Leontaritis, K.J., Mansoori, G.A., 1988. Asphaltene deposition: a survey of field experiences and research approaches. J. Pet. Sci. Eng. 1, 229.
Lichaa, P.M., 1977. Asphaltene deposition problems in venezuelan crudes—usage of asphaltene constituents in emulsion stability. Can. Pet. Technol. J. Oil Sands June, 609–624.
Lira-Galeana, C., Firoozabadi, A., Prausnitz, J.M., 1996. Thermodynamics of wax precipitation in petroleum mixtures. AIChE J. 42 (1), 239–248.
Long, R.B., 1979. The concept of asphaltene constituents. Div. Petrol. Chem. Am. Chem. Soc. 24 (4), 891. (Preprints).
Long, R.B., 1981. The concept of asphaltene constituents. In: Bunger, J.W., Li, N. (Eds.), The Chemistry of Asphaltene Constituents American Chemical Society, Washington, DC. Advances in Chemistry Series No. 195.
Long, R.B., Speight, J.G., 1989. Studies in petroleum composition. I: Development of a compositional map for various feedstocks. Révue Institut Français du Petrole 44, 205.
Long, R.B., Speight, J.G., 1998. The composition of petroleum. In: Speight, J.G. (Ed.), Petroleum Chemistry and Refining Taylor & Francis, Washington, DC. (Chapter 2).
Luo, K., Li, S., Zheng, X., Liu, H., Zhong, T., Zhu, Y., 2001. Phase behavior of highly waxy gas condensate systems. Paper No. 68228. Proceedings. SPE Middle East Oil Show, Bahrain, March 17–20. Society for Petroleum Engineers, Richardson, TX.
Mansoori, G.A., 1996. Asphaltene, resin, and wax deposition from petroleum fluids. Arabian J. Sci. Eng. 21 (48), 707–723.
Minssieux, L., 1997. Core damage from crude asphaltene deposition. Paper SPE 37250 presented at the SPE International Symposium on Oilfield Chemistry, Houston, TX, February 18–21.
Mitchell, D.L., Speight, J.G., 1973. The solubility of asphaltenes in hydrocarbon solvents. Fuel 52, 149.

Mitra-Kirtley, S., Mullins, O.C., van Elp, J., George, S.J., Chen, J., Cramer, S.P., 1993. Determination of the nitrogen chemical structures in petroleum asphaltene constituents using XANES spectroscopy. J. Am. Chem. Soc. 115, 252–258.
Moschopedis, S.E., Fryer, J.F., Speight, J.G., 1976. An investigation of asphaltene molecular weights. Fuel 55, 227.
Moschopedis, S.E., Speight, J.G., 1976a. An investigation of hydrogen bonding by oxygen functions in Athabasca bitumen. Fuel 55, 187.
Moschopedis, S.E., Speight, J.G., 1976b. Oxygen functions in asphaltene constituents. Fuel 55, 334.
Moschopedis, S.E., Speight, J.G., 1976c. An investigation of asphalt preparation from Athabasca bitumen. Proc. Assoc. Asphalt Paving Technol. 45, 78.
Moschopedis, S.E., Speight, J.G., 1978. The influence of metal salts on bitumen oxidation. Fuel 57, 235.
Moschopedis, S.E., Speight, J.G., 1979. Investigation of nitrogen types in Athabasca bitumen. Div. Petrol. Chem. Am. Chem. Soc. 24 (4), 1007. (Preprints).
Mousavi-Dehghania, S.A., Riazi, M.R., Vafaie-Sefti, M., Mansoori, G.A., 2004. An analysis of methods for determination of onsets of asphaltene phase separations. J. Petrol. Sci. Eng. 42 (2004), 145–156.
Mushrush, G.W., Speight, J.G., 1995. Petroleum Products: Instability and Incompatibility. Taylor & Francis Publishers, Washington, DC.
Nagarajan, N.R., Honarpour, M.M., Sampath, K., 2007a. Reservoir-fluid sampling and characterization—key to efficient reservoir management. J. Petrol. Technol. 59, 80–91.
Nagarajan, N.R., Honarpour, M.M., Sampath, K., 2007b. Reservoir-fluid sampling and characterization—key to efficient reservoir management. Paper No. SPE 103501 and 101517. Proceedings. Abu Dhabi International Petroleum Exhibition & Conference, Abu Dhabi, November 5–8. Society of Petroleum Engineers, Richardson, TX.
Nghiem, L.X., Hassam, M.S., Nutakki, R., George, A.E.D., 1993. Efficient modeling of asphaltene precipitation. Paper No. SPE 26642. Annual Technical Conference and Exhibition, Houston, TX, October 3–6. Society of Petroleum Engineers, Richardson, TX.
Otzisk, B., Kempen, H., 2009. Stabilization of asphaltenes. Petrol. Technol. Quart. Q1, 31–36.
Pan, H., Firoozabadi, A., Fotland, P., 1996. Pressure and composition effect on wax precipitation: experimental data and model results. Paper No. 36740. Proceedings. Technical Conference and Exhibition, Denver, October 6–9. Society for Petroleum Engineers, Richardson, TX.
Pan, H., Firoozabadi, A., Fotland, P., 1997. Pressure and composition effect on wax precipitation: experimental data and model results. SPE J. 12 (4), 250–258.
Park, S.J., Mansoori, G.A., 1988. Aggregation and deposition of heavy organics in petroleum crudes. Int. J. Energy Sources 10, 109–125.
Pedersen, K.S., Christensen, P.L., 2006. Phase Behavior of Petroleum Reservoir Fluids. CRC Press, Taylor & Francis Group, Boca Raton, FL.
Pedersen, K.S., Rønningsen, H.P., 2003. Influence of wax inhibitors on wax appearance temperature, pour point, and viscosity of waxy crude oils. Energy Fuels 17 (2), 321–328.
Por, N., 1992. Stability Properties of Petroleum Products. Israel Institute of Petroleum and Energy, Tel Aviv.
Power, A.J., Mathys, G.I., 1992. Characterization of distillate fuel sediment molecules: functional group derivatization. Fuel 71, 903–908.
Ramirez-Jaramillo, E., Lira-Galeana, C., Manero, O., 2006. Modeling asphaltene deposition in production pipelines. Energy Fuels 20 (3), 1184–1196.
Redelius, P.G., 2000. Solubility parameters and bitumen. Fuel 79, 27–35.
Redelius, P.G., 2004. Bitumen solubility model using hansen solubility parameter. Energy Fuels 18, 1087–1092.

Reerink, H., 1973. Size and shape of asphaltene particles in relationship to high temperature viscosity. Ind. Eng. Chem. Prod. Res. Dev. 12, 82–88.
Reynolds, J.G., 1998. Metals and heteroatoms in heavy crude oils Petroleum Chemistry and Refining. Taylor & Francis Publishers, Washington, DC63–102, (Chapter 3).
Ritchie, R.G.S., Roche, R.S., Steedman, W., 1979. Phenols from asphaltene constituents. Fuel 58, 523.
Rose, K.D., Francisco, M.A., 1987. Characterization of acidic heteroatoms in heavy petroleum fractions by phase-transfer methylation and NMR spectroscopy. Energy Fuels 1, 233.
Rose, K.D., Francisco, M.A., 1988. A two-step chemistry for highlighting heteroatom species in petroleum material using ^{13}C NMR spectroscopy. J. Am. Chem. Soc. 110, 637.
Sadeghazad, A., Christiansen, R.L., 1998. The effect of cloud point temperature on wax deposition. Paper No. 49467. Proceedings. 8th Abu Dhabi International Petroleum Exhibition and Conference, Abu Dhabi, United Arab Emirates, October 11–14.
Sanchez, J.H.P., Mansoori, G.A., 1998. Prediction of the phase behavior of asphaltene micelle/aromatic hydrocarbon systems. J. Pet. Sci. Technol. 16 (3/4), 377–394.
Schmitter, J.M., Garrigues, P., Ignatiadis, I., De Vazelhes, R., Perin, F., Ewald, M., et al., 1984. Occurrence of tetra-aromatic Aza-Arenes in petroleum. Org. Geochem. 6, 579–586.
Schucker, R.C., Keweshan, C.F., 1980. Reactivity of cold lake asphaltene constituents. Div. Fuel Chem. Am. Chem. Soc. 25, 155. (Preprints).
Singh, A., Benavides, M.S., Marcoux, J.F.S., Barrufet, M.A., 2004. Dilution strategies for wax management and control for deep water development from a flow assurance perspective: part 1—current practice and perspective. Paper No. 90688. Proceedings. Annual Technical Conference and Exhibition, Houston, TX, September 26–29. Society for Petroleum Engineers, Richardson, TX.
Singh, P., Venkatesan, R., Fogler, H.S., Nagarajan, N., 2000. Formation and aging of incipient thin film wax-oil gels. AIChE J. 46 (5), 1059–1074.
Singh, P., Venkatesan, R., Fogler, H.S., Nagarajan, N.R., 2001. Morphological evolution of thick wax deposits during aging. AIChE J. 47 (1), 6–18.
Soltani Soulgani, B., Rashtchian, D., Tohidi, B., Jamialahmadi, M., 2010. A novel method for mitigation of asphaltene deposition in the wellstring. Iran. J. Chem. Chem. Eng. 29 (2), 131–142.
Soltani Soulgani, B., Tohidi, B., Jamialahmadi, M., Rashtchian, D., 2011. Modeling formation damage due to asphaltene deposition in the porous media. Energy Fuels 25, 753–761.
Speight, J.G., 1970a. A structural investigation of the constituents of Athabasca bitumen by proton magnetic resonance spectroscopy. Fuel 49, 76.
Speight, J.G., 1970b. Thermal cracking of Athabasca bitumen, athabasca asphaltene constituents, and athabasca deasphalted heavy oil. Fuel 49, 134.
Speight, J.G., 1971. Structural analysis of athabasca asphaltene constituents by proton magnetic resonance spectroscopy. Fuel 50, 102.
Speight, J.G., 1972. The application of spectroscopic techniques to the structural analysis of coal and petroleum. Appl. Spectrosc. Rev. 5, 211.
Speight, J.G., 1979. Studies on bitumen fractionation: (a) Fractionation by a cryoscopic method, (b) effect of solvent type on asphaltene solubility. Information Series No. 84, Alberta Research Council, Edmonton, Alberta, Canada.
Speight, J.G., 1984. The chemical nature of petroleum asphaltene constituents. In: Tissot, B. (Ed.), Characterization of Heavy Crude Oils and Petroleum Residues Editions Technip, Paris.
Speight, J.G., 1986. Polynuclear aromatic systems in petroleum. Div. Petrol. Chem. Am. Chem. Soc. 31 (4), 818. (Preprints).
Speight, J.G., 1992. A chemical and physical explanation of incompatibility during refining operations. Proceedings. 4th International Conference on the Stability and Handling of Liquid Fuels. U.S. Department of Energy (DOE/CONF-911102), p. 169.

Speight, J.G., 1994. Chemical and physical studies of petroleum asphaltenes In: Yen, T.F. Chilingar, G.V. (Eds.), Asphaltenes and Asphalts, I. Developments in Petroleum Science, vol. 40 Elsevier, Amsterdam. (Chapter 2).

Speight, J.G., 2001. Handbook of Petroleum Analysis. John Wiley & Sons Inc., New York.

Speight, J.G., 2014a. The Chemistry and Technology of Petroleum, fifth ed. CRC Press, Taylor & Francis Group, Boca Raton, FL.

Speight, J.G., 2014b. Oil and Gas Corrosion Prevention. Gulf Professional Publishing, Elsevier, Oxford.

Speight, J.G., 2015. Handbook of Petroleum Product Analysis, second ed. John Wiley & Sons Inc., Hoboken, NJ.

Speight, J.G., Moschopedis, S.E., 1977. Asphaltene molecular weights by a cryoscopic method. Fuel 56, 344.

Speight, J.G., Moschopedis, S.E., 1981. Anomalous effects in the character of the carbonyl absorption in the infrared spectra of petroleum resins and asphaltene constituents. Div. Petrol. Chem. Am. Chem. Soc. 26 (4), 907. (Preprints).

Speight, J.G., Ozum, B., 2002. Petroleum Refining Processes. Marcel Dekker, Inc., New York.

Speight, J.G., Pancirov, R.J., 1984. Structural types in asphaltene constituents as deduced from pyrolysis-gas chromatography-mass spectrometry. Liquid Fuels Technol. 2, 287.

Speight, J.G., Wernick, D.L., Gould, K.A., Overfield, R.E., Rao, B.M.L., Savage, D.W., 1985. Molecular weights and association of asphaltenes: a critical review. Révue de l'Institut Français du Pétrole 40, 27.

Stavinoha, L.L., Henry, C.P. (Eds.), 1981. Distillate Fuel Stability and Cleanliness. Special Technical Publication No. 751 American Society for Testing and Materials, Philadelphia, PA.

Swanson, J., 1942. A contribution to the physical chemistry of the asphalts. J. Phys. Chem. 46, 141–150.

Taft, R.W., Berthelot, M., Laurence, C., Leo, A.J., 1996. Hydrogen bonds and molecular structure. Chemtech July, 20.

Tiwary, R., Mehrotra, A.K., 2009. Deposition from wax–solvent mixtures under turbulent flow: effects of shear rate and time on deposit properties. Energy Fuels 23 (3), 1299–1310.

Tronov, J.P., 1969. Effect of flow rate and other factors on the formation of paraffin deposits. Tr. Tatar. Neft. Nauch-Issled Inst. 13, 207.

Udourioh, G.A., Ibezim-Ezeani, M.I., Ofodile, S.E., 2014. Comparative investigation of heavy organics precipitation from crude oil using binary mixtures and single n-alkane organic solvents. J. Petrol. Gas Explor. Res. 4 (4), 53–59.

Venkatesan, R., Creek, J.L., 2007. Wax deposition during production operations: SOTA. Paper No. OTC 18798. Proceedings. Offshore Technology Conference, Houston, TX, April 30–May 3.

Wang, S., Civan, F., 2001. Productivity decline of vertical and horizontal wells by asphaltene deposition in petroleum reservoirs. Paper No SPE 64991. Proceedings. International Symposium on Oilfield Chemistry, Houston, TX, February 13–16. Society of Petroleum Engineers, Richardson, TX.

Waxman, M.H., Deeds, C.T., Closmann, P.J., 1980. Thermal alterations of asphaltene constituents in Peace River Tars. Paper No. 9510. Society of Petroleum Engineers, Richardson, TX.

Weingarten, J.S., Euchner, J.A., 1988. Method for predicting wax precipitation and deposition. Paper No. 15654. Proceedings. Annual Technical Conference and Exhibition, New Orleans, Louisiana, February. Society for Petroleum Engineers, Richardson, TX.

Weispfennig, K., 2006. Recent advances in wax deposition modeling. Proceedings. Advances in Flow Assurance Technology Conference, Milan, Italy, February 23.

Wiehe, I.A., 1993. A phase-separation kinetic model for coke formation. Ind. Eng. Chem. Res. 32, 2447–2454.
Wiehe, I.A., 1994. The pendant-core building block model of petroleum residua. Energy Fuels 8, 536–544.
Wiehe, I.A., 1995. Polygon mapping with two-dimensional solubility parameters. Ind. Eng. Chem. Res. 34, 661–673.
Yarranton, H.W., 2005. Asphaltene self-association. J. Dispersion Sci. Technol. 26, 5–8.
Yen, T.F., 1974. The structure of asphaltene and its significance. Energy Sources 1, 447.
Yen, T.F., 1975. The Role of Trace Metals in Petroleum. Ann Arbor Science Publishers Inc., Ann Arbor, MI.
Yen, T.F., 1994. Multiple structural orders of asphaltene constituents. In: Yen, T.F., Chilingarian, G.V. (Eds.), Asphaltene Constituents and Asphalt, 1. Developments in Petroleum Science, 40 Elsevier, Amsterdam.
Zhu, T., Walker, J.A., Liang, J., 2008. Evaluation of wax deposition and its control during production of Alaska North Slope oils. Final Report. DOE Award No.: DE-FC26-01NT41248. Office of Fossil Energy, United States Department of Energy, Washington, DC.

CHAPTER 5

Evaluation of Reservoir Fluids

5.1 INTRODUCTION

Heavy oil and tar sand bitumen are finding wide use in complementing crude oil as feedstocks for liquid fuels, as well as fuels for industry and heating and source materials for a variety of other products, plus the ubiquitous plastics. From a chemical standpoint, heavy oil and bitumen, are extremely complex mixtures of hydrocarbon compounds, with substantial amounts of nitrogen-containing, oxygen-containing, and sulfur-containing constituents, as well as trace amounts of metal-containing constituents. As a result, a variety of production processes are being used and developed to recover heavy oil and bitumen. However, technologies and services used for the recovery of conventional crude oil face limitations with heavy oil and tar sand bitumen. Mathematical and engineering models developed for application to the predictability of conventional oil recovery are typically not applicable to the recovery of heavy oil and bitumen and such models require more than the often quoted *tweaking*. For example, the problems associated with paraffin deposition during conventional oil recovery are replaced by problems related to asphaltene deposition during heavy oil and bitumen recovery—the deposition of paraffin constituents and asphaltene constituents require different mathematical treatment according to the conditions during which deposition occurs. In addition, the high molecular weight fraction of heavy oil and bitumen are comprised of compounds (not only the asphaltene constituents) with high melting points and high pour points that greatly contribute to the fluid properties of these viscous feedstocks and hence to reduced mobility, compared to conventional petroleum, under a variety of conditions. It is typically this poor mobility, as opposed to the disposition of asphaltene constituents in the pore throats of the reservoir rock or in production lines that are usually the cause of production problems.

Fluid typing in terms of heavy oil and bitumen composition, specific gravity, viscosity, solution gas, asphaltene constituents, and mineral content are important in planning the production method. This requires reliable data from laboratory test methods derived from *representative* fluid samples

Introduction to Enhanced Recovery Methods for Heavy Oil and Tar Sands.
DOI: http://dx.doi.org/10.1016/B978-0-12-849906-1.00005-9

at representative (of the reservoir parameters) temperatures and pressures are needed, possibly also in the presence of solvents, or during the changes initiated by combustion processes.

From such investigations, flow assurance (which is needed to guarantee the transport of heavy oil possibly mixed with sand, water, natural gas, and/or solvents, in pipelines or horizontal wells) must be predictable and guaranteed. As an example of adverse effects that affect flow assurance: (1) suspended sand may deposit from the fluid and block flow in pipelines or long horizontal wells; (2) the deposition of asphaltene constituents in the well tubing or pipeline is also possible; and (3) lower temperatures that are prevalent in the overburden, in arctic regions, subsea, and northern pipelines may require heating elements, the addition of solvents, insulated tubing, and other flow assurance methods. There is also the need for monitoring the composition of the heavy oil or bitumen as a means of assessing the potential for in situ upgrading (Chapter 9). One approach is to test the produced fluids or to use a method for monitoring the fluids in situ, either with cased-hole logging, downhole fluid sampling, or permanent sensors.

Furthermore, it is essential to recognize that properties such as viscosity (sometimes referred to as the *heaviness*) of heavy oil and tar sand bitumen are primarily the result of an internal balance between a relatively high proportion of complex, high molecular weight, nonparaffinic compounds and a low proportion of volatile, lower molecular weight compounds. The problems of producing heavy oil from the reservoir or bitumen from the deposit are typically a result of disturbing the internal balance which, in turn, influences the mobility of the material, in addition to the deposition of the asphaltene constituents. Success with the recovery of heavy oil and bitumen depends as much on understanding the fluid properties of these materials as it does on knowing the geology of the reservoir itself. The reason is that the chemical differences between heavy oil (and tar sand bitumen) and conventional oil ultimately affect their viscosity and, hence, their mobility and recovery.

New developments for the recovery of heavy oil and tar sand bitumen require some form of enhanced oil recovery (EOR), which generally means steam, solvents, or a combination of both. Thermal recovery options in some reservoirs include the use of cyclic steam injection (also known as the *huff "n" puff* process), downhole heaters, or the recently developed steam assisted gravity drainage (SAGD) process. The use of steam influences the mobility of the heavy oil or bitumen which influences recovery rates.

Strategies for recovering heavy oil and tar sand bitumen (excluding the use of mining for bitumen recovery) emphasize the difference in properties that influences the choice of methods or combinations thereof for conversion to various products. Furthermore, the availability of processes that can be employed to produce heavy oil from the reservoir has increased significantly in recent years (Speight, 2014a). Thus, in order to determine the potential recoverability of heavy oil or bitumen a series of consistent and standardized characterization procedures are required. These procedures can be used with a wide variety of feedstocks to develop a general approach to predict behavior during recovery (Speight, 2014a).

Recovering heavy oil and bitumen can be designed in an optimal manner by performing diligent evaluation of the quality of these materials using standard test methods that lead to an understanding of their chemical and physical properties (Speight, 2015). The evaluation schemes do not need to be complex, but must focus on key parameters that affect recovery and can be determined by a number of characteristics that affect the recovery of the heavy oil or bitumen. The two most common measurements of quality are the specific gravity (which is also presented in degrees API) and the sulfur content. Acid content is also a factor in determining the corrosive properties of the heavy oil or bitumen and the effects on recovery equipment as well as refinery units (Speight, 2014b, 2014c).

For example, the identification of the important features can be made with a saturates–aromatics–resins–asphaltene separation (Fig. 5.1), with the focus on the potential for asphaltene deposition. Subsequent analysis of the fractions also provides further information about the intramolecular relationships of the oil, again with the focus on the potential for asphaltene deposition and the potential for any thermal changes that might affect the stability of the heavy oil or bitumen and cause a dramatic effect on solids deposition during recovery. In addition, useful information can be gained from knowledge of elemental composition and molecular weight, and size exclusion chromatography and high performance liquid chromatography (HPLC) profiles.

Finally, understanding the properties and behavior of heavy oil and tar sand bitumen is important for any preliminary work related to reservoir/deposit simulation and production response predictions. In heavy oil reservoirs and tar sand deposits, the fluid viscosity and density are commonly reported, but there is little experimental data in the literature reporting how oil properties change with pressure. This information would be especially useful for production companies seeking to understand and

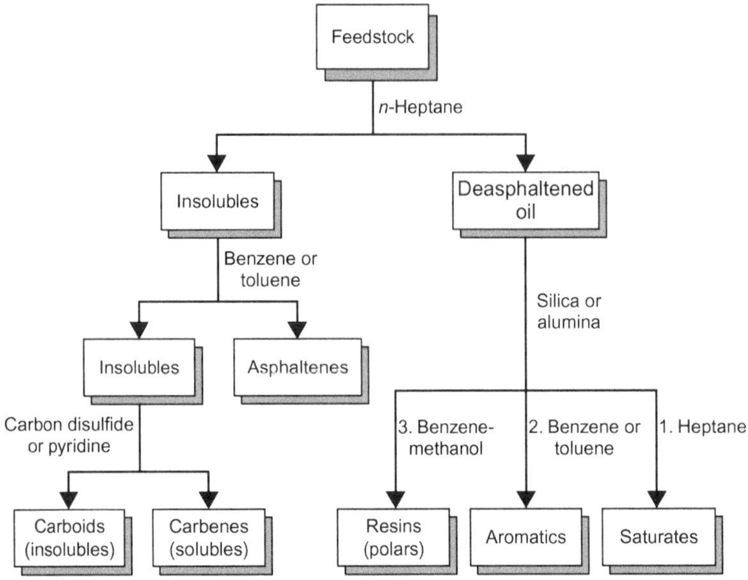

Figure 5.1 Schematic of the separation of heavy oil into various bulk fractions—the carbenes and carboids are generally considered to be the organic product of thermal treatment and are not typically found in untreated crude oil, heavy oil, or bitumen.

improve their primary (cold production) response (Goodarzi et al., 2005). Furthermore since there are many effects that can influence the recovery process—such as the presence of foamy oil—it is valuable, if not essential, to investigate the bulk fluid properties of heavy oil and bitumen prior to initiation of any (laboratory or field) process. If the behavior of the material, such as the response of various properties to incremental pressure reduction, can be established, this can be compared with fluid expansion during the process, especially during pressure depletion. This applies specifically to the determination of density, formation value factor, and solution gas–oil ratio as a function of pressure. The incremental benefit of the these measurement is that they provide more detailed information about that heavy oil or bitumen, compared to conventional pressure–volume–temperature (PVT) measurements. In particular, viscosity is one of the most important pieces of information that has to be obtained. The high viscosity of heavy oil and bitumen presents a significant obstacle to the technical and economic success of an EOR option. As a result, in situ oil viscosity measurement techniques (as an example of several other property options) and the ability to predict such properties through meaningful modeling programs would be of considerable benefit to defining the recovery process (Gabitto and Barrufet, 2003; Goodarzi et al., 2005).

5.2 EVALUATION OF RESERVOIR FLUIDS AND FLUID ASSAY

The term reservoir fluids covers a considerable range of fluid properties from conventional (light) oil, heavy oil, extra heavy oil, and tar sand bitumen. In the context of this book, connate water (Chapter 4) must also be acknowledged as a reservoir fluid but the analysis of the water is not dealt with here. Similarly, natural gas, which is also recognized generally as a reservoir fluid is omitted from this text being the subject of several other major texts (Mokhatab et al., 2006; Speight, 2007, 2014a).

5.2.1 Fluid Evaluation

Turning to the organic fluids, which are the subject of this book, when the low-boiling constituents are absent, the nonvolatile material (heavy oil, extra heavy oil, and tar sand bitumen) becomes viscous with a high proportion of asphaltic molecules, and with substitution in the carbon network of heteroatoms, such as nitrogen, sulfur, and oxygen. Therefore, heavy oil, extra heavy oil, and tar sand bitumen (regardless of source) always contain the heavy asphaltic fractions, which consist of resin constituents and asphaltene constituents (Speight, 2014a, 2014b, 2014c and references cited therein). No known heavy oil or bitumen fails to incorporate asphaltenes (Speight, 2014a, 2014b, 2014c). The resin constituents and the asphaltene constituents play an important role in the accumulation, recovery, processing, and utilization of petroleum.

Thus, the recovery of oil from a reservoir requires an understanding of the behavior of oil during displacement from the rock and flow through the reservoir to the wellbore. Oil–rock interactions and flow through the reservoir rock (often referred to in the laboratory as a porous medium) is complicated, but important for many processes, but in the current context it is essential to consider fluid properties that are relevant to petroleum reservoirs and oil recovery. Inside a reservoir there can be displacements and miscible and/or immiscible flow, with one, two, or sometimes three mobile phases (oil, gas, and water). An understanding of the chemistry and physics of displacement is important for the correct interpretation of laboratory core data, for the assessment of quantities and position of residual oil and for reservoir simulation.

Generally, oil recovery involving miscible displacement (eg, solvent displacing oil in front of an immiscible displacement front or water displacing water behind it) occurs when oil flows to the production wellbore from deep within the reservoir displacing the oil near the wellbore, or water moving within the aquifer. For miscible displacement then the streamlines generally follow the permeability changes. If the viscosity of the initial and displacing fluids are different then viscous effects are imposed on top of the heterogeneity effects (Dawe, 2004).

Oil can also move to the wellbore by immiscible displacement when the oil is displaced by water or gas. There can also be mass transport across phase boundaries (eg, gas from liquid to vapor in solution gas processes, liquid dropout from gas in condensate reservoirs, multicontact processes in miscible gas EOR processes, and transfer of chemicals, such as in surfactant EOR). In addition, there can be temperature changes with corresponding physical changes, such as in thermal EOR and cooling during long-term waterflooding of a reservoir with cold water. If immiscible displacements occur, interpretation becomes even more problematic, as not only are there permeability and viscous effects but also effects due to capillary forces, wettability, and saturation which can become dominant, due in no small part to the presence of oil–rock interaction caused by such polar constituents as the asphaltene constituents (Buckley et al., 1997; Speight, 2014a, 2014b, 2014c). The saturation effects are particularly complex since they are affected by pore scale wettability, which modifies the relative permeability and capillary pressures, which themselves modify the displacement patterns (Dawe, 2004).

Reservoir-fluid PVT properties are critical for efficient reservoir management throughout the life of the reservoir, from discovery to abandonment (Honarpour et al., 2006; Nagarajan et al., 2007a, 2007b). In fact, reliable data related to the properties of in situ fluids are essential for the determination of in-place volumes and recovery-factor calculations as well as for technical evaluation of reservoir–development–depletion plans. Fluid characterization and distribution within the reservoir help in defining reservoir continuity and communication among various zones.

In fact, reservoir-fluid properties play a key role in the design and optimization of injection/production strategies and surface facilities for efficient reservoir management (Nagarajan et al., 2007a, 2007b). Inaccurate fluid characterization often leads to significant errors in estimation of the in-place volume of oil as well as predictions of the recovery efficiency. Reservoir-fluid PVT characterization begins with acquisition of fluid samples which must be representative of the fluid in the reservoir followed by measurement of the PVT parameters with strict protocols for quality assurance and quality control (QA and QC) as well as modeling of the phase behavior through reliable methods.

Reservoir-fluid characterization consists of several key steps: (1) acquisition of representative samples; (2) identification of reliable service laboratories to perform PVT measurements; (3) implementation of QA/QC procedures to ensure data quality; and (4) development of mathematical models to capture fluid-property changes accurately as functions of pressure, temperature, and composition. The fluid type and production processes dictate the type and the volume of required fluid data.

Heavy oil, which is a subcategory of petroleum, and bitumen (the primary organic component of tar sand) are often defined (loosely and incorrectly) in terms of API gravity (Speight, 2014a, 2014b, 2014c). However, by legal definition, a more appropriate definition of bitumen, which sets it aside from heavy oil and conventional petroleum, is based on the definition offered by the US government as the "extremely viscous hydrocarbon which is not recoverable in its natural state by conventional oil well production methods including currently used enhanced recovery techniques" (Chapter 1). As already noted, by inference, conventional petroleum and heavy oil (recoverable by *conventional oil well production methods including currently used enhanced recovery techniques*) are different to tar sand bitumen. Be that as it may, at some stage of production, conventional petroleum (in the later stages of recovery) and heavy oil (in the earlier stages of recovery) may require the application of enhanced oil recovery methods for recovery (Speight, 2014a, 2014b, 2014c).

Thus, even though heavy oil is a subcategory of petroleum, and is not recognized as being petroleum-related, and bitumen must both be treated with caution when handled for analysis. In the first instance, sampling must be beyond reproach insofar as the method takes a representative sample of the more viscous (than crude oil) heavy oil and the near solid bitumen.

5.2.2 Fluid Assay

An assay, as the term is used in the petroleum industry, is the application of a series of tests to petroleum, heavy oil, extra heavy oil, or tar sand bitumen that present data upon which the character and behavior of the feedstock during recovery (and refining) operations can be estimated (Table 5.1). Since crude oil, heavy oil, and bitumen are complex mixtures of hydrocarbons and hydrocarbonaceous constituents, that include

Table 5.1 Recommended Inspection Data Required for Petroleum and Heavy Feedstocks, Which Include Heavy Oil, Extra Heavy Oil, and Tar Sand Bitumen

Petroleum	Heavy Feedstocks
Density, specific gravity	Density, specific gravity
API gravity	API gravity
Carbon, wt%	Carbon, wt%
Hydrogen, wt%	Hydrogen, wt%
Nitrogen, wt%	Nitrogen, wt%
Sulfur, wt%	Sulfur, wt%
	Nickel, ppm
	Vanadium, ppm
	Iron, ppm
Pour point	Pour point
Wax content	
Wax appearance temperature	
Viscosity (various temperatures)	Viscosity (various temperatures)
Carbon residue of residuum	Carbon residue
	Ash, wt%
Distillation profile:	Fractional composition:
All fractions plus vacuum residue	Asphaltenes, wt%
	Resins, wt%
	Aromatics, wt%
	Saturates, wt%

compounds of sulfur, nitrogen, oxygen, metals, and other elements (ASTM D4175), it is not surprising that crude oil, heavy oil, and bitumen vary in composition properties and produce wide variations in recovery behavior. Thus, recovery operations require analytical test methods that can adequately evaluate the material and monitor any changes in composition and behavior (Speight, 2015). Furthermore, upgrading heavy oil and bitumen during recovery (Chapter 9) involves processing (usually partial conversion) into a higher value product. But before emerging into the detailed aspects of petroleum product analysis, it is necessary to understand the nature and character of crude oil, heavy oil, and bitumen as well as the methods used to recover these materials from the reservoir or deposit.

Crude oil, heavy oil, and tar sand bitumen exhibit wide variations in composition and properties within each category and between the categories. These variations occur not only in crude oil and heavy oil from different fields but also in oils taken from different production depths in the same well, as well as the vertical and horizontal variations in the properties of bitumen from a specific deposit. Historically, physical properties such as boiling point, density (gravity), and viscosity have been used to describe petroleum but the needs are even more extensive as increased amounts of heavy oil and bitumen are being recovered and sent to refineries as feedstocks (Table 5.1). Furthermore, because of the differences in petroleum composition, the importance of the correct sampling of crude oil that contains light hydrocarbons cannot be overestimated. Properties have to be selected with care and suitable sampling methods have to be used and care taken during the subsequent handling of the material in order to avoid the loss of any volatile constituents or oxidation during exposure to the air. In addition, adequate records of the circumstances and conditions during sampling have to be made.

Hence, the production of data focuses on (1) measurement, (2) accuracy, (3) precision, and (4) method validation. All of which depend upon the sampling protocols that were used to obtain the material to be analyzed. Without strict sampling protocols, variation and loss of accuracy (or precision) must be anticipated. For example, correct sampling of the product in storage or carrier tanks is important in order to obtain a representative sample for the laboratory tests which are essential in converting measured quantities to the standard volume.

Elemental analyses of crude oil, heavy oil, and bitumen show that the major constituents are carbon and hydrogen with smaller amounts of sulfur (0.1–8%, w/w), nitrogen (0.1–1.0%, w/w), and oxygen (0.1–3%, w/w) and trace elements such as vanadium, nickel, iron, and copper present at the part-per-million (ppm) level. Of the nonhydrocarbon (heteroelements) elements, sulfur is the most abundant and often considered the most important by refiners and by regulators (Speight, 2014a, 2014b, 2014c). However, compounds containing nitrogen and the trace metals also have deleterious effects on refinery catalysts and are deleterious to the environment at the time of recovery and after recovery and must not be discounted because of relative abundance. Thus, an efficient assay is derived from a series of test data that give an accurate description of petroleum quality and petroleum product quality and allow an indication of its behavior during refining or service. The first step is, of course, to assure adequate (correct) sampling by use of the prescribed protocols (ASTM D4057).

Thus, analyses are performed to determine whether each batch of crude oil, heavy oil, or bitumen in the reservoir or deposit is suitable for a particular recovery process after which an assay of the material recovered from the reservoir or deposit is suitable for transportation purposes and, ultimately, for refining. The tests are also applied to determine if there has been any contamination during storage or transportation that may increase the processing difficulty (cost). The information required is generally specific to a particular refinery and is also a function of refinery operations and desired product slate. To obtain the necessary information, two different analytical schemes are commonly used and these are: (1) an inspection assay, and (2) a comprehensive assay.

Inspection assays usually involve the determination of several key bulk properties of the sample (such as API gravity, sulfur content, pour point, and distillation range) as a means of determining if *major* changes in characteristics have occurred since the last comprehensive assay was performed. For example, a more detailed inspection assay might consist of the following tests: API gravity (or density or relative density), sulfur content, pour point, viscosity, salt content, water and sediment content, trace metals (or organic halides). The results from these tests—as well as the archived data from a comprehensive assay—provide an estimate of any changes that have occurred in the crude oil, heavy oil, or bitumen during the recovery operations. Furthermore, inspection assays are routinely performed on all feedstocks received at a refinery (Speight, 2014a, 2014b, 2014c, 2015).

On the other hand, the comprehensive (or full) assay is more complex (as well as time-consuming and costly) and is usually performed only when a new field (reservoir) or deposit comes on stream, or when the inspection assay indicates that significant changes in the composition of the crude oil have occurred. Except for these circumstances, a comprehensive assay of a particular crude oil stream may not (unfortunately) be updated for several years.

5.2.3 Data Acquisition and QA/QC

The objective of the data-gathering phase is to obtain reliable high-quality data for reservoir evaluation and development. The data requirement depends on the fluid type and the expected development and production strategies (Whitson and Brule, 2000). In addition, some heavy oils require customized PVT cells and experimental procedures to accelerate the time needed for attaining equilibrium conditions because of the slow (sometime, nonexistent) gas liberation. On the other hand, more complex near-critical fluids and miscible-gas injection processes need special PVT tests and precise measurement techniques to capture the complex phase behavior exhibited by these fluids. Gas condensates in the presence of water require PVT cells that can handle three-phase mixtures of gas, water, and condensate.

A heavy oil sampling program requires extra steps to obtain adequate volumes of representative single-phase oil samples for laboratory analysis. This includes adequate near-wellbore cleaning to minimize sample contamination by drilling-mud filtrate and optimal drawdown to minimize sand production and avoid two-phase flow while mobilizing the oil from the reservoir into the sample chamber (Reddie and Robertson, 2004). During surface sampling, measurement uncertainty in the producing gas–oil ratio (GOR) is a concern because of large drawdown and incomplete gas separation from the oil. Another issue with surface samples is the slow dissolution of gas while recombining them to prepare reservoir fluid. The main advantage of bottomhole sampling over surface sampling is that the former offers a viable means to capture single-phase samples and eliminate uncertainties associated with surface samples.

The C7+ fraction of the reservoir fluid contains numerous compounds of different homologs (paraffinic, naphthenic, and aromatic) and plays a dominant role in determining the PVT behavior of the fluid. For example, in a gas-condensate fluid, the dew point pressure is (as anticipated) a strong function of C7+ molecular weight and its relative amount in the fluid. On the other hand, in heavy oils, the C7+ components dictate the viscosity behavior and control the asphaltene deposition and wax deposition characteristics of the oil. Similarly, in volatile oils and rich condensates, the oil volumes and other properties below the saturation pressure are determined by the amounts of intermediate and heavy components.

Therefore, it is important to characterize them accurately and several methods are used to lump these components into pseudo-components for equation-of-state models (Whitson, 1983) in which the C7+ distribution is represented by a continuous gamma distribution that is optimally discretized into a specific number of fractions (ie, pseudo-components) in which fluid type and the production process involved further guides the component selection.

Because of slow gas liberation and dissolution in heavy oil, special care should be exercised in selecting equipment and procedures for sample preparation and PVT measurements (Cengiz et al., 2004). It is essential to measure the true bubble point pressure as well as the viscosity by means of, for example, a capillary-flow viscometer. Because the oil is saturated at each pressure step in the differential-liberation experiment, small pressure drops in the capillary viscometer caused by the flow will liberate the gas.

Therefore, it may be necessary to conduct several viscosity measurements above the saturated pressure and use an extrapolation technique to determine the viscosity at the desired differential-liberation pressure.

PVT data interpretation and modeling for heavy oils require reliable treatment of C7+ components because a majority of components in heavy oils fall in this range (Ancheyta and Speight, 2007). Solid-forming compounds, such as resin constituents and asphaltene constituents, should be characterized properly for flow assurance needs. Thus, it is imperative that reservoir fluid characterization studies should relate where possible, considering the nature of heavy oil, to the following issues: (1) acquisition of representative samples at various depths to quantify initial fluid gradients and for PVT studies; (2) PVT measurements to capture near-critical behavior, evaluate gas-injection strategies, and design the surface-separator train; (3) fluid modeling to predict observed near-critical behavior and property changes during gas injection; and (4) development of thermodynamically consistent compositional-gradient models for use in reservoir studies.

Fluid characterization strongly affects in-place volume, recovery-factor, injectivity/productivity, and well-deliverability calculations. Accurate fluid characterization minimizes technical uncertainties and, thus, provides a reliable representation of the asset value. However, fluid-sampling programs must be tailored to the fluid type, reservoir-rock and -fluid conditions, and fluid distribution. The fluid type and production processes dictate PVT data requirements, measurement methods, and data accuracy. In addition, the C7+ components must be characterized accurately and rigorous modeling methods, such as energy minimization, and robust solution techniques are needed to model near-critical fluids and processes. Finally, sampling tar sand bitumen requires additional precautions to ensure that the acquired data represent the sample. The sample should, at first, be homogenized to fall into one of three categories: (1) low grade, approximately 8% (w/w) bitumen; (2) medium grade, approximately 8% (w/w) bitumen; or (3) high grade, approximately 13% (w/w) bitumen. After this, handling and analytical protocols should be similar to the handling and analytical protocols applied to the analysis of atmospheric and vacuum residua (Speight, 2015). As with crude oil and heavy oil, sampling is an important precursor to bitumen analysis and a standard method is available that provides guidance for the sampling of liquid and semisolid asphalt products.

Finally, in order to ensure high-quality data requires routine laboratory visits, evaluation of laboratory procedures and methods, and QC as the data become available. The QA/QC methods can range from simple graphical techniques to sophisticated material-balance calculations (Whitson and Brule, 2000).

Finally, heavy oil, extra heavy oil, and tar sand bitumen resources can be markedly different in their characteristics. Hence, each production method must be tailored for the particular resource and for its fluid properties. A method that is successful in one reservoir (or deposit) may fail utterly in a different reservoir (or deposit). Hence, it is essential that the properties of the resource be fully understood before selecting a production scheme. Essential properties include (1) the geological setting, (2) the depth, (3) areal extent, (4) thickness of the resource, (5) fluid composition, density, viscosity, and gas content, (6) the presence of bottom water or top gas zones, (7) petrophysical and geomechanical properties, such as porosity, permeability, and rock strength, (8) the presence of shale layers or shale lenses, (9) the presence of clay layers or clay lenses, (10) vertical and horizontal permeability, and (11) the variation of these (and any other relevant) properties across the reservoir or deposit.

In summary, fluid characterization strongly affects estimations of (1) the in-place volume, (2) the recovery-factor, (3) the injectivity and productivity, and (4) the well performance. Accurate fluid characterization minimizes technical uncertainties and, thus, provides a reliable representation of the asset value. Moreover, fluid-sampling programs must be tailored to (1) the fluid type, (2) reservoir-rock, (3) the fluid conditions, and (4) the fluid distribution. The fluid type and production processes dictate the requirements for the PVT data measurement methods as well as data accuracy. In addition, the laboratory test methods and procedures must be tailored to specific fluids coupled with quality assurance and quality control protocols.

5.3 CHEMICAL AND PHYSICAL PROPERTIES

The chemical and physical properties of heavy oil and tar sand bitumen lead to a variety of problems with respect to recovery, which is reflected in the increased cost of extraction and processing and physical limitations on production capacity. Due to the high viscosity of these materials, some form of EOR or innovative recovery is usually required for recovery. Steam injection has been common practice, in both vertical and lateral wells while mining was the only method of choice for recovery of bitumen from tar sand deposits. In terms of nonmining techniques for bitumen recovery, an addition to recovery technology has been a variety of innovative processes such as SAGD, combined with horizontal drilling (Chapter 8). In this method a horizontal steam-injection well is drilled several feet above a production well. Another recovery technique involves the injection of solvent rather than steam in the supra-adjacent well—the vapor extraction (VAPEX) process (Chapter 7).

In the VAPEX process, vaporized hydrocarbon solvent is injected into heavy oil and bitumen reservoirs; the solvent-diluted oil drains by gravity to a horizontal production well. The process is relatively energy efficient, environmentally friendly, and causes in situ upgrading (by deposition of asphaltic materials). The applicability of the VAPEX process may even surpass the performance of the SAGD in thin reservoirs, reservoirs underlain by aquifer, and even in offshore recovery operations. Another noteworthy recovery method is cold heavy oil production with sand (CHOPS). This process can be achieved through horizontal lateral wells in combination with electric submersible pumps and progressing cavity pumps. Efforts are also continuing to improve production of heavy oil (and bitumen) through downhole electrical resistance heating.

These processes aside, heavy oil and bitumen exhibit a wide range of physical properties and several relationships can be made between various physical properties (Speight, 2014a, 2015). Whereas the properties such as viscosity, density, and boiling range may vary widely, the ultimate or elemental analysis varies over a narrow range for a large number of samples. The carbon content is relatively constant, while the hydrogen content and the content of the heteroatoms are responsible for the major differences within the heavy oil family and also within bitumen from different tar sand deposits or even bitumen from different sections of the same deposit.

Heavy oil or bitumen containing 9.5% (w/w) (or more) heteroatoms may contain few pure hydrocarbon constituents insofar as the constituents contain *at least one or more* nitrogen, oxygen, and/or sulfur atoms within the molecular structure of the constituents. And it is the heteroelements that can have substantial effects on the recovery process. Coupled with the changes brought about to the feedstock constituents by disturbing the delicate intramolecular balance, it is not surprising that recovering heavy oil and bitumen can be a monumental task. Thus, initial inspection of the oil (conventional examination of the physical properties) is necessary. From this, it is possible to make deductions about the propensity of facile or difficult recovery. In fact, evaluation of heavy oil and tar sand from physical property data can indicate (nothing is ever certain until tried) as to which recovery sequences should be employed for heavy oil or bitumen and must be a predominant part of the initial examination of the material for any heavy oil.

The chemical composition of heavy oil or bitumen is a much truer indicator of behavior. Whether the composition is represented in terms of compound types or, more likely, in terms of generic compound classes, it can assist in determination of the nature and type of any potential interactions of the oil with, for example, the rock, or with changes in pressure and temperature. Hence, chemical composition can play a large part in determining the nature of the products that arise from the recovery operations. It can also play a role in determining the means by which a particular feedstock should be processed (Speight, 2014a) and this becomes particularly important when partial upgrading in the reservoir is considered as a serious option for recovery.

To satisfy specific needs with regard to heavy oil and bitumen recovery, as well as to the nature of the recovered product, various standard test methods are available from organizations, such as the ASTM International. A complete discussion of the large number of routine tests available for petroleum and heavy oil and bitumen, as well as petroleum residua is the subject of a number of texts (eg, Gwynn, 1971; Wood and Ritzma, 1972; Wallace, 1988; Subramanian et al., 1996; CCQTA, 2008; EIPA, 2015; Speight, 2015). However, it seems appropriate that in any discussion of the physical properties of heavy oil and bitumen, reference should be made to the corresponding test method and, accordingly, the various test numbers have been included in the text. Some of the tests have been abandoned but are still used in many laboratories, especially where a replacement test has not been developed and, for reference purposes, these test methods are also included here.

Thus, initial inspection of the nature of the heavy oil and bitumen will provide deductions about the most logical methods to be used for recovery. Indeed, careful evaluation from physical property data is a major part of the initial study of heavy oil or bitumen and *proper* interpretation of the data resulting from the inspection of these materials requires an understanding of their significance (Speight, 2014a). But before any analysis occurs, it is necessary to ensure that the sample is consistent with the reservoir and that the data can be reproduced.

5.3.1 Sampling and Sample Preparation

Reservoir fluids found in oil fields, heavy oil fields, and tar sand deposits worldwide vary greatly in composition within each category. The rocks which contain these fluids also vary considerably in composition

(Chapter 2), and in physical and flow properties. In certain cases, this can serve to complicate the sampling procedure but other factors, such as producing area, height of the column of hydrocarbon fluid, fracturing or faulting, and water production also serve to distinguish one reservoir or deposit from another. The combination of all these factors affects the choice of sampling methods and preparations for sampling.

When a reservoir or deposit is relatively small, a properly taken sample from a single well or borehole can be representative of the fluid throughout the entire reservoir (or deposit) whereas for reservoirs and deposits which are large and/or complex, samples from several wells and/or depths may be required. However, significant horizontal and vertical variations in fluid composition often occur in very thick formations and additional sampling during the later life of a reservoir or deposit is in common practice because production experience can show significant variations in fluid composition and properties and that the reservoir (or deposit) is more complex than earlier analytical information indicated.

The main objective of a successful sampling campaign is to obtain representative fluid samples for determining properties, and adequate volumes should be collected for analysis, geochemical analysis for fluid-source identification and reservoir continuity, as well as crude assay for refinery processes (Speight, 2001, 2014a, 2014b, 2014c, 2015; API, 2003; Nagarajan et al., 2007a, 2007b). The sampling program should focus on selecting an appropriate sampling method and developing sound sampling, sample-transfer, and QC procedures. In addition, sample character and specific sampling issues should be addressed in the form of a sample history that details the acquisition, storage, and tests carried out on the sample (Speight, 2001, 2014a, 2014b, 2014c, 2015).

Proper management of production from a reservoir can maximize the recovery of the hydrocarbon fluids (gas and/or crude oil) originally in the reservoir. Developing proper management strategies requires accurate knowledge of the characteristics of the reservoir fluid. Practices are recommended herein for obtaining samples of the reservoir fluid, from which the pertinent properties can be determined by subsequent laboratory tests. For example, near-wellbore cleaning is necessary in order to minimize sample contamination by drilling-mud filtrate as well as optimal drawdown to minimize sand production and avoidance of two-phase flow while transferring the sample from the reservoir into the sample chamber (Reddie and Robertson, 2004). During surface sampling, measurement uncertainty in the producing GOR may be an issue because of large drawdown and incomplete gas separation from the oil. Another potential issue that can influence the condition of the samples at the surface is the slow dissolution of gas while recombining them to prepare reservoir fluid.

Conditioning a well before sampling is always necessary—initial well testing or normal production operations often result in the fluid near the wellbore having a composition which has been altered from that of the original reservoir fluid (for reasons described later). The objective of conditioning the well is to remove this altered (nonrepresentative) fluid. Well conditioning consists of producing the well at a rate which will move the altered fluid into the wellbore and allow it to be replaced by unaltered (representative) fluid flowing in from further out in the reservoir. Well conditioning is especially important when the reservoir fluid is at or near its saturation pressure at the prevailing reservoir conditions because reduction in pressure near the wellbore, which inevitably occurs from producing the well, will alter the composition of the fluid flowing into the wellbore. The types of problems which can be encountered in conditioning a well, and methods for their control are discussed.

If the sampling procedure is incorrect or if samples are collected from an improperly conditioned well, the resulting samples may not be representative of the reservoir fluid. A nonrepresentative sample may not exhibit the same properties as the reservoir fluid. The use of fluid-property data obtained from nonrepresentative samples, however accurate the laboratory test methods, may result in errors in reservoir management. Poor planning can also result in incomplete data being taken during the sampling program. Incomplete data can make it difficult or impossible for laboratory personnel to perform and interpret tests that provide accurate and meaningful fluid-property information.

Reservoir fluids found in gas and oil fields around the world vary greatly in composition. In some fields, the fluid is in the gaseous state and in others it is in the liquid state, as is the case with heavy oil. The rocks which contain these reservoir fluids also vary considerably in composition, as well as in physical and flow properties. In certain cases, this can serve to complicate the sampling procedure. Other factors, such as producing area, height of the column of hydrocarbon fluid, fracturing or faulting, and water production also serve to distinguish one reservoir from another. The combination of all these factors affects the choice of sampling methods and preparations for sampling.

When a reservoir is relatively small, a properly taken sample from a single well can be representative of the fluid throughout the entire reservoir.

For reservoirs which are large or complex, samples from several wells and/or depths may be required. Significant variations in fluid composition often occur in very thick formations, in really large reservoirs, or in reservoirs subjected to recent tectonic disturbances. Additional sampling during the later life of a reservoir is not uncommon because production experience can show that the reservoir is more complex than earlier information indicated.

5.3.1.1 Sampling

As a result of the complexity of these materials, the importance of the correct *sampling* of heavy oil or bitumen (Wallace, 1988; Speight, 2000; CCQTA, 2008) cannot be overstressed. Properties such as elemental analysis, metals content, density (specific gravity), and viscosity (to mention only a few properties) are affected by the homogeneity (or heterogeneity) of the sample. In addition, adequate records of the circumstances and conditions during sampling have to be made, for example, in sampling from oilfield separators, the temperatures and pressures of the separation plant and the atmospheric temperature would be noted. An accurate sample handling and storage log should be maintained and should include information such as: (1) the precise source of the sample, that is, the exact geographic location or refinery locale from which the sample was obtained; (2) a description of the means by which the sample was obtained; (3) the protocols that have been used to store the sample; (4) chemical analyses, such as elemental composition; (5) physical property analyses, such as API gravity, pour point, and distillation profile; (6) the standard (ASTM or other recognized organization) test methods used to determine the properties in items 4 and 5; and (7) the number of times that the samples have been retrieved from storage to extract a portion, that is, indications of exposure to the air or oxygen.

Attention to factors such as these enables standardized comparisons to be made when subsequent samples are taken.

Methods for sampling reservoir fluids fall into two general categories (API, 2003). They are referred to as subsurface sampling or surface sampling and, as the names imply, each category reflects the location at which the sampling process occurs. Subsurface sampling may also be referred to as downhole or bottomhole sampling. The two commonly used sampling methods are bottomhole sampling and surface sampling. Bottomhole sampling attempts to capture samples close to reservoir conditions, while surface sampling aims at capturing gas and oil samples from the separator under stable flow conditions. Separator fluids then are recombined at a measured producing GOR to prepare representative reservoir fluid.

The choice of either the surface or downhole sampling method cannot be considered a simple or routine matter. Each reservoir usually presents certain constraints or circumstances peculiar to it. For example, field operation requirements can impose restrictions on the preparation and execution of a sampling program; sand production or downhole equipment in the well may limit the use of some of the equipment normally used in the sampling operation. Wells that exhibit rapid variations in production rate present special problems in making the necessary measurements with acceptable accuracy. Seasonal or daily weather changes can also influence the sampling operation. Thus, the details of a given sampling procedure often require modification to circumvent local problems.

In bottomhole-sampling operations, adequate cleaning of near-wellbore regions and controlled drawdown are critical for obtaining uncontaminated representative samples (Witt and Crombie, 1999). Controlled drawdown helps avoid two-phase flow in the reservoir. Downhole fluid analyzers are used to monitor sample contamination and ensure single-phase flow prior to sample capture. In surface sampling operations, proper well conditioning with minimum drawdown is the key to acquiring high-quality samples. Well conditioning requires that the well be flowed at an optimum rate for an extended period of time with a stable producing GOR but it must be recognized that sample quality and separator efficiency introduce uncertainties in the quality of the fluids (API, 2003).

Modern open-hole wireline formation testers now provide the means of recovering representative samples before the effects of subsequent production take place. Selection of one particular method over another is influenced by the type of reservoir fluid, the producing characteristics and mechanical condition of the well, the design and mechanical condition of the surface producing equipment, the relative expense of the various methods, and safety considerations. While bottomhole sampling has the advantage of capturing fluids at reservoir conditions, surface sampling operation has a potential for obtaining cleaner samples as a result of large volumes of fluid production before sampling.

These modifications are usually made based upon on-the-spot judgments. Some of the modifications which can be made to accommodate special situations are presented herein.

5.3.1.2 Sample Preparation

Once a sample has been obtained, there are several protocols involved in initial isolation and cleaning of the sample prior to analysis (Wallace, 1988; CCQTA, 2008; Speight, 2015). In fact, considerable importance attaches to the presence of *water* or *sediment* in crude oil, heavy oil, and tar sand bitumen (ASTM D1796; ASTM D4007) because they lead to difficulties during recovery and in the refinery, for example, corrosion of equipment and pipelines, uneven running on the distillation unit, blockages in heat exchangers and adverse effects on product quality.

The *sediment* consists of finely divided solids that may be dispersed in the oil or carried in water droplets. The solids may be drilling mud or sand or scale picked up during the transport of the oil, or may consist of chlorides derived from evaporation of brine droplets in the oil. In any event, the sediment can lead to serious plugging of the equipment, corrosion due to chloride decomposition, and a lowering of residual fuel quality.

Water may be found in the crude either in an emulsified form or in large droplets and can cause flooding of distillation units and excessive accumulation of sludge in tanks. The quantity is generally limited by pipeline companies and by refiners, and steps are normally taken at the wellhead to reduce the water content as low as possible. However, water can be introduced during shipment and, in any form, water and sediment are highly undesirable in a refinery feedstock and the relevant tests (ASTM D1796; ASTM D4006) are regarded as important for the determination of quality. Prior to assay it is often necessary to separate the water from the sample. In terms of water content, heavy oil and bitumen recovered by a steam-based process often form persistent emulsions which are difficult to separate, while in testing such emulsions for sediment and water care has to be taken to ensure that separated asphaltene constituents suspended in the sample are recognized and brought into solution prior to the test otherwise the organic deposit will be recorded as sediment. Thus, careful sampling is necessary when examination has to be made for suspended impurities of the type discussed.

There is a great variation in the *salt content* of heavy oil and bitumen depending mainly on the source, and possibly on the producing wells or zones within a field or deposit. In addition, at the refinery, salt water introduced during shipment by tanker may have contributed to this total salt content. These salts have adverse effects on refinery operations especially in increasing maintenance following corrosion in crude units and heat exchangers. It is common practice to monitor wells in a producing field or deposit for high salt content, and it is also general practice to desalt the heavy crude or bitumen (usually solvent-diluted bitumen) at the refinery. The determination of salt content is often made, but as with water and sediment tests, careful sampling is necessary. It would appear that further tests to determine the corrosiveness due to the individual chemical components of the salt (which is often a complex mixture of inorganic components) and also a determination of the extent of evolution of hydrogen chloride on heating would be desirable.

Finally, downhole sampling in which the fluids are recovered, as they exist in situ, is an important aspect of any sampling program. Recovering the fluids (oil, water, and natural gas) without contamination, loss of constituents, or degradation is needed for realistic measurement of the properties in the laboratory. Without downhole sampling of the fluids, properties such as PVT relationships, viscosity, and composition may be of limited value for planning leading to production. However, if the heavy oil may have limited mobility without an injection of steam, a solvent, or heat, downhole sampling becomes a more difficult challenge. For example, there can be very large variations in viscosity (several orders of magnitude) in heavy oil reservoirs, extra heavy oil reservoirs (or deposits), and in tar sand bitumen deposits, making the properties of such highly heterogeneous resources a challenge to measure leading to further challenges in production.

5.3.2 Elemental Analysis

The elemental (ultimate) analysis of heavy oil and bitumen for the percentages of carbon, hydrogen, nitrogen, oxygen, and sulfur is perhaps the first method used to examine the general nature, and perform an evaluation, of a feedstock. The atomic ratios of the various elements to carbon (ie, H–C, N–C, O–C, and S–C) are frequently used for indications of the overall character of the heavy oil and bitumen. It is also of value to determine the amounts of trace elements, such as vanadium and nickel, in a feedstock since these materials can have serious deleterious effects on catalyst performance during partial upgrading during recovery or even using a partial upgrading process at the surface before transportation.

There are procedures (standard test methods) for the ultimate analysis of petroleum and petroleum products but many such methods may have been designed for other materials. For example, *carbon content* can be determined by the method designated for coal and coke or by the method designated for municipal solid waste (ASTM E777). There are also methods designated for:

1. *Hydrogen content* (ASTM E777);
2. *Nitrogen content*;
3. *Oxygen content*; and
4. *Sulfur content*.

Of the data that are available, the proportions of the elements in heavy oil and bitumen vary only slightly over narrow limits (Speight, 2014a, 2015). Perhaps the more pertinent property in the present context is the *sulfur content*, which along with the API gravity represent the two properties that have the greatest influence on the value (in terms of the refinability) of heavy oil and bitumen. The sulfur content varies from about 0.1% to about 5% by weight (Speight, 2014a, 2015 and references cited therein).

In terms of classifying crude oil, heavy oil, and tars and bitumen by the sulfur content, the decision is purely arbitrary and virtually for an accurate classification method. Crude oil is defined as *sweet* if the sulfur content is 0.5% (w/w) or less and sour if the sulfur content is greater than 1.0% (w/w). Sulfur compounds in crude oil are chemically bonded within the hydrocarbon structures and very few crude oil actually have free sulfur—the Qayarah crude oil from Iraq in an example of crude oil that contains free sulfur (Speight, 1978, 2014a). Additional equipment in the refinery is required to remove the sulfur from crude oil, intermediate hydrocarbon feedstocks, and finished products. Transportation fuel specifications require extremely low sulfur contents, usually less than 80 ppm.

5.3.3 Density and Specific Gravity

Density is the mass of a unit volume of material at a specified temperature and has the dimensions of grams per cubic centimeter (a close approximation to grams per milliliter). *Specific gravity* is the ratio of the mass of a volume of the substance to the mass of the same volume of water and is dependent on two temperatures, those at which the masses of the sample and the water are measured. When the water temperature is 4°C (39°F), the specific gravity is equal to the density in the centimeter-gram-second (cgs) system, since the volume of 1 g of water at that temperature is, by definition, 1 mL. Thus the density of water, for example, varies with temperature, and its specific gravity at equal temperatures is always unity. The standard temperatures for a specific gravity in the petroleum industry in North America are 60/60°F (15.6/15.6°C).

The density dictates the buoyancy of an oil on water and influences spreading and natural dispersion. Density also provides a first indication as to volatility and distillation characteristics, the density of oil is one of many standard parameters used commercially to describe petroleum products. This information is also important to oil spill responders to predict evaporation loss and the development of other weathering processes. Specific gravity of heavy oil and bitumen falls within the range of 0.92–1.10 at 15°C (60°F). Because of this they can generally not be expected to float in water (specific gravity 1.00), which may have to be taken into account during recovery.

The *density* and *specific gravity* of heavy oil and bitumen is typically measured using the API and is the measure of the weight of crude oil in relation to the weight of water (water has an API gravity of 10°). Crude oil is arbitrarily characterized as heavy, intermediate, or light with respect to its API gravity (ASTM D287; ASTM D1298; ASTM D1217; ASTM D1555) which is a property that has found wide use for preliminary assessment of the character of the material. In particular, heavy oil and bitumen with a high content of asphaltene constituents and resin constituents and poor mobility at ambient temperature and pressure, thereby requiring vastly different processing sequences, may have a specific gravity (density) of about 0.95 (heavy oil) to 1.05 (bitumen).

The characterization of petroleum-related feedstocks and tar sand bitumen is arbitrary and is subject to variation depending upon the source of the data. For example, tar sand bitumen can be quoted as having an API gravity up to 12° while the often recognized more general upper limit is 10°. Heavy crude oil (which is viscous and resistant to flow) can be quoted as having an API gravity of as low as 18° or as high as 25°. On the other hand, intermediate gravity crude oil can be quoted as having an API gravity anywhere between 18° and 36°, while light crude oil can be quoted as having an API gravity of 36° or greater. The best default position is to use the definitions of tar sand bitumen, heavy oil, and conventional crude oil that are based on the recovery method (Chapter 1) since specific gravity (hence, density and API gravity) is influenced by chemical composition but quantitative correlation is difficult to establish (Speight, 2014a, 2015).

Nevertheless, it is generally recognized that increased amounts of aromatic compounds result in an increase in density, whereas an increase in saturated compounds results in a decrease in density. It is also possible to recognize certain preferred trends between the API gravity of heavy oils, bitumen, and residua and one or more of the other physical parameters. For example, a correlation exists between the API gravity and sulfur content, Conradson carbon residue, and viscosity (Speight, 2000 and references cited therein). However, the derived relationships between the density of heavy oil or bitumen and the fractional composition are valid only when applied to a certain type of heavy oil and may lose their significance when applied to heavy oil from different sources.

The values for density (and specific gravity) cover an extremely narrow range considering the differences in heavy oil behavior. In an attempt to inject a more meaningful relationship between the physical properties and processability of the various crude oils, the API devised a measurement of gravity devised upon the Baumé scale for industrial liquids. The Baumé scale for liquids lighter than water was used initially:

$$\text{Degrees Baumé} = 140/\text{sp gr}\ @\ 60/60°\text{F} - 130$$

However, a considerable number of hydrometers calibrated according to the Baumé scale were found at an early period to be in error by a consistent amount, and this led to the adoption of the equation:

$$\text{Degrees API} = 141.5/\text{sp gr}\ @\ 60/60°\text{F} - 131.5$$

The specific gravity of conventional crude oil usually ranges from about 0.8 (45.3° API) to approximately 1.0 (10° API) for heavy oil and heavier than 1.0 (<10° API) for bitumen. This is in keeping with the general trend that a lower atomic hydrogen–carbon ratio (increased aromaticity) leads to a decrease in API gravity (or, more correctly, an increase in specific gravity).

Density or specific gravity or API gravity may be measured, depending upon the properties of the heavy oil or bitumen sample, by means of a hydrometer (ASTM D287; ASTM D1298) or by means of a pycnometer (ASTM D1217). The variation of density with temperature, effectively the coefficient of expansion, is a property of great technical importance, since most crude oils (including heavy crude oil and bitumen) are sold by volume and specific gravity is usually determined at the prevailing temperature (21°C, 70°F) rather than at the standard temperature (60°F, 15.6°C). The tables of gravity corrections (ASTM D1555) are based on an assumption that the coefficient of expansion is a function (at fixed temperatures) of density only.

5.3.4 Viscosity

Viscosity is the most important single fluid characteristic governing the motion of heavy oil and bitumen and is actually a measure of the internal resistance to motion of a fluid by reason of the forces of cohesion between molecules or molecular groupings. Because of the high viscosity, heavy oil and tar sand bitumen are more difficult to pump without elevation of the temperatures to bring these fluids to a liquid state.

Crude oils are often assessed by their viscosity and API gravity. However, a note of caution is worthy insofar as the viscosity and the API gravity are not necessarily indicators of the producibility of the oil. From a subsurface perspective, it is the API gravity and the viscosity at reservoir temperature and pressure that are more relevant to properties of the oil. Reservoir temperature and the GOR have a very significant influence on the viscosity of the live oil, so that crude oils with a similar API gravity can have very different viscosities in the reservoir. A low API crude oil at a deeper horizon may be easily producible, on account of the lower in situ viscosity (Jayasekera and Goodyear, 1999).

By definition, *viscosity* is the force in dynes required to move a plane of 1 cm^2 area at a distance of 1 cm from another plane of 1 cm^2 area through a distance of 1 cm in 1 second. In the cgs system the unit of viscosity is the *poise* (P) or *centipoise* (cP) (1 cP = 0.01 P). Two other terms in common use are *kinematic viscosity* and *fluidity*. The kinematic viscosity is the viscosity in centipoise divided by the specific gravity. The unit of kinematic viscosity is the *stoke* (cm^2/s), although *centistokes* (0.01 cSt) is in more common usage; fluidity is simply the reciprocal of viscosity.

The viscosity (ASTM D445; ASTM D88; ASTM D2161; ASTM D341; ASTM D2270) of heavy oil and bitumen vary markedly over a very wide range. Values vary from several hundred centipoises at room temperature to many thousands of centipoises at the same temperature.

Many types of instruments have been proposed for the determination of viscosity but like density determination, the choice of an instrument depends upon the properties of the oil.

The simplest and most widely used are capillary types (ASTM D445), and the viscosity is derived from the equation:

$$\mu = \frac{Br^4P}{8nl}$$

In the equation, *r* is the tube radius, *l* the tube length, *P* the pressure difference between the ends of a capillary, *n* the *coefficient of viscosity*, and μ the quantity discharged in unit time. Not only are such capillary instruments the most simple, but when designed in accordance with known principles and used with known necessary correction factors, they are probably the most accurate viscometers available. It is usually more convenient, however, to use relative measurements, and for this purpose the instrument is calibrated with an appropriate standard liquid of known viscosity.

Batch flow times are generally used; in other words, the time required for a fixed amount of sample to flow from a reservoir through a capillary is the datum that is actually observed. Any features of technique that contribute to longer flow times are usually desirable. Some of the principal capillary viscometers in use are those of Cannon-Fenske, Ubbelohde, Fitzsimmons, and Zeitfuchs. The *Saybolt universal viscosity* (SUS; ASTM D88) is the time in seconds required for the flow of 60 mL of the sample, at constant temperature, through a calibrated orifice. The *Saybolt furol viscosity* (ASTM D88) is determined in a similar manner except that a larger orifice is employed.

As a result of the various methods for viscosity determination, it is not surprising that much effort has been spent on interconversion of the several scales, especially converting Saybolt to kinematic viscosity (ASTM D2161),

$$\text{Kinematic viscosity} = a \times \text{Saybolt}\ s + \frac{b}{\text{Saybolt}\ s}$$

In this equation, a and b are constants.

The SUS equivalent to a given kinematic viscosity varies slightly with the temperature at which the determination is made because the temperature of the calibrated receiving flask used in the Saybolt method is not the same as that of the oil. Conversion factors are used to convert kinematic viscosity from 2 to 70 cSt at 38°C (100°F) and 99°C (210°F) to equivalent SUS in seconds. Appropriate multipliers are listed to convert kinematic viscosity in excess of 70 cSt. For a kinematic viscosity determined at any other temperature the equivalent SUS is calculated by use of the Saybolt equivalent at 38°C (100°F) and a multiplier that varies with the temperature:

$$\text{Saybolt}\ s\ \text{at}\ 100°\text{F}(38°\text{C}) = \text{cSt} \times 4.635$$

$$\text{Saybolt}\ s\ \text{at}\ 210°\text{F}(99°\text{C}) = \text{cSt} \times 4.667$$

Various studies have also been made on the effect of temperature on viscosity since the viscosity of heavy oil and bitumen decreases as the temperature increases:

$$\log\log(n + c) = A + B\log T$$

In the equation, *n* is absolute viscosity, *T* is temperature, and *A* and *B* are constants. This equation has been sufficient for most purposes and has come into very general use. The constants *A* and *B* vary widely with different oils, but *c* remains fixed at 0.6 for all oils having a viscosity over 1.5 cSt; it increases only slightly at lower viscosity (0.75 at 0.5 cSt). The viscosity-temperature characteristics of any oil, so plotted, thus create a straight line, and the parameters *A* and *B* are equivalent to the intercept and slope of the line. To express the viscosity and viscosity-temperature characteristics of an oil, the slope and the viscosity at one temperature must be known; the usual practice is to select 38°C (100°F) and 99°C (210°F) as the observation temperatures.

Suitable conversion tables are available (ASTM D341), and each table or chart is constructed in such a way that for any given heavy oil or bitumen the viscosity-temperature points typically result in a straight line over the applicable temperature range. Thus, only two viscosity measurements need be made at temperatures far enough apart to determine a line on the appropriate chart from which the approximate viscosity at any other temperature can be read. However, to be absolutely certain of the viscosity-temperature relationship it is sternly recommended that at least three or four points should be used to determine the slope of the line.

5.3.5 Solubility

Although not truly a thermal property, the solubility parameter of crude oil and its constituent fractions is of interest during thermal methods of recovery.

The solubility parameter of a crude oil fraction, especially the asphaltene fraction, has been the subject of study with some interesting results emerging (Speight, 2014a, 2015). In fact, phase separation as can occur during thermal recovery of heavy oil and bitumen can be explained by use of the solubility parameter, δ, for petroleum fractions and for the solvents. As an extension of this concept, there is sufficient data to draw a correlation between the atomic hydrogen–carbon ratio and the solubility parameter for hydrocarbons and the constituents of the lower boiling fractions of petroleum. Recognition that hydrocarbon liquids can dissolve polynuclear hydrocarbons, a case in which there is usually less than a three-point difference between the lower solubility parameter of the solvent and the higher solubility parameter of the solute. Thus, a parallel, or near-parallel, line can be assumed that allows the solubility parameter of the asphaltenes and resins to be estimated (Fig. 5.2).

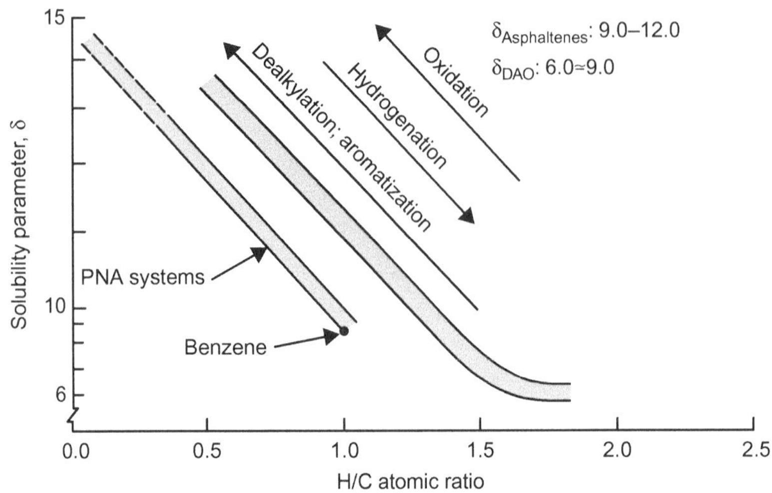

Figure 5.2 Representation of the variation of the solubility parameter of petroleum fractions through variation with the H–C atomic ratio and comparison to benzene and polynuclear aromatic systems.

By this means, the solubility parameter of asphaltenes can be estimated to fall in the range 9–12, which is in keeping with the asphaltenes being composed of a mixture of different compound types with an accompanying variation in polarity. As the thermal reaction proceeds (especially if the recovery process employs superheated steam of combustion) it causes removal of alkyl side chains from the asphaltenes, decreases the hydrogen-to-carbon atomic ratio (Fig. 5.3), and increases the solubility parameter. Concurrently changes occur to the oil medium but are of lesser effect thereby bringing about a higher solubility parameter differential between the reacted asphaltene constituents and resin constituents and the oil. As a result deposition ensues.

The deposited reacted material (Fig. 5.4) is usually a product of the action of the highest molecular weight and/or the highest polarity constituents in the asphaltene and resin fractions. This is of benefit to properties of the produced oil but may be a disadvantage insofar as the deposit causes blockage of the reservoir flow channels.

Another aspect of this reaction is the order of deposition relative to models applied to the system. The more polar constituents (eg, the amphoteric constituents) of the asphaltene and resin fractions are more thermally labile than the lower polarity constituents (eg, the neutral polar

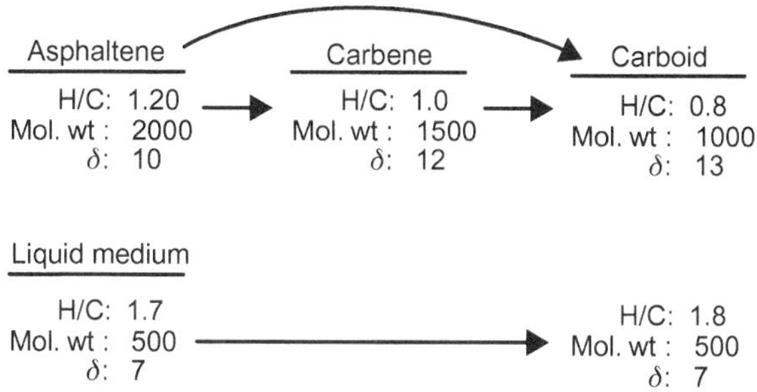

Figure 5.3 Variation of the solubility parameter of the asphaltene fraction and the oil with reaction progress.

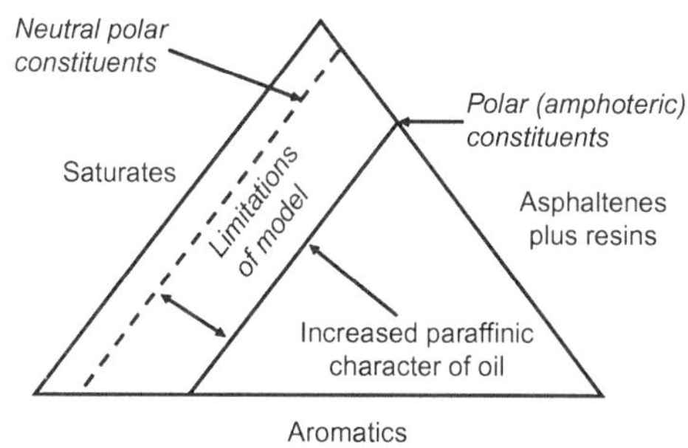

Figure 5.4 Order of deposition of asphaltene constituents during thermal changes.

constituents) (Fig. 5.4). As a result products from the amphoteric constituents will exceed the solubility parameter differential more quickly and will separate first from the oil medium first and at an earlier time than could be predicted if an average property is used for any model applied to the system.

5.3.6 Acid Number

Another characteristic of crude oil, heavy oil, and tar sand bitumen that can influence recovery is acid content, determined by the total acid number (TAN; Speight, 2014a, 2015). The TAN represents a composite of acids present in the oil and is measured in milligrams (mg) per gram of potassium hydroxide. Acid contents are usually monitored as a TAN determined by potentiometric titration (ASTM D664). This method is wrought with potential interferences and interpretative problems (Piehl, 1988; Speight,

2014c, 2015) but it remains a standard method by which oils are assayed and valued. TAN generally increases with increasing levels of biodegradation. The current activity of biodegrading organisms may be most important in determining organic acid contents because acids may dissipate rapidly owing to relatively high water solubility and reactivity.

The acids originate during the generation of petroleum, heavy oil, and bitumen is associated with the deposition of organic detritus. The detritus deposition occurs during the development of fine-grained sedimentary rocks that occur in marine, near-marine, or even nonmarine environments and the composition of petroleum, heavy oil, and bitumen is greatly influenced not only by the nature of the precursors in the protopetroleum but also by the relative amounts of these precursors (that are dependent upon the local flora and fauna) that occur in the source material (Speight, 2014a). Hence, it is not surprising that petroleum, heavy oil, and tar sand bitumen composition can vary with the location and age of the field or deposit in addition to any variations that occur with the depth of the individual well. The same rationale can apply to the occurrence and character of the various constituents of petroleum, not the least of which (in the current context) is the fraction known as *naphthenic acids*.

Naphthenic acid is the generic name used for all of the organic acids present in crude oils—most of the acids arise as biochemical markers of crude oil origin and maturation (Speight, 2014a). Most of these acids are believed to have the chemical formula $R(CH_2)_nCOOH$, where R is a cyclopentane or cyclohexane ring and n is typically greater than 12 which can lead to the representation of structural characteristics (Tables 5.2–5.4) (Hsu et al., 1994, 2000; Barrow et al., 2003; Petkova et al., 2009). In addition to $R(CH_2)_nCOOH$, a multitude of other acidic organic constituents are also present in crude oil(s) but not all of the species have been fully analyzed and identified. Although saturated carboxylic acids are the predominant compounds found in most crude oils, aromatic and even heterocyclic compounds have also been reported. For example, mono-aromatics and di-aromatics have also been identified within the naphthenic acid group (Derungs, 1956; Hsu et al., 2000).

Naphthenic acids are generally described by the formula $C_nH_{2n+Z}O_2$, where n is the number of carbons and Z is the *hydrogen deficiency* index. Z is either zero (for simple fatty acids with one carbon–oxygen double bond) or a negative even integer (acids with additional rings/double bonds) that specify homologous series. On a molecular basis, Z is

Table 5.2 Representative Structures of Naphthenic Acids Found in Crude Oils (Hsu et al., 2000; Barrow et al., 2003

1 ring	2 rings	3 rings	4 rings	5 rings	6 rings	C22–C33
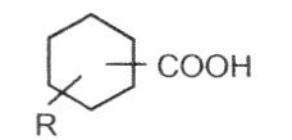	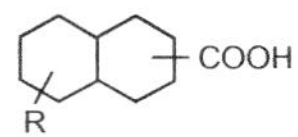	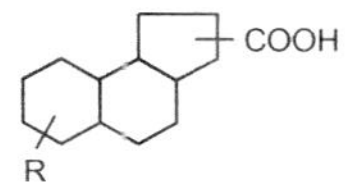	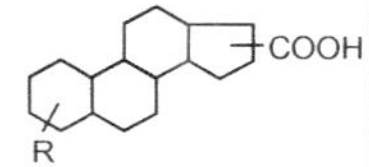	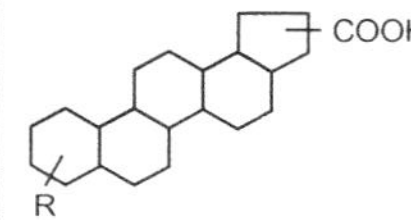		$C_nH_{2n-2}O_2$
	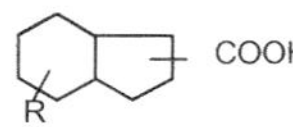	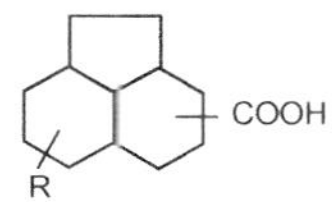	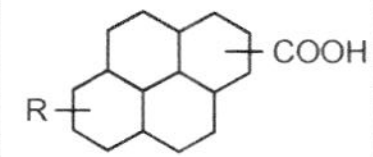	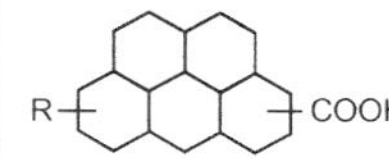	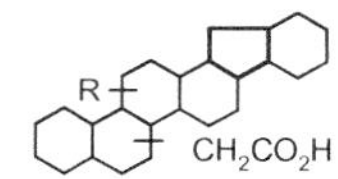	$C_nH_{2n-4}O_2$
		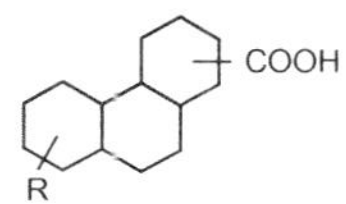				$C_nH_{2n-6}O_2$
						$C_nH_{2n-8}O_2$

In each case, *R* is an alkyl group of varying size and there may be more than one alkyl group per molecule.

Table 5.3 Predominant Naphthenic Acid Compounds Classes in a California Crude Oil

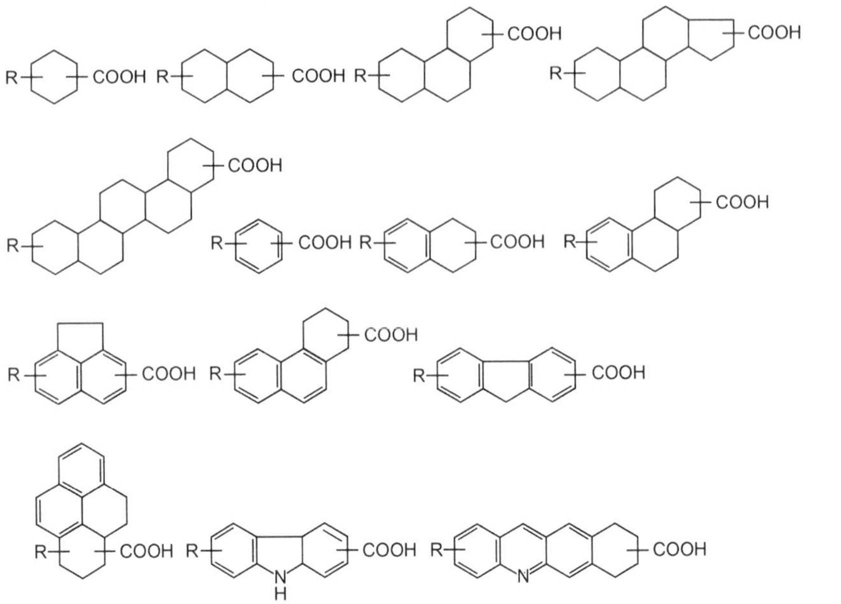

Table 5.4 Alternate Structural Types or Naphthenic Acids Where the Carboxylic Acid Function Is Attached to an Alkyl Side Chain

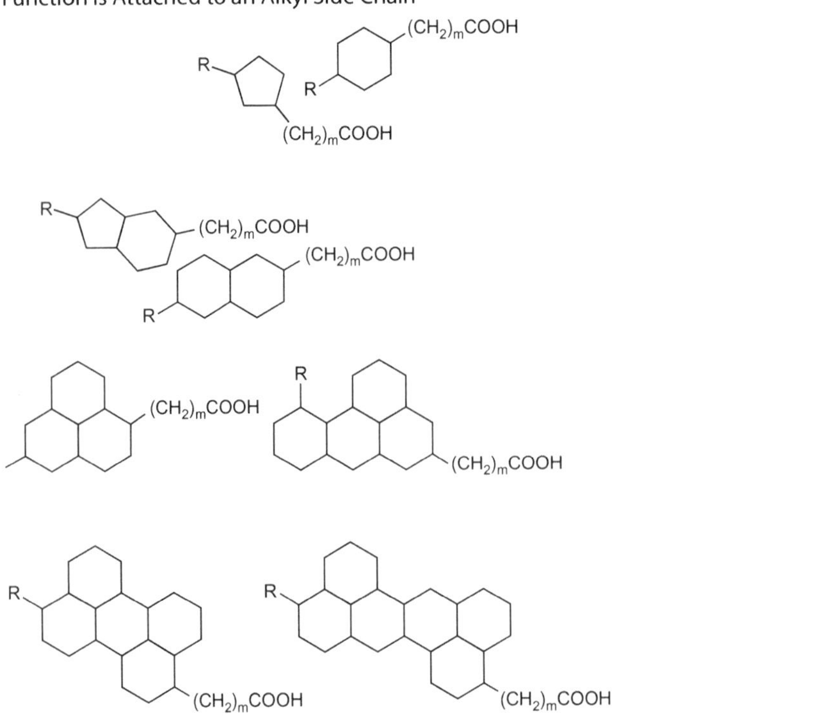

frequently referred to as the molecular *hydrogen deficiency* (Hughey et al., 2002, 2007; Qian et al., 2001, 2008).

Finally, *naphthenates* are the salts of naphthenic acids have the formula M(naphthenate)$_2$ (M = divalent metal) or are basic oxides with the formula M_3O(naphthenate)$_6$. Naphthenate salts form when naturally occurring naphthenic acids in the crude oil come in contact with metal ions (such as calcium) in the produced water under the right conditions of pH and temperature. Naphthenates (RCOO–) are formed during crude oil production due to pressure decrease and release of carbon dioxide, which leads to increase in pH and dissociation of naphthenic acid (RCOOH). Naphthenates can precipitate with metal cations present in brine and form deposits, mainly of calcium naphthenates, which can accumulate in topside facilities, desalters, and pipelines, leading to shutdowns and other serious problems during crude refining.

A TAN number greater than 0.5 mg is considered high. As an example, Wilmington and Kern crude oil have a TAN ranging from 2.2 to 3.2 mg, respectively. However, some acids are relatively inert. Thus, the TAN number does not always represent the corrosive properties of the crude oil. Further, different acids will react at different temperatures—making it difficult to pinpoint the processing units within the refinery that will be affected by a particular high TAN crude oil. Nonetheless, high TAN crude oils contain naphthenic acids, a broad group of organic acids that are usually composed of carboxylic acid compounds. These acids corrode the distillation unit in the refinery and form sludge and gum which can block pipelines and pumps entering the refinery.

The impact of corrosive, high TAN crude oils can be overcome by blending higher and lower TAN oils, installing or retrofitting equipment with anticorrosive materials, or by developing low temperature catalytic decarboxylation processes using metal catalysts such as copper. Many California refineries already process high TAN crude. High TAN oils are sold on the market at a discount compared to higher quality crude oils. High TAN oils account for an increasing percentage of the global crude oil market. Crude oil with a TAN greater than 1.0 mg increased in the world market from 7.5% in 1998 to 9.5% in 2003.

5.4 THERMAL PROPERTIES

The thermal properties of heavy oil and bitumen are not always considered to be necessary aspects of petroleum recovery. But these properties are a determinant of the behavior (including the thermal or reactivity) if the material when it is subjected to recovery by thermal (steam-based or combustion) processes as well as heat fluctuations over time. As with all properties of petroleum, heavy oil, and bitumen, a collection of standard test methods is instrumental in the evaluation and assessment of the thermal properties. These standards allow recovery engineers and other geological and chemical engineers to appropriately examine and recovery the heavy oil or bitumen in a safe and efficient manner.

5.4.1 Carbon Residue

The carbon residue (ASTM D189 and ASTM D524) of heavy oil or bitumen is a property that can be correlated with several other properties and may be used to evaluate the deposition of carbonaceous products from heavy oil or bitumen during thermal recovery operations. The carbonaceous products can either be reacted constituents such as (1) reacted asphaltene constituents, (2) reacted resin constituents (both of which may undergo phase separation during the recovery process), or (3) truly carbonaceous material that may be formed during a recovery process involving in situ combustion.

There are two older well-used methods for determining the carbon residue: the Conradson method (ASTM D189) and the Ramsbottom method (ASTM D524). Both are equally applicable to heavy oil and bitumen but the presence of metallic constituents in the sample will give erroneously high carbon residues. The metallic constituents must first be removed from the oil or they can be estimated as ash by complete burning of the coke after carbon residue determination. There is no exact correlation between the two methods so it is not possible to interconnect the data.

Another method (ASTM D4530) requires smaller sample amounts and was originally developed as a *thermogravimetric method*. The carbon residue produced by this method is often referred to as the *microcarbon residue*. Agreements between the data from the three methods are good, making it possible to interrelate all of the data from carbon residue tests (Long and Speight, 1989).

5.4.2 Specific Heat

Specific heat is the quantity of heat required to raise a unit mass of material through one degree of temperature (ASTM D2766). It is an extremely important engineering quantity in practice and is used in all calculations on heating and cooling heavy oil and bitumen. Many measurements have been made on various hydrocarbon materials, but the data for most purposes may be summarized by the general equation:

$$C = 1/d(0.388 + 0.00045t)$$

where C is the specific heat at t°F of an oil whose specific gravity 60/60°F is d; thus, specific heat increases with temperature and decreases with specific gravity.

5.4.3 Heat of Combustion

The gross heat of combustion of heavy oil and bitumen is given with a reasonable degree of accuracy by the equation:

$$Q = 12,400 - 2100d^2$$

In the equation, d is the 60/60°F specific gravity. Deviation is generally less than 1% although highly aromatic heavy oil or bitumen may show considerably higher values.

For thermodynamic calculations of equilibria, combustion data of extreme accuracy are required because the heats of formation of water and carbon dioxide are large in comparison with those in the hydrocarbons. Great accuracy is also required of the specific heat data for the calculation of free energy or entropy. Much care must be exercised in selecting values from the literature for these purposes, since many of those available were determined before the development of modern calorimetric techniques.

5.4.4 Volatility

The volatility of a liquid or liquefied gas may be defined as its tendency to vaporize, that is, to change from the liquid to the vapor or gaseous state. Because one of the three essentials for combustion in a flame is that the fuel be in the gaseous state, volatility is a primary characteristic of liquid fuels. The distillation profile is also a measure of the relative amounts of these liquid fuels (albeit small and unrefined) in heavy oil and certain samples of bitumen.

Similarly there must also be some estimate of the ability of the constituents of heavy oil or bitumen to distill, or steam distill (that can occur during the application of steam-based recovery processes), from the material during thermal methods of EOR. However, before any volatility tests are carried out it must be recognized that the presence of more than 0.5% water in test samples of heavy oil or bitumen can cause several problems during distillation procedures. Water has a high heat of vaporization, necessitating the application of additional thermal energy to the distillation flask. Water is relatively easily superheated and therefore excessive *bumping* can occur, leading to erroneous readings and the potential for destruction of the glass equipment is real. In addition, steam formed during distillation can act as a carrier gas and high boiling point components may end up in the distillate (often referred to as *steam distillation*).

Centrifugation can be used to remove water (and sediment) if the sample is not a tight emulsion. Other methods that are used to remove water include: (1) heating in a pressure vessel to control loss of light ends; (2) addition of calcium chloride as recommended in ASTM D1160; (3) addition of an azeotroping agent such as *iso*-propanol or *n*-butanol; (4) removal of water in a preliminary low-efficiency or flash distillation followed by reblending the hydrocarbon which codistills with the water into the sample; and (5) separation of the water from the hydrocarbon distillate by freezing.

The vaporizing tendencies of petroleum and petroleum products are the basis for the general characterization of the volatility of fossil fuels (ASTM D2715). A test (ASTM D6) also exists for determining the loss of material when heavy oil and bitumen are heated. Another standard test method (ASTM D20) is applied to determining the volatility of road tar (refinery-produced asphalt) that might also be applied to estimating the volatility of high molecular weight unknown residues.

For some purposes it is necessary to have information on the initial stage of vaporization and the potential hazards, even with heavy oil and bitumen, that such a property can cause. To supply this need, flash and fire, vapor pressure, and evaporation methods are available. The data from the early stages of the several distillation methods are also useful. For other uses it is important to know the tendency of a product to partially vaporize or to completely vaporize, and in some cases to know if small quantities of high-boiling components are present. For such purposes, a chief reliance is placed on the distillation methods.

The *flash point* of petroleum or a petroleum product is the temperature to which the product must be heated under specified conditions to give off sufficient vapor to form a mixture with air that can be ignited momentarily by a specified flame (ASTM D56; ASTM D92; ASTM D93; ASTM D1310; ASTM D3828).

The Pensky-Marten apparatus using a closed or open system (ASTM D93) is the standard instrument for flash points above 50°C. The Cleveland Open-Cup method (ASTM D92) is also used for the determination of the *fire point* (the temperature at which the sample will ignite and burn for at least 5 seconds). The Pensky-Marten apparatus consists of a brass cup, mounted in an air bath and heated by a gas flame. A propeller-type stirrer, operated by a flexible drive, extends from the center of the cover into the cup. The cover has four openings: one for a thermometer, and the others fitted with sliding shutters for the introduction of a pilot flame and for ventilation. The temperature of the oil in the cup is raised at 5–6°C/minute (9–11°F/minute). The stirrer is rotated at approximately 60 rpm. When the temperature has risen to approximately 15°C (27°F) from the anticipated flash point, the pilot flame is dipped into the oil vapor for 2 seconds for every 1°C (1.8°F) rise in temperature up to 105°C (221°F). Above 105°C (221°F), the flame is introduced every 2°C (3.6°F) rise in temperature. The flash point is the temperature at which a distinct flash is observed when the pilot flame meets the vapor in the cup.

The *open-flash point* is determined after the closed flash point by removing the cover and continuing the heating until a distinct flash occurs across the open cup. The Abel *closed-cup* apparatus (IP 170) consists of a brass cup sealed in a small water bath that is immersed in a second water bath. The cover of the brass cup is fitted in a manner similar to that in the Pensky-Marten apparatus. For conventional crude oil and crude oil products with flash point <30°C (<86°F), the outer bath is filled with water at 55°C (131°F) and is not heated further. The oil under test is then placed inside the cup. When the temperature reaches 19°C (66°F) the pilot flame is introduced every 0.5°C (1°F) until a flash is obtained. For oils with flash points in excess of 30°C (>86°F) and less than 50°C (<122°F), the inner water bath is filled with cold water to a depth of 35 mm. The outer bath is filled with cold water and heated at a rate of 1°C/minute (1.8°F/minute). The flash point is obtained as before.

The *fire point* is the temperature to which the product must be heated under the prescribed conditions of the method to burn continuously when the mixture of vapor and air is ignited by a specified flame (ASTM D92).

From the viewpoint of safety, information about the *flash point* is of most significance at or slightly above the maximum temperatures (30–60°C, 86–140°F) that may be encountered in storage, transportation, and use of heavy oil, bitumen, and their respective products, in either closed or open containers. In this temperature range the relative fire and explosion hazard can be estimated from the flash point. For products with flash point below 40°C (104°F) special precautions are necessary for safe handling. Flash points above 60°C (140°F) gradually lose their safety significance until they become indirect measures of some other quality.

The flash point of heavy oil and bitumen can also be used to detect contamination. A substantially lower flash point than expected for a product is a reliable indicator that a product has become contaminated with a more volatile product, such as gasoline. The flash point is also an aid in establishing the identity of a particular hydrocarbon contaminant.

A further aspect of volatility that receives considerable attention is the vapor pressure of heavy oil and bitumen, which may be close to zero. The *vapor pressure* is the force exerted on the walls of a closed container by the vaporized portion of a liquid. Conversely, it is the force that must be exerted on the liquid to prevent it from vaporizing further (ASTM D323). The vapor pressure increases with temperature and the temperature at which the vapor pressure of a liquid, either a pure compound of a mixture of many compounds, equals 1 atmosphere pressure (14.7 psi, absolute) is designated as the boiling point of the liquid.

Heavy oil (and on occasion) tar sand bitumen can, in some cases, be subdivided by distillation into a variety of fractions of different *boiling ranges* (*cut points*) (the lower boiling fractions are de-emphasized because of the nature of heavy oil and bitumen) using a variety of standard test methods specifically designed for this task (ASTM D86; ASTM D1160; ASTM D2887; ASTM D2892; ASTM D5236).

Distillation involves the general procedure of vaporizing the petroleum liquid in a suitable flask either at *atmospheric pressure* (ASTM D86; ASTM D2892) or at *reduced pressure* (ASTM D1160). There are also test methods for the distillation of pitch (ASTM D2569) and cutback asphalt (ASTM D402) that can be applied to heavy oil and bitumen. However, most of the methods specify an upper atmospheric equivalent temperature (AET) limit of 360°C (680°F) and therefore are too limited to be of value in the analysis of low-volatility API gravity heavy oil and lower volatility bitumen.

In the simplest case, the distillation method involves using a standard round-bottom distillation flask of 250-mL capacity attached to a water-cooled condenser. The thermometer bulb is placed at the opening to the side arm of the flask. One hundred milliliters of sample are placed in the flask and heated by a small gas flame so as to produce 10 mL of distillate every 4 or 5 minutes. The temperature of initial distillation is recorded; the temperature at which each further 10 mL distills and the final boiling point are also recorded. The ASTM D2892 method describes the procedure for distilling crude petroleum up to 400°C (750°F) AET. This method is often referred to as the *true boiling point (TBP) distillation* method and is typically adequate for heavy oil and bitumen where an estimation of the volatile constituents and nonvolatile constituents is all that is required.

The ASTM D1160 method is used to determine the boiling ranges of petroleum products to a maximum liquid temperature of 400°C (752°F) at pressures as low as 1 mm Hg and is suitable for use with heavy oil and bitumen. In the method, a 200-mL sample is weighed to the nearest 0.1 g in a distillation flask. The distillation assembly is evacuated to the desired pressure and heat is applied to the flask as rapidly as possible using a 750-W heater. When refluxing liquid appears, the rate of heating is adjusted so that the distillate is recovered at 4–8 mL/minute until the distillation is complete. However, because of the thermal sensitivity of heavy oil and bitumen (being immature products of the various maturation processes), cracking will most likely occur before the liquid temperature reaches 400°C (752°F). An increase in distillation rate accompanied by a drop in head temperature, loss of vacuum in the system, which is restored when heat to the still is reduced, and production of vapor clouds in the system are all evidence of cracking.

The *vacuum pot still method* (ASTM D5236) is a procedure for the distillation of heavy oil samples having initial boiling points above 150°C (300°F). The method employs a pot still with a low-pressure drop entrainment separator. The method also provides for the determination of standard distillation profiles to the highest AET possible by conventional distillation. In the *spinning band* method, fractions of feedstocks, such as heavy oil and bitumen, with an initial boiling point above room temperature at atmospheric pressure can be prepared. The distillation is terminated at an AET of 524°C (975°F). For heavy oil and bitumen samples the distillation is terminated at a pot temperature of 360°C (680°F).

Samples are distilled under atmospheric and reduced pressures of 20.0 and 0.60 mm Hg (2.67 kPa and 80 Pa) in a distillation unit equipped with a spinning band column. The spinning band, which effectively provides a large contact area between the liquid phase and the vapor phase, increases the number of theoretical plates in the column and thus the fractionating efficiency of the column. Readings of vapor temperature (ie, convertible to AET) and distillate volume (ie, convertible to percent by volume) are used to plot a distillation curve. Distillate yields for naphtha, light gas oil, heavy gas oil, and residue fractions are determined on a gravimetric basis.

When attempting to fractionate heavy oil or bitumen by distillation, the pot temperature cannot exceed 360°C (680°F) or thermal decomposition will ensue. In the method, samples are distilled under atmospheric and reduced pressures in a still equipped with a spinning band column. Vapor temperatures are converted to atmospheric equivalent temperatures and can be plotted as a function of volume or weight percent distilled to yield a distillation profile.

Short path distillation produces a single distillate and a single residue fraction defined by the operating temperature and pressure of the still. This procedure is used to generate high boiling point fractions with endpoints up to 700°C (1290°F) for further analysis. Since only one cut temperature is used in each run, generation of a distillation curve using this equipment would be time-consuming. In the method, the material to be fractionated is introduced at a constant rate onto the hot inner wall of the evaporator under high vacuum. Rotating (Teflon) rollers ensure that the film on the wall is kept thin. The feedstock is progressively distilled at the fixed conditions of temperature and pressure. The distillate vapors condense on a concentric cold surface (60°C, 140°F) placed a short distance from the hot wall inside the still. The condensate then drains by gravity to the base of the *cold finger* where it is collected. The residue drains down the hot wall and is collected through a separate port.

Short path distillation permits collection of deeper cut distillate and residue fractions than conventional vacuum distillation methods. AET endpoints of up to 660°C (1220°F) can be attained without thermal cracking of the material. The distillation, however, is conducted under nonequilibrium conditions so the cut point is not as specific as in other vacuum distillation methods.

Generally, the distillation tests are planned so that the data are reported in terms of one or more of the following items: (1) initial boiling point; (2) distillation temperatures; (3) end point or maximum distillation temperature; (4) dry point; (5) recovery; (6) total recovery; and (7) percent evaporated.

The *initial boiling point* is the thermometer reading in the neck of the distillation flask when the first drop of distillate leaves the tip of the condenser tube. This reading is materially affected by a number of test conditions, namely, room temperature, rate of heating, and condenser temperature. *Distillation temperatures* are usually observed when the level of the distillate reaches each 10% mark on the graduated receiver, with the temperatures for the 5% and 95% marks often included. Conversely, the volume of the distillate in the receiver, that is, the percentage recovered, is often observed at specified thermometer readings. The *end-point* or *maximum temperature* is the highest thermometer reading observed during distillation. In most cases it is reached when all of the sample has been vaporized. If a liquid residue remains in the flask after the maximum permissible adjustments are made in heating rate, this is recorded as indicative of the presence of very high boiling compounds. The *dry point* is the thermometer reading at the instant the flask becomes dry and is for special purposes, such as for solvents and for relatively pure hydrocarbons. For these purposes the dry point is considered more indicative of the final boiling point than end point or maximum temperature.

The *recovery* is the total volume of distillate recovered in the graduated receiver and *residue* is the liquid material, mostly condensed vapors, left in the flask after it has been allowed to cool at the end of distillation. The residue is measured by transferring it to an appropriate small graduated cylinder. Low or abnormally high residues indicate the absence or presence, respectively, of high-boiling components. The *total recovery* is the sum of the liquid recovery and residue; *distillation loss* is determined by subtracting the total recovery from 100%. It is, of course, the measure of the portion of the vaporized sample that does not condense under the conditions of the test. Like the initial boiling point, distillation loss is affected materially by a number of test conditions, namely, condenser temperature, sampling and receiving temperatures, barometric pressure, heating rate in the early part of the distillation, and others. Provisions are made for correcting high distillation losses for the effect of low barometric pressure because of the practice of including distillation loss as one of the items in some specifications for motor gasoline. The *percent evaporated* is, the percentage recovered at a specific thermometer reading or other distillation temperatures, or the converse. The amounts that have been evaporated are usually obtained by plotting observed thermometer readings against the corresponding observed recoveries plus, in each case, the distillation loss. The initial boiling point is plotted with the distillation loss as the percentage evaporated. Distillation data are reproducible, particularly for the more volatile products.

The actual procedure involves the distillation of the sample in one to five gallon quantities in a column equivalent to 14 theoretical plates and with reflux adjusted to obtain a TBP versus yield relationship. Narrow fractions can be taken throughout the distillation and the still pressure can be reduced as necessary to avoid decomposition in the higher boiling ranges. The fractionating distillation is continued to a relatively high temperature such as 370°C (700°F) equivalent temperature at 760 mm of mercury absolute. The actual operating pressure of the still at this stage would be in the region of 2 mm of mercury absolute. The distillate fractions up to this point cover the light and middle distillates, and after the analysis of the required fractions in this range, curves relating the various properties to yield on fraction or on crude oil can be derived. A series of residues can also be prepared by appropriate blending of distillates with the final residue, and after analysis, property relationships with yield can also be prepared for the residues.

From the TBP curve, the yields of products of a required boiling range can be deduced and fractions can be blended to match required products for further analysis or treatments. The residue (b.p. >370°C, >700°F) obtained from the TBP distillation quoted, contains the main lubricating oil stock, and by further distillation under high vacuum a series of distillates can be prepared to cover the various lubricating oil grades.

It is often useful in this type of assay to be able to extend the boiling point data to higher temperatures than are possible in the fractionating distillation method previously described, and for this purpose a vacuum distillation in a simple still, with no fractionating column (similar to the ASTM D1160) can be carried out. This distillation, which is done under fractionating conditions equivalent to one theoretical plate, allows the boiling point data to be extended to about 600°C (1110°F) (corrected to 760 mm of mercury absolute) with many crude oils, heavy oil, and bitumen. Bearing in mind the limitations in accuracy of the temperature measurements in the high boiling ranges of crude it is possible to construct a wide range boiling point curve for the crude by combining the results of the simple distillation with those of the main distillation. This method gives useful comparative and reproducible results that are often accurate enough for engineering purposes.

It is advantageous in any analytical, research, or development organization to collect information by the above methods on a standard series of distillates or residues from crude oils, heavy oils, and bitumen so that valuable comparisons become available as the data accumulate. These distillates should be prepared as near as possible to the boiling ranges that are known to be suitable for marketable products together with a TBP yield curve: The residue (>370°C, >700°F) can be further fractionated, if required, using however a high vacuum still of sufficient capacity to give fractions in suitable quantity for subsequent analysis.

There is also another method that is increasing in popularity for application to a variety of feedstocks and that is the method commonly known as *simulated distillation* (ASTM D2887). This is a low-resolution, temperature-programmed gas chromatographic analysis intended to simulate the time-consuming TBP distillation method. It is based on the observation that hydrocarbons generally are eluted from a nonpolar column in the order of their boiling point, with the retention time being converted to equivalent distillation temperatures (Green, 1976). The accumulated detector response is related to the amount of sample distilled. In practice, the nonpolar stationary phases do discriminate to some extent among different classes of hydrocarbon. The retention time is a function of the adsorptive property of the stationary phase, the vapor pressure, and the heat of vaporization. The benefits of the technique include good comparisons with other ASTM distillation data as well as the application to higher boiling fractions. In fact, data output includes the provision of the corresponding Engler profile (ASTM D86) as well as the prediction of other properties such as vapor pressure and flash point.

For a more detailed distillation analysis of feedstocks and products, a low-resolution, temperature-programmed gas chromatographic analysis has been developed to simulate the time-consuming TBP distillation. The method relies on the general observation that hydrocarbons are eluted from a nonpolar adsorbent in the order of their boiling points. The regularity of the elution order of the hydrocarbon components allows the retention times to be equated to distillation temperatures and the term *simulated distillation by gas chromatography* (or *simdis*) is used throughout the industry to refer to this technique (Worman and Green, 1965; Green, 1976; MacAllister and DeRuiter, 1985; Neer and Deo, 1995).

The quality of crude oil is (in the current context) among the most important determinants for recovery operations. The quality and value of a crude oil depends on the TBP curve that is the content of light fractions (boiling up to 350°C/660°F), 350–510°C (vacuum gas oil) fraction, and the bottom of the crude oil barrel (510°C+/950°F; vacuum residue), and the level of impurities such as sulfur, nitrogen, and metals. The light oil fractions are generally used for production of transportation fuels. The vacuum gas oil fraction can be used as a feedstock for catalytic cracking or hydrocracking units which produce motor fuels, or for production of base lubricating oils. The crude oil bottom (black-oil fraction) is generally used for production of fuel oil.

Depending on the quality of the heavy oil (from data such as the carbon residue, asphaltene content, and metals content), there is the potential for conversion to high value transportation fuels by thermal conversion or hydrocracking processes. The TBP analysis is not typically part of the assay for heavy oil or tar sand bitumen. However, for recovery operations, especially where the presence of any volatile material needs to be known because of the potential for steam distillation to play an active role in a steam-based recovery process since the lack of information about the heavy oil quality could negatively impact the optimum recovery operation and, hence, there is an opening for determining the distillation analysis of, at least, heavy oil (ASTM D2887; ASTM D5236; ASTM D7169; Riazi, 1989; Nedelchev et al., 2011; Stratiev et al., 2011, 2014; Nikolaychuk et al., 2015).

Finally, heavy oil and tar sand bitumen have minimal constituents in the naphtha range (0–200°C, 32, 50, 239°F) (Subramanian et al., 1996; Speight, 2000, 2014a). This is especially true in the case of tar sand bitumen. Nevertheless, whether the bitumen is produced by surface recovery or by in situ processes, the characterization of these materials is important for the development of recovery processes and of kinetic and thermodynamic models of the processes employed to convert them in situ or to refinery feedstocks.

5.4.5 Liquefaction and Solidification

The liquefaction and solidification of heavy oil and bitumen seems to draw little attention in the standard textbooks of petroleum science and yet, both properties are very important in handling heavy oil and bitumen at the recovery site prior to transportation. In fact, since heavy oil and bitumen can both be borderline liquid or near-solid at ambient temperature, problems may arise from solidification during normal use or storage.

The *melting point* is a test (ASTM D87; ASTM D127) that is widely used by suppliers of wax and by the wax consumers, which can also be applied to heavy oil and bitumen but it is the *softening point* (ASTM D36), which is the temperature at which a disk of the material softens and sags downward a distance of 25 mm under the weight of a steel ball under strictly specified conditions, that finds wider use for heavy oil and bitumen. The *dropping point* (ASTM D566) is the near-equivalent test that is used for lubricating greases.

Recently, more emphasis has been placed on the *pour point* which, in conjunction with the reservoir temperature, can give indications of the fluidity and mobility of the heavy oil in the reservoir or bitumen in the deposit. The *pour point* is the temperature below which the oil becomes a semisolid and will not flow. In addition to being viscous, some heavy oils and especially bitumen are semisolids to solid at reservoir or deposit ambient temperatures. At temperatures below the pour point, heavy oil and bitumen will not flow. For this reason, these material are usually transported in heated pipeline or in tanks and if the heating system fails, the oil may cool to temperatures below its pour point and thus solidify. The change in state, from liquid to semisolid, arises from internal changes in the physical structure of the oil and is a key variable in reservoir or deposit behavior and recovery operations. An alternate option for transportation of high pour point heavy oil or bitumen is dilution with aromatic naphtha—use of paraffinic naphtha will likely may cause the separation of asphaltene constituents as an insoluble solid phase.

Briefly, determination of the pour point was originally applied to crude oil that had a high wax content. More recently, the pour point, like the viscosity, is determined principally for use evaluating mobility or immobility at reservoir or deposit temperature as well as in pumping arid pipeline design calculations. In addition, the pour point has also found use as an indicator of the temperature at which heavy oil will flow during in situ recovery operations (Fig. 5.5). For example, for asphaltic crude oils where paraffin precipitation will not occur, if 21°C (70°F) is the pour point of heavy oil in a reservoir where the temperature is 38°C (100°F), the oil is liquid under reservoir conditions and will be mobile and will flow under those conditions. On the other hand, tar sand bitumen (pour point: 60°C, 140°F) in a deposit (temperature: 10°C, 50°F) will be solid and immobile. This state of the oil in the reservoir can also have consequences on the ability of gases and liquids (eg, steam, hot water) used for recovery operations to penetrate the reservoir/deposit. Although pressure can have some influence on the pour point, the effect is not large and is unlikely to affect any general conclusions. Indeed, there is a relationship between API gravity and pour point. Thus, any increase in pour point due to an increase in pressure (surface compared to reservoir or deposit pressure) will be most likely be negated as the API gravity decreases with increase in temperature (60°F compared to reservoir temperature).

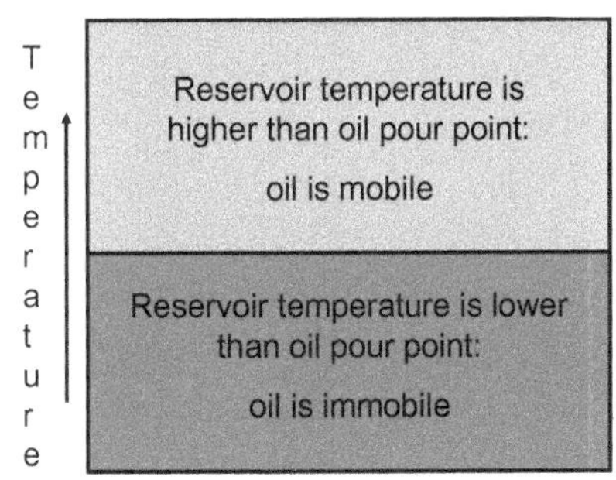

Figure 5.5 Relationship of pour point and reservoir temperature.

To determine the *pour point* (ASTM D97), the sample is first heated to 46°C (115°F) and cooled in air to 32°C (90°F) before the tube is immersed in the same series of coolants as used for the determination of the *cloud point*. The sample is inspected at temperature intervals of 2°C (3°F) by withdrawal and holding horizontal for 5 seconds until no flow is observed during this time interval.

Whether or not the relationship of pour point and reservoir temperature is generally accepted remains to be seen but the concept of withdrawal from the use of a single parameter to predict oil behavior is a necessity. The use of two parameters, whatever they may be, gives a more realistic view of heavy oil and bitumen behavior.

5.4.6 Wax Analysis and Wax Appearance Temperature

Many techniques have been developed to quantify the composition of crude, but by far the most utilized method is by gas chromatography by which a complete characterization up to carbon number C_{100} can be performed. Detection levels of ppb (parts per billion) can be repeatedly obtained with small sample sizes. In the modern equipment, the inlet, chromatographic column, and the detectors can be changed to suit the desired analysis. However, some detectors will not detect nitrogen, carbon dioxide, or anything else that will not burn.

Precipitated wax can be detected in different forms by different techniques, such as (1) the form of the quantity of wax precipitated, relative to the oil quantity or composition, (2) the size of the wax crystals, and (3) the number of wax crystals, but all of these affect the sensitivity of the measurement techniques. In addition, some techniques detect wax crystals at the microscopic level (nucleation stage of wax crystal formation) while others detect wax crystals at the early stage of growth. However, it has been suggested that none of the available techniques is able to measure the true wax appearance temperature, where the first crystal appears under thermal equilibrium (Hammami et al., 2003).

Standard test methods are available for determination of the cloud point (ASTM D2500; ASTM D5551; ASTM D5771; ASTM D5772; ASTM D5773; ASTM D7397; ASTM D7683; ASTM D7689) and the methods are unique for detection of cloudiness of a fluid sample in a glass container as the temperature is reduced. Other methods that are also available and under development and include: (1) cold finger testing; (2) viscometry; (3) light transmittance; (4) cross-polarization microscopy; and (5) flow loop methods.

The *cold finger testing* technique uses a cold surface (cold finger) that is placed in a sample of heated crude oil, and cooling fluid is circulated through the interior of the cold finger. The cooling fluid is provided by a controlled circulating heating and cooling bath. In the method, the crude oil is maintained at a temperature above the wax appearance temperature and is gently agitated about the cold finger with a magnetic stirrer. With time, a deposit forms on the cold finger's surface which gives a measure of the problem that can be encountered in the field that is producing the oil. The time parameter is not standard and is individually dependent.

The *viscometry* technique employs the linear relationship between fluid viscosity and temperature. Wax formation changes the crude oil from Newtonian to non-Newtonian fluid behavior. The wax appearance temperature is determined from a plot of natural log of viscosity (Ln viscosity) versus the inverse of absolute temperature ($1/T$) to be the point at which deviation from linearity occurs as temperature is lowered. Viscometry can be used for opaque fluids. The method requires a considerable volume fraction of wax crystals to be able to detect the phase transition, which gives the cloud point. The wax appearance temperature that is determined depends on the sharpness of the deviation from linearity of the viscosity versus temperature plot. However, cautions have been expressed that viscometry technique should not be used independently for wax appearance temperature measurement, but along with other techniques of greater certainty and accuracy. Results inferred from viscometry could be misleading especially when the change in fluid behavior does not produce a distinctive point.

The *light transmittance* technique relies on the variation of light transmission by different phases such as solid and liquid. The intensity of light transmitted through or scattered by a sample as the temperature varies is measured. When this is applied to oil samples in a cooling process, the variation in light transmission due to appearance of crystals indicates the wax appearance temperature. The light transmission versus temperature curve shows a marked decrease in light transmission at the wax appearance temperature (Kruka et al., 1995). The method requires the formation of a certain quantity of crystals before a detectable reduction in transmitted light can be obtained and, thus, the measured cloud point would be lower than the true cloud point. The cloud point is detected at the growth stage of wax formation. The light transmittance technique can be used for all fluids, both opaque and transparent, depending on light source and also can be used for high pressure cloud point measurement, which implies that it can be used for determination of the wax appearance temperature of live oil (Coutinho and Daridon, 2005).

The *cross-polarization microscopy* technique is based on the ability of wax crystals to rotate the plane of transmitted polarized light, whereas liquid hydrocarbons do not. The basic components of the microscope are the polarizer and the analyzer. In this method, the slide containing the sample is placed on a thermal microscope stage, brought into focus, and viewed under polarized light, where wax crystals appear as bright spots on a dark background. A digital camera or video camera is employed to capture the appearance of wax crystals, thereby enhancing accuracy. The method has also been stated to be satisfactory for use with dark colored crude oils (Hammami and Raines, 1997, 1999).

Flow loops are used for wax deposition studies under dynamic conditions (Bidmus and Mehrotra, 2009; Lashkarbolooki et al., 2010). In this method, the oil is usually circulated in closed loops designed to suit the aim/objective of the research (Leontaritis and Leontaritis, 2003). Flow loops can be used for a variety of crude oils and for both opaque and transparent fluids. In dynamic test loops, the active condition of the crude oil can be simulated, which has the potential to generate results that depict the actual field situation.

5.5 METALS CONTENT

Heteroatoms (*nitrogen*, *oxygen*, *sulfur*, and *metals*) are found in every crude oil, heavy oil, and bitumen and the concentrations of these elements have to be reduced to convert the oil to transportation fuel. The reason is that if nitrogen and sulfur are present in the final fuel during combustion, nitrogen oxides (NO_x) and sulfur oxides (SO_x) form, respectively. In addition, metals affect many upgrading processes adversely, poisoning catalysts in refining and causing deposits in combustion.

Heteroatoms do affect every aspect of recovery and refining. Sulfur is usually the most concentrated and is fairly easy to remove; many commercial catalysts are available that routinely remove 90% of the sulfur. Nitrogen is more difficult to remove than sulfur, and there are fewer catalysts that are specific for nitrogen. Metals cause particular problems because they poison catalysts used for sulfur and nitrogen removal as well as other processes such as catalytic cracking. In addition, the levels of nickel and vanadium in heavy oil and bitumen need to be considered by the geochemist when the origins of these two natural products are considered.

Heavy oil and bitumen contain relatively high proportions of *metals* either in the form of salts or as organo-metallic constituents (such as the metallo-porphyrins), which are extremely difficult to remove from the feedstock. Indeed, the nature of the process by which residua are produced virtually dictates that all the metals in the original crude oil are concentrated in the residuum (Speight, 2000). Those metallic constituents that may actually *volatilize* under the distillation conditions and appear in the higher boiling distillates are the exceptions here. The deleterious effect of metallic constituents on the catalyst is known particularly through their ability to modify the selectivity of Zeolite catalysts thereby causing an increase in the formation of coke at the expense of the more desirable liquid products. Thus, serious attempts have been made to develop catalysts that can tolerate a high concentration of metals without serious loss of catalyst activity or catalyst life. However, for the most part, the metals concentrate in the coke formed during thermal processes.

A number of the heavy metals such as nickel, vanadium, copper, and iron can also be effectively bound in large organic molecules characteristic of those found in the asphaltene fraction (pentane- or heptane-insoluble portion of the feedstock) and resins (ASTM D3279; ASTM D4124). *Nickel* and *vanadium* porphyrins are commonly found and show high thermal stability allowing them to pass through the extraction process into the upgrading process. Porphyrins are the major, but certainly not the only, organo-metallic complexes present. Metals may simply be entrapped or loosely bound in the very large molecules present in the asphaltenes and resins. Although *iron* is present as organo-metallic compounds, it occurs mostly in the form of process-accumulated rust or is scrounged from pipelines during shipping and pipelining. Thus, worrying about the geochemical significance of a metal, such as iron, in crude oil may not be a top priority of recovery operators or to refinery operators. Without doubt these metals are present in recovered heavy oil and bitumen and since catalysts are used extensively in upgrading and are readily poisoned by such metals, it is important to know the amounts present. Significant levels typically found in these materials include nickel 20–200 ppm; vanadium 50–500 ppm; and iron 150–1000 ppm.

Metals (particularly *vanadium* and *nickel*) are found in every heavy oil or bitumen sample (Reynolds, 1998; Speight, 2014a). These materials contain relatively high proportions of metals either in the form of salts or as organo-metallic constituents (such as the metallo-porphyrins), which are extremely difficult to remove from the feedstock. The metallic constituents that may actually *volatilize* under the thermal recovery operations appear in the reservoir or in the production lines.

Thus, a variety of tests have been designated for the determination of metals in crude oil, crude oil products (such as residua and asphalt, heavy oil, and bitumen) (ASTM D1318; ASTM D3340; ASTM D3341; ASTM D3605) (Speight, 2014a, 2015). Determination of metals in whole feeds can be accomplished by combustion of the sample so that only inorganic ash remains. The ash can then be digested with an acid and the solution examined for metal species by atomic absorption (AA) spectroscopy or by inductively coupled argon plasma (ICAP) spectrometry.

The analysis of heavy oil and bitumen can be accomplished by several instrumental techniques: ICAP Spectrometry; AA Spectrometry; and X-ray Fluorescence (XRF) Spectrometry. Each technique has limitations in terms of sample preparation, sensitivity, sampling, time for analysis, and overall ease of use. Thus a variety of tests (ASTM D482; ASTM D1318; ASTM D3340; ASTM D3341; ASTM D3605) either directly or as the constituents of combustion ash have been designated to determine metals in oil based on a variety of techniques. However, at the time of writing, the specific test for the determination of metals in whole feeds has not been designated but this task can be accomplished by combustion of the sample so that only inorganic ash remains (ASTM D482). The ash can then be digested with an acid and the solution examined for metal species by AA spectroscopy or by ICAP spectrometry.

AA provides very high sensitivity but requires careful subsampling, extensive sample preparation, and detailed sample-matrix corrections. XRF requires little in terms of sample preparation but suffers from low sensitivity and the application of major matrix corrections. ICAP spectrometry provides high sensitivity and few matrix corrections, but requires a considerable amount of sample preparation depending on the process stream to be analyzed.

Whenever, a technique requires conversion of any mineral matter in the sample to mineral ash, one of the issues that arises is the potential loss of volatile nickel and vanadium compounds. Both sulfuric acid and elemental sulfur have been employed to prevent loss of these metals. In fact, because of this a wet ashing method is often preferred over dry ashing (Wallace, 1988). The analytical method should be selected depending on the sensitivity required, the compatibility of the sample matrix with the specific analysis technique, and the availability of facilities. Sample preparation, if it is required, can present problems. Significant losses can occur, especially in the case of organo-metallic complexes, and contamination of environmental samples is of serious concern.

Analysis for metals in an organic matrix is most easily accomplished using XRF or neutron activation techniques for quantitative analyses. These particular techniques are also applicable for the direct analyses of oil sand solids. Spark techniques can be used for qualitative analyses. The direct aspiration of metal-containing hydrocarbons into the plasma instrument also provides quantitative information, with the assumption that any particulates present are included in the analyses. The precision of the analysis depends on the metal itself, the method used, and the standard used for calibration of the instrument. Relative standard deviations range from 1% to more than 20%.

In the ICAP emission spectrometer method, nickel, iron, and vanadium content of gas oil samples range from 0.1 to 100 mg/kg. Thus, a 10 g sample of gas oil is charred with sulfuric acid and subsequently combusted to leave the ash residue. The resulting sulfates are then converted to their corresponding chloride salts to ensure complete solubility. A barium internal standard is added to the sample which before analysis.

The ICAP spectrometer is an instrument routinely used for the analysis of the dissolved metal composition of aqueous samples. Solid samples are prepared in such a way (ashing then fusion) that their final analysis matrix is in the aqueous dissolved form. The analysis of metals in a hydrocarbon matrix, potentially containing solids important to the analysis, is a problem. The ashing of a hydrocarbon sample containing metals (whether in organic-complexes, elemental, or inorganic complexes) can lead to analysis errors in that some of the complexes, specifically the organo-metallic compounds, are in a volatile form at typical ashing temperatures (700–800°C, 1290–1470°F). The metals in organic complexes tend to originate in the high molecular weight hydrocarbons found naturally in the feed or generated in the reactor. The inorganic metal complexes originate naturally in the feed, or from the breakdown of catalyst, or the formation of inorganic complexes under the reactor conditions. Without ashing and subsequent fusion, any solids present may not reach the analytical region of the instrument resulting in analytical errors.

Using the ICAP method for the analysis of nickel, vanadium, and iron present counteracts the two basic issue arising from metals analysis. The most serious issue is the fact that these metals are partly or totally in the form of volatile chemically stable porphyrin complexes and extreme conditions are needed to destroy the complexes without losing the metal through volatilization of the complex. The second issue is that the alternate direct aspiration of the sample introduces large quantities of carbon into the plasma. This carbon causes marked and somewhat variable background changes in all direct measurement techniques.

5.6 BULK COMPOSITION AND MOLECULAR WEIGHT

The bulk (physical) composition of reservoir fluids is a subset of fluid characterization (Chapter 5) and distribution within the reservoir which helps in defining reservoir continuity and communication among various zones. Interpretation of well-test data and the design of surface facilities and processing plants require accurate fluid information and its variation with time. In addition to initial reservoir-fluid samples, periodic sampling is necessary for reservoir surveillance.

Reservoir-fluid characterization consists of several key steps: (1) acquisition of representative samples; (2) identification of reliable service laboratories to perform PVT measurements; (3) implementation of QA/QC procedures to ensure data quality; and (4) development of mathematical models to capture fluid-property changes accurately as functions of pressure, temperature, and composition. The fluid type and production processes dictate the type and the volume of required fluid data. This paper outlines recommended sampling techniques, PVT data-acquisition strategies, and modeling methods and presents field examples covering a wide range of fluid types from heavy oils to lean gas condensates and production processes such as depletion, pressure maintenance, and miscible recovery.

The term *physical composition* (or *bulk composition*) refers to the composition of crude oil as determined by various physical techniques. For example, the separation of petroleum using solvents and adsorbents (Speight, 2014a, 2014b, 2014c, 2015) into various bulk fractions determines the physical composition of crude oil. These methods of separation are not always related to chemical properties and the terminology applied to the resulting fractions if often a *terminology of convenience*.

Proper management of production from a heavy oil reservoir or bitumen for a tar sand deposit can maximize the recovery of the oil originally in the reservoir. Developing proper management strategies requires accurate knowledge of the characteristics of the reservoir fluid as long as fluid samples obtained from the reservoir fluid reflect the pertinent properties of the fluid, as determined by subsequent laboratory tests.

As already stated, the objective of reservoir fluid sampling is to collect a sample that is representative of the fluid present in the reservoir at the time of sampling (Speight, 2001, 2015). Adequate volumes should be collected for plant and process analysis, geochemical analysis for fluid-source identification and reservoir continuity, and crude assay for refinery processes. The critical steps in any successful sampling program are avoiding two-phase flow in the reservoir, minimizing fluid contamination introduced by drilling and completion fluids, and preserving sample integrity. A sampling program should focus on the key issues of selecting an appropriate sampling method and associated tools, customizing the tool string, and developing sound sampling, sample-transfer, and QC procedures. In addition, specific sampling issues should be addressed related to fluid type and condition, saturated vs undersaturated, and fluids with nonhydrocarbon components or fluids. If the sampling procedure is incorrect or if samples are collected from an improperly conditioned well, the resulting samples may not be representative of the reservoir fluid.

A nonrepresentative sample may not exhibit the same properties as the reservoir fluid. The use of fluid-property data obtained from nonrepresentative samples, however accurate the laboratory test methods, may result in errors in reservoir management. Poor planning can also result in incomplete data being taken during the sampling program. Incomplete data can make it difficult or impossible for laboratory personnel to perform and interpret tests that provide accurate and meaningful fluid-property information.

When a reservoir is relatively small, a properly taken sample from a single well can be representative of the fluid throughout the entire reservoir. For reservoirs which are large or complex, samples from several wells and/or depths may be required. Significant variations in fluid composition often occur in very thick formations, in really large reservoirs, or in reservoirs subjected to recent tectonic disturbances. Additional sampling during the later life of a reservoir is not uncommon because production experience can show that the reservoir is more complex than earlier information indicated.

Conditioning a well before sampling is almost always necessary. Initial well testing or normal production operations often result in the fluid near the wellbore having a composition which has been altered from that of the original reservoir fluid (for reasons described later). The objective of conditioning the well is to remove this altered (nonrepresentative) fluid.

Well conditioning consists of producing the well at a rate which will move the altered fluid into the wellbore and allow it to be replaced by unaltered (representative) fluid flowing in from further out in the reservoir. Well conditioning is especially important when the reservoir fluid is at or near its saturation pressure at the prevailing reservoir conditions because reduction in pressure near the wellbore, which inevitably occurs from producing the well, will alter the composition of the fluid flowing into the wellbore.

Once the sample has been obtained, storage protocols must be observed because of the potential for oxidation and tests can then commence (Chapter 4) (Mushrush and Speight, 1995; Speight, 2014a, 2014b, 2014c).

5.6.1 Asphaltene Separation

Heavy oil evaluation by separation into various fractions has been used successfully for several decades. The knowledge of the bulk fractions of heavy oil (Fig. 5.1) on a *before recovery* (core sample analysis) and *after recovery* (well fluid analysis) basis, as well as variations over time, has been a valuable aid to recovery process development. But first, the asphaltene fraction and the resin fractions are not chemical classes. Each fraction is, by definition, a *separation class* and the composition of each class is dictated by the method of separation.

In the case of the asphaltene fraction, it is the collection of constituents that is precipitated from feedstocks by the addition of 40 volumes of the liquid low-boiling hydrocarbon (Girdler, 1965; Mitchell and Speight, 1973; Speight et al., 1982, 1984; Andersen and Birdi, 1990, 1991; Andersen, 1994; Speight, 1994, 2001, 2014a, 2014b, 2014c, 2015; Gholoum et al., 2003; Gholoum and Oskui, 2004). The fraction is separated as dark brown to black friable solid that has no definite melting point and usually foams and swells on heating to leave a carbonaceous residue. The composition of the asphaltene fraction varies with the hydrocarbon employed for the separation (Girdler, 1965; Mitchell and Speight, 1973; Speight, 2014a, 2014b, 2014c, 2015) and, in fact, propane is used commercially in processing petroleum residua for asphalt production—the asphalt is a mixture of resin and asphaltene constituents with some heavy gas oil constituents (Speight and Ozum, 2002; Hsu and Robinson, 2006; Gary et al., 2007; Speight, 2014a, 2014b, 2014c).

The complexity of the asphaltene fraction is reflected in the separation of the fraction by functional group composition (Chapter 4) (Francisco and Speight, 1984) but it cannot be defined by the separation method and the solvent employed. In fact, there is no one parameter that is operational in the separation of the asphaltene fraction (Speight, 1994; Chukwu et al., 2011). The relevant parameters for asphaltene separation are physical and chemical in nature and include:

1. Polarity (the presence of functional groups derived from the presence of heteroatoms in the asphaltene constituents) (Long, 1979, 1981; Speight, 1994).
2. Aromaticity (the presence of polynuclear aromatic systems in the asphaltene constituents) (Girdler, 1965; Mitchell and Speight, 1973).
3. Molecular weight (molecular size) (Long, 1979, 1981).
4. Three-dimensional structure (the micelle) of the asphaltene constituents as they exist in relationship with the other constituents of crude oil (Speight, 1992; Andersen and Speight, 1992, 1994).
5. Solvent power of the precipitating/extracting liquid used for the separation (Girdler, 1965; Mitchell and Speight, 1973).
6. The time required to allow the precipitating/extracting liquid to penetrate the micelle which is dependent upon the ability of the hydrocarbon liquid to penetrate the micelle, indicating that the process is diffusion-controlled (Speight et al., 1984).
7. The ratio of the precipitating/extracting liquid to crude oil that dictates the yield and character of the asphaltene product (Girdler, 1965; Mitchell and Speight, 1973; Speight et al., 1984).
8. The temperature, which may have an effect on the induction period that is a requirement of diffusion-controlled processes (Mitchell and Speight, 1973; Hirschberg et al., 1984; Andersen and Birdi, 1990; Hassanvand et al., 2012).
9. The pressure, as employed in several refinery processes as a means of maintaining the low-boiling liquid hydrocarbon in the liquid phase (Speight and Ozum, 2002; Hsu and Robinson, 2006; Gary et al., 2007; Speight, 2014a, 2014b, 2014c).

Other parameters may be defined as subsets of those enumerated above. It is also worthy of note that, in order to remove entrained resin material, precipitation of the asphaltene constituents from benzene or toluene is often necessary (Speight et al., 1984). But, none of these parameters applied to the separation of petroleum can be related to the separation of distinct chemical types.

Briefly, the separation of the asphaltene fraction can be ascribed to changes in crude oil composition caused by the addition of lower boiling components that the complex equilibrium keeping the asphaltene constituents dispersed in a *peptized state*. In the case of addition of low-boiling liquid hydrocarbons to the feedstock, the hydrocarbon causes a change in the solubility parameter of the oil medium that, in turn, changes the tolerance of the medium for the complex micelle structure. As this occurs, the lower molecular weight and less polar constituents of the micelle (the resin constituents) are extracted into the liquid, leaving the asphaltene constituents without any surrounding (dispersing) sheath—separation then ensues. If the process is thermal (ie, at a temperature in excess of the thermal decomposition temperature) a different sequence of events ensues leading to the formation of an insoluble phase and ultimately to coke (Chapter 3) (Speight, 2014a, 2014b, 2014c). Although the thermal decomposition of asphaltene constituents has been represented as a linear path, the actual path is multifaceted (Figs. 5.6 and 5.7) and, considering the complexity of the asphaltene fraction, adds an additional dimension to the pathways for the thermal decomposition of asphaltene constituents since each chemical type will decomposed at different rates and with a different product slate. Similar pathways for thermal decomposition of the other bulk fractions (Fig. 5.1) may be estimated with the multifaceted pathway being the more likely to be operative.

Resin constituents are soluble in the liquids that precipitate asphaltene constituents and are usually soluble in most organic liquids, except in the lower alcohols and acetone, but they are precipitated by liquid propane and liquid butanes. The resin constituents often coprecipitate with the asphaltene constituents in controlled propane deasphalting procedures, and the product, called *propane asphalt*, contains appreciable amounts of adsorbed resin constituents and has the properties of a low-melting-point asphalt. The resin fraction is dark, semisolid or solid, and adhesive (Koots and Speight, 1975; Andersen and Speight, 2001). The composition of the fraction can vary depending on the kind of precipitating liquid and on the temperature of the liquid system. In addition, the fraction becomes fluid on heating but often shows pronounced brittleness when cold.

The *oils fraction* (comprising the *saturates fraction* plus the *aromatics fraction*) comprises the lowest molecular weight fraction of petroleum and is

Asphaltenes → Carbenes → Carboids → Coke

Asphaltenes :	Benzene-soluble, pyridine-soluble
Carbenes :	Benzene-insoluble, pyridine-soluble
Carboids :	Benzene-insoluble, pyridine-insoluble
Coke :	Insoluble

Figure 5.6 Linear path for thermal decomposition of asphaltene constituents.

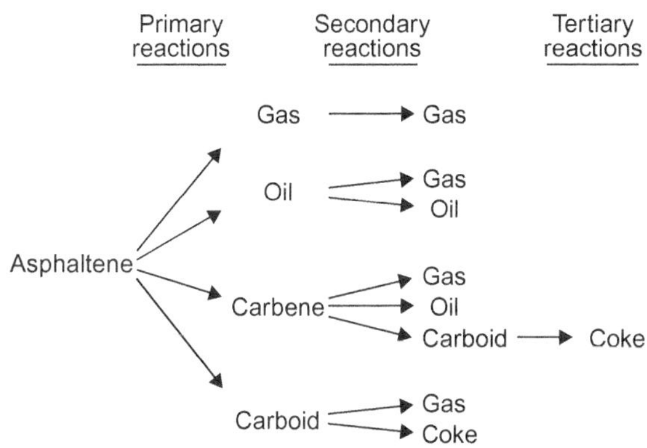

Figure 5.7 Multireaction sequence for the thermal decomposition of asphaltene constituents.

the dispersion medium for the peptized asphaltene constituents. Although the *oils fraction* is sometimes colored, fractions obtained by chromatography are colorless and are similar to white medicinal oils or lubricating oils of high purity. The oils fraction may be quite viscous because of the presence of paraffin waxes that vary over a wide range for crude oil and other feedstocks. The oils fraction is miscible with low molecular weight liquid hydrocarbons (such *n*-pentane and *n*-heptane), as well as most organic solvents, and may be separated into various hydrocarbon subfractions by suitable solvent or adsorbent methods.

Thus, the *asphaltene fraction* is that portion of heavy oil feedstock that is precipitated when a large excess (40 volumes) of a low-boiling liquid hydrocarbon (eg, *n*-pentane or *n*-heptane) is added to the crude oil (1 volume) (Speight, 1994, 2014a, 2014b, 2014c, 2015). *n*-Heptane is the preferred hydrocarbon with *n*-pentane still being used although hexane is used on occasion (Speight, 2014a, 2014b, 2014c, 2015 and references cited therein). Although, *n*-pentane and *n*-heptane are the solvents of choice in the laboratory other solvents can be used (Speight, 1979) and cause the separation of the asphaltene fraction as a brown-to-black powdery solid material. In the refinery, supercritical low molecular weight hydrocarbons (eg, liquid propane, liquid butane, or mixtures of both) are the solvents of choice and the product is a semisolid (tacky) to solid asphalt. The amount of asphalt that settles out of the paraffin/residuum mixture depends on the size of the paraffin, the temperature, and the paraffin-to-feedstock ratio (Girdler, 1965; Mitchell and Speight, 1973; Speight et al., 1984).

When pentane and the lower molecular weight hydrocarbon solvents are used in large excess, the quantity of precipitate and the composition of the precipitate changes with *increasing temperature* (Mitchell and Speight, 1973). At ambient temperature, the quantity of precipitate first increases with increasing *ratio of solvent to feedstock* and then reaches a maximum. In fact, for many heavy oils, there are indications that when the proportion of solvent in the mix is <35% little or no asphaltene constituents are precipitated.

Contact time between the hydrocarbon and the feedstock also plays an important role in asphaltene separation. Yields of the asphaltenes reach a maximum after approximately 8 hours, which may be ascribed to the time required for the asphaltene particles to agglomerate into particles of a *filterable size* as well as the diffusion-controlled nature of the process. Heavier feedstocks also need time for the hydrocarbon to penetrate their mass.

For example, if the precipitation method (deasphalting) involves the use of solvent and heavy oil, it is essentially a leaching of the soluble constituents form the insoluble constituents and may be referred to as *extraction*. However, under the prevailing conditions now in laboratory use, the term *precipitation* is perhaps more correct and descriptive of the method. Variation of *solvent type* also causes significant changes in asphaltene yield. Thus, the contact time between the feedstock and the hydrocarbon liquid can have an important influence on the yield and character of the asphaltene fraction.

At this point, a mention of the phase behavior of asphaltenes and fluids containing asphaltenes is worthwhile.

The phase behavior of fluid containing asphaltenes is complex. Chemically, asphaltenes are difficult to define in general, and the physics and chemistry underlying the definition of this fraction is not open to debate; the fraction is a solubility fraction and is, in reality, an artifact of the separation method (Speight, 2014a, 2014b, 2014c, 2015). In fact, asphaltene fractions possessing similar constituents may exhibit different properties in their native fluids and in solvents/nonsolvent mixtures. Asphaltenes constituents intra-act and interact with one another and with solvent media (Speight, 2014a, 2014b, 2014c, 2015). Phase behavior and precipitation models must capture the relevant physics and chemistry if derived models are to be truly predictive. But the issue is the use of *average parameters* rather than the recognition the asphaltene fraction is collection of different molecular types (Fig. 4.3) that vary from crude oil to crude oil as well as from crude oil to heavy oil to bitumen (Fig. 4.4), because of the complexities of the maturation process (Speight, 2014a, 2014b, 2014c), and should be represented as such in any models. In fact, analytical methods such as HPLC have shown conclusively that the asphaltene fraction is a mix of unknown (at best *speculatively known*) molecular types (Fig. 4.5) (Speight, 2001, 2014a, 2014b, 2014c, 2015).

In spite of this, *average parameters* and *average properties* continue to be used for derivation of predictive models when the behavioral characteristics for any average structure for such a complex mixture cannot in any way be truly representative of the behavior of the different constituents. Thus, it is not surprising that many models for heavy oil behavior are of little value in terms of predictive capability of heavy oil behavior and the models generally flounder on the rocky shores of asphaltene science (Andersen and Speight, 1999).

5.6.2 Fractionation

Fractionation of heavy oil into components other than asphaltenes (Fig. 4.2) has also been of interest in following recovery procedures. By careful selection of a characterization scheme it may be possible to obtain a detailed overview of oil composition that can be used for process predictions. Thus, fractionation methods also play a role, along with the physical testing methods, of evaluating heavy oils and the various recovery processes, especially when determining whether or not in situ upgrading occurs. For example, by careful selection of an appropriate technique it is possible to obtain a detailed *map* of heavy oil that can be used for predictions of behavior (Long and Speight, 1989, 1998; Speight, 2014a, 2014b, 2014c, 2015 and references cited therein).

After removal of the asphaltene fraction, further fractionation of petroleum is also possible by variation of the hydrocarbon solvent. For example, liquefied gases, such as propane and butane, precipitate as much as 50% by weight of the residuum or bitumen. The precipitate is a black, tacky, semi-solid material, in contrast to the pentane-precipitated asphaltenes, which are usually brown, amorphous solids. Treatment of the propane precipitate with pentane then yields the insoluble brown, amorphous asphaltenes and soluble, near-black, semisolid resins, which are, as near as can be determined, equivalent to the resins isolated by adsorption techniques.

Separation by adsorption chromatography essentially commences with the preparation of a porous bed of finely divided solid, the adsorbent. The adsorbent is usually contained in an open tube (column chromatography); the sample is introduced at one end of the adsorbent bed and induced to flow through the bed by means of a suitable solvent. As the sample moves through the bed the various components are held (adsorbed) to a greater or lesser extent depending on the chemical nature of the component. Thus, those molecules that are strongly adsorbed spend considerable time on the adsorbent surface rather than in the moving (solvent) phase, but components that are slightly adsorbed move through the bed comparatively rapidly.

There are three standard test methods that provide for the separation of heavy oil into four or five constituent fractions (Speight, 2001, 2014a, 2014b, 2014c, 2015). It is interesting to note that as the methods have evolved there has been a change from the use of pentane (ASTM D2007) to heptane (ASTM D3279; ASTM D4124) to separate asphaltenes. This is, in fact, in keeping with the production of a more consistent fraction that represents these higher molecular weight, more complex constituents of petroleum (Girdler, 1965; Speight et al., 1984).

Two of the methods (ASTM D2007; ASTM D4124) use adsorbents to fractionate the deasphaltened heavy oil, but the third method advocates the use of various grades of sulfuric acid to separate the material into compound types. Caution is advised in the application of this method since the method does not work well with all feedstocks. For example, when the *sulfuric acid* method is applied to the separation of heavy oil, complex emulsions can be produced (Speight, 2014a, 2015).

Obviously, there are precautions that must be taken when attempting to separate heavy oil into constituent fractions. The disadvantages in using ill-defined adsorbents are that adsorbent performance differs with the same feed and, in certain instances, may even cause chemical and physical modification of the feed constituents. The use of a chemical reactant like sulfuric acid should only be advocated with caution since feeds react differently and may even cause irreversible chemical changes and/or emulsion formation. These advantages may be of little consequence when it is not, for various reasons, the intention to recover the various product fractions *in toto* or in the original state, but in terms of the compositional evaluation of different feedstocks the disadvantages are very real.

Methods used for the separation of petroleum and heavy oil into various fractions are often identified by the acronyms for the names: PONA (paraffins, olefins, naphthenes, and aromatics), PIONA (paraffins, *iso*-paraffins, olefins, naphthenes, and aromatics), PNA (paraffins, naphthenes, and aromatics), PINA (paraffins, *iso*-paraffins, naphthenes, and aromatics), or SARA (saturates, aromatics, resins, and asphaltenes). However, it must be recognized that the fractions produced by the use of different adsorbents will differ in content and will also be different from fractions produced by solvent separation techniques. However, for heavy oil fractions, the absence of paraffins in the sample usually precludes many of these acronyms and the most common method is the SARA method.

HPLC has found great utility in separating different hydrocarbon group types and identifying specific constituent types. Of particular interest, is the application of the HPLC technique to the identification of the molecular types in the heavy oil, especially the molecular constituents of the asphaltene fraction. The general advantages of HPLC are: (1) each sample may be analyzed *as received* even though the boiling range may vary over a considerable range; and (2) the total time per analysis is usually on the order of minutes.

5.6.3 Molecular Weight

Even though recovery processes, in general, do not affect the quality of the oil, there is still the need to determine the molecular weight of the original constituents as well as the molecular weights of the products as a means of understanding the process. For those original constituents and products, for example, resins and asphaltenes, that have little or no volatility, *vapor pressure osmometry* (*VPO*) has been proven to be of considerable value.

There have been numerous attempts made to measure the molecular weight of heavy oil, particularly the asphaltene fraction, using a variety of different methods but there appears to be a noticeable lack of consensus on the value obtained for a specific sample. This has been irrevocably traced to structural aspects of heavy oil and to the behavior of the asphaltene constituents in molecular dispersion in their own maltene fraction (the heptane-soluble or pentane-soluble constituents, depending on the hydrocarbon that has been used for separation of the asphaltene fraction). In particular, the reliability of the method, the meaning of the term *average molecular weight*, and the interpretation or usefulness of the data has received considerable attention.

Currently, of the methods available, several standard methods are recognized as being useful for determining the molecular weight of petroleum fractions (ASTM D2224; ASTM D2502; ASTM D2503; ASTM D2878). However, before attempting to measure molecular weights care should be taken to ensure that the oil sample does not contain water or volatile components (boiling below 200°C, 392°F) because this would invalidate the basic calculation. Methods for molecular weight measurement are also included in other more comprehensive standards (ASTM D128; ASTM D3712) and there are several indirect methods have been proposed for the estimation of molecular weight by correlation with other, more readily measured physical properties. They are satisfactory when dealing with the conventional type of crude oils or their fractions and products and when approximate values are desired.

VPO (ASTM D2503), also called vapor phase osmometry, is a relatively simple and cheap method for the determination of molecular weight. Most osmometers can operate over a range of temperature through the use of probes that cover specific temperature ranges. This gives the number average molecular weight and not the molecular weight distribution.

The measuring elements are two temperature-sensitive thermistors placed in a closed, heat-insulated chamber. By means of two syringes, a drop of solvent is placed on one and a drop of the sample in solution in the same solvent is placed on the other. The chamber is saturated with solvent vapor and carefully temperature controlled. Because the solution has a lower vapor pressure than the solvent, solvent from the chamber atmosphere will condense on it. The difference in vapor pressure is proportional to its sample concentration. The heat of condensation warms the solution drop until its vapor pressure is the same as that of the surroundings. From then on, a steady state of condensation and warming is established. The temperature increase is measured and recorded. The solvent drop on the other thermistor is in equilibrium with the solvent in the chamber and, theoretically, gives no temperature effect. In reality, convection and other effects cause minute disturbances in the system, the effect of which can be minimized by subtracting the voltage of the solvent thermistor from that of the solution thermistor.

The temperature difference ΔT, is related to concentration and molecular weight at infinite dilution by the equation:

$$\Delta T = \frac{K_1 c}{\text{MW}}$$

In the equation, K_1 is a constant determined by calibration and c is the solute concentration. The effect is measured at several concentrations, and the results are plotted versus the reciprocal of the molecular weight (1/MW) and extrapolated to zero concentration (also referred to as unlimited dilution). However, one of the issues that arises through the use of this method for determining the *number average molecular weight* is that there can be a shift to the low values by the presence of low-molecular weight contaminants, as might be used in preparing heavy oil for transport.

A common solvent for VPO is toluene, which is satisfactory for hydrocarbons and moderately polar compounds. However, for the highly polar fractions, such as asphaltenes, more polar solvents such as pyridine are required. The molecular weight of such fractions measured by VPO in pyridine is distinctly lower than those measured in toluene (Speight, 2015 and references cited therein) indicating a lower degree of aggregation, assuming that contamination with trace amounts of previously used solvent in the separation of the samples can truly be excluded.

The observance of lower than true molecular weights though the influence of solvent and/or adsorbed lower molecular weight species on the asphaltenes constituents is not specific to the VPO method. It applies to any method producing number average results. Only further fractionation or other suitable measurements can provide the molecular weight distribution of these polar fractions. Reprecipitation or further fractionation of asphaltenes mitigates the preferential solvation of large polar molecules by small ones (Speight, 2015 and references cited therein).

One technique for reducing the effect of aggregation is to measure the molecular weight of polar species (asphaltenes) at different concentrations and different temperatures to illustrate and negate any aggregation effects. Thus, the molecular weight is measured at three different concentrations at each of three different temperatures (Speight, 1994). The data are extrapolated to determine the molecular weight at zero concentration at each temperature after which the data can be extrapolated to room temperature to negate concentration and temperature effects. This, of course, assumes that the extrapolation line is straight!

The *aggregation effect* has, on occasion, been incorrectly referred to as an *error*, in that the accuracy of the molecular weight value is influenced by *aggregation* of the molecular species in solution. In some instances, this may not be classed as an error insofar as the aggregation is recognized and the method is used (along with molecular weight determination in a polar solvent) to determine the maximum aggregation that can occur in the nonpolar solvent. Such effects are magnified when measuring the molecular weight of the asphaltenes that consist primarily of polar (heteroatom) constituents. Indeed, from such experiments the aggregation effect can be determined as the ratio of the molecular weight in a nonpolar solvent (eg, toluene) to the molecular weight on a polar solvent (eg, pyridine or nitrobenzene). Thus:

$$\text{Aggregation factor}, N = \frac{\text{MW}_{\text{toluene}}}{\text{MW}_{\text{pyridine}}}$$

$$\text{Aggregation factor}, N = \frac{\text{MW}_{\text{toluene}}}{\text{MW}_{\text{nitrobenzene}}}$$

The method using the concept of *freezing point depression* determines the average molecular weight of various fractions using simple equipment. The samples must be completely soluble and chemically inert in the solvent (benzene). Molecular weight values represent a number average molecular weight for the sample components.

The freezing points for benzene and for a benzene solution containing a known weight of sample are determined from their cooling curves. From the depression of the freezing point due to the sample, the average molecular weight is calculated by application of Raoult's law, viz. the vapor pressure lowering of a solvent is directly proportional to the concentration of the solute:

$$\frac{p_o - p}{p_o} = \frac{n_2}{n_1 + n_2}$$

In the equation, n_1 and n_2 are the number of moles of solvent and solute, respectively. Thus, if w_2 g of solute of molecular weight M_2 is dissolved in w_1 g of solvent of molecular weight M_1, Raoult's law gives:

$$\frac{p_o - p}{p_o} = \frac{w_2/M_2}{w_1M_1 + w_2M_2}$$

For a dilute solution, n_2 may be neglected in favor of n_1.

Thus, the calculation for molecular weight determination by cryoscopy is:

$$M = \frac{W(K_f - \Delta T)}{\Delta T \cdot S}$$

M is the number average molecular weight, W is the weight of sample in g, K_f is the cryoscopic constant for the solvent (K_f for benzene = 65.6), ΔT is the freezing point depression in °C (ie, the freezing point of benzene minus the freezing point of the solution), and S is the moles of solvent. Many solvents have been tried, but with the best technique an accuracy of 1–2% is to be considered good.

Boiling point elevation (*ebullioscopic*) methods are, in general use, more rapid and equally accurate but tend to fall short when applied to the higher molecular weight fractions of petroleum, heavy oil, and bitumen. Molecular weights in the low range (<500 Da) are readily determined by vapor-density methods. For any one sample the boiling point elevation is determined at a series of concentrations of solute. Such determinations were carried out in practice by comparing the boiling point of pure solvent, measured in an ebulliometer, with the boiling points, determined in a second ebulliometer, of a series of solutions prepared by adding successive portions of the sample to the solvent. The first ebulliometer serves as the control experiment.

In order to determine the ebullioscopic constant K—which depends not only on the solvent used, but also to a certain extent on the construction of the apparatus and the procedure followed—a series of determinations is carried out on a pure hydrocarbon of known molecular weight, preferably of the same order of magnitude as the molecular weight under investigation. The ebullioscopic constant calculated for each concentration is plotted on a graph, just as in the case of a molecular weight determination, and is extrapolated to zero concentration. These methods all give number average molecular weights, defined by:

$$\frac{\Sigma(n_i M_i)}{\Sigma n_i} = M_n$$

M_n is the number average molecular weight and n_i is the number of molecules having a molecular weight M_i.

There is a method for determining the molecular weight of petroleum, heavy oil, bitumen, and their constituent fraction using *size exclusion (gel permeation) chromatography* (ASTM D5296). The principle of size exclusion chromatography is the exclusion of larger sample molecules from smaller pores in the packing. As a result, larger molecules cannot reside in the entire column volume but are restricted to smaller regions. In the extreme, the largest ones are restricted to the interstitial volume, that is, the space between particles, whereas the smallest ones can penetrate the entire open column volume, that is, the interstitial and all the pore volume. As a consequence, the large molecules elute first and the smallest ones last.

For the higher molecular weight petroleum fractions, such as the asphaltene fraction of heavy oil, column packing of nominal pore sizes between 50–100 Å and 5000–10,000 Å are used for complete resolution. The bimodal column combinations give good linear calibration curves and well-characterized polymer samples as well as narrow polymer fractions, for example, polystyrenes, are available for calibrating column sets a cross the entire molecular weight range.

However, these polymers are structurally different from petroleum constituents and cannot be directly used for accurate calibration. In fact, petroleum fractions, such as the resin fraction and the asphaltene fraction, differ from polymers in three ways: (1) the hydrocarbon skeletal structure; (2) the polar nature; and (3) the varying composition with molecular weight. Therefore, it is preferable to employ a set of *narrow* subfractions obtained by preparative size exclusion chromatography from the same or a similar petroleum fraction for calibration. Calibration curves obtained by polymers should at least be checked with such subfractions and adjusted as necessary.

It is important to remember that size exclusion chromatography separates by molar volume rather than by molecular weight. The method will, therefore, differentiate by structure in addition to molecular weight. In principle, size exclusion chromatography is a very powerful method for separating petroleum fractions by molecular weight. It is used frequently in petroleum analysis despite the tendency of the components toward adsorption and aggregation, and other potential problems.

REFERENCES

Ancheyta, J., Speight, J.G., 2007. Hydroprocessing of Heavy Oils and Residua. CRC Press, Taylor & Francis Group, Boca Raton, FL.

Andersen, S.I., 1994. Dissolution of solid boscan asphaltene constituents in mixed solvents. Fuel Sci. Technol. Int. 12, 1551–1577.

Andersen, S.I., Birdi, K.S., 1990. Influence of temperature and solvent on the precipitation of asphaltene constituents. Fuel Sci. Technol. Int. 8, 593–615.
Andersen, S.I., Birdi, K.S., 1991. Aggregation of asphaltenes as determined by calorimetry. J. Colloid Interface Sci. 142, 497.
Andersen, S.I., Speight, J.G., 1992. Asphaltene precipitation and incipient flocculation in mixed solvents. Div. Fuel Chem. Am. Chem. Soc. 37 (3), 1335. (Preprints).
Andersen, S.I., Speight, J.G., 1994. Observations on the critical micelle concentration of asphaltene constituents. Fuel 72, 1343–1344.
Andersen, S.I., Speight, J.G., 1999. Thermodynamic models for asphaltene solubility and precipitation. J. Petrol. Sci. Eng. 22, 53.
Andersen, S.I., Speight, J.G., 2001. Petroleum reins: separation, character, and role in petroleum. Petrol. Sci. Technol. 19, 1.
API, 2003. Sampling Petroleum Reservoir Fluids. Report API RP 44. American Petroleum Institute, Washington, DC.
ASTM D6, 2014. Standard test method for loss on heating of oil and asphaltic compounds Annual Book of Standards. ASTM International, West Conshohocken, PA.
ASTM D20, 2014. Standard test method for distillation of road tars Annual Book of Standards. ASTM International, West Conshohocken, PA.
ASTM D36, 2014. Standard test method for softening point of bitumen (Ring-and-Ball Apparatus) Annual Book of Standards. ASTM International, West Conshohocken, PA.
ASTM D56, 2014. Standard test method for flash point by tag closed cup tester Annual Book of Standards. ASTM International, West Conshohocken, PA.
ASTM D86, 2014. Standard test method for distillation of petroleum products at atmospheric pressure Annual Book of Standards. ASTM International, West Conshohocken, PA.
ASTM D87, 2014. Standard test method for melting point of petroleum wax (Cooling Curve) Annual Book of Standards. ASTM International, West Conshohocken, PA.
ASTM D88, 2014. Standard test method for saybolt viscosity Annual Book of Standards. ASTM International, West Conshohocken, PA.
ASTM D92, 2014. Standard test method for flash and fire points by cleveland open cup tester Annual Book of Standards. ASTM International, West Conshohocken, PA.
ASTM D93, 2014. Standard test methods for flash point by pensky-martens closed cup tester Annual Book of Standards. ASTM International, West Conshohocken, PA.
ASTM D97, 2014. Standard test method for pour point of petroleum products Annual Book of Standards. ASTM International, West Conshohocken, PA.
ASTM D127, 2014. Standard test method for drop melting point of petroleum wax, including petrolatum Annual Book of Standards ASTM International, West Conshohocken, PA.
ASTM D128, 2014. Standard test methods for analysis of lubricating grease Annual Book of Standards. ASTM International, West Conshohocken, PA.
ASTM D189, 2014. Standard test method for conradson carbon residue of petroleum products Annual Book of Standards. ASTM International, West Conshohocken, PA.
ASTM D287, 2014. Standard test method for API gravity of crude petroleum and petroleum products (Hydrometer Method) Annual Book of Standards. ASTM International, West Conshohocken, PA.
ASTM D323, 2014. Standard test method for vapor pressure of petroleum products (Reid Method) Annual Book of Standards. ASTM International, West Conshohocken, PA.
ASTM D341, 2014. Standard test method for viscosity-temperature charts for liquid petroleum products Annual Book of Standards. ASTM International, West Conshohocken, PA.
ASTM D402, 2014. Standard test method for distillation of cut-back asphaltic (Bituminous) products Annual Book of Standards. ASTM International, West Conshohocken, PA.
ASTM D445, 2014. Standard test method for kinematic viscosity of transparent and opaque liquids (and the Calculation of Dynamic Viscosity) Annual Book of Standards. ASTM International, West Conshohocken, PA.

ASTM D482, 2014. Standard test method for ash from petroleum products Annual Book of Standards. ASTM International, West Conshohocken, PA.
ASTM D524, 2014. Standard test method for ramsbottom carbon residue of petroleum products Annual Book of Standards. ASTM International, West Conshohocken, PA.
ASTM D566, 2014. Standard test method for dropping point of lubricating grease Annual Book of Standards. ASTM International, West Conshohocken, PA.
ASTM D664, 2014. Standard test method for the acid number of petroleum products by potentiometric titration Annual Book of Standards. ASTM International, West Conshohocken, PA.
ASTM D1160, 2014. Standard test method for distillation of petroleum products at reduced pressure Annual Book of Standards. ASTM International, West Conshohocken, PA.
ASTM D1217, 2014. Standard test method for density and relative density (Specific Gravity) of liquids by bingham pycnometer Annual Book of Standards. ASTM International, West Conshohocken, PA.
ASTM D1298, 2014. Standard test method for density, relative density (Specific Gravity), or API gravity of crude petroleum and liquid petroleum products by hydrometer method Annual Book of Standards. ASTM International, West Conshohocken, PA.
ASTM D1310, 2014. Standard test method for flash point and fire point of liquids by tag open-cup apparatus Annual Book of Standards. ASTM International, West Conshohocken, PA.
ASTM D1318, 2014. Standard test method for sodium in residual fuel oil (Flame Photometric Method) Annual Book of Standards. ASTM International, West Conshohocken, PA.
ASTM D1555, 2014. Standard test method for calculation of volume and weight of industrial aromatic hydrocarbons and cyclohexane Annual Book of Standards. ASTM International, West Conshohocken, PA.
ASTM D1796, 2014. Standard test method for water and sediment in fuel oils by the centrifuge method (Laboratory Procedure) Annual Book of Standards. ASTM International, West Conshohocken, PA.
ASTM D2007, 2014. Standard test method for characteristic groups in rubber extender and processing oils and other petroleum-derived oils by the clay-gel absorption chromatographic method Annual Book of Standards. ASTM International, West Conshohocken, PA.
ASTM D2161, 2014. Standard practice for conversion of kinematic viscosity to saybolt universal viscosity or to saybolt furol viscosity Annual Book of Standards. ASTM International, West Conshohocken, PA.
ASTM D2224, 2014. Test method for mean molecular weight of mineral insulating oils by the Cryoscopic Method (discontinued in 1989 but still used by some laboratories for determining the molecular weight of petroleum fractions up to an including gas oil).
ASTM D2270, 2014. Standard practice for calculating viscosity index from kinematic viscosity at 40 and 100°C Annual Book of Standards. ASTM International, West Conshohocken, PA.
ASTM D2500, 2014. Standard test method for cloud point of petroleum products Annual Book of Standards. ASTM International, West Conshohocken, PA.
ASTM D2502, 2014. Test method for estimation of molecular weight (Relative Molecular Mass) of petroleum oils from viscosity measurements Annual Book of Standards. ASTM International, West Conshohocken, PA.
ASTM D2503, 2014. Standard test method for relative molecular mass (molecular weight) of hydrocarbons by thermoelectric measurement of vapor pressure Annual Book of Standards. ASTM International, West Conshohocken, PA.
ASTM D2569, 2014. Standard test method for distillation of pitch Annual Book of Standards. ASTM International, West Conshohocken, PA.
ASTM D2715, 2014. Standard test method for volatilization rates of lubricants in vacuum Annual Book of Standards. ASTM International, West Conshohocken, PA.

ASTM D2766, 2014. Standard test method for specific heat of liquids and solids Annual Book of Standards. ASTM International, West Conshohocken, PA.
ASTM D2878, 2014. Method for estimating apparent vapor pressures and molecular weights of lubricating oils Annual Book of Standards. ASTM International, West Conshohocken, PA.
ASTM D2887, 2014. Standard test method for boiling range distribution of petroleum fractions by gas chromatography Annual Book of Standards. ASTM International, West Conshohocken, PA.
ASTM D2892, 2014. Standard test method for distillation of crude petroleum (15-Theoretical Plate Column) Annual Book of Standards. ASTM International, West Conshohocken, PA.
ASTM D3279, 2014. Standard test method for n-Heptane insolubles Annual Book of Standards. ASTM International, West Conshohocken, PA.
ASTM D3340, 2014. Standard test method for lithium and sodium in lubricating greases by flame photometer Annual Book of Standards. ASTM International, West Conshohocken, PA.
ASTM D3341, 2014. Standard test method for lead in gasoline-iodine monochloride method Annual Book of Standards. ASTM International, West Conshohocken, PA.
ASTM D3605, 2014. Standard test method for trace metals in gas turbine fuels by atomic absorption and flame emission spectroscopy Annual Book of Standards. ASTM International, West Conshohocken, PA.
ASTM D3712, 2014. Standard test method of analysis of oil-soluble petroleum sulfonates by liquid chromatography Annual Book of Standards. ASTM International, West Conshohocken, PA.
ASTM D3828, 2014. Standard test methods for flash point by small scale closed cup tester Annual Book of Standards. ASTM International, West Conshohocken, PA.
ASTM D4006, 2014. Standard test method for water in crude oil by distillation Annual Book of Standards. ASTM International, West Conshohocken, PA.
ASTM D4007, 2014. Standard test method for water and sediment in crude oil by the centrifuge method (Laboratory Procedure) Annual Book of Standards. ASTM International, West Conshohocken, PA.
ASTM D4057, 2014. Standard practice for manual sampling of petroleum and petroleum products Annual Book of Standards. ASTM International, West Conshohocken, PA.
ASTM D4124, 2014. Standard test method for separation of asphalt into four fractions Annual Book of Standards. ASTM International, West Conshohocken, PA.
ASTM D4175, 2014. Standard terminology relating to petroleum, petroleum products, and lubricants Annual Book of Standards. ASTM International, West Conshohocken, PA.
ASTM D4530, 2014. Standard test method for determination of carbon residue (Micro Method) Annual Book of Standards. ASTM International, West Conshohocken, PA.
ASTM D5236, 2014. Standard test method for distillation of heavy hydrocarbon mixtures (Vacuum Potstill Method) Annual Book of Standards. ASTM International, West Conshohocken, PA.
ASTM D5296, 2014. Standard test method for molecular weight averages and molecular weight distribution of polystyrene by high performance size-exclusion chromatography Annual Book of Standards. ASTM International, West Conshohocken, PA.
ASTM D5551, 2014. Standard test method for determination of the cloud point of oil Annual Book of Standards. ASTM International, West Conshohocken, PA.
ASTM D5771, 2014. Standard test method for cloud point of petroleum products (Optical Detection Stepped Cooling Method) Annual Book of Standards. ASTM International, West Conshohocken, PA.
ASTM D5772, 2014. Standard test method for cloud point of petroleum products (Linear Cooling Rate Method) Annual Book of Standards. ASTM International, West Conshohocken, PA.

ASTM D5773, 2014. Standard test method for cloud point of petroleum products (Constant Cooling Rate Method) Annual Book of Standards. ASTM International, West Conshohocken, PA.
ASTM D7169, 2014. Standard test method for boiling point distribution of samples with residues such as crude oils and atmospheric and vacuum residues by high temperature gas chromatography Annual Book of Standards. ASTM International, West Conshohocken, PA.
ASTM D7397, 2014. Standard test method for cloud point of petroleum products (Miniaturized Optical Method) Annual Book of Standards. ASTM International, West Conshohocken, PA.
ASTM D7683, 2014. Standard test method for cloud point of petroleum products (Small Test Jar Method) Annual Book of Standards. ASTM International, West Conshohocken, PA.
ASTM D7689, 2014. Standard test method for cloud point of petroleum products (Mini Method) Annual Book of Standards. ASTM International, West Conshohocken, PA.
ASTM E777, 2014. Standard test method for carbon and hydrogen in the analysis sample of refuse-derived fuel Annual Book of Standards. ASTM International, West Conshohocken, PA.
Barrow, M.P., McDonnell, L.A., Feng, X., Walker, J., Derrick, P.J., 2003. Determination of the nature of naphthenic acids present in crude oils using nanospray fourier transform ion cyclotron resonance mass spectrometry: the continued battle against corrosion. Anal. Chem. 75 (4), 860–866.
Bidmus, H.O., Mehrotra, A.K., 2009. Solids deposition during cold flow of wax-solvent mixtures in a flow-loop apparatus with heat transfer. Energy Fuels 23 (6), 3184–3194.
Buckley, J.S., Liu, Y., Xie, X., Morrow, N.R., 1997. Asphaltenes and crude oil wetting—the effect of oil composition. Soc. Petrol. Eng. J. 2, 107–119.
CCQTA, 2008. Heavy oil and bitumen analytical methods: understanding their capabilities and limitations. In: Fuhr, B. (Ed.), Canadian Crude Quality Technical Association (CCQTA) Alberta Research Council, Edmonton, Alberta.
Cengiz, S., Robertson, C., Kalpacki, B., Gupta, D., 2004. A study of heavy oil solution gas drive for hamaca field: depletion studies and interpretations. Paper SPE 86967-MS. SPE International Thermal Operations and Heavy Oil Symposium and Western Regional Meeting, Bakersfield, California, March 16–18.
Chukwu, O.C., Ofodile, S.E., Okoye, I.P., 2011. The effect of different solvent polarity on the precipitation of heavy organics from a crude oil deposit in the Niger delta. Scientia Afr. 10 (2), 9–15.
Coutinho, J.A.P., Daridon, J.L., 2005. The limitations of the cloud point measurement techniques and the influence of the oil composition on its detection. Petrol. Sci. Technol. 23, 1113–1128.
Dawe, R.A., 2004. Miscible displacement in heterogeneous porous media. Proceedings. 6th Caribbean Congress of Fluid Dynamics, University of the West Indies, St. Augustine, Trinidad, Republic of Trinidad and Tobago, January 22–23.
Derungs, W.A., 1956. Naphthenic acid corrosion—an old enemy of the petroleum industry. Corrosion 12 (12), 617–622.
EIPA, 2015. AOSTRA/AERI Publications. Alberta Innovates, Energy and Environment Solutions, Government of Alberta, Edmonton, Alberta, Canada. <http://eipa.alberta.ca/resources/publications.aspx> (accessed 16.09.15.).
Francisco, M.A., Speight, J.G., 1984. Asphaltene characterization by a non-spectroscopic method. Div. Fuel Chem. Am. Chem. Soc. 29 (1), 36. (Preprints).
Gabitto, J., Barrufet, M., 2003. Experimental and theoretical determination of heavy viscosity under reservoir conditions. Final Report, Contract No. DE-FG26-99FT40615. United States Department of Energy, Washington, DC.
Gary, J.G., Handwerk, G.E., Kaiser, M.J., 2007. Petroleum Refining: Technology and Economics, fifth ed. CRC Press, Taylor & Francis Group, Boca Raton, FL.

Gholoum, E.F., Oskui, G.P., 2004. Investigation of the asphaltene precipitation process for Kuwaiti reservoirs. Petrol. Sci. Technol. 22 (7, 8), 1165–1185.
Gholoum, E.F., Oskui, G.P., Salman, M., 2003. Investigation of asphaltene onset conditions for Kuwaiti reservoirs. SPE Paper No. 81571. Proceedings. SPE 13th Middle East Oil Show Conference, Bahrain, April 5–8. Society of Petroleum Engineers, Richardson, TX.
Girdler, R.B., 1965. Constitution of asphaltene constituents and related studies. Proc. Assoc. Asphalt Paving Technol. 34, 45.
Goodarzi, N., Bryan, J., Mai, A., Kantzas, A., 2005. Heavy-oil fluid testing with conventional and novel techniques. Paper No SPE 97803. Proceedings. SPE/PS-CIM/CHOA International Thermal Operations and Heavy Oil Symposium, Calgary, Alberta, Canada, November 1–3.
Green, L.E., 1976. Chromatography gives boiling point. Hydrocarb. Process. 55 (5), 205.
Gwynn, J.W., 1971. Instrumental analysis of tars and their correlations in oil-impregnated sandstone beds, Uintah and Grand Counties, Utah. Special Studies No. 37. Utah Geological and Mineralogical Survey, Salt Lake City, Utah, October.
Hammami, A., Raines, M.A., 1997. Paraffin deposition from crude oils: comparison of laboratory results to field data. Paper No. SPE-38776. Proceedings. SPE Annual Technical Conference and Exhibition, San Antonio, TX, USA, October 5–8.
Hammami, A., Raines, M.A., 1999. Paraffin deposition from crude oils: comparison of laboratory results with field data. SPE J 4 (1), 9–18.
Hammami, A., Ratulowski, J., Coutinho, J.A.P., 2003. Cloud points: can we measure or model them? Petrol. Sci. Technol. 21 (3–4), 345–358.
Hassanvand, M., Shahsavani, B., Anooshe, A., 2012. Study of temperature effect on asphaltene precipitation by visual and quantitative methods. J. Petrol. Technol. Altern. Fuels 3 (2), 8–18.
Hirschberg, A., deJong, L., Schipper, B., Maijer, J., 1984. Influence of temperature and pressure on asphaltene flocculation. SPE J. June, 283–293.
Honarpour, M.M., Nagarajan, N.R., Sampath, K., 2006. Rock/fluid characterization and their integration—implication on reservoir management. J. Pet. Tech. 58 (9), 120.
Hsu, C.S., Dechert, G.J., Robbins, W.K., Fukuda, E.K., 2000. Naphthenic acids in crude oils characterized by mass spectrometry. Energy Fuels 14, 217–223.
Hsu, C.S., Robinson, P.R., 2006. Practical Advances in Petroleum Processing, vols. 1 and 2. Springer, New York.
Hsu, J.J.C., Santamaria, M.M., Brubaker, J.P., 1994. Wax deposition of waxy live crudes under turbulent flow conditions. Paper No 28480. Proceedings. 69th Annual Technical Conference and Exhibition, New Orleans, Louisiana, September 25–28. Society for Petroleum Engineers, Richardson, TX.
Hughey, C.A., Galasso, S.A., Zumberge, J.E., 2007. Detailed compositional comparison of acidic NSO compounds in biodegraded reservoir and surface crude oils by negative ion electrospray fourier transform ion cyclotron resonance mass spectrometry. Fuel 86 (5–6), 758–768.
Hughey, C.A., Rodgers, R.P., Marshall, A.G., Qian, K., Robbins, W.K., 2002. Identification of acidic NSO compounds in crude oils of different geochemical origins by negative ion electrospray fourier transform ion cyclotron resonance mass spectrometry. Org. Geochem. 33 (7), 743–759.
Jayasekera, A.J., Goodyear, S.G., 1999. The development of heavy oil fields in the uk continental shelf: past, present, and future. Paper No. SPE 54623. Proceedings. SPE Western Regional Meeting, Anchorage, Alaska, May 26–28. Society of Petroleum Engineers, Richardson, TX.
Koots, J.A., Speight, J.G., 1975. The relation of petroleum resins to asphaltene constituents. Fuel 54, 179.

Kruka, V.R., Cadena, E.R., Long, T.E., 1995. Cloud point determination for crude oils. J. Petrol. Technol. 47 (8), 681–687.
Lashkarbolooki, M., Seyfaee, A., Esmaeilzadeh, F., Mowla, D., 2010. Experimental investigation of wax deposition in Kermanshah crude oil through a monitored flow loop apparatus. Energy Fuels 24 (2), 1234–1241.
Leontaritis, K.J., Leontaritis, J.D., 2003. Cloud point and wax deposition measurement techniques. Paper No. 80267. Proceedings. SPE International Symposium on Oil Field Chemistry held in Houston, TX, February 5–7. Society for Petroleum Engineers, Richardson, TX.
Long, R.B., 1979. The concept of asphaltene constituents. Div. Petrol. Chem. Am. Chem. Soc. 24 (4), 891. (Preprints).
Long, R.B., 1981. The concept of asphaltene constituents. In: Bunger, J.W., Li, N. (Eds.), The Chemistry of Asphaltene Constituents American Chemical Society, Washington, DC. Advances in Chemistry Series No. 195.
Long, R.B., Speight, J.G., 1989. Studies in petroleum composition. I: development of a compositional map for various feedstocks. Révue Institut Français du Petrole 44, 205.
Long, R.B., Speight, J.G., 1998. The composition of petroleum. In: Speight, J.G. (Ed.), Petroleum Chemistry and Refining Taylor & Francis, Washington, DC. (Chapter 2).
MacAllister, D.J., DeRuiter, R.A., 1985. Further development and application of simulated distillation for enhanced oil recovery. Paper No. SPE 14335. 60th Annual Technical Conference, September 22–25. Society of Petroleum Engineers, Las Vegas.
Mitchell, D.L., Speight, J.G., 1973. The solubility of asphaltene constituents in hydrocarbon solvents. Fuel 52, 149.
Mokhatab, S., Poe, W.A., Speight, J.G., 2006. Handbook of Natural Gas Transmission and Processing. Elsevier, Amsterdam.
Mushrush, G.W., Speight, J.G., 1995. Petroleum Products: Instability and Incompatibility. Taylor & Francis Publishers, Washington, DC.
Nagarajan, N.R., Honarpour, M.M., Sampath, K., 2007a. Reservoir-fluid sampling and characterization—key to efficient reservoir management. J. Petrol. Technol. 59, 80–91.
Nagarajan, N.R., Honarpour, M.M., Sampath, K., 2007b. Reservoir-fluid sampling and characterization—key to efficient reservoir management. Paper No. SPE 103501 and 101517. Proceedings. Abu Dhabi International Petroleum Exhibition & Conference, Abu Dhabi, November 5–8. Society of Petroleum Engineers, Richardson, TX.
Nedelchev, A., Stratiev, D., Ivanov, A., Stoilov, G., 2011. Boiling point distribution of crude oils based on TBP and ASTM D-86 distillation Data. Petrol. Coal 53 (4), 275–290.
Neer, L.A., Deo, M.D., 1995. Simulated distillation of oils with a wide carbon number distribution. J. Chromatogr. Sci. 33, 133–138.
Nikolaychuk, E., Stratiev, D., Velkov, L., Veli, A., Sotirov, S., Mitkova, M., 2015. Conversion of heavy oil distillation data from ASTM D1160 to ASTM D5236. Petrol. Coal 57 (3), 266–279.
Petkova, N., Angelova, M., Petkov, P., 2009. Establishing the reasons and type of enhanced corrosion in the crude oil atmospheric distillation unit. Petrol. Coal 51 (4), 286–292.
Piehl, R.L., 1988. Naphthenic corrosion in crude distillation units. Mater. Perform. 27 (1), 37.
Qian, K., Edwards, K.E., Dechert, G.J., Jaffe, S.B., Green, L.A., Olmstead, W.N., 2008. Measurement of total acid number (TAN) and TAN boiling point distribution in petroleum products by electrospray ionization mass spectrometry. Anal. Chem. 2008 (80), 849–855.
Qian, K., Robbins, W.K., Hughey, C.A., Cooper, H.J., Rodgers, R.P., Marshall, A.G., 2001. Resolution and identification of elemental composition for more than 3000 crude acids in heavy petroleum by negative-ion microelectronspray high-field fourier transform ion cyclotron resonance mass spectrometry. Energy Fuels 15 (6), 1505–1511.

Reddie, D.R., Robertson, C.R., 2004. Innovative reservoir fluid sampling systems. Paper SPE 86951. Proceedings. SPE International Thermal Operations and Heavy Oil Symposium and Western Regional Meeting, Bakersfield, CA, March 16–18. Society of Petroleum Engineers, Richardson, TX.

Reynolds, J.G., 1998. Metals and heteroatoms in heavy crude oils. In: Speight, J.G. (Ed.), Petroleum Chemistry and Refining Taylor & Francis Publishers, Washington, DC, pp. 63–102. (Chapter 3).

Riazi, M.R., 1989. Distribution model for properties of hydrocarbon-plus fractions. Ind. Eng. Chem. Res. 28, 1831–1835.

Speight, J.G., 1978. The use of bitumen as an adhesive and building material at the ancient cities of Babylon, Calah, Nineveh, and Ur. Report to the Dean, College of Science, University of Mosul, Mosul, Iraq.

Speight, J.G., 1979. Studies on bitumen fractionation: (a) fractionation by a Cryoscopic Method, (b) effect of solvent type on asphaltene solubility. Information Series No. 84. Alberta Research Council, Edmonton, Alberta, Canada.

Speight, J.G., 1992. A chemical and physical explanation of incompatibility during refining operations. Proceedings. 4th International Conference on the Stability and Handling of Liquid Fuels. Report No. (DOE/CONF-911102). United States Department of Energy US, vol. 1, Washington, DC, p. 169.

Speight, J.G., 1994. Chemical and physical studies of petroleum asphaltene constituents. In: Yen, T.F. Chilingarian, G.V. (Eds.), Asphaltene Constituents and Asphalts. I. Developments in Petroleum Science, vol. 40 Elsevier, Amsterdam. (Chapter 2).

Speight, J.G., 2000. The Desulfurization of Heavy Oils and Residua, second ed. Marcel Dekker, Inc., New York.

Speight, J.G., 2001. Handbook of Petroleum Analysis. John Wiley & Sons, Inc., Hoboken, NJ.

Speight, J.G., 2007. Natural Gas: A Basic Handbook. GPC Books, Gulf Publishing Company, Houston, TX.

Speight, J.G., 2014a. The Chemistry and Technology of Petroleum, fifth ed. CRC Press, Taylor and Francis Group, Boca Raton, FL.

Speight, J.G., 2014b. Oil and Gas Corrosion Prevention. Gulf Professional Publishing, Elsevier, Oxford.

Speight, J.G., 2014c. High Acid Crudes. Gulf Professional Publishing, Elsevier, Oxford.

Speight, J.G., 2015. Handbook of Petroleum Product Analysis, second ed. John Wiley & Sons, Inc., Hoboken, NJ.

Speight, J.G., Long, R.B., Trowbridge, T.D., 1982. On the definition of asphaltene constituents. Div. Fuel Chem. Am. Chem. Soc. 27 (3/4), 268. (Preprints).

Speight, J.G., Long, R.B., Trowbridge, T.D., 1984. Factors influencing the separation of asphaltene constituents from heavy petroleum feedstocks. Fuel 63, 616.

Speight, J.G., Ozum, B., 2002. Petroleum Refining Processes. Marcel Dekker, Inc., New York.

Stratiev, D.S., Marinov, I., Nedelchev, A., Velkov, I., Stratiev, D.D., Veli, A., et al., 2014. Evaluation of approaches for conversion of ASTM into TBP distillation data of oil fractions. Oil Gas Eur. Mag. 40 (4), 216–221.

Stratiev, D.S., Nedelchev, A., Dinkov, R., Batchvarov, A., 2011. It's possible to derive TBP from partial distillation data. Oil Gas J. 109 (17), 114–122.

Subramanian, M., Deo, M.D., Hanson, F.V., 1996. Compositional analysis of bitumen and bitumen-derived products. J. Chromatograph. Sci. 34, 20–26.

Wallace, E.D. (Ed.), 1988. A review of analytical methods for bitumens and heavy oils. AOSTRA Technical Publication Series No. 5. Alberta Oil Sands Technology and Research Authority, Edmonton, Alberta, Canada.

Whitson, C.H., 1983. Characterizing hydrocarbon-plus fractions. SPE J. 23 (4), 683.

Whitson, C.H., Brule, M., 2000. Phase Behavior. Monograph Series. Society for Petroleum Engineers, Richardson, TX.

Witt, C.J., Crombie, A., 1999. A comparison of wireline and drillstem test fluid samples from a deep water gas-condensate exploration well. Paper SPE 56714-MS. SPE Annual Technical Conference and Exhibition, Houston, October 3–6.

Wood, R.E., Ritzma, H.R. 1972. Analysis of oil extracted from oil-impregnated sandstone deposits in Utah. Special Studies No. 39. Utah Geological and Mineralogical Survey, Salt Lake City, Utah, January.

Worman, J.C., Green, L.E., 1965. Simulated distillation of high boiling petroleum fractions. Anal. Chem. 37, 1620–1621.

PART II

Oil Recovery

CHAPTER 6

General Methods of Oil Recovery

6.1 INTRODUCTION

The initial production of crude oil from an underground reservoir is achieved by the use of the natural energy of the reservoir (Alvarado and Manrique, 2010). As soon as the reservoir is opened, the natural energy comes into play and, whether through gas pressure or water pressure, forces the fluids to the surface facility (primary production). In the process, the actual sources of natural reservoir energy that lead to primary production include the swelling of reservoir fluids, the release of solution gas as the reservoir pressure declines, nearby communicating aquifers, and gravity.

When the natural reservoir energy has been depleted through production, it becomes necessary to supplement or even substitute the natural reservoir energy with energy from an external source. This is usually accomplished by the injection of fluids, such as natural gas or water (secondary recovery). The main purpose of either a natural gas injection process (gas flooding) or a water injection process (water flooding) is to repressurized the reservoir and then to maintain the reservoir at a high pressure. The term *pressure maintenance* is sometimes used to describe a secondary recovery process.

When gas is used as the pressure maintenance agent, it is usually injected into a zone of free gas (ie, a gas cap) to maximize recovery by gravity drainage. The injected gas is usually natural gas produced from the reservoir in question which defers the sale of that gas until the gas flooding operation is completed and the gas can be recovered. On the other hand, other gases, such as nitrogen, can be injected to maintain reservoir pressure, which allows the natural gas to be sold as it is produced.

The situation is different for heavy oil and tar sand bitumen where the reservoirs of deposits have little to no reservoir energy in the form of the pressure exerted within the reservoir by the presence of or water. Heavy oil reservoir and tar sand deposits are found in the microscopic pores of sedimentary rocks such as sandstone and limestone. Not all of the pores

Introduction to Enhanced Recovery Methods for Heavy Oil and Tar Sands.
DOI: http://dx.doi.org/10.1016/B978-0-12-849906-1.00006-0

in a reservoir rock or deposit will contain heavy oil or bitumen—some will be filled with water or brine—the latter is water that is saturated with minerals. Seismic surveys are used to try to predict where fields may be found but the only way of making certain is by drilling.

Once found, the events leading to recovery start to unfold (Drew, 1997). Production rates from reservoirs depend on a number of factors, such as reservoir geometry (primarily formation thickness and reservoir continuity), reservoir pressure, reservoir depth, rock type and permeability, fluid saturations and properties, extent of fracturing, number of wells and their locations, and the ratio of the permeability of the formation to the viscosity of the heavy oil or bitumen and what it will take to modify the viscosity to enable recovery methods to be effective (Taber and Martin, 1983; DOE, 1996; Jayasekera and Goodyear, 1999). Operators can increase production over that which would naturally occur by such methods as fracturing the reservoir to open new channels for flow, injecting gas and water to increase the reservoir pressure, or lowering oil viscosity with heat or chemicals. These supplementary techniques are expensive, and the extent to which they are used depends on such external factors as the operator's economic condition, sales prospects, and perceptions of future prices.

The extraordinary geological variability of different reservoirs means that production profiles differ from field to field. Heavy oil reservoirs (see chapter: Microbial Enhanced Oil Recovery) can be developed to significant levels of production and maintained for a period of time by supplementing natural drive force, while gas reservoirs normally decline more rapidly. On this basis an oil reservoir with the seemingly large reserve of a million barrels might produce only 200 to 400 barrels per day during the best production years.

The two important concepts involved in oil recovery are *mobility ratio*, M, and the *capillary number*, N_c. If the *mobility ratio* is greater than unity (ie, $M > 1$), the displacing fluid will flow past much of the displaced fluid, displacing it inefficiently. Thus the mobility ratio influences *displacement efficiency*, that is, the (microscopic) efficiency of oil displacement within the pores. When the mobility far exceeds unity, the displacing fluid will channel past oil clusters (often referred to as *viscous fingering*). For maximum displacement efficiency, the mobility should be less than unity ($M < 1$) and is usually denoted as a favorable mobility ratio. If $M > 1$ (unfavorable mobility ratio), then, in the absence of viscous fingering, it merely means that more fluid will have to be injected to attain a given residual oil saturation in the pores. The mobility ratio can be reduced by increasing the viscosity of the displacing fluid (such as by the addition of a suitable polymer to the water), or by reducing the viscosity of the displaced fluid (such as by heating the oil or the addition of a solvent to the oil).

Briefly and by way of explanation, *viscous fingering* is a condition whereby the interface of two fluids, such as oil and water, bypasses sections of a reservoir as it moves along, creating an uneven, or fingered, profile. Fingering is a relatively common condition in reservoirs with water-injection wells and the result is an inefficient sweeping action that can bypass significant volumes of recoverable oil and, in severe cases, an early breakthrough of water into adjacent production wellbores.

The *capillary number* (N_c) represents the relative effect of viscous forces versus surface tension acting across an interface between a liquid and a gas or between two immiscible liquids. The capillary number is defined as:

$$N_c = \mu V/\gamma$$

In the equation, μ is the dynamic viscosity of the liquid, V is a characteristic velocity, and γ is the surface or interfacial tension between the two fluid phases.

The residual oil saturation as a function of the capillary number can be increased thereby decreasing the residual oil saturation by reducing viscosity of the oil or by increasing pressure gradient.

Finally, heavy oil, extra heavy oil, and tar sand bitumen resources can be markedly different in their characteristics. Hence, each production method must be tailored for the particular resource and for its fluid properties. A method that is successful in one reservoir (or deposit) may fail utterly in a different reservoir (or deposit). Hence, it is essential that the properties of the resource be fully understood before selecting a production scheme. Essential properties include the geological setting; the depth, areal extent, and thickness of the resource; oil composition, density, viscosity, and gas content; the presence of bottom water or top gas zones; petrophysical and geomechanical properties such as porosity, permeability, and rock strength; the presence of shale layers; vertical and horizontal permeability; and the variation of these properties across the reservoir.

This chapter, for the most part, deals with those recovery methods that are applied to recovery of conventional crude oil and, in some cases, to recovery of heavy oil and tar sand bitumen.

6.2 PRIMARY RECOVERY METHODS

The initial stage in producing oil from a reservoir (primary recovery, primary production) is the stage in which oil is forced to the surface by such natural forces as: (1) expansion of oil; (2) expansion of the contained gas; (3) displacement by migration of naturally pressurized water from a communicating zone such as in natural water drive; and (4) drainage downward from a high elevation in a reservoir to wells penetrating lower elevations. The natural expulsive forces present in a given reservoir depend on rock and fluid properties, geologic structure and geometry of the reservoir, as well as on the rate of oil and gas production—several of the forces may be present in a given reservoir. Recovery efficiency of conventional (light) crude oil varies in the primary stage from 10% to approximately 50% v/v of the oil in place. However, these numbers do not apply to heavy oil or to tar sand bitumen which are generally immune to recovery by primary methods.

Briefly, the overall recovery efficiency E of any fluid displacement process is given by the product of the macroscopic, or volumetric, displacement efficiency E_v and the microscopic displacement efficiency E_d:

$$E = E_v E_d.$$

The macroscopic displacement efficiency is a measure of how well the displacing fluid has come into contact with the oil-bearing parts of the reservoir. The microscopic displacement efficiency is a measure of how well the displacing fluid mobilizes the residual oil once the fluid has come into contact with the oil. Another relationship gives the fraction of oil that has been recovered from a zone swept by a water flood or other displacement process as:

$$E = (V_{oi} - V or)/V_{oi}$$

In this equation, V_{oi} is the volume of oil at start of flood, V_{or} is the volume of oil remaining after flood.

The macroscopic displacement efficiency is made up of two other terms, the areal sweep efficiency, E_s, and the vertical sweep efficiency, E_i. Thus,

$$E_v = E_s E_i$$

Primary recovery, in long accepted practice, is defined as production by natural reservoir pressure, or pumping, until depletion. Typically, conventional crude oil is produced from the reservoir by drilling wells into the formation and recovery of the oil by any one of several possible methods (Fig. 6.1). In this recovery method, crude oil is driven from the formation up through the wells (production wells) by energy stored in the formation, such as the pressure of water, dissolved natural gas (*primary recovery, natural recovery*)—often referred to as the *natural energy of the reservoir* or *reservoir energy*. When the natural energy of the reservoir is expended, energy must be injected into the reservoir in order to stimulate production using methods such as adding water and gas (*secondary recovery*), followed by other more energy intensive methods of recovery (*enhanced recovery*) (Chakma et al., 1991; Islam et al., 1994; Speight, 2014a). Furthermore, crude oil recovery depends upon several factors that, in turn, are site specific and a variety of selection criteria are involved in terms of fluid properties, production method, and formation geology (Figs. 6.2 and 6.3).

In most reservoir traps, the initial reservoir pressure (natural pressure or reservoir energy) is sufficient to push the oil to the surface of the production well with only minimal help from a downhole pump. But, with declining well pressures, it becomes more difficult to get the hydrocarbon to the surface. Sometimes, artificial oil lift is needed. Thus, if the

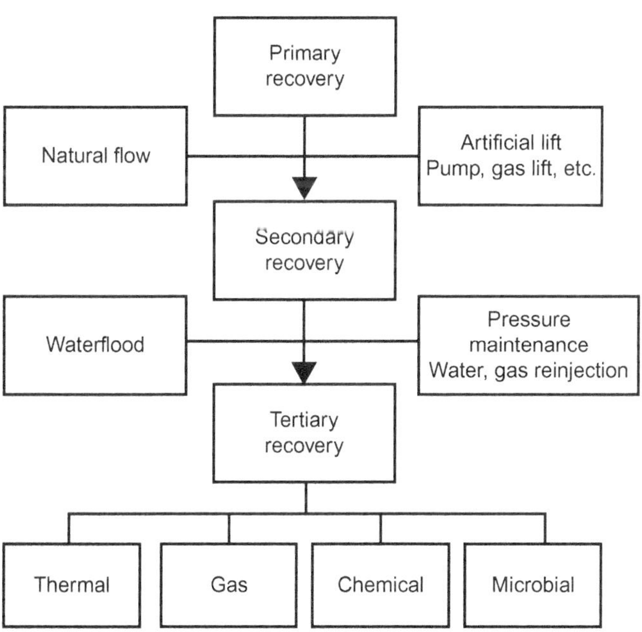

Figure 6.1 Methods for oil recovery.

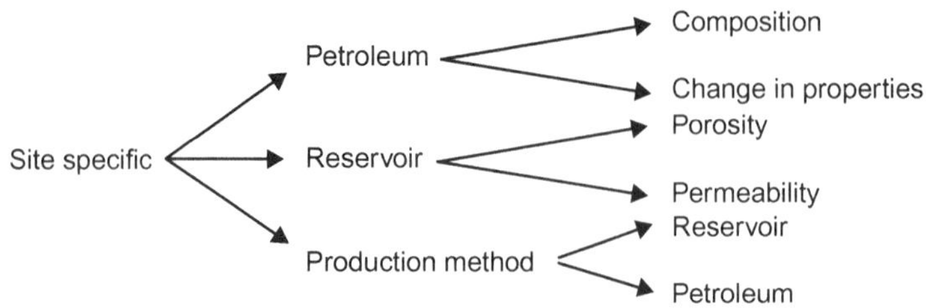

Figure 6.2 Parameters that contribute to site specificity.

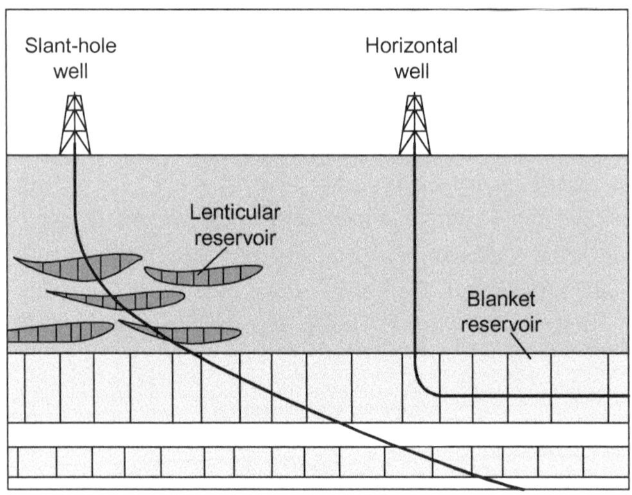

Figure 6.3 Directional drilling.

underground pressure in the oil reservoir is sufficient, then this pressure will force the oil to the surface. Gaseous fuels, natural gas, or water are usually present, which also supply needed underground pressure. In this situation, it is sufficient to place a complex arrangement of valves (the Christmas tree) (Fig. 6.4) on the wellhead to connect the well to a pipeline network for storage and processing.

Petroleum reservoirs usually start with a formation pressure high enough to force crude oil into the well and sometimes to the surface through the tubing (Lake and Walsh, 2004). However, since production is invariably accompanied by a decline in reservoir pressure, *primary recovery* through natural drive soon comes to an end. In addition, many oil reservoirs enter production with a formation pressure high enough to push the oil into the well but not up to the surface through the tubing. In these

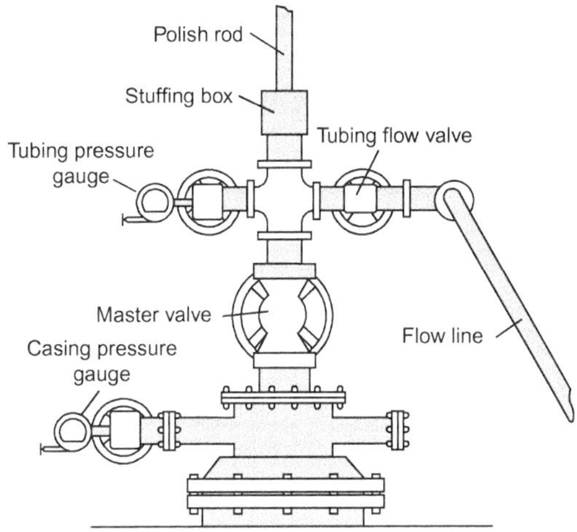

Figure 6.4 The Christmas Tree.

cases, some means of *artificial lift* must be installed. The most common installation uses a pump at the bottom of the production tubing that is operated by a motor and a *walking beam* (an arm that rises and falls like a seesaw) on the surface. A string of solid metal *sucker rods* connects the walking beam to the piston of the pump. Another method, called gas lift, uses gas bubbles to lower the density of the oil, allowing the reservoir pressure to push it to the surface. Usually, the gas is injected down the annulus between the casing and production tubing and through a special valve at the bottom of the tubing. In a third type of artificial lift, produced oil is forced down the well at high pressure to operate a pump at the bottom of the well.

With the artificial lift methods described above, oil may be produced as long as there is enough nearby reservoir pressure to create flow into the wellbore. Inevitably, however, a point is reached at which commercial quantities no longer flow into the well. In most cases, less than one-third of the oil originally present can be produced by naturally occurring reservoir pressure alone, and in some cases (eg, where the oil is quite viscous and at shallow depths) primary production is not economically possible at all.

There are several means that serve to drive the petroleum fluids from the formation, through the well, and to the surface, and these methods are classified as either natural or applied flow. If the underground pressure in the oil reservoir is sufficient, then the oil will be forced to the surface under this pressure. Gaseous fuels or natural gas are usually present, which also supplies needed underground pressure. In this situation it is sufficient to place a complex arrangement of valves (the Christmas tree) at the well head to connect the well to a pipeline network for storage and processing. This is called *primary oil recovery*.

Primary oil production (*primary oil recovery*) is the first method of producing oil from a well and depends upon natural reservoir energy to drive the oil through the complex pore network to producing wells. If the pressure on the fluid in the reservoir (reservoir energy) is great enough, the oil flows into the well and up to the surface. Such driving energy may be derived from liquid expansion and evolution of dissolved gases from the oil as reservoir pressure is lowered during production, expansion of free gas, or a gas cap, influx of natural water, gravity, or combinations of these effects.

Thus, once the well is completed, the flow of oil into the well is commenced. For limestone reservoir rock, acid is pumped down the well and out the perforations. The acid dissolves channels in the limestone that lead oil into the well. For sandstone reservoir rock, a specially blended fluid containing *proppants* (sand, walnut shells, aluminum pellets) is pumped down the well and out the perforations. The pressure from this fluid makes small fractures in the sandstone that allow oil to flow into the well, while the proppants hold these fractures open. Once the oil is flowing, the oil rig is removed from the site and production equipment is set up to extract the oil from the well.

A well is always carefully controlled in its flush stage of production to prevent the potentially dangerous and wasteful *gusher*. This is actually a dangerous condition, and is (hopefully) prevented by the blowout preventer and the pressure of the drilling mud. In most wells, acidizing or fracturing the well starts the oil flow.

For a newly opened formation and under ideal conditions the proportions of gas may be so high that the oil is, in fact, a solution of liquid in gas that leaves the reservoir rock so efficiently that a core sample will not show any obvious oil content. A general rough indication of this situation is a high ratio of gas to oil produced. This ratio may be zero for fields in which the rock pressure has been dissipated. The oil must be pumped out to as much as 50,000 ft^3 or more of gas per barrel of oil in the so-called condensate reservoirs, in which a very light crude oil (0.80 specific gravity or lighter) exists as vapor at high pressure and elevated temperature. Crude oil moves out of the reservoir into the well by one or more of three processes. These processes are: *dissolved gas drive*, *gas cap drive*, and *water drive*.

Early recognition of the type of drive involved is essential to the efficient development of an oil field.

In *solution gas drive* (*dissolved gas drive*) (Fig. 6.5) the propulsive force is the gas in solution in the oil, which tends to come out of solution because of the pressure release at the point of penetration of a well. Dissolved gas drive is the least efficient type of natural drive as it is difficult to control the gas-oil ratio; the bottom-hole pressure drops rapidly, and the total eventual recovery of petroleum from the reservoir may be less than 20%. If gas overlies the oil beneath the top of the trap, it is compressed and can be utilized (*gas cap drive*) to drive the oil into wells situated at the bottom of the oil-bearing zone (Fig. 6.6). By producing oil only from below the gas cap, it is possible to maintain a high gas–oil ratio in the reservoir until almost the very end of the life of the pool. If, however, the oil deposit is not systematically developed so that bypassing of the gas occurs, an undue proportion of oil is left behind. The usual recovery of petroleum from a reservoir in a gas cap field is 40% to 50%.

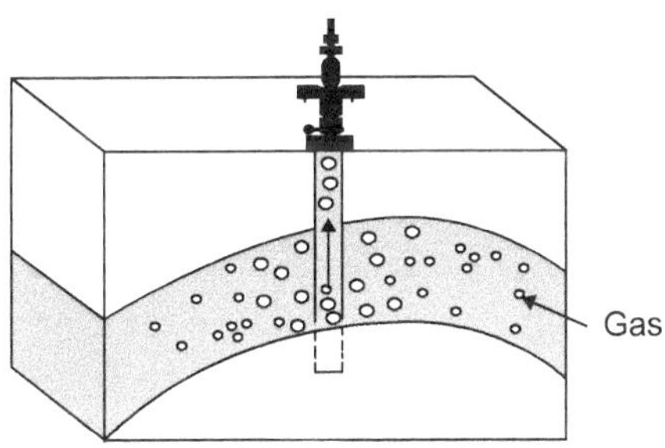

Figure 6.5 Solution-gas drive.

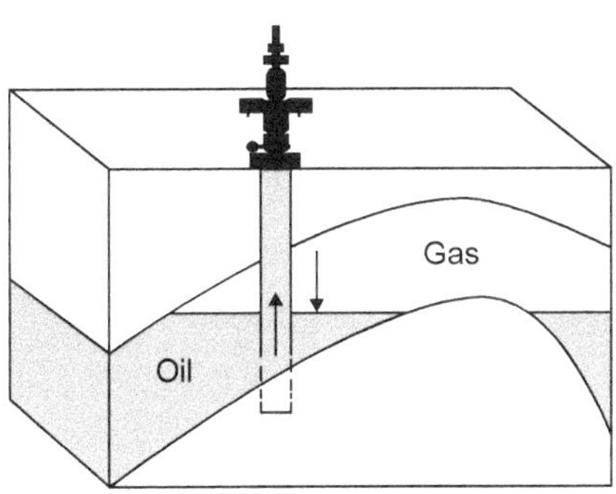

Figure 6.6 Gas-cap drive.

Usually the gas in a gas cap (*associated natural gas*) contains methane and other hydrocarbons that may be separated out by compressing the gas. A well-known example is *natural gasoline* that was formerly referred to as *casing head gasoline* or *natural gas gasoline*. However at high pressures, such as those existing in the deeper fields, the density of the gas increases and the density of the oil decreases until they form a single phase in the reservoir. These are the so-called retrograde condensate pools because a decrease (instead of an increase) in pressure brings about condensation of the liquid hydrocarbons. When this reservoir fluid is brought to the surface and the condensate is removed, a large volume of residual gas remains. The modern practice is to cycle this gas by compressing it and inject it back into the reservoir, thus maintaining adequate pressure within the gas cap, and condensation in the reservoir is prevented. Such condensation prevents recovery of the oil, for the low percentage of liquid saturation in the reservoir precludes effective flow.

The most efficient propulsive force in driving oil into a well is natural *water drive*, in which the pressure of the water forces the lighter recoverable oil out of the reservoir into the producing wells (Fig. 6.7). In anticlinal accumulations, the structurally lowest wells around the flanks of the dome are the first to come into water. Then the oil-water contact plane moves upward until only the wells at the top of the anticline are still producing oil; eventually these also must be abandoned as the water displaces the oil.

In a water drive field it is essential that the removal rate be adjusted so that the water moves up evenly as space is made available for it by the removal of the hydrocarbons. An appreciable decline in bottom-hole pressure is necessary to provide the pressure gradient required to cause water influx. The pressure differential needed depends on the reservoir

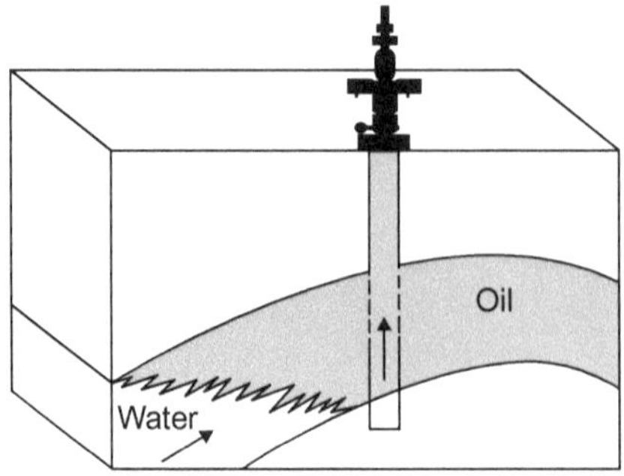

Figure 6.7 Water drive.

permeability; the greater the permeability, the less the difference in pressure necessary. The recovery of petroleum from the reservoir in properly operated water drive pools may run as high as 80%. The force behind the water drive may be hydrostatic pressure, the expansion of the reservoir water, or a combination of both. Water drive is also used in certain submarine fields.

Gravity drive is an important factor when oil columns of several thousands of feet exist, as they do in some North American fields. Furthermore, the last bit of recoverable oil is produced in many pools by gravity drainage of the reservoir. Another source of energy during the early stages of withdrawal from a reservoir containing undersaturated oil is the expansion of that oil as the pressure reduction brings the oil to the bubble point (the pressure and temperature at which the gas starts to come out of solution).

For primary recovery operations, no pumping equipment is required. If the reservoir energy is not sufficient to force the oil to the surface, then the well must be pumped. In either case, nothing is added to the reservoir to increase or maintain the reservoir energy or to sweep the oil toward the well. The rate of production from a flowing well tends to decline as the natural reservoir energy is expended. When a flowing well is no longer producing at an efficient rate, a pump is installed.

The recovery efficiency for primary production is generally low when liquid expansion and solution gas evolution are the driving mechanisms. Much higher recoveries are associated with reservoirs with water and gas cap drives and with reservoirs in which gravity effectively promotes drainage of the oil from the rock pores. The overall recovery efficiency is related to how the reservoir is delineated by production wells. Thus for maximum recovery by primary recovery it is often preferable to sink several wells into a reservoir, thereby bringing about recovery by a combination of the methods outlined here.

When production begins to drop off, it may be time for the well to receive a workover (a major repairing and cleaning out of all pipes). Producing wells are like anything else; they require periodic maintenance. Corrosion can roughen pipe walls or cause failure, allowing product to leak onto the surface. Pieces of rock from the side of the well may break off and fall into the well clogging it. Natural gas pipes tend to accumulate paraffin (hydrate compounds that build up inside the pipe causing restrictions). Maintenance can result in everything from cleaning fluids being injected into the pipes to wire brushes being inserted to brush the pipes clean. Residues are flushed from the system before it is reconnected.

But workover is not restricted only to the hardware; it may also be applied to the down-hole portion of the rock formation. Often, the formation through which the hydrocarbons are flowing becomes clogged, which diminishes the volume of product reaching the well. Two processes used to improve formation characteristics are acidizing and fracturing. *Acidizing* involves injecting an acid into a soluble formation, such as a carbonate, where it dissolves rock. This process enlarges the existing voids and increases permeability. *Hydraulic fracturing* (*fracking*) involves injecting a fluid into the formation under significant pressure that makes existing small fractures larger and creates new fractures.

Thus, in conventional oil production, the concept of applying more than one recovery technology, one after the other, to a reservoir is well established. When primary production declines and become less economic, producers investigate the opportunity to water flood the reservoir as a secondary recovery technology. Finally, tertiary methods may be applied when water floods yield diminishing returns. Heavy oil and oil sands have a shorter history and generally reservoirs have only been subject to only one recovery technology. In the case of oil sands, primary and secondary recovery technologies, as defined for conventional oil, are not applicable because bitumen is not mobile at reservoir conditions. Therefore, oil sands developments generally start with a thermal recovery technology which would be considered a tertiary method or enhanced recovery method for conventional oil. However, as the development of heavy oil and oil sands matures, the concept of applying more than one recovery technology in a specific order is likely to also be applied to heavy oil and bitumen reservoirs. In particular, in the Lloydminster area, researchers and producers have already been investigating for several years the concept of follow-up recovery technologies once primary production is no longer economic.

Over the lifetime of the well the pressure will fall, and at some point there will be insufficient underground pressure to force the oil to the surface. If economical, and it often is, the remaining oil in the well is extracted using secondary oil recovery methods.

6.3 SECONDARY RECOVERY METHODS

Secondary recovery processes are necessary because most of the crude oil in a reservoir remains in place after the natural energy pressurizing the reservoir has been dissipated. Several techniques are available for repressuring the reservoir by injecting fluids into the reservoir to augment the natural forces have been widely used for many years. Fluids, most commonly water (water flooding) and (although not truly a fluid but considered to be a reservoir fluid (see Chapter: Reservoir Fluids)) natural gas (gas flooding) are injected into the reservoir through a series of wells (injection wells) to force the crude oil toward another series of wells (production wells). The pattern of injection and production wells most appropriate to a reservoir are a technical matter and depend upon the properties of the reservoir and the properties of the crude oil. Furthermore, it is frequently desirable to initiate such processes as soon as sufficient knowledge is available of the geology of the reservoir and the type of natural expulsive forces that are operative. And there is no reason (in fact, it is often disadvantageous) to delay the fluid injection process until the natural energy in a reservoir is exhausted.

Thus, secondary recovery methods are methods in which the reservoir is repressurized and the remaining oil (that is no longer amenable to recovery by primary methods) is driven into the wellbore (or production zone). When a large part of the crude oil remaining in the reservoir cannot be recovered by primary means, a method for supplying extra energy must be found. Thus the success of secondary recovery processes depends on the mechanism by which the injected fluid displaces the oil (displacement efficiency) and on the volume of the reservoir that the injected fluid enters (conformance or sweep efficiency).

Most often, secondary recovery is accomplished by injecting gas or water into the reservoir to replace produced fluids and thus maintain or increase the reservoir pressure. When gas alone is injected, it is usually put into the top of the reservoir, where petroleum gases normally collect to form a gas cap. Gas injection can be a very effective recovery method in reservoirs where the oil is able to flow freely to the bottom by gravity. When this gravity segregation does not occur, however, other means must be sought.

Water, which is usually readily available and the least expensive of the fluids used, constitutes the oldest and most widely used secondary recovery method. In the process (waterflooding), water is pumped through the injection well into the reservoir. The water is forced from injection wells through the rock pores, sweeping the oil ahead of it toward production wells. This is practical for light to medium crudes. Over time, the percentage of water in produced fluids—the water cut—steadily increases. Some wells remain economical with a water cut as high as 99. But at some point, the cost of removing and disposing of water exceeds the income from oil production, and secondary recovery is then halted.

Secondary oil recovery methods use various techniques to aid in recovering oil from depleted or low-pressure reservoirs. Sometimes pumps on the surface or submerged (electrical submersible pumps), are used to bring the oil to the surface. Other secondary recovery techniques increase the reservoir's pressure by water injection and gas injection, which injects air or some other gas into the reservoir.

There are two main objectives in secondary crude oil production. One objective is to supplement the depleted reservoir energy pressure, and the second objective is to sweep the crude oil from the injection well toward and into the production well. In fact, secondary oil recovery involves the introduction of energy into a reservoir to produce more oil. For example, the addition of materials to reduce the interfacial tension of the oil results in a higher recovery of oil.

The most common follow-up, or *secondary recovery*, operations usually involve the application of pumping operations or of injection of materials into a well to encourage movement and recovery of the remaining petroleum. The pump, generally known as the *horsehead pump* (*pump jack*, *nodding donkey*, or *sucker rod pump*) (Fig. 6.8) provides mechanical lift to the fluids in the reservoir.

The pump is powered by an electric motor but some older, less producing wells use propane as an alternative source of power rather than a large power grid. Some of these wells can even use the natural gas from the casing as fuel, and the well can be completely self-sufficient. The

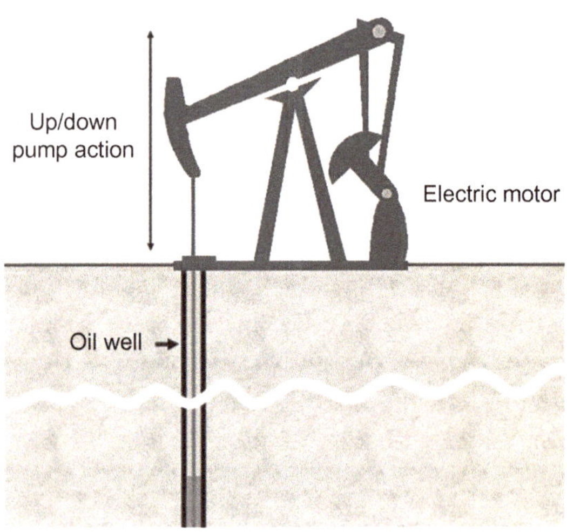

Figure 6.8 A horsehead pump.

engine of the pump runs a set of pulleys to the transmission which in turn drives a pair of cranks, generally with counter weights on them to assist the motor in lifting the heavy string of sucker rods. The cranks in turn raise and lower one end of the "I" beam which is free to move on an "A" frame. A metal or fiberglass cable called a (what else!) a bridle, connects the horse head to the polished rod.

The bridle follows the curve of the horsehead section of the pump as it lowers and rises to create a completely vertical stroke. The up-and-down movement of the sucker rods forces the oil up the tubing to the surface. A walking beam powered by a nearby engine may supply this vertical movement, or it may be brought about through the use of a pump jack, which is connected to a central power source by means of pull rods. Depending on the size of the pump, it generally produces up to one third of a barrel of an oil–water emulsion at each stroke. The size of the pump is also determined by the depth and weight of the oil to be removed, with deeper extraction requiring more power to move the heavier lengths of polish rod. Electrically powered centrifugal pumps and submersible pumps (both pump and motor are in the well at the bottom of the tubing) have proven their production capabilities in numerous applications.

There are also *secondary oil recovery* operations that involve the injection of water or gas into the reservoir. When water is used the process is called a *waterflood*; with gas, a *gas flood*. Separate wells are usually used for injection and production. The injected fluids maintain reservoir pressure or repressure the reservoir after primary depletion and displace a portion of the remaining crude oil to production wells. In fact, the first method recommended for improving the recovery of oil was probably the reinjection of natural gas, and there are indications that gas injection was utilized for this purpose before 1900 (Craft and Hawkins, 1959; Frick, 1962). These early practices were implemented to increase the immediate productivity and are therefore classified as pressure maintenance projects.

Waterflooding recovers oil by the water moving through the reservoir as a bank of fluid and *pushing* oil ahead of it. The recovery efficiency of a waterflood is largely a function of the sweep efficiency of the flood and the ratio of the oil and water viscosities. The sweep efficiency is a measure of how well the water has come in contact with the available pore space in the oil-bearing zone. Heterogeneity in the rock matrix (such as fractures, changes in minerology, and faults) are examples of rock heterogeneity and lead to low sweep efficiencies. Fractures, high-permeability streaks, and faults are examples of gross heterogeneities while homogeneous rock formations provide the optimum setting for high sweep efficiencies.

With respect to heavy oil production, when the injected water is much less viscous than the oil it is meant to displace, the water could begin to finger, or channel, through the reservoir (*viscous fingering*). This leads to significant bypassing of residual oil or heavy oil and lower production efficiency and the bypassing phenomenon is an important issue in applying enhanced oil recovery (EOR) techniques as well as in waterflooding.

The wells to be used for injecting water are usually located in a pattern that will best push oil toward the production wells. Water injection often increases oil recovery to twice that expected from primary means alone. Some oil reservoirs (eg, the East Texas Field) are connected to large, active water reservoirs, or aquifers, in the same formation. In such cases it is necessary only to reinject water into the aquifer in order to help maintain reservoir pressure.

In secondary recovery, the injected fluid must dislodge the oil and propel it toward the production wells. The term miscible means the mixing of two fluids—for instance, oil and a solvent such as carbon dioxide (CO_2)—into a single-phase fluid and may also apply to a continuity between the oil and injected gas, due to a multiphase transition zone between the two. Carbon dioxide can be injected into an oil reservoir to recover oil left behind during waterflooding. This EOR technique is commonly referred to as carbon dioxide miscible displacement. During a miscible displacement project, carbon dioxide dynamically develops miscibility as it mixes with the oil in the porous media. This process is conducted at or slightly above the *minimum miscibility pressure (MMP)* to ensure that the solvent strength of the carbon dioxide is sufficiently high to attain the requisite degree of solvency for the oil that it contacts. As the reservoir fluids are produced from the reservoir, the carbon dioxide can be readily separated from the oil and brine via pressure reduction. Thus, the primary objective of miscible carbon dioxide-based oil recovery is to remobilize and dramatically reduce the after waterflooding residual oil saturation in the reservoir's pore space. In addition, the reservoir energy must also be increased to displace the oil. Using techniques such as gas and water injection, there is no change in the state of oil. Similarly, there is no change in the state of the oil during miscible fluid displacement technologies.

Another recognized secondary recovery technique is injection of a hydrocarbon-based gas into an existing gas cap or directly into the oil itself. Gas may be injected over a considerable period of time-up to a year-while producing wells are hut in, until reservoir pressure is restored and production resumed. Another method is injection of gas to maintain pressure during production. Gas injection requires a nearby source of inexpensive gas in sufficient volume.

In most proposed secondary projects, water does both these things more effectively than gas. It must be decided if the use of gas offers any economic advantages because of availability and relative ease of injection. In reservoirs with high permeability and high vertical span, the injection of gas may result in high recovery factors as a result of gravity segregation, as described in a later section. However, if the reservoir lacks either adequate vertical permeability or the possibility for gravity segregation, a frontal drive similar to that used for water injection can be used (dispersed gas injection). Thus, dispersed gas injection is anticipated to be more effective in reservoirs that are relatively thin and have little dip. Injection into the top of the formation (or into the gas cap) is more successful in reservoirs with higher vertical permeability (200 milliDarcys or more) and enough vertical relief to allow the gas cap to displace the oil downward.

Vaporization is another recovery mechanism used to inject gas into oil reservoirs. A portion of the oil affected by the *dry* injection gas is vaporized into the oil and transported to the production wells in the vapor phase. In some instances this mechanism has been responsible for a substantial amount of the secondary oil produced.

During the withdrawal of fluids from a well, it is usual practice to maintain pressures in the reservoir at or near the original levels by pumping either gas or water into the reservoir as the hydrocarbons are withdrawn. This practice has the advantage of retarding the decline in the production of individual wells and considerably increasing the ultimate yield. It also may bring about the conservation of gas that otherwise would be wasted, and the disposal of brines that otherwise might pollute surface and near-surface potable waters.

In older fields it was not the usual practice to maintain the reservoir pressure, and it is now necessary to obtain petroleum from these fields by means of secondary recovery projects. Thus, several methods have been developed to obtain oil from reservoirs where previous economic policies dictated that ordinary production systems were no longer viable.

Considerable experimentation has been carried on in the use of different types of input gas. Examples are *wet casinghead gas*, *enriched gas*, *liquefied petroleum gas* (*LPG*), such as butane and propane, high-pressure gas, and even nitrogen. High-pressure gas not only pushes oil through the reservoir but may also produce a hydrocarbon exchange so that the concentration of liquid petroleum gases in the oil is increased.

Waterflooding is a form of oil recovery wherein some of the energy required to move the oil from the reservoir rock into a producing well is supplied from the surface by means of water injection. Waterfloods are essentially artificial water drives and are normally the most efficient of the various drive mechanisms, recovering 40% to 80% of the original oil in place. Artificial water drives (waterfloods) typical average about 50% of the original oil in place.

In the waterflooding process, water is injected into a reservoir to obtain additional oil recovery through movement of reservoir oil to a producing well. Generally, the selection of an appropriate flooding pattern for the reservoir depends on the quantity and location of accessible wells. Frequently, producing wells can be converted to injection wells whereas in other circumstances it may be necessary or advantageous to drill new injection wells.

In designing a waterflood project, it is general practice to locate injection and producing wells in a regular geometric pattern so that a symmetrical and interconnected network is formed. However, the relative location of injectors and producers depends on: (1) reservoir geometry; (2) lithology; (3) reservoir depth; (4) porosity; (5) permeability; (6) continuity of reservoir rock properties; (7) magnitude and distribution of fluid saturations; and last, but certainly not least (8) fluid, that is, oil, properties. Overall, the goal is to increase the mobility of the oil.

Briefly, the mobility of oil is the effective permeability of the rock to the oil divided by the viscosity of the oil.

$$\lambda = k/\mu$$

In the equation, λ is the mobility, md/cP, k is the effective permeability of reservoir rock to a given fluid, md, and μ is the fluid viscosity, cp. Thus the mobility ratio (M) is the mobility of the water divided by the mobility of oil:

$$M = K_{rw}\mu_o/K_{ro}\mu_w$$

K_{rw} is the relative permeability to water, K_{ro} is the relative permeability to oil, μ_o is the viscosity of the oil, and μ_w if the viscosity of water. The mobility ratio (M) infers that K_o is the mobility of oil ahead of the front (measured at S_{wc}) while K_w is the mobility of water at average water saturation in the water-contacted portion of the reservoir.

The mobility ratio of a waterflood will remain constant before breakthrough, but will increase after water breakthrough corresponding to the increase in water saturation and relative permeability to water in the water-contacted portion of the reservoir. Furthermore, the mobility ratio at water breakthrough is the term that is of significance in describing relative mobility ratio, that is, $M < 1$ indicates a favorable displacement as oil moves faster than water and $M = 1$ indicates a favorable displacement as both oil and water move at equal speed whereas $M > 1$ indicates an unfavorable displacement as water moves faster than oil.

Whenever it is feasible, the injection design should take advantage of gravity, that is to say, dipping or inclined reservoirs, underlying aquifers, or gas caps. A variety of injection-production patterns have been employed in waterflooding operations. In pattern flooding, the injectors are distributed among the producers in some repeating fashion. Pattern selection is important because it can affect the area swept by the injected fluid. Well locations frequently do not coincide with the ideal well locations associated with the well patterns used. Generally the choice of pattern (Fig. 6.9) for waterflooding must be consistent with the existing wells. The objective is to select the proper pattern that will provide the injection fluid with the maximum possible contact with the crude oil to minimize bypassing by the water.

In a *four-spot pattern* the distance between all like wells is constant. Any three injection wells form an equilateral triangle with a production well at the center. The four spot may be used when the injectivity is high or the heterogeneity is minimal. In a *five-spot pattern* the distance between all like wells is constant. Four injection wells form a square with a production

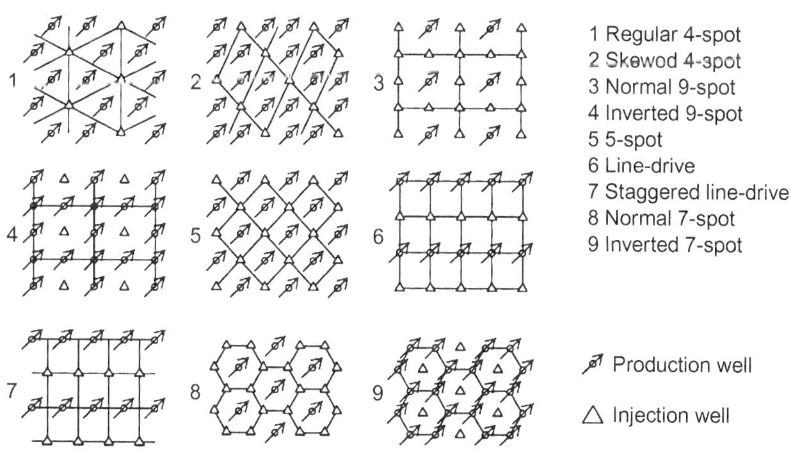

Figure 6.9 Well patterns.

well at the center. If existing wells were drilled on square patterns, five-spot patterns (as well as nine-spot patterns) are most commonly used since they allow easy conversion to a five-spot water flood. If the injectivity is low or the heterogeneity is large the five-spot pattern may not be the best choice and a nine-spot may be more efficient.

In the *seven-spot pattern* the injection wells are located at the corner of a hexagon with a production well at its center. If the reservoir characteristics yield lower than preferred injection rates, either a seven-spot (or a nine-spot) pattern should be considered because there are more injection wells per pattern than producing wells. In the *nine-spot pattern* the arrangement is similar to that of the five spot but with an extra injection well drilled at the middle of each side of the square. The pattern essentially contains eight injectors surrounding one producer. If existing wells were drilled on square patterns, nine-spot patterns (as well as five-spot patterns) are most commonly used. If the reservoir characteristics yield lower injection rates than those desired, one should consider using either a nine-spot pattern (or a seven-spot pattern) where there are more injection wells per pattern than producing wells.

In the *inverted seven-spot pattern* the arrangement is similar to the normal seven-spot pattern except where the position of the producer well was in the normal seven-spot pattern there is now an injector well. Likewise where the injector wells were in the normal seven-spot pattern, there are now producer wells. The inverted seven-spot pattern may be used when the injectivity is high or the heterogeneity is minimal. In this pattern, the arrangement of the wells is similar to the normal nine-spot pattern except the position of the producer well in the normal nine-spot pattern is occupied by an injector well. Likewise where the positions of the injector wells were in the normal nine-spot, there are now producer wells. If the reservoir is fairly homogenous and the mobility ratio is unfavorable, the inverted nine-spot pattern may be promising.

In the *direct line-drive pattern* the lines of injection and production are directly opposite to each other. If the injectivity is low or the heterogeneity is large direct line drive is a good option. Anisotropic permeability, permeability trends, or oriented fracture systems favor line drive patterns. In the *staggered line-drive pattern* the wells are in lines as in the direct line, but the injectors and producers are no longer directly opposed but laterally displaced by a specified distance that is dependent upon the distance between wells of the same type and the distance between the lines of injector wells and producer wells. The staggered line-drive pattern is also effective for reservoirs where there is anisotropic permeability or where there are permeability trends or oriented fracture systems.

Reservoir uniformities also dictate the choice of pattern and mobility ratio has an important influence on pattern selection. If the ratio is unfavorable, the injectivity of an injector will exceed the productivity of a producer and water injection will be supersede oil production. Hence, to balance the production with the water injection, more producers than injectors are required. On the other hand, if the mobility ratio is favorable, the injectivity is impaired, and the pattern should have more injectors than producers. Consideration should be given to the desired oil production level after waterflooding, which will influence the ratio of injectors to producers (Table 6.1) as an aid to increasing the areal sweep efficiency.

In the ideal case, the displacement of oil by water takes place in the form of a homogenous front and, depending on the mobility ratio of oil and water, the front starts to generate fingers resulting in bypassing oil. After the waterfront breaks through at the producer, high water-cut and low oil production characterizes the production behavior of this producer. The velocity of the frontal movement depends on the amount of water injected in the wells surrounding the producer and the amount of total fluid that is produced from the well in the center of the water injectors. Furthermore, the recovery of oil through water displacement can be considered as: (1) displacement, that is, the fraction of oil displaced from each pore by the physical contact between the displacing water and oil; and (2) conformance, that is, the fraction of the total volume (area × thickness) that is contacted by the injected water.

Displacement efficiency is the fraction of the oil displaced from each pore, which is contacted by the encroaching water. Areal sweep efficiency

Table 6.1 The Ratio of Injectors to Producers for Various Well Patterns

Pattern	Ratio of Producing Wells to Injection Wells	Drilling Pattern Required
Four-spot	2	Equilateral triangle
Five-spot	1	Square
Seven-spot	½	Equilateral triangle
Inverted seven-spot	2	Equilateral triangle
Nine-spot	⅓	Square
Inverted nine-spot	3	Square
Direct line drive	1	Rectangle
Staggered line drive	1	Offset lines of wells

is the fraction of oil displaced from the area physically contacted by the encroaching water. The vertical sweep efficiency is the fraction of oil displaced from the vertical cross-sectional coverage by the encroaching water. Hence, the recovery efficiency is the product of displacement efficiency and sweep efficiency (areal efficiency × vertical efficiency).

Water flooding remains a predominantly secondary recovery process (*waterflood*). Probably the principal reason for this is that reservoir formation water is ordinarily not available in volume during the early years of an oil field and pressure maintenance water from outside the field may be too expensive. When a young field produces considerable water, it may be injected back into the reservoir primarily for the purpose of nuisance abatement, but reservoir pressure maintenance is a valuable by-product.

Nevertheless, some passages in the formation are larger than others, and the water tends to flow freely through these, bypassing smaller passages where the oil remains. A partial solution to this problem is possible by *miscible fluid flooding*. Liquid butane and propane are pumped into the ground under considerable pressure, dissolving the oil and carrying it out of the smaller passages; additional pressure is obtained by using natural gas.

Finally, various mechanisms can interfere with the recovery of much of the original-oil-in-place after secondary recovery. Geologic heterogeneity in the reservoir may cause a large volume of mobile oil to be bypassed and remain within a field as a result of poor sweep efficiency when injected displacement water moves preferentially through higher permeability zones toward the production well. Even in regions of the reservoir that have been swept by large quantities of water, residual, immobile oil can remain in the pore spaces (of the reservoir rock) by capillary forces.

Finally, the waterflood-alternating-gas injection (WAG) process is used to increase the extent of reservoir contacted by the injected gas, the water-alternating-gas process is the most commonly employed commercial field gas injection process. Conceptually, the WAG process is meant to break-up the continuous slug of gas into smaller slugs by alternating them with water. In the process, the counter tendencies of gas to rise upward and water to descend within the reservoir are supposed to compensate each other to provide a more uniform reservoir sweep of the entire reservoir. The WAG process attempts to combine the good microscopic displacement arising from gas injection with improved macroscopic efficiency by injection water to improve the flood mobility ratio.

Currently, the WAG process is applied to many fields and regarded as the default process for commercial gas injection projects. The large-scale

WAG applications have been driven by proven improved performances over continuous gas injection (Christensen et al., 1998; Kulkarni and Rao, 2005).

6.4 ENHANCED OIL RECOVERY METHODS

Processes that inject fluids other than natural gas and water to augment the depleted reservoir energy have been designated improved oil recovery (IOR) techniques, tertiary recovery techniques, and EOR techniques. For the purposes of this text, the term EOR is preferred.

Primary and secondary recovery methods typically leave oil in place—as much as 50% to 67% v/v of the original oil in place—even in the most amenable conventional (light) crude oil reservoirs. The oil that remains—residual oil—is not the same as the residuum produced as a result of application of the refinery distillation process (Speight and Ozum, 2002; Parkash, 2003; Hsu and Robinson, 2006; Gary et al., 2007; Speight, 2014a). Residual oil, by definition, is the oil that remains in exhausted or energy-depleted reservoirs. It is similar in composition to the recovered oil but this is dependent upon the recovery processes so there might be some dissimilarities to the composition of the recovered oil and the residual oil cannot be recovered by conventional primary or secondary techniques. Since the primary and secondary techniques can recover approximately 33% to 50% v/v of the original oil in place, the amount of residual oil can be considerable.

Similar to the initiation of secondary recovery in conjunction with the cessation of primary recovery, at some point before secondary recovery ceases to remain feasible EOR methods must be applied if further oil is to be recovered. In the case of heavy oil recovery, which offers very little oil by primary and secondary recovery methods (again this is oil and reservoir dependent), it is often decided to apply EOR (tertiary oil recovery) methods at the onset of reservoir development. Thus, while waterflooding is effective in nearly all reservoirs, no single EOR technique is a cure-all. Most reservoir are complex, as are most EOR processes. Efficient reservoir management treats EOR as a high-cost, high-risk but critical component of a comprehensive plan that plans primary recovery through abandonments.

EOR, the third step (tertiary oil recovery) in the recovery of oil from a reservoir, is not an exact term but it is the term applied to the recovery of oil from a reservoir after primary and secondary recovery methods have been exhausted. Another term—IOR has come into the petroleum recovery lexicon and is often applied to the description of all recovery methods other than natural (primary) production, and the designation EOR is used for those processes beyond simple waterflood and gas flood—basically or, in summary, recovery by injection of anything not originally in the reservoir. The three major methods used for EOR are (1) thermal methods, that is, the application of heat, (2) miscible methods, that is, use of a solvent that is miscible with the oil, and (3) chemical flood methods, that is, the use of chemicals to minimize the oil–rock interactions and assist in the movement of oil from the reservoir into the production well (or production zone).

Thus, the intent of EOR is to increase the effectiveness of oil removal from pores of the rock (*displacement efficiency*) and to increase the volume of rock contacted by injected fluids (*sweep efficiency*) (Schumacher, 1980). *EOR* (*tertiary oil recovery*) was developed to recover the incremental ultimate oil (the residual oil left after application of primary and secondary recovery methods) from a petroleum reservoir (Fig. 6.10) (Lake, 1989;

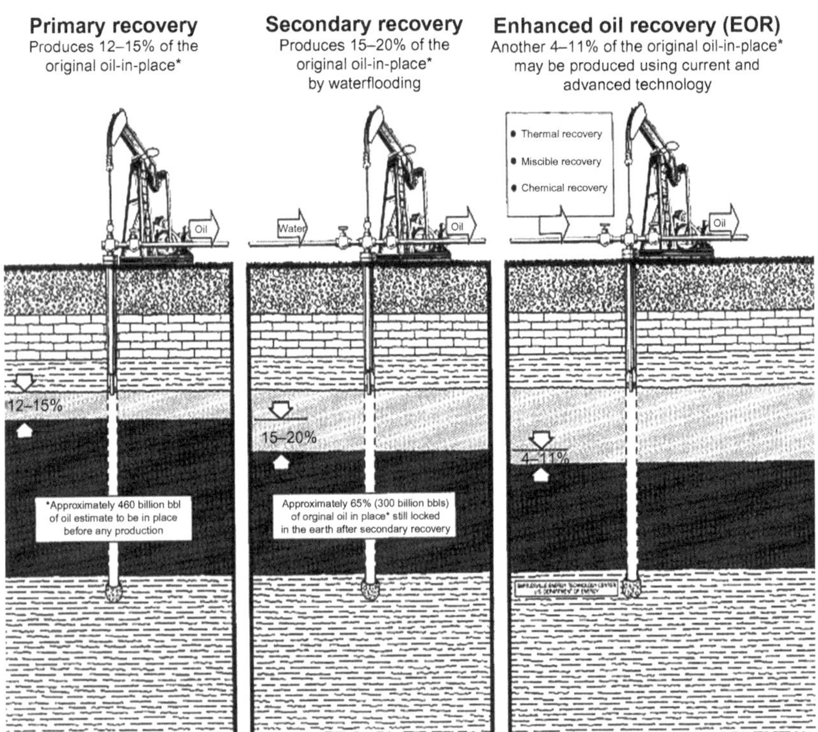

Figure 6.10 Oil production methods.

Arnarnath, 1999). EOR is often synonymous with *tertiary recovery* and, to some extent *IOR* as well as *advanced oil recovery*, although these terms also apply to primary and secondary methods. The intent of EOR (in the current context) is to increase the effectiveness of the removal of heavy oil or bitumen from pores of the rock (displacement efficiency). In addition, EOR processes use *thermal*, *chemical*, or *fluid phase behavior* effects to reduce or eliminate the capillary forces that trap oil within pores, to thin the oil or otherwise improve its mobility or to alter the mobility of the displacing fluids. The various processes differ considerably in (1) complexity and (2) the physical mechanisms. The degree to which the EOR methods are applicable in the future will depend on the development of improved process technology. It will also depend on improved understanding of fluid chemistry, phase behavior, and physical properties; and on the accuracy of geology and reservoir engineering in characterizing the physical nature of individual reservoirs (Venuto, 1989). Thus, EOR methods are designed to reduce the viscosity of the crude oil (ie, to reduce the pour point of the crude oil relative to the temperature of the reservoir), thereby increasing oil production.

Each EOR process is suited to a particular type of reservoir because unexpected or unknown reservoir characteristics can cause most process failure. EOR begins with a thorough geologic study. Technical rule-of-thumb screening criteria are available to aid preliminary evaluation of a reservoir's suitability for EOR (Taber and Martin, 1983) and, after these criteria are applied to a prospect, stringent economic analysis follows, generally through repeated reservoir simulations.

Once preliminary reservoir information has been assembled and used to select options for the EOR process to be applied to the reservoir, project design requires investigation of laboratory simulation of the process, fluid flow simulation, followed laboratory simulation, and then a pilot test. In the laboratory, the proposed processes are testing in core floods using samples of the reservoir rock and fluids. These tests—often referred to as one-dimensional flow tests in relatively homogeneous media—do not always successfully scale up to reservoir dimensions. Nevertheless, it is usually assumed that if the process simulation fails in the laboratory, the process is likely to fail in the field.

Fluid-flow simulations are based on a geologic reservoir model and can start with an assessment of the primary and secondary recovery methods and their respective production history to determine residual oil (oil not recovered during the application of the primary and secondary recovery methods). Care should be taken not to confuse the tem residual oil with the term residual material, which is the nonvolatile material that is left at the bottom of the atmospheric and vacuum, distillation units (Speight and Ozum, 2002; Parkash, 2003; Hsu and Robinson, 2006; Gary et al., 2007; Speight, 2014a). After this, the process-variable sensitivities can be calculated, followed by prediction of oil recovery by an EOR method.

Following from work on the reservoir geologic model, a pilot test of the proposed EOR process is usually carried out to confirm expected performance before an expensive, full-scale implementation. At this stage, the pilot test is performed in an area that is geologically similar to the field and large enough to be statistically representative of overall heterogeneity. Monitoring and data acquisition throughout pilot testing provide information needed to plan a full-scale commercial operation. Tests may also be carried out to investigate the feasibility of novel enhanced recovery techniques. Moving beyond the pilot test and a full-scale commercial test (demonstration test), for commercial operations, important considerations are secure and plentiful sources of water and other injectants, storage and transportation facilities (such as pipelines), surface processing, separation, recycling, and upgrading facilities, and environmental and safety requirements.

As stated above, in terms of the choice of an EOR process, there are three options: (1) *thermal processes*; (2) *chemical processes*; or (3) *fluid phase behavior* process, also called *miscible processes* which reduce or eliminate the capillary forces that trap oil within pores, to thin the oil or otherwise improve its mobility or to alter the mobility of the displacing fluids. In some cases, the effects of gravity forces, which ordinarily cause vertical segregation of fluids of different densities, can be minimized or even used to advantage. The various processes differ considerably in complexity, the physical mechanisms responsible for oil recovery, and the amount of experience that has been derived from field application. The degree to which the EOR methods are applicable in the future will depend on the development of improved process technology. It will also depend on improved understanding of fluid chemistry, phase behavior, and physical properties; and on the accuracy of geology and reservoir engineering in characterizing the physical nature of individual reservoirs (Borchardt and Yen, 1989).

Polymer flooding (*Polymer augmented waterflooding*) is waterflooding in which organic polymers are injected with the water to improve horizontal and vertical sweep efficiency. The process is conceptually simple and inexpensive, and its commercial use is increasing despite relatively small potential incremental oil production. Surfactant flooding is complex and requires detailed laboratory testing to support field project design but does have an excellent ability to improve the recovery of low-viscosity to moderate-viscosity oil (Bondor, 1991).

6.4.1 Chemical Processes

Chemical flooding involves the addition of one or more chemical compounds to an injected fluid either to reduce the interfacial tension between the reservoir oil and the injected fluid or to improve the sweep efficiency of the injected fluid. There are three general methods in chemical flooding technology: (1) alkaline flooding; (2) polymer flooding; and (3) micellar-polymer flooding. Alkaline flooding and micellar-polymer flooding uses chemicals that reduce the interfacial tension between an oil and a displacing fluid. Polymer flooding uses a macromolecule to increase the displacing fluid viscosity which leads to improved sweep efficiency in the reservoir.

Technical issues that have arisen with the use of chemical processes include: (1) screening chemicals to optimize the microscopic displacement efficiency; (2) making contact with the oil in the reservoir; and (3) maintaining good mobility in order to lessen the effects of viscous fingering. The requirements for screening of chemicals vary with the type of process and the chemicals must also be able to tolerate the environment in which they are used. Also, high temperature and salinity may limit the chemicals that could be used.

6.4.1.1 Alkaline Flooding

Alkaline flooding involves the use of aqueous solutions of certain chemicals, such as sodium hydroxide, sodium silicate, and sodium carbonate, that are strongly alkaline. These solutions will react with constituents present in some crude oils or present at the rock/crude oil interface to form detergent-like or surfactant-type materials which reduce the ability of the formation to retain the oil. These chemicals enhance oil recovery by one or more of the following mechanisms: interfacial tension reduction; spontaneous emulsification; or wettability alteration. These mechanisms rely on the in situ formation of surfactants during the neutralization of petroleum acids in the crude oil by the alkaline chemicals in the displacing fluids. Alkaline substances that have been used include sodium hydroxide, sodium orthosilicate, sodium metasilicate, sodium carbonate, ammonia, and ammonium hydroxide. Sodium hydroxide has been the most popular.

Sodium orthosilicate has some advantages in brines with a high divalent ion content.

When an alkaline solution is mixed with certain crude oils (high-acid crude oils are an example of such oils), surfactant molecules are formed—formation of the surfactants in situ raised the possibility that the interfacial tension between the brine and oil phases could be reduced. The reduction of interfacial tension causes the microscopic displacement efficiency to increase, which thereby increases oil recovery. There are optimum concentrations of alkaline and salt and optimum pH where the interfacial tension values experience a minimum.

Several mechanisms have been identified that aid oil recovery in the alkaline process. These include the following: (1) lowering of the interfacial tension; (2) emulsification of the oil; and (3) wettability changes in the rock formation. All three mechanisms can affect the microscopic displacement efficiency, and emulsification can also affect the macroscopic displacement efficiency. If a wettability change is desired, a high (2.0% to 5.0% w/w) concentration of alkali should be used, otherwise an alkali concentration on the order of 0.5% to 2.0% w/w of alkaline is used. The emulsification mechanism has been suggested to work by either of two methods: (1) emulsion formation and the emulsion becomes mobile and later trapped in downstream pores—the emulsion "blocks" the pores, which thereby diverts flow and increases sweep efficiency; or (2) emulsion formation and the mobile emulsion carries oil droplets that it has entrained to downstream production sites. The wettability changes that sometimes occur with the use of alkaline affect relative permeability characteristics, which in turn affect mobility and sweep efficiencies. Mobility control is an important consideration in the alkaline process as it is in all EOR processes. Often, it is necessary to include polymer in the alkaline solution in order to reduce the tendency of viscous fingering to take place.

As promising as this method may be, not all crude oils are amenable to alkaline flooding. The surfactant molecules are formed with the higher molecular weight acidic constituents of the crude oil. Standard test methods are available to determine the susceptibility of a given crude oil to alkaline flooding such as determining the acid number which is the number of milligrams of potassium hydroxide (KOH) required to neutralize one gram of the oil (Speight, 2014b). The higher the acid number, the more reactive the oil will be and the more readily it will form surfactants.

6.4.1.2 Polymer Flooding

The addition of polymers to an injected water can often increase the effectiveness of a conventional waterflood. Polymers are usually added to the water in concentrations ranging from 250 to 2000 parts per million (ppm) and a polymer solution is typically more viscous than a brine without polymer. In a flooding application, the increased viscosity changes the mobility ratio between the injected fluid and the reservoir oil leading to better vertical and areal sweep efficiencies and thus higher oil recoveries. Polymers have also been used to alter gross permeability variations in some reservoirs. In this application, polymers form a gel-like material by cross-linking with other chemical species and the polymer gel deposits in large permeability streaks and fractures diverting the flow of any injected fluid to a different location. Polymer flooding has seen success in moderately heterogeneous reservoirs and reservoirs containing oils with low viscosity (<100 cP).

In the process, a polymer-thickened water solution process is introduced after the chemicals are injected to help obtain a more uniform movement (sweep) through the reservoir. Fresh water is then injected behind the polymer solution to prevent contamination from the final drive water which may be salty or otherwise incompatible with the chemicals. Alkaline flooding is usually more efficient if the acid content of the reservoir oil is relatively high. A new modification to the process is the addition of surfactant and polymer to the alkali, giving rise to an alkaline-surfactant-polymer enhanced oil recovery method. Although emulsification in alkaline flooding processes decreases injection fluid mobility to a certain degree, emulsification alone may not provide adequate sweep efficiency. Sometimes polymer is included as an ancillary mobility control chemical in an alkaline waterflood to augment any mobility ratio improvements due to alkaline-generated emulsions.

Two general types of polymers have been used: (1) polyacrylamides; and (2) biologically produced polysaccharides. Polyacrylamides are long molecules with a small effective diameter and are susceptible to mechanical shear. High rates of flow through valves will sometimes break the polymer into smaller entities and reduce the viscosity of the solution, this can also occur as the polymer solution passes through the pore openings on the sand face of the injection well. However, polyacrylamides are also sensitive to salt (NaCl) and high concentrations of salt (in excess of 1% to 2% w/w) have an adverse effect on the viscosity-building effect of the polymers. On the other hand, polysaccharides are less susceptible to both mechanical shear and salt and, since they are produced biologically, care must be taken to prevent biological degradation in the reservoir.

Typically, the use of polymers does not affect the microscopic displacement efficiency and the improvement in oil recovery is due to an improved sweep efficiency over what is obtained during a conventional waterflood. Typical oil recoveries from polymer flooding applications are varied and are more likely to be successful if the process is initiated early in the producing life of the reservoir.

6.4.1.3 Micellar-Polymer Flooding

The micellar-polymer process uses a surfactant to lower the interfacial tension between the injected fluid and the reservoir oil. The surfactant migrates to the interface between the oil and water phases and helps make the two phases more miscible. Interfacial tensions can be reduced from approximately 30 dynes/cm, found in typical waterflooding applications, to 10^{-4} dynes/cm with the addition of as little as 0.1% to 5.0% w/w surfactant to water–oil systems. As the interfacial tension between an oil phase and a water phase is reduced, the capacity of the aqueous phase to displace the trapped oil phase from the pores of the rock matrix increases. The reduction of the interfacial tension results in a shifting of the relative permeability such that the oil will flow much more readily at lower oil saturations.

When surfactants are mixed above a critical saturation in a water-oil system, the result is a stable micellar solution which is made up of microemulsions that are homogeneous, transparent, and stable to phase separation. A microemulsion consists of external and internal phases sandwiched around one or more layers of surfactant molecules. The external phase can be either aqueous or hydrocarbon in nature, as can the internal phase. The microemulsions can exist in several shapes, which depend on the concentrations of surfactant, oil, water, and other constituents. Spherical microemulsions have a typical size ranging from 10^{-6} to 10^{-4} mm.

There are, in general, two types of micellar–polymer processes: (1) one which uses a low-concentration surfactant solution (<2.5% w/w) but a large injected volume (up to 50% pore volume); and (2) another which involves use of a high-concentration surfactant solution (5% to 12% w/w) and a small injected volume (5% to 15% v/v of the pore volume). Either type of process has the potential of achieving low interfacial tensions with a wide variety of brine–crude oil systems. Whichever system is selected, the system is made up of several components. The multicomponent facet leads to an optimization problem, since many different combinations could be chosen. Because of this, a detailed laboratory screening procedure is necessary and typically involves three types of tests for: (1) phase behavior; (2) interfacial tension; and (3) oil displacement.

Phase behavior tests are typically conducted in small (up to 100 mL) vials in order to determine what type, if any, of microemulsion is formed with a given micellar– crude oil system. The salinity of the micellar solution is usually varied around the salt concentration of the field brine where the process will be applied. Besides the microemulsion type, other factors examined could be oil uptake into the microemulsion, ease with which the oil and aqueous phases mix, viscosity of the microemulsion, and phase stability of the microemulsion. Interfacial tension tests are conducted with various concentrations of micellar solution components to determine the optimal concentration ranges.

The oil displacement tests are usually conducted using two or more types of porous media. Often initial screening experiments are conducted in unconsolidated sand packs and then in Berea sandstone. The last step in the sequence is to conduct the oil displacement experiments in actual cored samples of reservoir rock. Frequently, core samples are placed end to end in order to obtain a core of reasonable length since the individual core samples may be only 5 to 7 inches long.

6.4.2 Miscible Processes

Miscible processes are those in which an injected fluid dissolves in the oil by contact, forming a single oil-like liquid that can flow through the reservoir more easily than the original crude. A variety of such processes have been developed using different fluids that can mix with oil, including alcohols, carbon dioxide, petroleum hydrocarbons, such as propane or propane-butane mixtures, and petroleum gases rich in ethane, propane, butane, and pentane. The fluid must be carefully selected for each reservoir and type of crude to ensure that the oil and injected fluid will mix. The cost of the injected fluid is usually quite high in all known processes, and therefore either the process must include a supplementary operation to recover expensive injected fluid, or the injected material must be used sparingly. In this process, a *slug*, which varies from 5% to 50% of the reservoir volume, is pushed through the reservoir by gas, water (brine), or chemically treated brine to contact and displace the mixture of fluid and oil (OTA, 1978).

Miscible processes involve only moderately complex technology compared with other EOR processes. Although many miscible fluids have been field tested, much remains to be determined about the proper formulation of various chemical systems to effect complete volubility and to maintain this volubility in the reservoir as the solvent slug is pushed through it. Because of the high value of hydrocarbons and chemicals derived from hydrocarbons, it is generally felt that such materials would not make desirable injection fluids under current or future economic conditions. For this reason, attention has turned to carbon dioxide as a solvent. Conditions for complete mixing of carbon dioxide with crude oil depend on reservoir temperature and pressure and on the chemical nature and density of the oil. When insufficient reservoir pressure is available or the reservoir's oil composition is less favorable (heavier), the injected carbon dioxide will not become miscible with the oil in the reservoir.

Carbon dioxide (CO_2) can sometimes be used to enhance the displacement of oil from a reservoir. Carbon dioxide occurs naturally in some reservoirs, either with natural gas or as a nearly pure compound. It can also be obtained as a by-product from chemical and fertilizer plants, or it can be manufactured or separated from power plant stack gas. Thus, when pressure in a candidate reservoir has been depleted through primary production and possibly waterflooding, it must be restored before carbon dioxide injection can begin. To do this, normally water is pumped into the reservoir through injection wells until pressure reaches a desired level, then carbon dioxide is introduced into the reservoir through these same injection wells. Even though carbon dioxide is not miscible with oil on first contact, when it is forced into a reservoir a miscible front is generated by a gradual transfer of smaller, lighter hydrocarbon molecules from the oil to the carbon dioxide. This miscible front is in essence a bank of enriched gas that consists of carbon dioxide and low-boiling hydrocarbons. Under favorable conditions of pressure and temperature, this front will be soluble with the oil, making it easier to move toward production wells. This initial carbon dioxide slug is typically followed by alternate water and carbon dioxide injection—the water serves to improve sweep efficiency and to minimize the amount of carbon dioxide required for the flood. Production will be from an oil bank that forms ahead of the miscible front. As reservoir fluids are produced through production wells, the carbon dioxide reverts to a gaseous state and provides a *gas lift* similar to that of original reservoir natural gas pressure. On the surface, the carbon dioxide can be separated from the produced fluids and may be reinjected, helping to reduce the amount of new carbon dioxide required for the project; thus, the carbon dioxide can be recycled.

Thus, another oil displacement mechanism—immiscible carbon dioxide flooding—occurs in which the main mechanisms involved in immiscible carbon dioxide flooding are: (1) oil phase swelling, as the oil becomes saturated with carbon dioxide; (2) viscosity reduction of the swollen oil/carbon dioxide mixture; (3) extraction of lower molecular weight hydrocarbons into the carbon dioxide phase; and (4) fluid drive plus pressure. This combination of mechanisms enables a portion of the oil remaining in the reservoir to be mobilized and produced. In general, EOR based on the immiscibility of the oil and carbon dioxide is less efficient at recovering the oil remaining in the reservoir than EOR based on the miscibility of carbon dioxide.

In summary, oil swelling is an important oil recovery mechanism, for both the miscible and the immiscible carbon dioxide processes. Viscosity reduction is also an important aspect of the carbon dioxide-based processes.

6.4.3 Miscible Fluid Displacement Processes

Miscible fluid displacement (*miscible displacement*) processes are oil displacement processes in which an alcohol, a refined hydrocarbon, a condensed petroleum gas, carbon dioxide, liquefied natural gas, or even exhaust gas is injected into an oil reservoir at pressure levels such that the injected gas or alcohol and reservoir oil are miscible; the process may include the concurrent, alternating, or subsequent injection of water (Stalkup, 1983).

In the miscible process, a fluid is injected into the formation and mixes with the crude oil, which forms an oil bank. A mixing zone develops between the fluid and the oil bank and will grow due to dispersion. A second fluid, which is miscible with the first fluid but not generally miscible with the oil and which is much cheaper than the first fluid, is then injected into the formation. A mixing zone will also be created at the interface of the two injected fluids—it is important that the amount of the first injected fluid is of sufficient quantity so that the two mixing zones do not come in contact. If the front of the interface between the two injected fluids mixing zone reaches the rear of the first fluid–oil mixing zone, viscous fingering of the second fluid through the oil could occur.

In general, there are two types of miscible processes: (1) the single-contact miscible process, which involves such injection fluids as LPGs and alcohols and the injected fluids are miscible with residual oil immediately on contact; and (2) the multiple-contact process also called the dynamic miscible process in which the injected fluids are usually methane, inert fluids, or an enriched methane gas supplemented with a low-boiling (C_2-C_6) hydrocarbon faction. Also, in the multiple contact process, the injected fluid and oil are usually not miscible on first contact but rely on a process of chemical exchange between phases to achieve miscibility.

Miscible displacement occurs when oil flows to the production wellbores from deep within the reservoir displacing the oil near the wellbore, or water moving within the aquifer. Oil can also move to the wellbore by immiscible displacement when the oil is displaced by water or gas (Caruana and Dawe, 1996a; Dawe, 2004). There can also be mass transport across phase boundaries (eg, gas from liquid to vapor in solution gas processes, liquid dropout from gas in condensate reservoirs, multicontact processes in miscible gas EOR processes and transfer of chemicals such as in surfactant EOR). Additionally, there can be temperature changes with corresponding physical changes as in thermal EOR and cooling during long-term waterflooding of a reservoir with cold water (Dawe, 2004).

For miscible displacement, for example, solvent displacing oil in front of an immiscible displacement front or water displacing water behind it, then the flow lines generally follow the permeability changes. If the flow lines are not parallel to the heterogeneity some interesting consequences can occur and the predictions can become difficult. If immiscible displacements occur, interpretation becomes even more problematic as not only are there permeability and viscous effects but also effects due to capillary forces, wettability, and saturation which can become dominant. The saturation effects are particularly difficult as they are affected by pore scale wettability, which modifies the relative permeability and capillary pressures, which themselves modify the displacement patterns.

When carbon dioxide is used for oil displacement, the carbon dioxide enters the oil-bearing porous media at reservoir temperature, which is usually between 26°C and 120°C (80°F and 250°F). The carbon dioxide is introduced at a pressure that is high enough to enable miscible displacement to be achieved, referred to as the MMP. The foremost disadvantage of carbon dioxide as an oil displacement fluid is its low viscosity, 0.03 to 0.10 cP at these reservoir conditions. The brine in a reservoir has a viscosity on the order of 1 cP, while the reservoir oil viscosity varies between 0.1 and 50 cp. The carbon dioxide slug therefore has a much higher mobility (the ratio of a fluid permeability in porous media to viscosity) than the fluid it is displacing, resulting in an unfavorably low mobility ratio (the ratio of the carbon dioxide mobility to the fluid it is displacing is much less than unity). As a result, the areal sweep efficiency of the flood can be very low as the carbon dioxide *fingers* toward the production wells rather than uniformly displacing the oil ahead of it toward the production wells. Even though the displacement efficiency of the carbon dioxide may be very high *for the oil it contacts*, these fingers result in the carbon dioxide bypassing *much of the oil* in the reservoir. Consequently, if the carbon dioxide viscosity could be elevated to a level comparable with the oil it is displacing, typically a one to two order of magnitude increase, substantial improvements in sweep efficiency and oil recovery could result.

The procedures for miscible displacement are the same in each case and involve the injection of a slug of solvent that is miscible with the reservoir oil followed by injection of either a liquid or a gas to sweep up any remaining solvent. It must be recognized that the miscible *slug* of solvent becomes enriched with oil as it passes through the reservoir and its composition changes, thereby reducing the effective scavenging action. However, changes in the composition of the fluid can also lead to wax deposition (Weingarten and Euchner, 1986; Majeed et al., 1990; Pedersen et al., 1991; Erickson et al., 1993; Pan and Firoozabadi, 1996; Calange et al., 1997) as well as deposition of asphaltene constituents (Leontaritis et al., 1987 and references cited therein; Leontaritis, 1989; Chung, 1992; Nghiem et al., 1993; Nor-Aziam and Adewumi, 1993; Kamath et al., 1994; Deo et al., 1995; Rassamdana et al., 1999). Therefore, caution is advised.

Microscopic observations of the leading edge of the miscible phase have shown that the displacement takes place at the boundary between the oil and the displacing phase (Koch and Slobod, 1957). The small amount of oil that is bypassed is entrained and dissolved in the rest of the slug of miscible fluids; mixing and diffusion occur to permit complete recovery of the remaining oil. If a second miscible fluid is used to displace the first, another zone of displacement and mixing follows. The distance between the leading edge of the miscible slug and the bulk of pure solvent increases with the distance traveled, as mixing and reservoir heterogeneity cause the solvent to be dispersed.

Since the microscopic displacement efficiency is largely a function of interfacial forces acting between the oil, rock, and displacing fluid (see Introduction section), if the interfacial tension between the trapped oil and the displacing fluid could be lowered to 10^{-2} to10^{-3} dynes/cm, the oil droplets could be deformed and could squeeze through the pore constrictions. A miscible process is one in which the interfacial tension is zero;

that is, the displacing fluid and the residual oil mix to form one phase. If the interfacial tension is zero, then the capillary number N_{vc} becomes infinite and the microscopic displacement efficiency is maximized.

Other parameters affecting the miscible displacement process are reservoir length, injection rate, porosity, and permeability of reservoir matrix, size and mobility ratio of miscible phases, gravitational effects, and chemical reactions. Miscible floods using carbon dioxide, nitrogen, or hydrocarbons as miscible solvents have their greatest potential for enhanced recovery of low-viscosity oils, but also suffer from disadvantages (Table 6.2). In fact, each reservoir has unique fluid and rock properties, and specific chemical systems must be designed for each individual application. The chemicals used, their concentrations in the slugs, and the slug sizes depend upon the specific properties of the fluids and the rocks involved and upon economic considerations. During a miscible displacement project, carbon dioxide dynamically develops miscibility as it mixes with the oil in the porous media. This process is conducted at or slightly above the *MMP* to ensure that the solvent strength of the carbon dioxide is great enough to attain a high degree of solvency for the oil that it contacts. As the reservoir fluids are produced from the reservoir, the carbon dioxide

Table 6.2 Recovery process parameters and the potential adverse effects leading to sludge and/or sediment formation

Method	Comment
Carbon dioxide injection	Lowers pH; can change oil composition leading to phase separation of sludge or sediment and blocking of channels
Miscible flooding	Hydrocarbon-rich gases lower the solubility parameter and solvent power of the oil and cause separation of asphaltene material
Organic chemicals	Able to lower the solubility parameter and solvent power of the oil and cause separation of asphaltene material; blocking of channels
Acidizing	Interaction of crude oil constituents upsetting molecular balance and deposition of sludge or sediment; blocking of channels
Pressure decrease	Able to change the composition of the oil medium leading to phase separation of asphaltene material as sludge or sediment; blocking of channels
Temperature decrease	Able to change the composition of the oil medium leading to phase separation of asphaltene material as sludge or sediment; blocking of channels

can be readily separated from the oil and brine via pressure reduction. The availability of large volumes of low cost carbon dioxide, along with its low-toxicity, nonflammability, and classification as a nonvolatile organic compound combine to make it an attractive fluid for EOR (Enick, 1998). Commercial hydrocarbon-miscible floods have been operated since the 1950s, but carbon dioxide-miscible flooding on a large scale is relatively recent and is expected to make the most significant contribution to miscible enhanced recovery in the future.

Carbon dioxide is capable of displacing many crude oils, thus permitting recovery of most of the oil from the reservoir rock that is contacted (*carbon dioxide-miscible flooding*). The carbon dioxide is not initially miscible with the oil. However, as the carbon dioxide contacts the in situ crude oil, it extracts some of the hydrocarbon constituents of the crude oil into the carbon dioxide and carbon dioxide is also dissolved in the oil. Miscibility is achieved at the displacement front when no interfaces exist between the hydrocarbon-enriched carbon dioxide mixture and the carbon dioxide-enriched oil. Thus, by a *dynamic* (*multiple-contact*) process involving interphase mass transfer, miscible displacement overcomes the capillary forces that otherwise trap oil in pores of the rock.

The reservoir operating pressure must be kept at a level high enough to develop and maintain a mixture of carbon dioxide and extracted hydrocarbons that, at reservoir temperature, will be miscible with the crude oil. Impurities in the carbon dioxide stream, such as nitrogen or methane, increase the pressure required for miscibility. Mixing due to reservoir heterogeneity and diffusion tends to locally alter and destroy the miscible composition, which must then be regenerated by additional extraction of hydrocarbons. In field applications both miscible and near-miscible displacements may actually proceed simultaneously in different parts of the reservoir.

The volume of carbon dioxide injected is specifically chosen for each application and usually ranges from 20% to 40% of the reservoir pore volume. In the later stages of the injection program, carbon dioxide may be driven through the reservoir by water or a lower cost inert gas (Meszaros et al., 1990). To achieve higher sweep efficiency, water and carbon dioxide are often injected in alternate cycles.

In some applications, particularly in carbonate (limestone, dolomite, and chert) reservoirs where it is likely to be used most frequently, carbon dioxide may prematurely break through to producing wells. When this occurs, remedial action using mechanical controls in injection and production wells may be taken to reduce carbon dioxide production. However, substantial carbon dioxide production is considered normal. Generally this produced carbon dioxide is reinjected, often after processing to recover valuable light hydrocarbons.

For some reservoirs, miscibility between the carbon dioxide and the oil cannot be achieved and is dependent upon the oil properties. However, carbon dioxide can still be used to recover additional oil. The carbon dioxide swells crude oils, thus increasing the volume of pore space occupied by the oil and reducing the quantity of oil trapped in the pores. It also reduces the oil viscosity. Both effects improve the mobility of the oil. Carbon dioxide-immiscible flooding has been demonstrated in both pilot and commercial projects, but overall it is expected to make a relatively small contribution to EOR.

The solution gas–oil ratio for carbonated crude oil should be measured in the normal way and plotted as gas–oil ratio in volume per volume versus pressure. The greater the solubility of carbon dioxide in the oil, the larger is the increase in the solution gas–oil ratio. In fact, the increase in the gas–oil ratio usually parallels the increase in the oil formation volume factor due to swelling. It should be noted that the gas in any gas–oil ratio experiment is not carbon dioxide but contains hydrocarbons that have vaporized from the liquid phase. Consequently, whether the gas–oil ratio is measured in a pressure–volume–temperature cell or from a slim tube experiment, compositional analysis must be carried out to obtain the composition of the gas as well as that of the equilibrium liquid phase. If actual measured values are not available, the correlation developed for crude oil containing dissolved gases can be used but give only approximate values at best. Since the density of pure gases is a function of pressure and temperature, for crude oil saturated with gases, the density in the mixing zone must be specified as a function of pressure and mixing zone composition.

Hydrocarbon gases and condensates have been used for over 100 commercial and pilot miscible floods. Depending upon the composition of the injected stream and the reservoir crude oil, the mechanism for achieving miscibility with reservoir oil can be similar to that obtained with carbon dioxide (dynamic or multiple-contact miscibility), or the miscible solvent and in situ oil may be miscible initially (first-contact miscibility). Except in special circumstances, these light hydrocarbons are generally too valuable to be used commercially.

Nitrogen and flue gases have also been used for commercial miscible floods. MMPs for these gases are usually higher than for carbon dioxide, but in high-pressure, high-temperature reservoirs where miscibility can be achieved these gases may be a cost-effective alternative to carbon dioxide. Gaseous nitrogen is attractive for flooding this type of reservoir because it can be manufactured on site at less cost than other alternatives. Since it can be extracted from air by cryogenic separation, there is an unlimited source, and being completely inert it is noncorrosive. Thus, nitrogen flooding can be a viable EOR method if certain conditions exist in the candidate reservoir, such as: (1) the reservoir oil must be rich in ethane through hexane (C_2-C_6) or lower-boiling hydrocarbons—these crude oils are characterized as *light oils* having an API gravity in excess of 35°; (2) the oil should have a high formation volume factor, or the capability of absorbing added gas under reservoir conditions; (3) the oil should be undersaturated or low in methane (C_1); and (4) the reservoir should be sufficiently deep (on the order of 5000 feet deep or greater) to withstand the high injection pressure (in excess of 5000 psi) necessary for the oil to attain miscibility with nitrogen without fracturing the producing formation.

In general, when nitrogen is injected into a reservoir, it forms a miscible front by vaporizing some of the lighter components from the oil. This gas, now enriched to some extent, continues to move away from the injection wells, contacting new oil and vaporizing more components, thereby enriching itself still further. As this action continues, the leading edge of this gas front becomes so enriched that it goes into solution, or becomes miscible, with the reservoir oil. At this time, the interface between the oil and gas disappears, and the fluids blend as one. Continued injection of nitrogen pushes the miscible front (which continually renews itself) through the reservoir, moving a bank of displaced oil toward production wells. Water slugs are injected alternately with the nitrogen to increase the sweep efficiency and oil recovery. At the surface, the produced reservoir fluids may be separated, not only for the oil but also for natural gas liquids and injected nitrogen.

Finally, in terms of disadvantages of the miscible fluid process, the differences in density and viscosity between the injected fluid and the reservoir fluid(s), the miscible process often suffers from poor mobility. Viscous fingering and gravity override frequently occur and there may be separation between the miscible agent and brine because of density differences, with the miscible agent flowing along the top of the porous medium and the brine along the bottom. In order to overcome these disadvantages, process variations of the simultaneous injection scheme have been suggested and researched and typically involve the injection of a miscible agent followed by brine or the alternating of miscible agent–brine injection. The latter variation (WAG process, water alternate gas injection process) has become the most popular but a balance between amounts of injected water and gas must be achieved. For example, too much gas will lead to viscous fingering and gravity override of the gas, whereas too much water could lead to the trapping of reservoir oil by the water. The addition of foam-generating substances to the brine phase has been suggested as a way to aid in reducing the mobility of the gas phase.

Operational problems involving miscible processes include transportation of the miscible flooding agent, corrosion of equipment and tubing, and separation and recycling of the miscible flooding agent.

6.4.4 Thermal Processes

Primary production processes and secondary production processes from reservoirs containing heavy, low-gravity, viscous crude oils is usually a small fraction of the initial oil in place. This is due to the fact that these types of oils are very thick and viscous and as a result do not migrate readily to producing wells. However, the typical relationship between the viscosity of a heavy, viscous crude oil and temperature shows that viscosity decreases by orders of magnitude with an increase in temperature of 38°C to 95°C (100°F to 200°F) which, in the case of a heavy oil reservoir, will reduce the oil viscosity significantly and will flow much more easily to a producing well (Speight, 2013a, b). The temperature of a reservoir can be raised by injecting a hot fluid (such as hot water or steam) or by generating thermal energy in situ by combusting the oil.

Most of the oil that has been produced by EOR processes to date has been as a result of thermal processes, especially steam-based processes. There is a practical reason for this, as well as several technical reasons. In order to produce more than 1% to 2% v/v of the initial oil in place from a heavy-oil reservoir, operators had to employ thermal methods. As a result, thermal methods were investigated much earlier than either miscible or chemical methods, and the resulting technology was developed much more rapidly.

Viscosity, a measure of a liquid's ability to flow, varies widely among crude oils—some crudes flow easily (water-like) while others have more difficulty in flow properties. However, the viscosity of most crude oils dramatically decreases as temperature increases, and the purpose of all thermal oil recovery processes is to apply heat to the oil to enable flow under the impetus of injected fluids, which may be steam or hot water (steam injection), or air (combustion processes). Thus, *thermal processes* for oil recovery have found most use when the oil in the reservoir has a high viscosity. For example, heavy oil is usually highly viscous (hence the use of the adjective *heavy*), with a viscosity ranging from approximately 100 centipoises to several million centipoises at the reservoir conditions. In addition, oil viscosity is also a function of temperature and API gravity (Speight, 2000 and references cited therein). Thus for heavy crude oil samples with API gravity ranging from 4 to 21°API (1.04 to 0.928 kg/m^3):

$$\log\log(\mu\sigma + \alpha) = A - B\log(T + 460)$$

In this equation, $\mu\sigma$ is oil viscosity in cP, T is temperature in °F, A and B are constants, and α is an empirical factor used to achieve a straight-line correlation at low viscosity. This equation is usually used to correlate kinematic viscosity in centistokes, in which case an alpha (α) value of 0.6 to 0.8 is suggested (dynamic viscosity in cP equals kinematic viscosity in cSt multiplied by density in g/mL).

An alternative equation for correlating viscosity data is:

$$\mu = ae^{b/T\star}$$

In this equation, a and b are constants, and $T\star$ is the absolute temperature.

Thermal EOR processes add heat to the reservoir to reduce oil viscosity and/or to vaporize the oil. In both instances, the oil is made more mobile so that it can be more effectively driven to producing wells. In addition to adding heat, these processes provide a driving force (pressure) to move oil to producing wells.

The main technical problems that have been associated with thermal techniques are (1) poor sweep efficiency, (2) loss of heat energy to unproductive zones underground, and (3) poor injectivity of steam or air. The poor sweep efficiency is due to the density differences between the injected fluids and the curd oil in the reservoir. The lighter steam or air tends to rise to the top of the formation and bypass large portions of crude oil and this may be reflected in significant differences in residual oil saturation in the top of the swept formation and the residual oil saturation in the bottom part of the swept formation. Thus, there is a continuing need to reduce the tendency for the injected fluids to override the reservoir oil.

Large heat losses are also associated with thermal processes although the forward wet combustion process has lowered these losses for the higher-temperature combustion techniques but, depending upon the properties of the oil and the geological character of the reservoir or deposit, the losses may be sufficiently high to prohibit the combustion process. Typically, the heat (energy) losses are not as significant with the steam processes because of the lower temperature and the development and use of feasible downhole steam generators (see chapter: Thermal Methods of Recovery) has been able to reduce significantly the losses associated with steam injection processes and improve process efficiency.

The poor injectivity found in thermal processes is largely a result of the nature of the reservoir crudes. Application of fracturing technology in connection with the injection of fluids in thermal processes has corrected the difficulties associated with many heavy oil reservoirs. In addition, other issues include: (1) the formation of emulsions; (2) the corrosion of injection and production tubing and facilities; and (3) the creation of adverse effects on the environment. When emulsions are formed with heavy crude oil, they are very difficult to break. Also in the high-temperature environments created in the combustion processes when water and stack gases mix in the production wells and facilities, corrosion becomes a serious problem. Stack gases from both steam and combustion applications must be cleaned to prevent environmental damage. In addition, stack gases are formed when steam is generated by either coal- or oil-fired generators and, of course, during the combustion process as the crude is burned.

6.4.4.1 Steam Drive Injection

The steam drive process (or processes) are considerably different in performance from hot-water drives due to the presence and effects of the steam which causes low molecular weight constituents in the oil to be distilled and carried along as hydrocarbon constituents in the gas phase. When and wherever the steam condenses, the condensable hydrocarbon components also condense and reduce the viscosity of the crude oil at the condensation front. Moreover, the condensing steam makes the displacement process more efficient and improves the sweep efficiency. Thus, the net effect is that recovery from steam drives is significantly higher than from hot-water drives.

An important additional phenomenon affecting displacement in steam drives is the steam distillation of the relatively low-boiling fractions in the oil. Distillation causes the vapor phase to be composed not only of steam but also of condensable hydrocarbon vapors. Some hydrocarbon vapors will condense along with the steam, mixing with the original crude and increasing the amount of relatively light fractions in the residual oil trapped by the advancing condensate water ahead of the front. Dilution by the low-boiling constituents causes some of the trapped oil to be displaced by the condensed water. The remainder of the oil is stripped by the steam of all the remaining low boilers to leave the higher boiling residuum. The lower-boiling constituents help to regenerate and maintain a solvent bank just downstream of the condensation front. During the process, the composition of the produced oil generally does not change until the steam zone is relatively near, at which point the volatile content increased markedly.

Steam drive injection (steam injection) has been commercially applied since the early 1960s. The process occurs in two steps: (1) steam stimulation of production wells, that is, direct steam stimulation; and (2) steam drive by steam injection to increase production from other wells (indirect steam stimulation). Steam drive differs considerably in performance from hot-water drive—the difference in performance is due to the presence and effects of the condensing vapor. The presence of the gas phase causes low-boiling constituents present in the crude oil to be distilled and carried along as hydrocarbon components in the gas phase. When the steam condenses, the condensable hydrocarbon constituents of the vapor also condense and reduce the viscosity of the crude oil at the condensation front. In addition, the condensing steam makes the displacement process more efficient and improves the sweep efficiency and the net effect of the steam drive is that crude oil recovery is significantly higher than the crude oil recovery from hot-water drives.

However, both steam and hot-water drives (1) improve oil mobility by reducing viscosity and (2) reduce residual oil at high temperatures. They achieve those effects to such a degree that the recovery of viscous crudes is better than could be attained by waterflooding. Moreover, all phenomena encountered in hot-water displacement are also present in steam displacement operations, not only because the arrival of the condensation front is preceded by a hot-water drive but also because a hot-water phase is present along with steam in all cases of practical interest. However, the magnitudes of these similar forces may be different as a result of the vapor phase.

The steam drive process is much like a conventional waterflood insofar as when a pattern arrangement is established, steam is injected into several injection wells while the oil is produced from other wells. This is different from the steam stimulation process, whereby the oil is produced from the same well into which the steam is injected. As the steam is injected into the formation, the thermal energy is used to heat the reservoir oil but, unfortunately, the energy also heats the entire environment such as formation rock as well as water and some energy is also lost to the underburden and overburden. Once the oil viscosity is reduced by the increased temperature, the oil can flow more readily to the producing wells. The steam moves through the reservoir and comes in contact with cold oil, rock, and water. As the steam comes in contact with the cold environment, it condenses and a hot water bank is formed. This hot water bank acts as a waterflood and pushes additional oil to the producing wells.

Typically, a hot-water drive involves the flow of only two phases—water and oil—whereas steam-based processes and combustion processes, on the other hand, always involve a third phase—gas. Thus, the hot water drive is a displacement process in which oil is displaced by both the immiscible hot water and immiscible cold water. Except for temperature effects and the fact that they generally are applied to relatively viscous crude oils, hot-water floods have many elements in common with conventional waterfloods. Moreover, because of the presence of water in all petroleum reservoirs, displacement by hot water must occur to some extent in all thermal recovery processes and is known as one of the processes that contribute to the displacement of oil in the zones downstream of both steam drives and combustion drives.

Several mechanisms have been identified that are responsible for the production of oil from a steam drive. These include thermal expansion of the crude oil, viscosity reduction of the crude oil, changes in surface forces as the reservoir temperature increases, and steam distillation of the lighter portions of the crude oil. In fact, an important phenomenon affecting displacement in steam drives is the steam distillation of the relatively light fractions in the crude. Distillation causes the vapor phase to be composed not only of steam but also of condensable hydrocarbon vapors. Some hydrocarbon vapors will condense along with the steam, mixing with the original crude and increasing the amount of relatively light fractions in the residual oil trapped by the advancing condensate water ahead of the front. Dilution by low-boiling constituents causes some of the trapped oil to be displaced by the condensed water while the remainder is stripped by the steam of all the remaining light ends, thus leaving less volatile residue. The lower-boiling constituents that are stripped from the bypassed oil help to regenerate and maintain a solvent bank just downstream of the condensation front. The produced crude oil may not change in composition until the steam zone is relatively close, at which point the volatile content of the produced crude oil will increase. In summary, the light ends in the vapor contribute to the high volatile content of the oil produced after steam breakthrough.

When there is some natural reservoir energy, steam stimulation normally precedes steam drive. In steam stimulation, heat is applied to the reservoir by the injection of high-quality steam into the produce well. This cyclic process, also called *huff and puff* or *steam soak*, uses the same well for both injection and production. The period of steam injection is followed by production of reduced viscosity oil and condensed steam (water). One mechanism that aids production of the oil is the flashing of hot water (originally condensed from steam injected under high pressure) back to steam as pressure is lowered when a well is put back on production.

Mechanisms of oil recovery due to this process include (1) reduction of flow resistance near the well bore by reducing the crude oil viscosity and (2) enhancement of the solution gas drive mechanism by decreasing the gas solubility in an oil as temperature increases. Often, in heavy-oil reservoirs, the steam stimulation process is applied to develop injectivity around an injection well and, once injectivity has been established, the steam stimulation process is converted to a continuous steam drive process. In some projects, producing wells are periodically steam stimulated to maintain high production rates. Normally, stream drive projects are developed on relatively close well spacing to achieve thermal communication between adjacent injection and production wells. To date, steam methods have been applied almost exclusively in relatively thick reservoirs containing viscous crude oil.

6.4.4.2 Cyclic Steam Injection

Cyclic steam injection is the alternating injection of steam and production of oil with condensed steam from the same well or wells. Thus, steam generated at surface is injected in a well and the same well is subsequently put back on production. A cyclic steam injection process includes three stages. The first stage is injection, during which a measured amount of steam is introduced into the reservoir. The second stage (*the soak period*) requires that the well be shut in for a period of time (usually several days) to allow uniform heat distribution to reduce the viscosity of the oil (alternatively, to raise the reservoir temperature above the pour point of the oil). Finally, during the third stage, the now-mobile oil is produced through the same well. The cycle is repeated until the flow of oil diminishes to a point of no returns.

Cyclic steam injection clearly reduces the flow resistance near the wellbore, but it also (1) enhances the depletion mechanism by causing gas dissolved in the crude to become less soluble as the temperature increases and (2) increases the amount of water retained in the reservoir through a combination of temperature-induced wettability changes and hysteresis in the relative permeability curves in a manner analogous to a countercurrent imbibition displacement process. In addition, the process may eliminate damage present in the neighborhood of the wellbore, for example, by circulating hot solvent to dissolve asphaltene constituents and other organic precipitates (see Chapter: Reservoir Fluids) from perforations. Alternatively (or in addition) the process may reduce the flow resistance into an undamaged wellbore by reducing the viscosity of the oil by means of a heated wellbore. Several of the more common thermal stimulation techniques do change the driving forces in the reservoir and should be considered a combination of stimulation and displacement processes. Furthermore, recognition of some of the mechanisms that occur naturally during cyclic steam injection have led to modifications to enhance them, for example, the addition of a noncondensable gas to the injected steam.

The relatively small portion of the oil that remains after these displacement mechanisms have acted becomes the fuel for the in situ combustion process. Production is obtained from wells offsetting the injection locations. In some applications, the efficiency of the total in situ combustion operation can be improved by alternating water and air injection. The injected water tends to improve the utilization of heat by transferring heat from the rock behind the combustion zone to the rock immediately ahead of the combustion zone.

6.4.4.3 In Situ Combustion

Early attempts at in situ combustion involved the *forward dry combustion process* in which the crude oil was ignited downhole, and then a stream of air or oxygen-enriched air was injected in the well where the combustion was originated. The flame front was then propagated through the reservoir but much energy was lost to the overburden and underburden with this process. To reduce the heat losses, the *reverse combustion process* was devised in which the oil is ignited as in forward combustion but the airstream is injected in a different well. The air is then pushed through the flame front as the flame front moves in the opposite direction. However, in many instances the flame would be shut off because there was no oxygen supply and that where the oxygen was being injected, the oil would self-ignite and the process would then revert to the forward combustion process. As a result of the failure of the reverse combustion process, a new technique—the forward wet combustion process—was introduced which begins as a forward dry combustion does, but once the flame front is established, the oxygen stream is replaced by water. As the water comes in contact with the hot zone left by the combustion front, it flashes to steam (thus is an in situ steam generating process), using energy that otherwise would have been wasted. The steam moves through the reservoir and aids the displacement of oil.

The performance of in situ combustion is predominantly determined by the four following factors: (1) the quantity of oil that initially resides in the rock to be burned; (2) the quantity of air required to burn the portion of the oil that fuels the process; (3) the distance to which vigorous combustion can be sustained against heat losses; and (4) the mobility of the air or combustion product gases.

Using combustion to stimulate oil production is regarded as attractive for deep reservoirs (Terwilliger, 1975) and, in contrast to steam injection, usually involves no loss of heat. The duration of the combustion may be short (<30 days) or more prolonged (approximately 90 days), depending upon requirements. In addition, backflow of the oil through the hot zone must be prevented or coking occurs.

Both forward and reverse combustion methods have been used with some degree of success when applied to tar sand deposits. The forward combustion process has been applied to the Orinoco deposits (Terwilliger et al., 1975) and in the Kentucky sands (Terwilliger, 1975). The reverse combustion process has been applied to the Orinoco deposit (Burger, 1978) and the Athabasca (Mungen and Nichols, 1975). In tests such as these it is essential to control the airflow and to mitigate the potential for spontaneous ignition (Burger, 1978). A modified combustion approach has been applied to the Athabasca deposit (Mungen and Nichols, 1975). The technique involved a heat-up phase and a production (or blowdown phase) followed by a displacement phase using a fireflood-waterflood (COFCAW) process.

Not all crude oils are amenable to the combustion process and for the process to function efficiently, the crude oil has to have enough heavy components to serve as the fuel source for the combustion. Usually this requires a heavy oil of low API gravity and, as the high molecular weight constituents of the heavy oil are combusted, lower molecular weight constituents (including gaseous products) are formed. The gases are produced with the oil and, if soluble in the produced oil, raise the effective API gravity of the produced oil.

6.4.4.4 Modified In Situ Extraction

In the *modified in situ extraction* process, a combinations of in situ and mining techniques are used to access the reservoir when the heavy oil proves too difficult to move to the production well. Such processes are amenable to recovery of bitumen from tar sand deposits (Fig. 6.11). A portion of the reservoir rock must be removed to enable application of the in situ extraction technology. The most common method is to enter the reservoir through a large-diameter vertical shaft, excavate horizontal drifts from the bottom of the shaft, and drill injection and production wells horizontally from the drifts. Thermal extraction processes are then applied through the wells. When the horizontal wells are drilled at or near the base of the tar sand reservoir, the injected heat rises from the injection wells through the reservoir, and drainage of produced fluids to the production wells is assisted by gravity.

Technologies that upgrade value, drive down costs, and reduce environmental impacts will have the greatest effect on increasing the

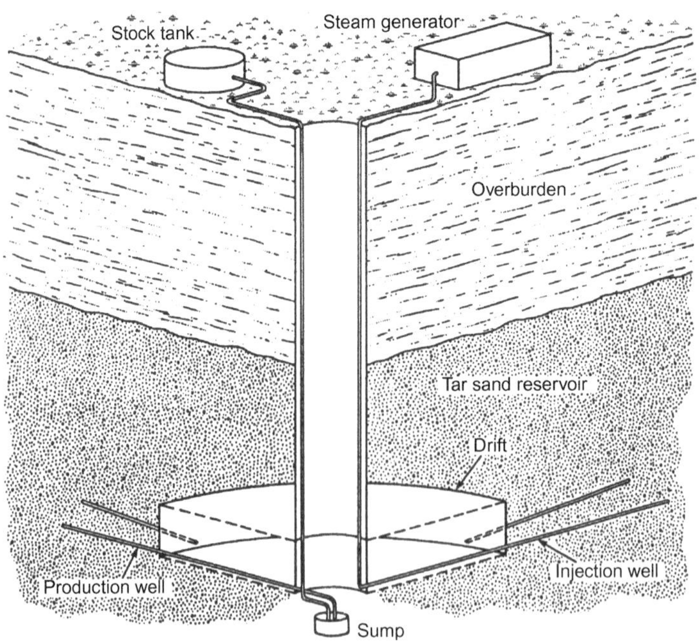

Figure 6.11 Modified in situ extraction.

production of heavy oil. There are a large number of technologies that can have an impact, but there is no single technology that can be generally applied owing to the variability of heavy oil properties as well as reservoir properties. In recent years there has been a renewed interest in the in-stir combustion technologies that have the potential to partially upgrade the oil and produce a product at the surface that is more valuable than the original oil.

No single EOR technique is the cure-all for oil recovery. Most reservoirs are complex and the oil-reservoir system must be considered as a whole rather than as individual, but equally complex, entities. Thus, the recovery technologies involve different concepts, some of which can cause changes to the heavy oil or bitumen during production. Technologies such as alkaline flooding, microemulsion (micellar/emulsion) flooding, polymer-augmented water flooding, and carbon dioxide miscible/immiscible flooding do not require or cause any change to the heavy oil or bitumen. The steaming technologies may cause steam distillation of the lower molecular weight constituents that can augment the process when the steam-distilled material moves with the steam front and acts as a solvent for the heavy oil or bitumen that lies ahead of the steam front. This may be reflected in favorable compositional changes to the recovered oil insofar as lower-boiling constituents are recovered and higher boiling (and higher molecular weight) constituents remain in the reservoir or deposit.

In fact, steam distillation is one of the principal mechanisms responsible for oil recovery during steam injection: (1) thermal expansion of the oil; (2) viscosity reduction; and (3) steam distillation. Steam distillation is the main mechanism that reduces the residual oil saturation behind the hot-water front during steam flood particularly for light and medium oils. The vaporizing effect of hydrocarbons is induced by increasing the system temperature, and it is reflected by the increase in the system pressure. Steam will evolve from the aqueous phase and strip some hydrocarbon component from the oleic phase. Since the pressure of the system is increased; the apparent bubble point of the water is consequently increased. This will cause the higher boiling point hydrocarbons to vaporize at the elevated saturation temperatures and steam distillation yields appear to be independent of the porous media and the steam injection rate, but dependent on the reservoir crude oil composition (Volek and Pryor, 1972; Wu and Brown, 1975; Duerksen and Hsueh, 1983; Wu and Elder, 1983). Several basic crude oil properties can be used to predict steam distillation yields reasonably well. A correlation using oil viscosity in centistokes at 38°C (100°F) can be used to predict the steam distillation yield as can the API gravity (Wu and Elder, 1983). Therefore, understanding the distillation effect of steam facilitates the prediction of fluid behavior in steam injection processes.

However, in any solvent-based process the separation (precipitation and deposition) of high-boiling constituents—such as asphaltene constituents—within the reservoir or deposit is a major concern. The probability of asphaltene precipitation and deposition occurring during any EOR operation and the subsequent effects on reservoir performance should be anticipated at earlier stages of any development project. Once the asphaltene deposition occurs, it causes reduction in the permeability and porosity as well as and wettability alteration. In addition, to changing the relative permeability of the reservoir rock there is always the potential for plugging the wellbore and surface facilities. Therefore, caution is advised when applying such processes to heavy oil and bitumen recovery.

The technology where changes do occur to heavy oil and bitumen involves combustion in situ. The concept of any combustion technology requires that the oil be partially combusted and that thermal decomposition occurs to (at least) part of the heavy oil or bitumen. This is sufficient to cause irreversible chemical and physical changes to the extent that the recovered product is markedly different to the heavy oil or bitumen in place and, in fact, represents partial upgrading during the process. In fact, in any oilfield where primary production is followed by a secondary or enhanced recovery method, there is the potential for noticeable differences in properties between the fluids produced but the changes may not always be beneficial. Significant differences may render the product outside of the range of acceptability for the usual refining options and force a higher demand for thermal processes, such as coking units.

If the reservoir contains heavy oil, it is often desirable to initiate *EOR* operations as early as possible. In tar sand deposits that contain bitumen, it is often desirable to initiate EOR operations as early as possible. This may mean considerably abbreviating conventional secondary recovery operations or bypassing them altogether. Thermal floods using steam and controlled in situ combustion methods are also used. Thermal methods of recovery reduce the viscosity of the crude oil by heat so that it flows more easily into the production well (Prats, 1986). Thus advanced techniques are usually variations of secondary methods with a goal of improving the *sweeping* action of the invading fluid.

Thermal recovery methods include cyclic steam injection, steam flooding (Fig. 6.12), use of detergents or surfactants (Fig. 6.13), and in situ

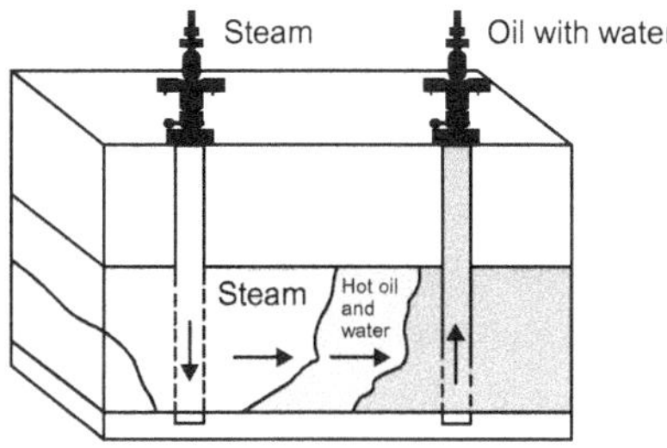

Figure 6.12 Steam flooding.

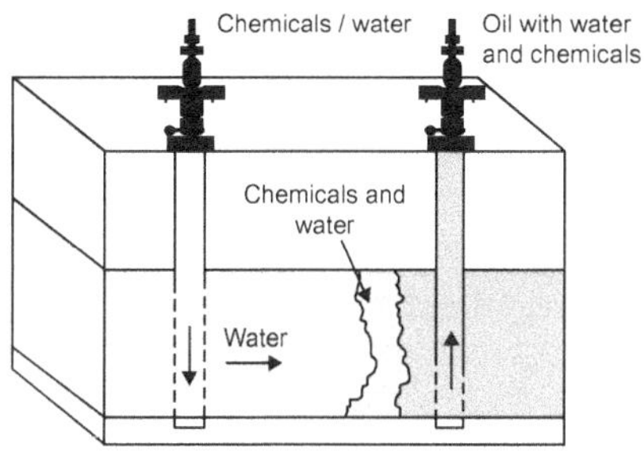

Figure 6.13 Use of detergents.

combustion (Prats, 1986). The steam processes are the most advanced of all EOR methods in terms of field experience and thus have the least uncertainty in estimating performance, provided that a good reservoir description is available. Steam processes are most often applied in reservoirs containing heavy oil, usually in place of rather than following secondary or primary methods. Commercial application of steam processes has been underway since the early 1960s. In situ combustion has been field tested under a wide variety of reservoir conditions, but few projects have proven economical and advanced to commercial scale.

6.4.5 Process Definition by the US Tax Code

Finally, for tax purposes, the Internal Revenue Service of the United States has listed the projects that qualify as EOR projects (CFR 1.43-2) and these are:

1. *Thermal recovery methods*:
   a. *Steam drive injection*—the continuous injection of steam into one set of wells (injection wells) or other injection source to effect oil displacement toward and production from a second set of wells (production wells);

   b. *Cyclic steam injection*—the alternating injection of steam and production of oil with condensed steam from the same well or wells; and
   c. *In situ combustion*—the combustion of oil or fuel in the reservoir sustained by injection of air, oxygen-enriched air, oxygen, or supplemental fuel supplied from the surface to displace unburned oil toward producing wells. This process may include the concurrent, alternating, or subsequent injection of water.
2. *Gas Flood recovery methods*:
   a. *Miscible fluid displacement*—the injection of gas (eg, natural gas, enriched natural gas, a liquefied petroleum slug driven by natural gas, carbon dioxide, nitrogen, or flue gas) or alcohol into the reservoir at pressure levels such that the gas or alcohol and reservoir oil are miscible;
   b. *Carbon dioxide augmented waterflooding*—the injection of carbonated water, or water and carbon dioxide, to increase waterflood efficiency;
   c. *Immiscible carbon dioxide displacement*—the injection of carbon dioxide into an oil reservoir to effect oil displacement under conditions in which miscibility with reservoir oil is not obtained; this process may include the concurrent, alternating, or subsequent injection of water; and
   d. *Immiscible nonhydrocarbon gas displacement*—the injection of nonhydrocarbon gas (eg, nitrogen) into an oil reservoir, under conditions in which miscibility with reservoir oil is not obtained, to obtain a chemical or physical reaction (other than pressure) between the oil and the injected gas or between the oil and other reservoir fluids; this process may include the concurrent, alternating, or subsequent injection of water.
3. *Chemical flood recovery methods*: Three EOR processes involve the use of chemicals: (1) surfactant/polymer; (2) polymer flooding; and (3) alkaline flooding (OTA, 1978).

Surfactant flooding is a multiple-slug process involving the addition of surface-active chemicals to water (Reed and Healy, 1977). These chemicals reduce the capillary forces that trap the oil in the pores of the rock. The surfactant slug displaces the majority of the oil from the reservoir volume contacted, forming a flowing oil–water bank that is propagated ahead of the surfactant slug. The principal factors that influence the surfactant slug design are interfacial properties, slug mobility in relation to the mobility of the oil-water bank, the persistence of acceptable slug properties and slug integrity in the reservoir, and cost.

Microemulsion flooding also known as *surfactant-polymer flooding* involves injection of a surfactant system (eg, a surfactant, hydrocarbon, cosurfactant, electrolyte, and water) to enhance the displacement of oil toward producing wells; and *caustic flooding*—the injection of water that has been made chemically basic by the addition of alkali metal hydroxides, silicates, or other chemicals. The terms *microemulsion* and *micellar solution* are used to describe concentrated, surfactant-stabilized dispersions of water and hydrocarbons that are used to enhance oil recovery. At concentrations above a certain critical value, the surfactant molecules in solution form aggregates called micelles. These micelles are capable of solubilizing fluids in their cores and are called swollen micelles. Spherical micelles have size ranges from 10^{-6} to 10^{-4} mm. The micellar solution or microemulsion is homogeneous, transparent or translucent, and stable to phase separation. The term *soluble oil* is often used to describe an oil external system having little or no dispersed water.

Microemulsion flooding (*micellar/emulsion flooding*) refers to a fluid injection process in which a stable solution of oil, water, and one or more surfactants along with electrolytes of salts is injected into the formation and is displaced by a mobility buffer solution (Reed and Healy, 1977; Dreher and Gogarty, 1979). *Micellar-polymer flooding* uses the injection of a micellar slug containing a mixture of a surfactant, cosurfactant, alcohol, brine, and oil that moves through the oil-bearing formation, releasing much of the oil trapped in the rock. This method is one of the most efficient methods for EOR but is also one of the most costly to implement.

Injecting water in turn displaces the mobility buffer. Depending on the reservoir environment, a preflood may or may not be used (Venuto, 1989). The microemulsion is the key to the process. Oil and water are displaced ahead of the microemulsion slug, and a stabilized oil and water bank develops. The displacement mechanism is the same under secondary and tertiary recovery conditions. In the secondary case, water is the primary produced fluid until the oil bank reaches the well.

In *microemulsion flooding*, two approaches have been developed to EOR (Gogarty, 1976). In the first process, a relatively low-concentration surfactant microemulsion is injected at large pore volumes of 15% to 60% to reduce the interfacial tension between the water and oil, thereby increasing oil recovery. In the second process, a relatively small pore volume, from 3% to 20% of a high-concentration surfactant microemulsion, is injected. With the high concentration of surfactant in the microemulsion, the micelles solubilize the oil and water in the displacing microemulsion. Consequently, the high-concentration system may initially displace the oil in a miscible-like manner. However, as the high-concentration slug moves through the reservoir it is diluted by the formation fluids and the process ultimately or gradually reverts to a low-concentration flood. However, this initial displacement forms an oil bank, which is very important in establishing displacement efficiency. Low-concentration systems typically contain 2% to 4% surfactant, whereas high-concentration systems contain 8% to 12% w/w.

Mobility control is important to the success of the process. The mobility of the microemulsion can be matched to that of the stabilized water–oil bank by controlling the microemulsion viscosity. The mobility buffer following the microemulsion slug prevents rapid slug deterioration from the rear and thus minimizes the slug size required for efficient oil displacement. Water external emulsions and aqueous solutions of high-molecular-weight polymers have been used as mobility buffers.

Microemulsion flooding can be applied over a wide range of reservoir conditions. Generally, wherever a waterflood has been successful, microemulsion flooding may also be applicable. In cases in which waterflooding was a failure because of poor mobility relationships, microemulsion flooding might be technically successful because of the required mobility control. If waterflooding was a failure because of certain reservoir conditions, such as fracturing or very high permeability streaks, microemulsion flooding will most likely also fail.

Polymer augmented waterflooding—the injection of polymeric additives with water to improve the areal and vertical sweep efficiency of the reservoir by increasing the viscosity and decreasing the mobility of the water injected; polymer augmented waterflooding does not include the injection of polymers for the purpose of modifying the injection profile of the wellbore or the relative permeability of various layers of the reservoir, rather than modifying the water–oil mobility ratio.

Conventional waterflooding can often be improved by the addition of polymers (*polymer flooding*) to injection water to improve the mobility ratio between the injected and in-place fluids (Sorbie, 1991). The polymer solution affects the relative flow rates of oil and water and sweeps a larger fraction of the reservoir than water alone, thus contacting more of the oil and moving it to production wells. Polymers currently in use are produced both synthetically (polyacrylamides) and biologically (polysaccharides).

The polymers may also be cross-linked in situ to form highly viscous fluids that will divert the subsequently injected water into different reservoir strata.

Polymer flooding has its greatest utility in heterogeneous reservoirs and those that contain moderately viscous oils. Oil reservoirs with adverse waterflood mobility ratios have a potential for increased oil recovery through better horizontal sweep efficiency. Heterogeneous reservoirs may respond favorably as a result of improved vertical sweep efficiency. Because the microscopic displacement efficiency is not affected, the increase in recovery over waterflood will likely be modest and limited to the extent that sweep efficiency is improved, but the incremental cost is also moderate. Currently, polymer flooding is being used in a significant number of commercial field projects. The process may be used to recover oils of higher viscosity than those for which a surfactant flood might be considered.

Polymer solutions must be stable for a prolonged period at reservoir conditions. Mechanical, chemical, thermal, and microbial effects can degrade polymers. However, degradation can be minimized or even prevented by using specific equipment or methods. Stability problems may occur as a result of oxygen contamination of the polymer solutions. Such contamination can lower the screen factor of polyacrylamide solutions by as much as 30%. In field operations, the loss of mobility reduction due to oxygen may be more serious since control of the reservoir fluid composition can be difficult. Sodium hydrosulfite in low concentrations is an effective oxygen collector for polyacrylamide solutions. However, sodium hydrosulfite tends to catalyze polymer deterioration when free oxygen and decomposed polymers are present. Therefore, the proper use of sodium hydrosulfite is imperative to avoid severe polymer degradation. In addition, caution is necessary to prevent oxygen from reentering the system once sodium hydrosulfite has been added to the makeup water.

Certain reservoir types, such as those with very viscous crude oils (heavy crude oils and extra heavy oils) and some low-permeability carbonate (limestone, dolomite, or chert) reservoirs, respond poorly to conventional secondary recovery techniques. The viscosity (or the API gravity) of the material is an important factor that must be taken into account when heavy oil is recovered from a reservoir.

In these reservoirs it is desirable to initiate *EOR* operations as early as possible. This may mean considerably abbreviating conventional secondary recovery operations or bypassing them altogether. Thermal floods using steam and controlled in situ combustion methods are also used. Thermal methods of recovery reduce the viscosity of the crude oil by heat so that it flows more easily into the production well. Thus tertiary techniques are usually variations of secondary methods with a goal of improving the *sweeping* action of the invading fluid.

Thermally-enhanced oil recovery methods are tertiary recovery techniques that heat the oil and make it easier to extract. Steam injection is the most common form of this process and is used extensively to increase oil production. In situ combustion is another form of thermally EOR but instead of using steam to reduce the crude oil viscosity, some of the oil is burned to heat the surrounding oil. Detergents are also used to decrease oil viscosity.

The same principles of EOR planning and engineering (as presented above) may not apply to offshore oil fields (Speight, 2015a). Generally, offshore wells tend to be highly deviated and the distance between the wells is often greater than between onshore wells. This extends the time between initiation of an enhanced recovery operations and meaningful results. Furthermore, such effects can complicate process control and limit the number of enhanced recovery techniques that may be applicable. Greater spacing between wells also increases the likelihood of any undetectable heterogeneity between the wells thereby impairing simulations of well behavior. Because the number of wells that can be drilled from a platform is fixed, infill drilling, often an important strategy for both secondary recovery and enhanced recovery, is not always possible. In fact, offshore EOR projects must be planned and started early enough so that production increases incrementally before primary and secondary production begins to decline. There is also the risk that there may be insufficient reservoir description at the early stages of field production and lack of time to acquire pilot test results to evaluate the enhanced recovery process.

6.4.6 Screening Criteria for Enhanced Oil Recovery Processes

A large number of variables are associated with a given oil reservoir, for instance, pressure and temperature, crude oil type and viscosity, and the nature of the rock matrix and connate water. Because of these variables, not every type of EOR process can be applied to every reservoir. An initial screening procedure would quickly eliminate some EOR processes from consideration in particular reservoir applications. This screening procedure involves the analysis of both crude oil and reservoir properties. This section presents screening criteria or guidelines for each of the general types of processes previously discussed in this article except microbial flooding.

A detailed geological study is usually desirable, since operators have found that unexpected reservoir heterogeneities have led to the failure of many EOR field projects. Reservoirs that are found to be highly faulted or fractured typically yield poor recoveries from EOR processes. Candidate reservoirs should contain sufficient recoverable oil and be large enough for the project to be potentially profitable. Also, deep reservoirs could involve large drilling and completion expenses if new wells are to be drilled.

The miscible process requirements are characterized by a low-viscosity crude oil and a thin reservoir. A low-viscosity oil will usually contain enough of the intermediate-range components for the multicontact miscible process to be established. The requirement of a thin reservoir reduces the possibility that gravity override will occur and yields a more even sweep efficiency.

In general, the chemical processes require reservoir temperatures of less than 95°C (200°F), a sandstone reservoir, and enough permeability to allow sufficient injectivity. The chemical processes will work on oils that are more viscous than what the miscible processes require, but the oils cannot be so viscous that adverse mobility ratios are encountered. Limitations are set on temperature and rock type so that chemical consumption can be controlled to reasonable values. High temperatures will degrade most of the chemicals that are currently being used in the industry.

In applying the thermal methods, it is critical to have a large oil saturation. This is especially pertinent to the steam flooding process, because much of the produced oil will be used on the surface as the source of fuel to fire the steam generators. In the combustion process, crude oil is used as fuel to compress the airstream on the surface. The reservoir should be of significant thickness in order to minimize heat loss to the surroundings.

6.5 FLUIDS PRODUCTION

The essence of fluids production from reservoirs is the success of the production process, which is often defined as the displacement efficiency, which is the fraction of oil that has been recovered from a zone swept by a water flood or other displacement process and the displacement efficiency, E, is represented by the equation:

$$E = (V_{oi} - V_{or})/V_{oi}$$

V_{oi} is the volume of oil at start of flood and V_{or} is the volume of oil remaining after water flood or displacement process flood.

The displacement efficiency is, in turn, defined as: (1) the microscopic displacement efficiency, which is a measure of the efficiency of the displacing fluid in mobilizing the residual oil once the fluid has come in contact with the oil, and (2) the macroscopic displacement efficiency, which is a measure of the efficiency of the displacing fluid in contacting the oil-bearing parts of the reservoir.

6.6 MICROSCOPIC DISPLACEMENT EFFICIENCY

The microscopic displacement efficiency is affected by the following factors: (1) interfacial tension and surface tension; (2) wettability; (3) capillary pressure; and (4) relative permeability. When a drop of one immiscible fluid is immersed in another fluid and comes to rest on a solid surface, the surface area of the drop will take a minimum value due to the forces acting at the fluid–fluid and rock–fluid interfaces. The force per unit length acting at the fluid–fluid and rock–fluid interface (interfacial tension) represents the amount of work required to create a new unit of surface area at the interface between two fluids. The interfacial tension is also to be thought of as a measure of the immiscibility of two fluids—typical values for the oil–brine interfacial tension are on the order of 20 to 30 dynes/cm. When certain chemical agents are added to an oil-brine system, it is possible to reduce the interfacial tension by several orders of magnitude.

Wettability is the tendency for a solid to prefer one fluid over another and is a function of the chemical composition of both the fluids and the rock. Rock surfaces (ie, reservoir rock surfaces) can be either oil-wet or water-wet, depending on the chemical composition of the fluids and this is strongly affected by the adsorption or desorption of constituents in the oil phase. Large, polar constituents (such as asphaltene constituents and resin constituents) of the oil phase can absorb onto the solid surface which leaves an oleophilic (the rock is often classed as oleophobic) film that may alter the wettability of the surface.

The concept of wettability leads to another significant factor in the recovery of residual oil—the capillary pressure which is caused by the lower density of oil compared to the density of brine. The pressure in the oil phase immediately above the oil–brine interface in a capillary tube will be slightly greater than the pressure in the water phase just below the interface. The difference in pressure (the capillary pressure P_c) of the system occurs in the nonwetting phase and is related to the contact angle θ, the radius r of the capillary, the oil-brine interfacial tension γ_{wo}, and the capillary pressure:

$$P_c = (2\gamma_{wo}\cos\theta)/r. \quad (6.1)$$

Thus, the capillary pressure in a porous medium is a function of the chemical composition of the rock and fluids, the pore size distribution, and the saturation of the fluids in the pores. Capillary pressures have also been found to be a function of the saturation history and, because of this, different values will be obtained during the drainage process (ie, displacing the wetting phase with the nonwetting phase) than those obtained during the imbibition process (ie, displacing the nonwetting phase with the wetting phase).

Another factor affecting the microscopic displacement efficiency is the fact that two or more fluids are usually flowing in an EOR process. When two or more fluid phases are present, the saturation of one phase affects the permeability of the other(s), and relative permeability have to be considered.

6.6.1 Macroscopic Displacement Efficiency

The macroscopic displacement efficiency is influenced by (1) heterogeneity, (2) anisotropy, which is the property of being directionally dependent, as opposed to isotropy, which implies identical properties in all directions, (3) the mobility of the displacing fluids compared with the mobility of the displaced fluids, (4) the physical arrangement of injection and production wells, and (5) the type of rock matrix in which the crude oil is found.

The heterogeneity and anisotropy of an oil-bearing formation have a significant effect on the macroscopic displacement efficiency—the movement of fluids through the reservoir will not be uniform if there are large variations in such properties as porosity, permeability, and content of clay minerals. Limestone formations generally have wide fluctuations in porosity and permeability and, since many formations have a system of microfractures or large macrofractures, when a fracture occurs in a reservoir, fluids will be inclined to travel through the fracture because of the high permeability of the fracture. This may lead to substantial bypassing of residual oil and is a major reason for the poor recovery efficiency of a recovery project.

Mobility is a relative measure of how easily a fluid moves through porous media and the apparent mobility is the ratio of effective permeability to fluid viscosity. Since the effective permeability is a function of fluid saturations, several apparent mobility phenomena can be defined and the mobility ratio M is a measure of the relative apparent mobility in a displacement process:

$$M = (\text{mobility of displacing phase})/(\text{mobility of displaced phase})$$

The sweep efficiency is a function of the mobility ratio and the phenomenon known as viscous fingering can take place if the mobility of the displacing phase is much greater than the mobility of the displaced phase.

In terms of the arrangement of injection and production wells, the geology of the formation and the size (areal extent) of the reservoir are the prime factors. Thus, when a reservoir or deposit is being considered for an EOR, a decision must be made whether or not to use the existing well arrangement or to drill new wells in different locations. If the existing wells are to be used, there may be a need to consider converting production wells to injection wells or converting injection wells to production wells. Also, when a production well is converted to an injection well, the production capacity of the reservoir has been reduced. Furthermore, knowledge of any directional permeability effects and variations in heterogeneity can aid in the consideration of well arrangements. For example, directional permeability trends could lead to a poor sweep efficiency in a developed pattern and might suggest that the pattern be altered in one direction or that a different pattern be used. In addition, the presence of faults, fractures, and high-permeability lenses or streaks in the formation could dictate the shutting in of a well near to the occurrence of a heterogeneity.

Sandstone formations are characterized by a more uniform pore geometry than that of limestone formations which have large holes (*vugs*) and can have significant fractures that are often connected. Limestone formations are also associated with connate water that can have high levels of divalent ions such as calcium (Ca^{2+}) and magnesium (Mg^{2+}). Porosity due to the presence of vugs (vugular porosity) and high divalent ion content in their connate waters hinder the application of EOR processes in limestone reservoirs. Conversely, a sandstone formation can be composed of such small sand grain sizes and be so tightly packed (tight formation) that fluids will not readily flow through the formation.

6.6.2 Effect of Capillary Number

In a water-wet system, during the early stages of a waterflood, the brine exists as a film around the sand grains and the oil fills the remaining pore space. At an intermediate time during the flood, the oil saturation has been decreased and exists partly as a continuous phase in some pore channels but as discontinuous droplets in other channels. At the end of the flood, when the oil has been reduced to residual oil saturation (S_{or}) the oil exists primarily as a discontinuous phase of droplets or globules that have been isolated and trapped by the displacing brine.

Waterflooding of oil in an oil-wet system (as might be exhibited in some heavy oil reservoirs) yields a different fluid distribution at the point of residual oil saturation (S_{or}). Early in the waterflood, the brine forms continuous flow paths through the center portions of some of the pore channels and the brine enters more and more of the pore channels as the waterflood progresses. At residual oil saturation, the brine has entered a sufficient number of pore channels to diminish and even shut off the oil flow. The residual oil exists as a film around the sand grains and in the smaller flow channels this film may occupy the entire void space.

The mobilization of the residual oil saturation in a water-wet system requires that the discontinuous globules be connected to form a flow channel whereas in an oil-wet porous medium, the film of oil around the sand grains needs to be displaced to large pore channels and be connected in a continuous phase before it can be mobilized. The mobilization of oil is governed by the viscous forces (pressure gradients) and the interfacial tension forces that exist in the sand grain–oil–water system.

Correlations have been developed between the capillary number (a dimensionless parameter, N_{vc}) and the fraction of oil recovered have been developed. Thus:

$$N_{vc} = V\mu_w/\gamma_{ow} = K_0P/\varphi\gamma_{ow}L$$

In the equation, V is the Darcy velocity, μ_w the viscosity of displacing fluid, γ_{ow} the interfacial tension between displaced and displacing fluid, K_0 the effective permeability to displaced phase, φ the porosity, and P/L the pressure drop associated with Darcy velocity. The capillary number Increases as the viscous forces increase or as the interfacial tension forces decrease.

6.7 HYDRAULIC FRACTURING

EOR processes use *thermal*, *chemical*, or *fluid phase behavior* effects to reduce or eliminate the capillary forces that trap oil within pores, to thin the oil or otherwise improve its mobility or to alter the mobility of the displacing fluids. In some cases, the effects of gravity forces, which cause vertical segregation of fluids of different densities, can be minimized or even used to advantage. The various processes differ considerably in complexity, the physical mechanisms responsible for oil recovery, and the amount of experience that has been derived from field application. The degree to which the EOR methods are applicable in the future will depend on development of improved process technology. It will also depend on improved understanding of fluid chemistry, phase behavior, and physical properties; and on the accuracy of geology and reservoir engineering in characterizing the physical nature of individual reservoirs (Borchardt and Yen, 1989). However, no single EOR technique is the cure-all for oil recovery. Most reservoirs are complex and the oil-reservoir system must be considered as a whole rather than as individual, but equally complex, entities.

In instances where fluids production is not possible because of, say, the low permeability of the reservoir or deposit, hydraulic fracturing is the next step to encourage fluids production and is included here because it is an example of technologies that contribute to improvements in production of crude oil. Hydraulic fracturing is used to create additional passageways in the oil reservoir that can facilitate the flow of oil to a producing well. *Tight reservoirs* are oil-containing reservoirs with restricted pore volume and connectivity in the reservoir rock that impede the flow of fluids through the reservoir. Such reservoirs are commonly fractured by injecting a fluid containing sand or other proppant under sufficient pressure to create fractures in the rock through which the oil can more easily flow. The process may also be applied top heavy oil reservoirs and tar sand deposits in ore to encourage recovery of the fluid.

In the process, a viscous fluid containing a proppant, such as sand, is injected under high pressure until the desired fracturing is achieved. The pressure is then released allowing the fluid to flow to the well while the proppant remains in the fractures to prevent them from closing. Care is also taken to contain the fracturing within the oil reservoir to avoid intersecting adjoining aquifers that would introduce excess water into the oil-producing zone. Tailored pulse fracturing is employed to control the extent and direction of the produced fractures. Precise quantities of solid rocket fuel-like propellants are ignited in the wellbore to create a controlled pressure pulse which creates fractures in a more predictable pattern. Foam fracturing, using foam under high pressure in gas reservoirs, has the advantage over high-pressure water injection because it does not create as much damage to the formation, and well cleanup operations are less costly. Carbon dioxide-sand fracturing increases production by eliminating much of the inhibiting effects of pumped fluids such as plugging by solids, water retention, and chemical interactions.

EOR is not the same as hydraulic fracturing which is part of a group of methods known as *oil production intensification*. The aim of hydraulic fracturing is to locally increase oil flow by changing the physical nature of the underlying rock strata (Speight, 2015b). Thus, once the well is completed, the flow of oil into the well is commenced. For limestone reservoir rock, acid is pumped down the well (*acid fracturing*) and the acid dissolves channels in the limestone that lead oil into the well. For sandstone reservoir rock, a specially blended fluid containing *proppants* (sand, walnut shells, aluminum pellets) is pumped down the well and out the perforations. The pressure from this fluid makes small fractures in the sandstone that allow oil to flow into the well, while the proppants hold these fractures open. Once the oil is flowing, the oil rig is removed from the site and production equipment is set up to extract the oil from the well.

Hydraulic fracturing has also been applied to EOR processes. In one common technique liquids or steam and other gases are injected into a well and an underground explosive charge is set off, forcing the liquids into the surrounding rock, fracturing it, and opening myriad small flow channels into the main wellbore. However, unlike EOR operations which rely on changing the nature of water and oil which impacts the way it flows through the oil field, hydraulic fracturing changes the physical nature of the rock strata and, instead, The underlying structure of the rock is left in place and is unchanged.

Hydraulic fracturing can, and is, also used with oil wells, but it is less commonly used in limestone reservoirs. In limestone ($CaCO_3$) reservoirs, drillers preferentially use a process known as *acidizing* (acidization, acid fracturing) which may also use the high pressure/explosive techniques of hydraulic fracturing (see chapter: Microbial Enhanced Oil Recovery). The oil industry uses these terms ambiguously to refer to cleanup as well as stimulation of the wellbore.

Thus, after drilling and cementing of the wellbore, smallish pieces of cement may remain in the wellbore, threatening to impede oil flow. Relatively small amounts of (typically) hydrochloric acid are pumped into the well to dissolve this cement and pump it out of the well. For *stimulation* purposes, much larger quantities of acid are pumped into the well. Technically, acid fracturing stimulation does not fracture the oil-bearing rock surrounding the wellbore; it *dissolves* the limestone, effectively increasing the diameter of the well.

In *modified in situ extraction* processes (Fig. 6.11), combinations of in situ and mining techniques are used to access the reservoir. A portion of the reservoir rock must be removed to enable application of the in situ extraction technology. The most common method is to enter the reservoir through a large-diameter vertical shaft, excavate horizontal drifts from the bottom of the shaft, and drill injection and production wells horizontally from the drifts. Thermal extraction processes are then applied through the wells. When the horizontal wells are drilled at or near the base of the tar sand reservoir, the injected heat rises from the injection wells through the reservoir, and drainage of produced fluids to the production wells is assisted by gravity.

Although not technically a hydraulic fracturing process, in a *modified in situ extraction* process a combination of in situ and mining techniques are used to access the reservoir when the heavy oil proves too difficult to move to the production well. Such processes are amenable to recovery of bitumen from tar sand deposits and serves as the basis of the steam-assisted-gravity-drainage (SAGD) process and the variants of the process, which may also include a degree of fracturing especially when steam injection occurs at fracture pressure.

Operating the SAGD injection and production wells at approximately reservoir pressure eliminates the instability problems that plague all high-pressure steam processes and produces a smooth, even production that can be as high as 70% to 80% of oil in place in suitable reservoirs. The process is relatively insensitive to shale streaks and other vertical barriers to steam and fluid flow because, as the rock is heated, differential thermal expansion causes fractures in it, allowing steam and fluids to flow through. This allows recovery rates of 60% to 70% of oil in place, even in formations with many thin shale barriers. Thermally, SAGD is twice as efficient as the older cyclic steam stimulation (CSS) process, which is applied to heavy-oil reservoirs to boost recovery during the primary production phase. During this time it assists natural reservoir energy by thinning the oil so it will more easily move through the formation to the injection/production wells. It can also be used, however, as a single-well procedure.

To utilize the CSS method, a predetermined amount of steam is injected into wells that have been drilled or converted for injection purposes. These wells are then shut in to allow the steam to heat (soak period) the producing formation around the well. After a sufficient time has elapsed to allow adequate heating, the injection wells are back in production until the heat is dissipated with the produced fluids. This cycle of soak-and-produce, or huff-and-puff method, may be repeated until the response becomes marginal because of declining natural reservoir pressure and increased water production. At this time a continuous steam flood is usually initiated to continue the heating and thinning of the oil and to replace declining reservoir pressure so that production may continue. When the steam flooding is started, some of the original injection wells will be converted for use as production wells, along with the others drilled or designated for that purpose. Currently, the SAGD process is preferred as it results in far fewer wells being damaged by high pressure. Combined with the higher oil recovery rates achieved, this means that SAGD is much more economic than the pressure-driven steam process where the reservoir is reasonably thick.

Technologies that upgrade value, drive down costs, and reduce environmental impacts will have the greatest effect on increasing the production of heavy oil. There are a large number of technologies that can have an impact, but there is no single technology that can be generally applied owing to the variability of heavy oil properties as well as reservoir properties. In recent years there has been a renewed interest in the in-stir combustion technologies that have the potential to partially upgrade the oil and produce a product at the surface that is more valuable than the original oil.

In addition to the recovery processes described above, hydraulic fracturing may also be necessary to contribute to improvements in oil production. In the process, hydraulic fracturing is used to create additional passageways in the oil reservoir that can facilitate the flow of oil to a producing well. Tight reservoirs (shale formations), where the gas-containing and oil-containing rocks have restricted pore volume and connectivity that impede the flow of oil through the reservoir, are commonly fractured by injecting a fluid containing sand or other proppant under sufficient pressure to create fractures in the rock through which the oil can more easily flow (Cramer, 2008). Care is also taken to contain the fracturing within the oil reservoir to avoid intersecting adjoining aquifers that would introduce excess water into the oil-producing zone.

The focus of current-day attention in terms of crude oil and natural gas production has been the cured oil and natural gas in tight formation. In many of these tight formations, the existence of large quantities of crude oil and natural gas has been known for decades, usually dating to the earliest exploration efforts within any given basin. In some areas multiple attempts were made to commercially produce from these oil-rich but tight zones. While initial production rates were often promising, or even exciting, productivity would typically decrease significantly within months or even days of initial production. As a result, other anecdotal and fragmentary (no pun intended) information on other tight oil and gas plays in North America there is very little available data on the production history of most of the tight oil and gas formations. Unfortunately, for most states and provinces the reporting of production statistics on a formation-by-formation basis is not typically readily available to the public, which makes it difficult to quantify and/or verify reports of tight oil production in many of the identified plays. North Dakota is a notable exception, with annual production data for each oil-producing formation in the state being available through the State government (National Petroleum Council, 2011). Thus, during the first Williston Basin oil boom, many wells in the 1950s and early 1960s were perforated both in the Bakken Formation and more conventional carbonate reservoirs of the overlying Madison Group. These early attempts at Bakken exploitation were through vertical wells, resulting in initial production values ranging from 150 to 450 barrel per day and typical cumulative production of 85,000 barrel per well. These statistics indicate that most early Bakken wells had productive lifetimes of less than 2 to 3 years. As such, very little oil was produced from the Bakken throughout most of the productive history of the Williston Basin. In the early 1990s the advent of horizontal drilling technology brought new interest to the Bakken, and IP values were increased to 230 to 500 B/D, with typical cumulative production also increasing to 145,000 barrels per well. However these improvements in production were not enough to economically sustain wide-scale exploitation of the Bakken and by the late 1990s the Bakken was once again considered an economically unattractive resource. In the early to mid-2000s a combination of high oil price environment, improved understanding of the geology of the Bakken, and improvements in well drilling, completion, and stimulation technologies sparked renewed interest in Bakken exploitation. From the information that is available, the Bakken appears to account for a vast majority of the current tight oil production. However recent reports of success and active development within the Niobrara and Eagle Ford plays suggest that their productivity may be comparable to that of the Bakken within just a few years (National Petroleum Council, 2011).

More recently, several alternative fracturing techniques designed to accomplish specific tasks: (1) tailored pulse fracturing; (2) foam fracturing; and (3) carbon dioxide-sand fracturing. *Tailored pulse fracturing* is employed to control the extent and direction of the produced fractures. Precise quantities of solid rocket fuel-like propellants are ignited in the wellbore to create a controlled pressure pulse which creates fractures in a more predictable pattern. *Foam fracturing*, using foam under high pressure in gas reservoirs, has the advantage over high-pressure water injection because it does not create as much damage to the formation, and well cleanup operations are less costly. *Carbon dioxide-sand* fracturing increases production by eliminating much of the inhibiting effects of pumped fluids, such as (1) plugging by solids, (2) water retention, and (3) chemical interactions.

Finally, when sandstone rocks contain crude oil or natural gas in commercial quantities but the permeability is too low to permit good recovery, a process called fracturing may be used to increase permeability to a practical level. Basically, to fracture a formation, a fracturing service company pumps a specially blended fluid down the well and into the formation under great pressure. Pumping continues until the fractures occur in the formation after which specially blended fluid containing *proppants* (sand, walnut shells, aluminum pellets) is pumped down the well and out of the perforations. Since the fractures tend to close after the pressure on the well is released, the proppant is needed to hold or prop the fractures open. These propped-open fractures provide passages for oil or gas to flow into the well and thence to the surface.

REFERENCES

Alvarado, V., Manrique, E., 2010. Enhanced oil recovery: an update review. Energies 3, 1529–1575.

Arnarnath, A., 1999. Enhanced Oil Recovery Scoping Study. Report No. TR-113836. Electric Power Research Institute, Palo Alto, California.

Bondor, P.L., 1991. Dilute surfactant flooding for north sea applications: technical and economics considerations. Proceedings. 6th European Symposium on Improved Oil Recovery, Stavanger, Norway; May 21–23. vol. 2, pp. 719–738.

Borchardt, J.K., Yen, T.F., 1989. Oil field chemistry Proceedings. Symposium Series No. 396. American Chemical Society, Washington, DC.

Burger, J., 1978. In situ combustion. In: Chilingarian, G.V., Yen, T.F. (Eds.), Developments in Petroleum Science, No. 7, Bitumens, Asphalts and Tar Sands Elsevier, New York, pp. 191.

Calange, S., Ruffier-Meray, V., Behar, E., 1997. Onset of Crystallization temperature and deposit amount for waxy crudes: experimental determinations and thermodynamic modeling. Paper No. SPE 37239. Proceedings. Annual Technical Conference and Exhibition, Houston, Texas, February 18 to 21.

Caruana, A., Dawe, R.A., 1996a. Effect of heterogeneities on miscible and immiscible flow processes in porous media. Trends Chem. Eng. 3, 185–203.

CFR 1.43-2, 2004. Internal Revenue Service, Department of the Treasury, Government of the United States, Washington DC.

Chakma, A., Islam, M.R., Berruti, F. (Eds). 1991. Enhanced oil recovery. AIChE Symposium Series No. 280, vol. 87, American Institute of Chemical Engineers, New York, pp. 147.

Christensen, J.R., Stenby, E.H., Skauge, A., 1998. Review of the WAG field experience. Paper No. SPE 71203. Proceedings. SPE International petroleum conference and exhibition of Mexico, Villahermosa, March 3–5. Society of Petroleum Engineers, Richardson, Texas.

Chung, T-H. 1992. Thermodynamic modeling for organic solids precipitation. Paper No. SPE 24851. Proceedings. 67th Annual Technical Conference, Washington, DC, October 4 to 7.

Craft, B.C., Hawkins, M.F., 1959. Applied Petroleum Reservoir Engineering. Prentice-Hall, Englewood Cliffs, New Jersey.

Dawe, R.A., 2004. Miscible displacement in heterogeneous porous media. Proceedings. Sixth Caribbean Congress of Fluid Dynamics, University of the West Indies, St. Augustine, Trinidad, Republic of Trinidad and Tobago. January 22–23.

Deo, M.D., Miharia, A., Kumar, R., 1995. Solids precipitation in reservoirs due to non-isothermal injections. Paper No. 28967. San Antonio, Texas, February 14 to 17.

DOE. 1996. Maintaining oil production from marginal fields: a review of the department of energy's reservoir class program. Panel on the Review of the Oil Recovery Demonstration Program of the Department of Energy, Committee on Earth Resources, Board on Earth Sciences and Resources, Commission on Geosciences, Environment, and Resources, National Research Council. National Academy Press, Washington, DC.

Dreher, K.D., Gogarty, W.B., 1979. An overview of mobility control in micellar-polymer enhanced oil recovery. J. Rheol. 23 (2), 209–229.

Drew, L.J., 1997. Undiscovered Mineral and Petroleum Deposits: Assessment & Controversy. Plenum Publishing Corporation, New York, (Chapter 3).

Duerksen, J.H., Hsueh, L., 1983. Steam distillation of crude oils. Soc. Petrol. Eng. J. 23 (2), 265–269.

Enick, R.M., 1998. A literature review of attempts to increase the viscosity of dense carbon dioxide. Contract DE-AP26-97FT25356. US DOE, Federal Energy Technology Center, Morgantown, West Virginia, October 10.

Erickson, D.D., Nielsen, V.G., Brown, T.S., 1993. Thermodynamic measurement and prediction of paraffin precipitation in crude oil. Paper No. SPE 26604. Proceedings. 68th Annual Technical Conference and Exhibition, Houston, Texas, October 3 to 6.

Frick, T.C., 1962. Petroleum Production Handbook, vol. II. McGraw-Hill Publishers, New York.

Gary, J.G., Handwerk, G.E., Kaiser, M.J., 2007. Petroleum Refining: Technology and Economics, fifth ed. CRC Press, Taylor & Francis Group, Boca Raton, Florida.

Gogarty, W.B., 1976. Status of surfactant micellar models. J. Petrol. Technol., 93–102.

Hsu, C.S., Robinson, P.R., 2006. Practical advances in petroleum processing, vols. 1 and 2. Springer, New York.

Islam, M.R., Chakma, A., Jha, K.N., 1994. Heavy oil recovery by inert gas injection with horizontal wells. Petrol. Sci. Eng. 11, 213–226.

Jayasekera, A.J., Goodyear, S.G., 1999. The Development of Heavy Oil Fields in the UK Continental Shelf: Past, Present, and Future. SPE 54623.

Kamath, V.A., Kakade, M.G., Sharma, G.D., 1994. An improved molecular thermodynamic model for asphaltene equilibria. In: Sharma, M.K., Yen, T.F. (Eds.), Asphaltene Particles in Fossil Fuel Exploration, Recovery, Refining, and Production Processes Plenum Press, New York.

Koch, H.A., Slobod, R.L., 1957. Miscible slug process. Trans. Am. Inst. Mech. Eng. 210, 40.

Kulkarni, M.M., Rao, D.N., 2005. Experimental investigation of miscible and immiscible Water-Alternating-Gas (WAG) process performance. J. Petrol. Sci. Eng. 48 (1–2), 1–20.

Lake, L.W., 1989. Enhanced Oil Recovery. Prentice-Hall Inc., Englewood Cliffs, New Jersey.

Lake, L.W., Walsh, M.P., 2004. Primary Hydrocarbon Recovery. Elsevier, Amsterdam, Netherlands.

Leontaritis, K.J., 1989. Asphaltene Deposition: A Comprehensive Description of the Problem, Manifestations, and Modeling Approaches. Paper No. SPE 18892. Proceedings. Symposium on Production Operations. Oklahoma City, Oklahoma. March 13 and 14.
Leontaritis, K.J., Mansoori, G.A., Mansoori, U., 1987. Asphaltene flocculation during oil production and processing: a thermodynamic colloidal model. Paper No. SPE 16258. Proceedings. International Symposium on Oilfield Chemistry, San Antonio, Texas. February 4 to 6.
Majeed, A., Bringedal, B., Overå, S., 1990. Model calculates wax deposition for N. sea oils. Oil Gas J. June (18), 63–69.
Meszaros, G., Chakma, A., Jha, K.N., Islam, M.R., 1990 Scaled model studies and numerical simulation of inert gas injection with horizontal wells. Proceedings. SPE Annual Technical Conference, EOR Volume. New Orleans, LA, September 23–26, pp. 585–598.
Mungen, R., Nichols, J.H., 1975. Tar Sands of the world. Proceedings. 9th World Petroleum Congress, 5: 29.
Nghiem, L.X., Hassam, M.S., Nutakki, R., George, A.E.D., 1993. Efficient modeling of asphaltene precipitation. Paper No. SPE 26642. Proceedings. 68th Annual Technical Conference, Houston, Texas, October 4 to 7.
Nor-Aziam, N., Adewumi, M.A., 1993. Development of asphaltene phase equilibria predictive model. Paper No. SPE 26905. Proceedings. Eastern Regional Conference and Exhibition, Pittsburgh, Pennsylvania, November 2 to 4.
OTA. 1978. Enhanced Oil Recovery Potential in the United States. Office of Technology Assessment, Washington, DC. January. NTIS order #PB-276594.
Pan, H., Firoozabadi, A., 1996. Pressure and composition effects on wax precipitation: experimental data and model results. Paper No. SPE 36740. Proceedings. Annual Technical Conference and Exhibition, Denver, Colorado, October 6 to 9.
Parkash, S., 2003. Refining Processes Handbook. Gulf Professional Publishing, Elsevier, Amsterdam, Netherlands.
Pedersen, K.S., Skovborg, P., Rønningsen, H.P., 1991. Wax precipitation from North sea crude oils. 4: thermodynamic modeling. Energy Fuels 5, 924–932.
Prats, M., 1986. Thermal recovery, vol. 7. Society of Petroleum Engineers, New York.
Rassamdana, H., Farhani, M., Dabir, B., Mozaffarian, M., Sahimi, M., 1999. Asphalt flocculation and deposition. V. Phase behavior in miscible and immiscible injections. Energy Fuels 13, 176–187.
Reed, R.L., Healy, R.N., 1977.. In: Shah, D.O., Schechter, R.S. (Eds.), Improved Oil Recovery by Surfactant and Polymer Flooding Academic Press, New York.
Schumacher, M.M., 1980. Enhanced Recovery of Residual and Heavy Oils. Noyes Data Corp., Park Ridge, New Jersey.
Sorbie, K.S., 1991. Polymer-Improved Oil Recovery. Springer, New York.
Speight, J.G., 2000. Desulfurization of Heavy Oils and Residua, second ed. Marcel Dekker Inc., New York.
Speight, J.G., 2013a. Heavy Oil Production Processes. Gulf Professional Publishing, Elsevier, Oxford, United Kingdom.
Speight, J.G., 2013b. Heavy and Extra Heavy Oil Upgrading Technologies. Gulf Professional Publishing, Elsevier, Oxford, United Kingdom.
Speight, J.G., 2014a. The Chemistry and Technology of Petroleum, fifth ed. CRC Press, Taylor & Francis Group, Boca Raton, Florida.
Speight, J.G., 2014b. High Acid Crudes. Gulf Professional Publishing, Elsevier, Oxford, United Kingdom.
Speight, J.G., 2015a. Handbook of Offshore Oil and Gas Operations. Gulf Professional Publishing, Elsevier, Oxford, United Kingdom.
Speight, J.G., 2015b. Handbook of Hydraulic Fracturing. John Wiley & Sons Inc., Hoboken, New Jersey.

Speight, J.G., Ozum, B., 2002. Petroleum Refining Processes. Marcel Dekker Inc., New York.

Stalkup, F.I., 1983. Miscible Displacement, SPE Monograph No. 8. Society of Professional Engineers, Richardson, Texas.

Taber, J.J, Martin, F.D., 1983. Technical screening guides for the enhanced recovery of oil. SPE 12069. Proceedings. 58th SPE Annual Technical Conference and Exhibition, San Francisco, California. October 5–8, Society of Petroleum Engineers, Richardson, Texas.

Terwilliger, P.L., 1975. Fireflooding shallow tar sands - a case history. Paper No. 5568. Proceedings. 50th Annual Fall Meeting. Society of Petroleum Engineers, American Institute of Mechanical Engineers, Dallas, Texas, September 28-October 1.

Terwilliger, P.L., Clay, R.R., Wilson, L.A., Gonzalez-Gerth, E., 1975. Fireflood of the P2-3 sand reservoir in **the** miga field of eastern venezuela. J. Petrol. Technol. 27, 9–14.

Venuto, P.B., 1989. Tailoring EOR processes to geologic environments. World Oil 209 (November), 61–68.

Volek, C.W., Pryor, J.A., 1972. Steam distillation drive – brea field, California. J. Petrol. Technol. 24 (8), 899–906.

Weingarten, J.S., Euchner, J.A., 1986. Methods for predicting wax precipitation and deposition. SPE Paper No. 15654. Proceedings. 61st Annual Technical Conference and Exhibition, New Orleans, Louisiana, October 5 to 8.

Wu, C.H., Brown, A., 1975. A laboratory study on steam distillation in porous media. Paper No. 5569. Proceedings. SPE Annual Meeting, Dallas, Texas. September 28-October1. Society of Petroleum Engineers, Richardson, Texas.

Wu, C.H., Elder, R.B., 1983. Correlation of crude oil steam distillation yields with basic crude oil properties. Soc. Petrol. Eng. J. 23 (6), 937–945.

CHAPTER 7

Microbial Enhanced Oil Recovery

7.1 INTRODUCTION

Microbial enhanced oil recovery (MEOR) flooding involves the injection of microorganisms that react with reservoir fluids to assist in the production of residual oil. There are two general types of MEOR processes: (1) processes in which microorganisms react with reservoir fluids to generate surfactants; and (2) processes in which microorganisms react with reservoir fluids to generate polymers.

Although the concept of MEOR is not a new concept (Goodwin et al., 1983), development of this concept to a fully operational recovery process has lagged behind the driving force that powered the bioremediation of contaminated soils and aquifers by spilled petroleum. Once petroleum is released and comes into contact with water, air, the necessary salts, and microorganisms present in the environment, the natural process of petroleum microbial degradation begins (Antai, 1990; Davies et al., 2001; Okoh, 2006; Salam et al., 2011; Speight and Arjoon, 2012; Miller and Miller, 1981; Obayori et al., 2009).

Biodegradation is the transformation of a substance into new compounds through biochemical reactions or the actions of microorganisms such as bacteria. It is an environmentally acceptable naturally occurring process that takes place when all of the nutrients and physical conditions involved are suitable for growth. Biodegradation allows for the breakdown of a compound to either fully oxidized or reduced simple molecules, such as carbon dioxide/methane, nitrate/ammonium, and water. It should be noted that the products of biodegradation can be more harmful than the substance degraded. Temperature is an important variable; keeping a substance frozen can prevent biodegradation. Most biodegradation occurs at temperatures between 10°C and 35°C (50 and 95°F). Water is essential for biodegradation.

Thus, bioremediation—the use of living organisms to recover crude oil, heavy oil, extra heavy oil and tar sand bitumen—is an option that offers the possibility to partially upgrade or fully upgrade and recover crude oil and heavy feedstocks (in a somewhat beneficially-changed form) for the

Introduction to Enhanced Recovery Methods for Heavy Oil and Tar Sands.
DOI: http://dx.doi.org/10.1016/B978-0-12-849906-1.00007-2

reservoir or deposit. Furthermore, the lessons learned from the application of microbes as bioremedial agents (Speight and Arjoon, 2012) can be applied to the use of microorganisms to petroleum recovery and refining (Speight, 2014; El-Gendy and Speight, 2015).

In the context of this book, biodegradation offers to be the key process in MEOR insofar as various microbes are able to degrade crude oil and heavy feedstocks under appropriate conditions. However, the success of the process depends on the ability to determine the appropriate conditions and establish them in the in situ environment of the reservoir. Thus, important factors required for the success of such a recovery operation (with the accompanying process of partial upgrading or full upgrading of the crude oil and heavy feedstock constituents) (El-Gendy and Speight, 2015) include: (1) the presence of metabolically capable and sustainable microbial populations, (2) suitable environmental growth conditions, such as the presence of oxygen; (3) the presence of water; (4) appropriate levels of nutrients and contaminants; and (5) favorable acidity or alkalinity (Speight and Arjoon, 2012).

The biodegradability of any petroleum constituent is a measure of the ability of that constituent to be metabolized (or cometabolized) by bacteria or other microorganisms through a series of biological process, which include ingestion by organisms as well as microbial degradation (Payne and McNabb, 1984). The chemical characteristics of the contaminants influence biodegradability; in addition, the location and distribution of petroleum contamination in the subsurface can significantly influence the likelihood of success for bioremediation.

Furthermore, the biodegradability of petroleum is inherently influenced by the composition of the substrate upon which the bacteria are acting. In addition, microbial degradation of heavy feedstocks can be enhanced by use of a consortium of different bacteria compared to the activity of single bacterium species (Ghazali et al., 2004; Milić et al., 2009).

Specifically, the microbial degradation of petroleum typically: (1) raises the viscosity and decreases the API gravity, which adversely reduces the ability of the degraded product to flow; (2) decreases the hydrocarbon content, thereby increasing the residuum content; (3) increases the concentration of certain metals; (4) increases the sulfur content; and (5) increases oil acidity, and adds compounds such as carboxylic acids and phenols. All of these changes are seen in the product relative to the unchanged (nonbiodegraded) petroleum (Miiller et al., 1987). In fact, there are indications that petroleum microbial degradation involves more biological components than just the microorganisms that directly attack petroleum constituents (the primary degraders) and shows that the primary degraders interact with these components (Head et al., 2006).

Hydrocarbon degrading bacteria and fungi are mainly responsible for the mineralization (conversion of hydrocarbons to carbon dioxide and water) of petroleum-related pollutants and are distributed in diverse ecosystems (Leahy and Colwell, 1990). Furthermore, the population of microorganisms found in a polluted environment will degrade petroleum-related constituents differently and at a different rate than microorganisms in a relatively clean environment (Obire, 1990, 1993; Obire and Okudo, 1997; Obire and Nwaubeta, 2001).

However, it is uncommon to find organisms that could effectively degrade both aliphatic constituents and aromatic constituents, possibly due to differences in metabolic routes and pathways for the degradation of the two classes of hydrocarbons. There are indications of the existence of bacterial species with propensities for simultaneous degradation of aliphatic hydrocarbons and aromatic hydrocarbons (Amund et al., 1987; Obayori et al., 2009). This rare ability may be as a result of long exposure of the organisms to different hydrocarbon pollutants resulting in genetic alteration and *acquisition* of the appropriate degradative genes.

7.2 MICROBIAL ALTERATION OF PETROLEUM CONSTITUENTS

The principle of MEOR is to encourage the microbes to feast on the petroleum in situ and convert part of the petroleum to lower-boiling products (such as naphtha) that can act as a solvent to assist (through dissolution) in the recovery of the remaining oil. In this way, the oil is recovered from the formation and partial upgrading also occurs.

Crude oil, heavy oil, extra heavy oil, and tars and bitumen are complex mixtures of differing molecular species of hydrocarbons and the constituents of these molecular categories are present in varied proportions resulting in high variability in petroleum and petroleum products (Speight, 2014). In terms of bulk fractions, the resin constituents and the asphaltene constituents are of particular interest (or notoriety) because these constituents generally resist degradation. After a spill, the constituents of petroleum and petroleum products are subjected to physical and chemical processes such as evaporation or photochemical oxidation which produce changes in the composition of the spilled material (Taghvaei Ganjali et al., 2007; Speight, 2014).

7.2.1 Alkanes

Alkanes are major constituents of conventional petroleum but are less likely to occur on heavy oil, extra heavy oil, and tar sand bitumen. Conventional (light) petroleum contains 10% to 40% w/w normal alkane derivatives while higher molecular weight alkanes constitute 5% to 20% w/w of light oils and up to 60% w/w of the more viscous oils and tar sand bitumen. Of these, the normal alkane series (straight-chain alkane series) is the most abundant and the most quickly degraded. Compounds with chains of up to 44 carbon atoms can be metabolized by microorganisms, but those having 10 to 24 carbon atoms (C_{10} to C_{24}) are usually the easiest to metabolize. Shorter chains (up to approximately C_8) also evaporate relatively easily. Only a few species can use C_1 to C_4 alkanes and C_5 to C_9 alkanes are degradable by some microorganisms but toxic to others.

Branched alkanes are usually more resistant to microbial degradation than normal alkanes but less resistant than cycloalkanes (naphthenes)—those alkanes having carbon atoms in ring-like central structures. Branched alkanes are increasingly resistant to microbial attack as the number of branches increases. At low concentrations, cycloalkanes may be degraded at moderate rates, but some highly condensed cycloalkanes can persist for long periods after a spill.

7.2.2 Aromatic Hydrocarbons

Aromatic hydrocarbons are characterized by the presence of at least one benzene ring or substituted benzene within the molecule. Low-molecular-weight aromatic hydrocarbons are subject to evaporation and, although toxic to much marine life, are also relatively easily degraded. Conventional (light) petroleum typically contain between 2% and 20% w/w low-boiling aromatic compounds, whereas heavy oils contain less than 2% w/w aromatic compounds. As molecular weight and complexity increase, aromatics are less readily degraded. Thus, the degradation rate of poly-aromatic derivatives is slower than that of mono-aromatic derivatives.

Two thermophilic aerobic bacteria (*Thermus aquaticus* and *Thermus* sp.) have been reported to degrade benzene, toluene, ethyl benzene, and xylene (BTEX) fractions cometabolically (Chen and Taylor, 1995, 1997a). However, only small fractions of benzene and toluene were metabolized to carbon dioxide and microbial degradation was inhibited by higher BTEX concentrations but was enhanced if strains were pregrown on catechol and *o*-cresol—indicating that preconditioning can enhance the performance of microbes (Chen and Taylor, 1995).

Two anaerobic consortia, consisting of unidentified bacterial cocci, could grow on all BTEX compounds as sole carbon and energy sources at 45°C to 75°C (93°F to 167°F), 50°C (122°F) being the optimal temperature (Chen and Taylor, 1997b). Only a small fraction of toluene was mineralized to carbon dioxide. Microbial degradation was coupled by both consortia to sulfate reduction as well as to generation of hydrogen sulfide. No growth or BTEX metabolism occurred when sulfate was omitted. Thus, sulfate-reducing bacteria are most likely the principal species that carry out the microbial degradation, while other thermophilic species may use the early water-soluble BTEX metabolites.

The above-described aerobic and anaerobic thermophiles are used for a patented in situ thermally enhanced strategy (Taylor et al., 1998). The authors propose the coupling of bioremediation to in situ dynamic underground stripping, a primary decontamination effort for the rapid treatment of contaminated soil at significant depths. A targeted subsurface volume is heated (by steam injection and electrical resistance) to vaporize the trapped contaminants, these are then removed by vacuum extraction. However, even VOCs such as BTEX are not completely removed during vaporization. Due to the use of heat, the entire underground environment remains at elevated temperatures of 50°C to 70°C for at least 60 days after the heat treatment. This enables thermophilic bacteria to metabolize the residual contaminants.

The microbial degradation of alkyl tetralin derivatives has also been studied. However, tetralin has been shown to be biodegraded by both mixed cultures of microbes (Strawinski and Stone, 1940; Soli and Bens, 1972) and by some strains able to utilize the compound as a sole carbon and energy source (Schreiber and Winkler, 1983; Sikkema and Bont, 1991; Hernáez et al., 1999). Recently, it has been demonstrated that *Rhodococci* strains are able to degrade recalcitrant alkyltetralins (Frenzel et al., 2007). The identification of such bacteria capable of the microbial degradation of alkyl tetralin derivatives may be an important step toward the development of bioremediation strategies for sites contaminated by toxic aromatic hydrocarbons.

7.2.3 Polynuclear Aromatic Hydrocarbons

Polynuclear aromatic hydrocarbons, in the current context, are persistent organic compounds with two or more aromatic rings in various structural configurations. Polynuclear aromatic hydrocarbons constitute a large and diverse class of organic compounds. However, derivatives such as tetralin (1,2,3,4-tetrahydronaphthalene) and decalin (decahydronaphthalene, bicyclo[4.4.0]decane) are not included in this group but are included in the alkane group because of the saturated ring.

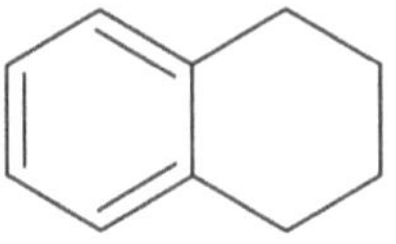

Tetralin

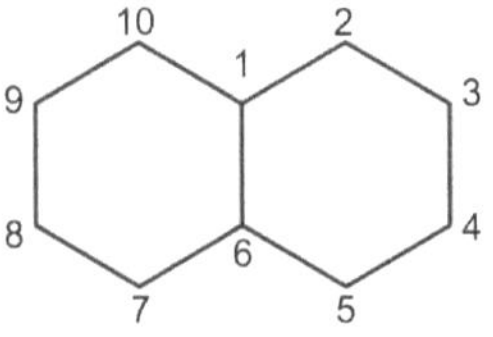

Decalin

The chemical properties, and hence the environmental fate, of polynuclear aromatic hydrocarbons are dependent in part upon molecular size (ie, the number of aromatic rings and the pattern of ring linkage). Ring linkage patterns (also known as molecular topology) in polynuclear aromatic hydrocarbons may occur such that the tertiary carbon atoms are centers of two or three interlinked rings, as in the linear kata-condensed polynuclear aromatic hydrocarbon anthracene or the peri-condensed polynuclear aromatic hydrocarbon pyrene. However, most polynuclear aromatic hydrocarbons occur as hybrids that encompass a variety of structural components, such as in the polynuclear aromatic hydrocarbon benzo[*a*]pyrene.

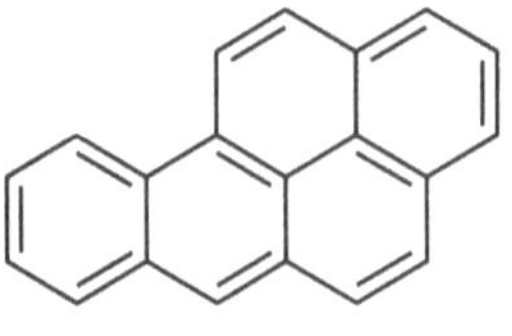

Benzo(a)pyrene

Generally, an increase in the size and angularity of a polynuclear aromatic hydrocarbon molecule results in a concomitant increase in hydrophobicity and electrochemical stability (Zander, 1983; Harvey, 1997). Polynuclear aromatic hydrocarbon molecule stability and hydrophobicity are two primary factors which contribute to their persistence during microbial attack.

Interest in the microbial degradation mechanisms and fate of polycyclic aromatic hydrocarbons (polynuclear aromatic hydrocarbons) is prompted by their ubiquitous distribution in crude oil and the heavy feedstocks. The degradation of polynuclear aromatic hydrocarbons by microorganisms is the subject of many excellent reviews and the microbial degradation of polynuclear aromatic hydrocarbons composed of three rings is well documented (Cerniglia, 1984, 1992; Gibson and Subramanian, 1984; Cerniglia and Heitkamp, 1989; Van der Meer et al., 1992; Shuttleworth and Cerniglia, 1995; Sutherland et al., 1995).

The microbial degradation of naphthalene (the simplest polynuclear aromatic hydrocarbon) process was optimized with preliminary experiments in slurry aerobic microcosms (Bestetti et al., 2003). From soil samples collected on a contaminated site, a *Pseudomonas putida* strain (designated as M8), capable of degrading naphthalene was selected. Microcosms were prepared with M8 strain by mixing noncontaminated soil and a mineral medium. Different experimental conditions were tested by varying naphthalene concentration, soil/water ratio, and inoculum density. The disappearance of hydrocarbon, the production of carbon dioxide, and the ratio of total heterotrophic and naphthalene-degrading bacteria were monitored at different incubation times. The kinetic equation that best fitted the disappearance of contaminant with time was determined. The results showed that the isolated strain enhanced the microbial degradation rate with respect to the natural microbial degradation.

Of the four-ring polynuclear aromatic hydrocarbons, fluoranthene, pyrene, chrysene, and benz[*a*]anthracene have been investigated to various degrees.

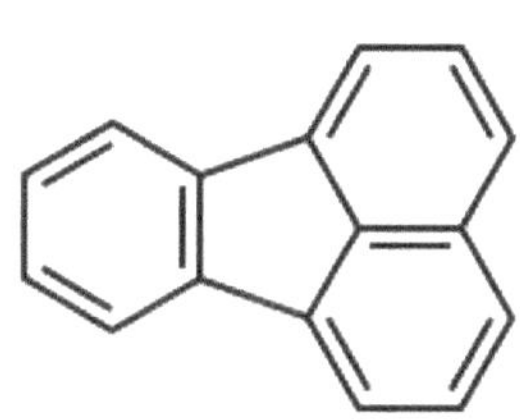

Fluoranthene

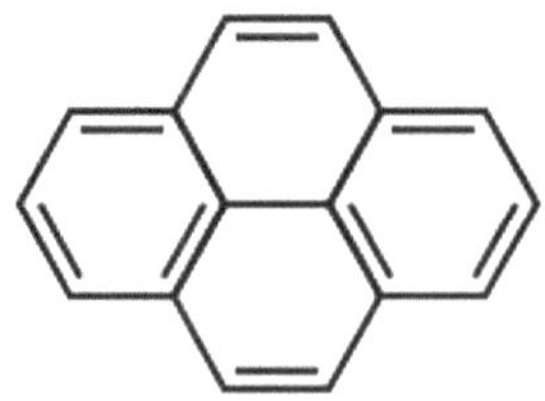

Pyrene

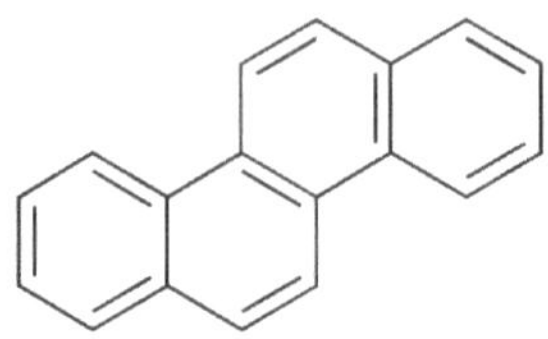

Chrysene

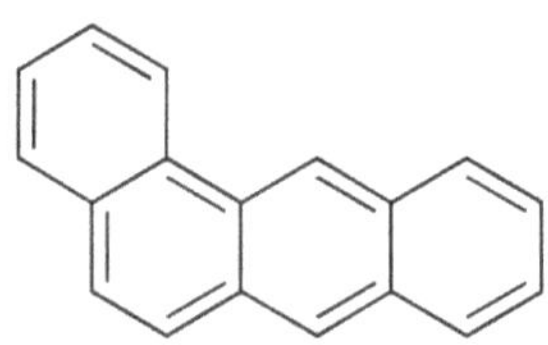

Benz(a)anthracene

Fluoranthene, a polynuclear aromatic hydrocarbon containing a five-membered ring, has been shown to be metabolized by a variety of bacteria, and pathways describing its microbial degradation have been proposed (Mueller et al., 1990; Weissenfels et al., 1990, 1991; Ye et al., 1996). Fluoranthene has been used as a model compound in studies which have investigated the effects of surface-active compounds on polynuclear aromatic hydrocarbon microbial degradation. Comparisons of the mineralization of Fluoranthene by four fluoranthene-degrading strains in the presence of the nonionic surfactants triton x-100 and tween 80 showed that responses differed between strains (Willumsen et al., 1998).

The bacterial degradation of pyrene, a peri-condensed polynuclear aromatic hydrocarbon, has been reported by a number of groups, and some have identified metabolites and proposed pathways (Cerniglia and

Heitkamp, 1990). Sediment microcosms inoculated with the mycobacterium showed enhanced mineralization of various polynuclear aromatic hydrocarbons, including pyrene and benzo(*a*)pyrene.

Measuring the success of microbial degradation petroleum-related constituents is based on several parameters, among them the degradation of polynuclear aromatic hydrocarbons in the crude oil. Though the lower *n*-alkanes are generally considered the most biodegradable compound class within crude oils (Leahy and Colwell, 1990; Prince, 1993), other studies point to exceptional conditions in which polynuclear aromatic hydrocarbons degrade preferentially to *n*-alkane derivatives. Also, it has been demonstrated (Jones et al., 1983) that the microbial degradation of alkyl aromatic hydrocarbons was preferential to that of *n*-alkanes in crude oil when oil-contaminated sediments were aerobically incubated. Preferential microbial degradation of the aromatic hydrocarbons was also observed in bitumen of the South Aquitaine Basin in which the *n*-alkane had not been completely biodegraded.

Currently, there is only limited information regarding the bacterial microbial degradation of polynuclear aromatic hydrocarbons with five or more rings. Most studies have focused on the five-ring benzo(*a*)pyrene due to the potential hazards to human health shown by this compound. Many studies have documented the recalcitrance of benzo(*a*)pyrene to microbial degradation (Cerniglia, 1992; Park et al., 1990; Wild and Jones, 1993; Goodin and Webber, 1995; Van Brummelen et al., 1996). Turnover times of in excess of 3 years in oil-contaminated freshwater sediments and possibly in excess of sixty years in uncontaminated sediments have been reported for the biotransformation of benzo(*a*)pyrene.

Increases in the understanding of the microbial degradation of polynuclear aromatic hydrocarbons and the mechanisms by which polynuclear aromatic hydrocarbon microbial degradation occur will prove helpful for predicting the potential for MEOR.

7.2.4 Microbial Degradation of Residua

A *residuum* (*pl. residua*, also shortened to *resid*, *pl. resids*) is the residue obtained from petroleum after nondestructive distillation has removed all the volatile materials (see chapter: Heavy Oil and Tar Sand Bitumen) (Speight and Ozum, 2002; Parkash, 2003; Hsu and Robinson, 2006; Gary et al., 2007; Speight, 2014). The temperature of the distillation is usually maintained below 350°C (660°F) since the rate of thermal decomposition of petroleum constituents is minimal below this temperature but the rate of thermal decomposition of petroleum constituents is substantial above 350°C (660°F). *Residua* are black, viscous materials and are obtained by distillation of a crude oil under atmospheric pressure (atmospheric residuum) or under reduced pressure (vacuum residuum). They may be liquid at room temperature (generally an atmospheric residuum) or almost solid (generally vacuum residua) depending upon the nature of the crude oil (Speight, 2014). Other terminology (often used incorrectly) that the investigator should be made aware of are the terms *tar* and *pitch* (see chapter: Heavy Oil and Tar Sand Bitumen) (Speight, 2014).

The properties of residua closely resemble heavy oil, extra heavy oil, and tar sand bitumen (Table 7.1) and is worth consideration when examining the potential for microbial enhanced recovery of heavy oils, extra heavy oils, and bitumen., The chemical composition of a residuum from an asphaltic crude oil is complex. Physical methods of fractionation usually indicate high proportions of asphaltene constituents and resin constituents, even in amounts up to 50% (or higher) of the residuum. In addition, the presence of ash-forming metallic constituents, including such organometallic compounds as those of vanadium and nickel, is also a distinguishing feature of residua and the heavier oils. Furthermore, the deeper the *cut* into the crude oil, the greater is the concentration of sulfur and metals in the residuum and the greater the deterioration in physical properties. Both the atmospheric residuum (which also contains a fraction known as vacuum gas oil) and the vacuum residuum (which is the atmospheric residuum without the vacuum gas oil) contain compounds that either is (1) nonbiodegradable or (2) degraded very slowly—because of the coke-forming tendencies of the two residua during refining, it might be advisable to leave these fractions in the reservoir or deposit.

In residua which are characterized by high viscosity and high density or low API gravity (Speight, 2014), asphaltene constituents, resin constituents, and the matrix compose a dynamic stable system, in which the non-asphaltene and nonresin fraction act as the medium for dispersion in which the asphaltene constituents are stabilized by the resin constituents (Speight, 2014). Resin constituents include petroleum compounds containing nitrogen, sulfur, and/or oxygen as constituents. If not highly condensed, they may be subject to limited microbial degradation. Asphaltene constituents and resin constituents are difficult to analyze and, to date, little information is available on the biodegradability of most compounds in these groups.

Table 7.1 Properties of various residua

Source/Residuum Cut Point	Gravity API	Sulfur % w/w	Nitrogen % w/w	Nickel, ppm	Vanadium, ppm	Asphaltene Fraction (heptane)	Carbon Residue (Conradson, % w/w)
Arabian light, >650°F	17.7	3.0	0.2	10.0	26.0	1.8	7.5
Arabian light, >1050°F	8.5	4.4	0.5	24.0	66.0	4.3	14.2
Arabian heavy, >650°F	11.9	4.4	0.3	27.0	103.0	8.0	14.0
Arabian heavy, >1050°F	7.3	5.1	0.3	40.0	174.0	10.0	19.0
Alaska, North Slope, >650°F	15.2	1.6	0.4	18.0	30.0	2.0	8.5
Alaska, North Slope, >1050°F	8.2	2.2	0.6	47.0	82.0	4.0	18.0
Lloydminster (Canada), >650°F	10.3	4.1	0.3	65.0	141.0	14.0	12.1
Lloydminster (Canada), >1050°F	8.5	4.4	0.6	115.0	252.0	18.0	21.4
Kuwait, >650°F	13.9	4.4	0.3	14.0	50.0	2.4	12.2
Kuwait, >1050°F	5.5	5.5	0.4	32.0	102.0	7.1	23.1
Tia Juana, >650°F	17.3	1.8	0.3	25.0	185.0		9.3
Tia Juana, >1050°F	7.1	2.6	0.6	64.0	450.0		21.6
Taching, >650°F	27.3	0.2	0.2	5.0	1.0	4.4	3.8
Taching, >1050°F	21.5	0.3	0.4	9.0	2.0	7.6	7.9

Light oils may contain about 1% to 5% of both asphaltene constituents and resin constituents whereas heavy oil or weathered oil may have up to 25% w/w asphaltene constituents and 20% w/w resin constituents.

The inherent biodegradability of these individual high-boiling constituents is a reflection of their chemical structure, but is also strongly influenced by the physical state of the constituents. Whilst *n*-alkanes as a structural group are the most biodegradable petroleum hydrocarbons, the higher molecular weight constituents become increasingly recalcitrant because not only the chemical structure but also the physical state of the constituents strongly influences biodegradation. In fact, asphaltene and resin constituents tend to increase during biodegradation in relative and sometimes absolute amounts (Okoh, 2006). This suggests that they not only tend to resist biodegradation but may also be formed by condensation reactions of biodegradation intermediates.

The biodegradation of residua requires a complex metabolic pathway which usually can be observed in a microbial community. Many studies have been carried out on the biodegradation of petroleum hydrocarbons using a consortium of microorganisms (Whyte et al., 1997; Foght et al., 1990; Morais and de Tauk-Tornisielo, 2009). During adaptation of microbial communities to hydrocarbon components, particularly complex ones, genes for hydrocarbon-degrading enzymes that are carried on plasmids or transposons may be exchanged between species and new catabolic pathways eventually may be assembled and modified for efficient regulation (Rabus et al., 2005). Other cell adaptations leading to new ecotypes may include modifications of the cell envelope to tolerate solvents (Ramos et al., 2002) and development of community level interactions that facilitate cooperation within consortia.

One of the limiting factors for biodegradation of petroleum hydrocarbons in oil is the bioavailability (Banat, 1995; Kim et al., 2001) of these components due to their chemical structure (Harvey, 1997; Cerniglia, 1992). One of the options to increase bioavailability of polynuclear aromatic hydrocarbons is the use of surfactants to increase desorption and apparent solubility in the aqueous phase and consequently enhance the oil mobility, improving the rate of biodegradation (Barathi and Vasudevan, 2001; Burd and Ward, 1996a, b; Rosenberg et al., 1988; Sar and Rosenberg, 1983) and, hence, the rate of recovery of a more valuable product.

The *asphaltene constituents* of heavy feedstocks have drawn considerable attention due to problems caused by their detrimental effects in the extraction, transportation, and processing of residua because of their viscous and flocculating nature and their relative resistance to biodegradation following spills (Speight, 2014). These constituents are the highest molecular weight and most polar fraction of crude oil. Although the structure of asphaltene constituents have not been fully elucidated, it is widely accepted (by speculation rather than by hard scientific foundation) that they are constituted by interacting systems of polynuclear aromatic sheets bearing alkyl side-chains. Asphaltene molecules have a high content of nitrogen, oxygen, and sulfur heteroatoms as well as metals (V, Ni, and Fe) (Speight, 2014). The problems associated with asphaltene constituents have increased due to the need to extract heavier crude oils, as well as the trend to extract larger amounts of light fractions out of crude oil by cracking and visbreaking.

The asphaltenic fraction is recognized as the most recalcitrant oil fraction. There is no clear evidence that asphaltene constituents can be degraded or transformed by microbial activity. Microorganisms have been found associated with bitumen (Wyndham and Costerton, 1981) and natural asphalt lake (Naranjo et al., 2007), which contain high amounts of asphaltene constituents. A molecular study (Kim and Crowley, 2007) revealed a wide range of phylogenetic groups within the Archaea and Bacteria domains in natural asphalt-rich tar pits; interestingly, genes encoding novel oxygenase derivatives were also detected in such samples.

Furthermore, most of the studies on asphaltene constituents biodegradation should be treated with caution since the asphaltene content was usually determined gravimetrically after *n*-alkane precipitation, and thus the reported changes may be attributed to the disruption of the asphaltenic matrix by the production of surfactants during bacterial growth, liberating trapped hydrocarbons. Other studies have reported that the asphaltene fraction does not support bacterial growth, and no changes in asphaltene content are found after bioconversion of heavy oil and asphaltene constituents (Lacotte et al., 1996; Thouand et al., 1999).

It has been reported (Pineda et al., 2004) that a bacterial consortium was able to grow in the asphaltene fraction as the sole carbon source. Mineralization of the asphaltene constituents was estimated by measuring carbon dioxide production. The authors found in two control experiments (inoculum without asphaltene constituents and noninoculated asphaltene constituents), carbon dioxide production equivalent to 39% and 26%, respectively, of that found in the consortium growing in the asphaltene fraction. Other evidence showed that enzymes were able to modify asphaltene constituents and involved the use of the fungus *Caldariomyces fumago* and a chemically modified *Cytochrome c* which were able to transform petroporphyrin derivatives and asphaltene constituents in reaction mixtures containing organic solvents (Fedorak et al., 1993; Mogollon et al., 1998; Garcia-Arellano et al., 2004). Notable spectral changes in the petroporphyrin-rich fraction of asphaltene constituents were observed and the enzymatic oxidation of petroporphyrin derivatives led to the removal of up to 74% of Ni and 95% of V.

It has also been reported (Uribe-Alvarez et al., 2011) that a fungus isolated from a natural asphalt lake is able to grow using asphaltene constituents as the sole source of carbon and energy. Thus, a fungal strain isolated from a microbial consortium growing in a natural asphalt lake is able to grow in purified asphaltene constituents as the only source of carbon and energy. The asphaltene constituents were rigorously purified in order to avoid contamination from other petroleum fractions. In addition, most petroporphyrin derivatives were removed. The 18S rRNA and b-tubulin genomic sequences, as well as some morphologic characteristics, indicate that the isolate is *Neosartorya fischeri*. After 11 weeks of growth, the fungus is able to metabolize 15.5% of the asphaltenic carbon, including 13.2% transformed to carbon dioxide. In a medium containing asphaltene constituents as the sole source of carbon and energy, the fungal isolate produces extracellular lactase activity, which is not detected when the fungus grows in a rich medium. The results obtained in this work clearly demonstrate that there are microorganisms able to metabolize and mineralize asphaltene constituents, which is considered the most recalcitrant petroleum fraction.

Furthermore, bacterial metabolites (especially polysaccharides) are of great value as enhancers of oil recovery due to their surfactant activity and bioemulsifying properties (Banat, 1995). Because the conditions in oil deposits are often saline, the use of salt-resistant metabolites may be advantageous to the recovery of oil. Furthermore, hypersaline water and soil are often contaminated with crude oils, heavy metals, or other toxic compounds from anthropogenic sources. However, conventional microbiological treatment processes do not function at high salt concentrations; therefore, the use of moderately halophilic bacteria should be considered (Hao and Lu, 2009).

In the past, biodegradation of asphaltene constituents through the use of a microbial consortium or mixed cultures isolated from soil samples, sediments contaminated with hydrocarbons, and oil wells have taken place but in low proportions of 0.55% to 3.5% v/v (Venkateswaran et al., 1995; Thouand et al., 1999). This is most likely due to the complex molecular structure of asphaltene constituents (Speight, 2014) which makes these molecular species resistant to biodegradation thereby causing their accumulation in ecosystems where petroleum and its refining byproducts are spilled in either accidental or purposeful ways (Guiliano et al., 2000).

In contrast to low-molecular-weight hydrocarbons, polycyclic aromatics and hydrocarbons included in the asphaltene fraction are usually considered as being only slightly biodegradable because of their insufficient availability to microbial attack (Gibson and Subramanian, 1984; Cerniglia, 1992; Kanaly and Harayama, 2000). Among the pentacyclic triterpane derivatives, the hopane constituents are so stable that they are commonly used as ubiquitous biomarkers for the assessment of biodegradation levels of crude oil (Ourisson et al., 1979). They were shown to be only slightly biodegraded by specialized microflorae under laboratory conditions (Frontera-Suau et al., 2002). The mechanism of the degradation is complex but is believed to be a sequential process in which *n*-alkanes moieties are generally removed first, followed by the degradation of *iso*-alkanes moieties, cycloalkanes moieties, one-to-three ring aromatics, and finally polyaromatics (Greenwood et al., 2008). However, the typical pattern of degradation varies with different bacteria as well as type and composition of oil (Greenwood et al., 2008; Zrafi-Nouira et al., 2009).

Microbes reduce the viscosity of crude oil and heavy feedstocks by degrading high molecular weight constituents into lower molecular weight constituents such as biological surface active substances, acids, and gases. In addition, anaerobic fermentation leads to the production of acids, carbon dioxide, hydrogen, and alcohols. Anaerobic bacteria produce acetate and butyrate during the initial growth phase (acidogenic phase) of the fermentation process. As the culture moves to the second phase of fermentation, the stationary growth phase, there is a shift in the metabolism of the cells to solvent production (solvent-generation phase). These gaseous and liquid metabolites dissolve into the oil resulting in reduced viscosity (Bryant et al., 1998). Moreover, the reaction of asphalt degradation within an acidic background is preferable as the proton (H^+) effectively interacts with the polar functionalities in the asphaltene constituents and resin constituents thereby efficiently reducing the polar interactions which results in breaking the intermolecular associations that exist in the raw residuum or asphalt.

Furthermore, when each of the fractions, namely aliphatic, aromatic, and asphaltic (asphaltene and resin) fractions, were treated with *Garciaella petrolearia* TERIG02 for 30 days it was found that maximum degradation was in the case of asphalt followed by aromatic fraction (Lavania et al., 2012). Indications were that *Garciaella petrolearia* TERIG02 could tolerate the toxicity of these compounds and was capable of utilizing them as a carbon and energy source.

7.2.5 Process Parameters

The composition and inherent biodegradability of the petroleum and heavy feedstock constituents is, perhaps, the first and most important consideration when determining the suitability of a MEOR process. Heavy feedstocks are generally much more difficult to biodegrade than conventional crude oil, just as heavier crude oils could be suitable for inducing increased selection pressure for the isolation of petroleum hydrocarbon degraders with enhanced efficiency. Also, the amount of heavy crude oil metabolized by some bacterial species increases with increasing concentration of the feedstock while degradation rates appear to be more pronounced within a specific concentration range (Okoh et al., 2002; Rahman et al., 2002).

Microbial degradation of petroleum-related and other feedstocks constituents by bacteria can occur under both aerobic (oxic) and anaerobic (anoxic) conditions (Zengler et al., 1999), usually by the action of different consortia of micro-organisms. In the subsurface, microbial degradation occurs primarily under anaerobic conditions, mediated by sulfate reducing bacteria in cases where dissolved sulfate is present (Holba et al., 1996), or methanogenic bacteria in cases where dissolved sulfate is low (Bennett et al., 1993). Although subsurface oil microbial degradation does not require oxygen, there is a requirement for the presence of essential nutrients (such as nitrogen, phosphorus, potassium), which can be provided by dissolution/alteration of minerals in the water layer. In the absence of nutrients, the potential for hydrocarbon degradation in anoxic sediments is markedly reduced (Dibble and Bartha, 1976).

7.2.5.1 Effect of Nutrients

Different types of nutrients (primarily nitrogen and phosphorus) have been applied to improve petroleum hydrocarbon degradation, including classic (water soluble) nutrients and oleophilic and slow-release fertilizers. Bioavailability is one main factor that influences the extent of microbial degradation of petroleum-related hydrocarbons which generally have low-to-poor solubility in water and, as a result, are adsorbed on to clay or humus fractions, so they pass very slowly to the aqueous phase where they are metabolized by microorganisms. Cyclodextrins are natural compounds that form soluble inclusion complexes with hydrophobic molecules and increase degradation rate of hydrocarbons in vitro.

In the perspective of an in situ application, β-cyclodextrin does not increase eluviation (the lateral or downward movement of the suspended material in soil through the percolation of water) of hydrocarbons through the soil and consequently does not increase the risk of groundwater pollution (Sivaraman et al., 2010). Furthermore, the combination of bioaugmentation and enhanced bioavailability due to β-cyclodextrin was effective for a full degradation (Bardi et al., 2003). Thus, in situ microbial degradation of polynuclear aromatic hydrocarbons can be improved by the augmentation of degrading microbial populations and by the increase of hydrocarbon bioavailability (Bardi et al., 2007).

Nutrients are very important ingredients for successful microbial degradation of hydrocarbon pollutants, especially nitrogen, phosphorus, and in some cases iron. Depending on the nature of the impacted environment, some of these nutrients could become limiting thus affecting the microbial degradation processes.

7.2.5.2 Effect of Temperature

Temperature influences rate of biodegradation by controlling the rate of enzymatic reactions within microorganisms. Generally, the rate of an enzymatic reaction approximately doubles for each 10°C (18°F) rise in temperature" (Nester et al., 2001) There is an upper limit to the temperature that microorganisms can withstand. Most bacteria found in soil, including many bacteria that degrade petroleum hydrocarbons, are mesophile organisms which have an optimum temperature ranging from 25°C to 45°C (77°F to 113°F). (Nester et al., 2001) Thermophilic bacteria (those which survive and thrive at relatively high temperatures) which are normally found in hot springs and compost heaps exist indigenously in cool soil environments and can be activated to degrade hydrocarbons with an increase in temperature to 60°C (140°F).

Thus, temperature plays an important role in the microbial degradation of petroleum-related hydrocarbons not only because of the direct effect on the chemistry of the pollutants but also because of the effect on the physiology and diversity of the microbial surroundings (Atlas, 1975). In short, temperature can play the role of increasing a microbial reaction or inhibiting a microbial reaction in a similar manner to the general rules for the influence of temperature on chemical reactions. Typically, microbial degradation of petroleum and petroleum products occurs at temperatures less than 80°C (<176°F) (Barnard and Bastow, 1991)—at higher temperatures (unless the microbes are of a specific thermophilic type) many of the microorganisms involved in subsurface oil microbial degradation cannot exist.

Although the microbial degradation of petroleum-related hydrocarbons can occur over a wide range of temperatures, the rate of microbial degradation generally decreases with decreasing temperature (Bossert and Bartha, 1984). In fact, the biodegradability of petroleum is highly dependent not only on composition but also on microbial incubation temperature (Atlas, 1975)—at 20°C (68°F) conventional petroleum has higher abiotic losses and is more susceptible to microbial degradation than heavy oil. As expected from petroleum chemistry and composition (Speight, 2014), the rate of mineralization for the heavy oil is significantly lower at 20°C (68°F) than for conventional oil. During microbial degradation, some preference is shown for removal of the paraffin constituents over the aromatic and asphaltic constituents, especially at low temperatures. Branched paraffins, such as pristane, are degraded at both 10°C and 20°C (50°F and 68°F) (Atlas, 1975).

7.2.5.3 Rates of Microbial Degradation

Different results for microbial activity measurements may be obtained in laboratory studies depending on pretreatment and size of the sample, even when the experimental conditions are strictly repeated. Thus, modeling chemical reactions to determine the rate of the reaction is a common practice in chemistry. However, blending of chemical reactants, which is a part of modeling studies, can have beneficial or adverse effects on chemical reactions. Blend time predictions are usually based on empirical correlations and when a competitive side reaction is present, the final product distribution may be suspect. In microbial degradation, the effect of the microbe in relation to the hydrocarbon stream on the reaction outcome is crucial. Also, the scale-up of such reactions from the laboratory to the field may not be straightforward. Thus, there is a need for comprehensive, optimistic caution when models of microbial degradation chemistry are used to predict information such as incubation period, product, by-products, and whether or not the microbes will survive the duration of the treatment. It is also equally important to recognize the potential for interference of one chemical species with another—a situation that is not often determined in the laboratory when simple model substrates (such as single chemical entities or mixtures of two-to-twelve compounds that are not truly representative of petroleum and petroleum products) are used for the experiments.

Evaporative loss is controlled by the composition and physical properties of the petroleum or heavy feedstock (Payne and McNabb, 1984). Derivation of a universal model for such a process may insert inaccuracies because of the complex and changing nature of petroleum and petroleum products. In addition, the material left behind is richer in metals (mainly nickel and vanadium), waxes, resin constituents, and asphaltene constituents than the original oil. With evaporation and stream distillation in the reservoir, the specific gravity and viscosity of the original oil also increase and a conventional crude oil may begin to resemble heavy oil in composition and properties (Mielke, 1990). In fact, biodegradation is a multivariable process and optimization through classical methods is subject to question. To overcome the disadvantages of the process, response surface methodology has been advocated for analyzing the effects of several independent variables on the bioremediation process in order to assess the optimum conditions for the process (Nasrollahzadeh et al., 2007; Huang et al., 2008; Pathak et al., 2009; Vieira et al., 2009; Mohajeri et al., 2010; Zahed et al., 2010).

7.3 PROCESSES TYPES

The MEOR method is very different from the traditional enhanced oil recovery methods (see chapter: General Methods of Oil Recovery) (Bailey et al., 2001; Sen, 2008; Gao and Zekri, 2011). The method is to inject live microorganisms and nutrients into a reservoir to mobilize the residual oil or heavy feedstock. If favorable bacteria already reside in reservoirs, it is feasible to inject nutrients only. The advantages of the method are: (1) it does not require modification of present water injection facilities; (2) it does not involve toxic chemicals; and (2) it is friendly to the environment.

7.3.1 General Principles

Microorganisms can be reacted with reservoir fluids to generate either surfactants or polymers in the reservoir. As with other recovery processes (see chapter: General Methods of Oil Recovery) (Alvarado and Manrique,

2010), the success of MEOR processes will be highly affected by reservoir characteristics. In fact, MEOR processes can be designed for reservoirs that have either a high or a low degree of channeling and, therefore, MEOR applications require a thorough knowledge of the reservoir since mineral content of the reservoir brine will also affect the growth of microorganisms. Once either the surfactant or the polymer is produced, the processes of mobilizing and recovering residual oil become similar to those discussed with regard to chemical flooding.

Most projects have involved near-well treatments of stripper wells in an application of the huff and puff process discussed with regard to thermal flooding. A solution of microorganisms is injected along with a nutrient, usually molasses and when the solution of microorganisms has been designed to react with the oil to form polymers, the injected solution will enter high-permeability zones and react to form the polymers that will then act as a permeability-reducing agent. When oil is produced during the huff stage, oil from lower permeability zones will be produced. Conversely, the solution of microorganisms can be designed to react with the residual crude oil to form a surfactant. The surfactant lowers the interfacial tension of the brine-water system, which thereby mobilizes the residual oil. The oil is then produced in the huff part of the process.

The reaction of the microorganisms with the reservoir fluids may also produce gases, such as carbon dioxide (CO_2), nitrogen (N_2), hydrogen (H_2), and methane (CH_4). The production of these gases will result in an increase in reservoir pressure, which will thereby enhance the reservoir energy. However, since microorganisms can be reacted to form either polymers or surfactants, a knowledge of the reservoir characteristics is critical. If the reservoir is heterogeneous, it is desirable to generate polymers in situ that could be used to divert fluid flow from high-to-low-permeability channels. If the reservoir has low injectivity, using microorganisms that produced polymers could be very damaging to the flow of fluids near the wellbore.

Reservoir brines can inhibit the growth of the microorganisms. Therefore, some simple compatibility tests can result in useful information as to the viability of the process. These can be simple test-tube experiments in which reservoir fluids and/or rock are placed in microorganism–nutrient solutions and growth and metabolite production of the microorganisms are monitored.

In summary, MEOR relies on microorganisms and their metabolic products to mobilize otherwise unrecoverable (residual oil) or difficult-to-recover oil (heavy oil, extra heavy oil, and tar sand bitumen) either in the original form, in a partially-upgraded form, or in a fully-upgraded form. The process is environmentally friendly and easy to operate and over the past two to three decades the method has attracted a great deal of attention in both research and field applications. The mechanisms of the processes are variable and can involve changes in interfacial tension and wettability alteration. On the downside (if liquids are the desired product), there is the potential for a high production of gas.

7.3.2 Mechanisms

The viability of the MEOR process are variable and are dependent upon the properties of (1) the microbes, (2) the feedstock, and (3) the characteristics of the reservoir (or deposit). For example, bacteria are able to produce either surfactants, or polymers, or gases, or solvents that can contribute to mobilization of the oil in reservoir.

The production of surfactants (*biosurfactants*) that reduce oil–water *interfacial tension* and any residual oil (or heavy oil) that is held in the porous reservoir rocks by capillary pressure starts to flow when the interfacial tension is reduced. However, the biosurfactants must reduce the interfacial tension below a critical value (that is dependent upon the oil-rock system) to have any effect on oil recovery.

The issue of *selective plugging* arises during MEOR because of the rock structure. A porous rock contains pores of various sizes and when a waterflooding project is in progress, the larger pores receive most of the injected water, while residual oil remains in small pores without being swept. Furthermore, when bacteria are introduced into the reservoir they also tend to enter large pores. Some bacteria can generate biopolymers that plug the high-permeability zones with large pores, thus forcing injected water to sweep the oil from the low permeability zones. Other bacteria produce gas (such as carbon dioxide) and solvents in the reservoir which can dissolve in the oil thereby reducing the oil viscosity. Lower oil viscosity leads to improved mobility ratio and oil recovery. The produced gas can also increase the reservoir pressure, which leads to higher production rate. In addition, other bacteria can degrade crude oil, especially the paraffin contents in crude oil thereby removing any paraffin deposit in the near wellbore region and improve the rock permeability and the crude oil production rate (Bailey et al., 2001).

The *wettability* of the reservoir rock has a major influence on the distribution of residual oil. In water-wet sandstone reservoirs, water is in contact with sand grains and the oil collects as droplets in the center of the pore space. On the other hand, for oil-wet rocks, the oil is in contact with sand-grain surfaces and remains in the small pores. In the former case (water wettability) the system is more favorable for oil recovery. However, there is some doubt about the influence of bacteria on the wettability of reservoir rock. There is a report (Shabani et al., 2009) that when Berea sandstone cores were treated by *Rhodococcus sp. 094* solution and that after injection of bacteria solution, the originally water-wet cores turned more water-wet. For mixed-wet cores, the change in wettability was insignificant. In another test (Zekri et al., 2003), water-wet limestone cores were submerged in a solution of bacteria and a reduction in contact angle was observed, indicating wettability alteration towards water wettability. In practice, the concentration of bacteria in a reservoir is low and, therefore, significant alteration of rock wettability is considered to be unlikely without any external influence.

7.3.3 Process Types

Most MEOR projects can be classified either as (1) huff and puff-type projects or as (2) bacterial flooding products and the differentiation between the two types of projects is made according to the types of the wells treated by bacteria. Bacteria can be injected through both oil and gas production well (*producer* well) and water injection well (*injector* well). The bacterial solution is delivered to the reservoir from production or injection wells through tubing or annulus.

7.3.3.1 Microbial Huff 'n' Puff

In the huff 'n' puff process, a bacterial solution is injected into the reservoir through a production well and the nutrient can be injected (1) simultaneously with the injection of the bacterial solution or (2) after the injection of the bacterial solution. The nutrient that is commonly used in practice is molasses (black treacle, a viscous product of the sugar refining industry). The system is then shut in for a predicated period of time, usually on the order of several days to several weeks. During this time, the bacteria produce solvent or surfactant that helps to eliminate the debris in the near wellbore region. Other bacteria can generate polymers to seal the high permeability channels in porous media. After the specified period of time, the system is brought back to production. The process is often repeated several times to maximize the recovery of oil. Typically, the oil is changed from its original from having been (most likely) partially or (less likely) fully upgraded to a less-viscous lower-API product.

7.3.3.2 Bacteria Flooding

In the bacterial flooding process (also called the *microbial flooding process*), the bacteria and a nutrient (or nutrients) are injected into a reservoir from an injector and normal water flooding operation is then resumed. As a result, the bacteria are carried deep into the reservoir with the injected water. During the transportation process in the interior of the reservoir, the bacteria produce surfactants that improve the recovery of oil as well as plug the high permeability zones with high permeability and force water to sweep the low permeability zones. The benefit of this process is that, compared to the huff 'n' puff process (in which the bacteria typically treat the near-wellbore region of the producing wells), the bacteria are transported deep into the reservoir and produce more reservoir per given bacterial culture.

In another option, use is made of the bacteria that already exist in the reservoir and which are able to enhance oil recovery, but these bacteria do not exist as a dominant colony. This being the case, a nutrient (and not a bacterial solution) is injected into the reservoir as a means of activating the indigenous bacteria. This option is less likely to be used compared with the two operations above—the bacterial flooding process and the huff 'n' puff process—unless favorable strains of bacteria have been indented in the reservoir in sufficient concentration to make the process efficient. Generally, bacterial strains that are capable of participating in an efficient process are not always in sufficiently high concentration and may not be able to compete favorably with other bacterial colonies even when nutrition is supplied. An option is to inject more of the indigent bacterial strain (with the nutrient) to maximize the effects of this strain and to assist the indigenous (favorable) bacteria to assume a dominant role in the produce and the effect the production of desirable fluids from the reservoir.

In summary, MEOR is on the verge of more general application in crude oil and heavy feedstocks reservoirs. With the development in biotechnology and petroleum technology, the understanding of the microbial enhanced recovery method will be improved.

REFERENCES

Alvarado, V., Manrique, E., 2010. Enhanced oil recovery: an update review. Energies 3, 1529–1575.

Amund, O.O., Adewale, A.A., Ugoji, E.O., 1987. Occurrence and characterization of hydrocarbon utilizing bacteria in nigerian soils contaminated with spent motor oil. Indian J. Microbiol. 27, 63–87.

Antai, S.P., 1990. Microbial degradation of Bonny light crude oil by Bacillus sp and Pseudomonas sp. Waste Manag. 10, 61–64.

Atlas, R.W., 1975. Effects of temperature and crude oil composition on petroleum microbial degradation. Appl. Microbiol. 30 (3), 396–403.

Bailey, S.A., Kenney, T.M., Schneider, D.R., 2001. Microbial enhanced oil recovery: diverse successful applications of biotechnology in the oil field. Paper No. SPE 72129. Proceedings. SPE Asia Pacific Improved Oil Recovery Conference, Kuala Lumpur, Malaysia. October 6–9. Society of Petroleum Engineers, Richardson, Texas.

Banat, I.M., 1995. Biosurfactants production and possible uses in microbial enhanced oil recovery and oil pollution remediation: a review. Biores. Tech. 51, 1–12.

Barathi, S., Vasudevan, N., 2001. Utilization of petroleum hydrocarbons by *Pseudomonas fluorescens* isolated from a petroleum-contaminated soil. Environ. Int. 26, 413–416.

Bardi, L., Ricci, R., Mario Marzona, M., 2003. In situ bioremediation of a hydrocarbon polluted site with cyclodextrin as a coadjuvant to increase bioavailability. Water Air Soil Pollution Focus 3, 15–23.

Bardi, L., Martini, C., Opsi, F., Bertolone, E., Belviso, S., Masoero, G., et al., 2007. Cyclodextrin-enhanced in situ bioremediation of polyaromatic hydrocarbons-contaminated soils and plant uptake. J. Incl. Phenom. Macrocycl. Chem. 57, 439–444.

Barnard, P.C., Bastow, M.A., 1991. Hydrocarbon generation, migration, alteration, entrapment and mixing in the Central and Northern North Sea. In: England, W.A., Fleet, A.J. (Eds.), Petroleum Migration Geological Society, London, United Kingdom, pp. 167–190. Special Publication.

Bennett, P.C., Siegel, D.E., Baedecker, M.J., Hult, M.F., 1993. Crude oil in a shallow sand and gravel Aquifer 1. Hydrogeology and inorganic geochemistry. Appl. Geochem. 8, 529–549.

Bestetti, G., Collina, E., Di Gennaro, P., Lasagni, M., Pitea, D., 2003. Kinetic study of naphthalene microbial degradation in aerobic slurry phase microcosms for the optimization of the process. Water Air Soil Pollution Focus 3, 223–231.

Bossert, I., Bartha, R., 1984. The fate of petroleum in soil ecosystems. In: Atlas, R.M. (Ed.), Petroleum Microbiology Macmillan, New York, pp. 453–473.

Bryant, R.S., Bailey, S.A., Step, A.K., Evans, D.B., Parli, J.A., 1998. Biotechnology for heavy oil recovery. Paper No. 36767 Proceedings. SPE/DOE Improved Oil Recovery Symposium. Society of Petroleum Engineers, Richardson, Texas, pp. 1–7.

Burd, G., Ward, O.P., 1996a. Physico-chemical properties of PM-factor, a surface active agent produced by *Pseudomonas marginalis*. Can. J. Microbiol. 42, 243–251.

Burd, G., Ward, O.P., 1996b. Bacterial degradation of polycyclic aromatic hydrocarbons on agar plates: the role of biosurfactants. Biotechnol. Tech. 10, 371–374.

Cerniglia, C.E., 1984. Microbial metabolism of polycyclic aromatic hydrocarbons. Adv. Appl. Microbiol. 30, 31–71.

Cerniglia, C.E., 1992. Microbial degradation of polycyclic aromatic hydrocarbons. Microbial. Degr. 3, 351–368.

Cerniglia, C.E., Heitkamp, M.A., 1989. Microbial degradation of polycyclic aromatic hydrocarbons in the aquatic environment. In: Varanasi, U. (Ed.), Metabolism of Polycyclic Aromatic Hydrocarbons in the Aquatic Environment CRC Press, Inc., Boca Raton, Florida, pp. 41–68.

Cerniglia, C.E., Heitkamp, M.A., 1990. Polycyclic aromatic hydrocarbon degradation by *Mycobacterium*. Methods Enzymol. 188, 148–153.

Cerniglia, C.E., 1992. Biodegradation of polycyclic aromatic hydrocarbons. Biodegradation 3, 351–368.

Chen, C.I., Taylor, R.T., 1995. Thermophilic microbial degradation of BTEX by two *Thermus* species. Biotechnol. Bioeng. 48, 614–624.

Chen, C.I., Taylor, R.T., 1997a. Batch and fedbatch bioreactor cultivations of a *Thermus* species with thermophilic BTEX-degrading activity. Appl. Microbiol. Biotechnol. 47, 726–733.

Chen, C.I., Taylor, R.T., 1997b. Thermophilic microbial degradation of BTEX by two consortia of anaerobic bacteria. Appl. Microbiol. Biotechnol. 48, 121–128.
Davies, L., Daniel, F., Swannell, R., Braddock, J., 2001. Biodegradability of chemically-dispersed oil. Report No. AEAT/ENV/R0421. AEA Technology Environment, Abingdon, Oxfordshire, United Kingdom.
Dibble, J.T., Bartha, R., 1976. The effect of iron on the microbial degradation of petroleum in seawater. Appl. Environ. Microbiol. 31, 544–550.
El-Gendy, N.S., Speight, J.G., 2015. Handbook of Refinery Desulfurization. CRC Press, Taylor & Francis Group, Boca Raton, Florida.
Fedorak, P.M., Semple, K.M., Vazquez-Duhalt, R., Westlake, D.W.S., 1993. Chloroperoxidase mediated modifications of petroporphyrins and asphaltenes. Enzyme Microb. Technol. 15, 429–437.
Foght, J.M., Fedorak, P.M., Westlake, D.W.S., 1990. Mineralization of [^{14}C] Hexadecane and [^{14}C] Phenanthrene in crude oil: specificity among bacterial isolates. Can. J. Microbiol. 36, 169–175.
Frenzel, M., James, P., Burton, S.K., Rowland, S.J., Lappin-Scott, H.M., 2007. Towards bioremediation of toxic unresolved complex mixtures of hydrocarbons: identification of bacteria capable of rapid degradation of alkyltetralins. J. Soil. Sediment. 9, 129–136.
Frontera-Suau, R., Bost, F., McDonald, T., Morris, P.J., 2002. Aerobic biodegradation of hopanes and other biomarkers by crude oil degrading enrichment cultures. Environ. Sci. Technol. 36, 4585–4592.
Gao, C.H., Zekri, A., 2011. Applications of microbial-enhanced oil recovery technology in the past decade. Energy Sour. Part A Recov. Util. Environ. Effects 33 (10), 972–989.
Garcia-Arellano, H., Buenrostro-Gonzalez, E., Vazquez-Duhalt, R., 2004. Biocatalytic transformation of petroporphyrins by chemical modified cytochrome c. Biotechnol. Bioeng. 85, 790–798.
Gary, J.G., Handwerk, G.E., Kaiser, M.J., 2007. Petroleum Refining: Technology and Economics, fifth ed. CRC Press, Taylor & Francis Group, Boca Raton, Florida.
Ghazali, F.M., Rahman, R.N.Z.A., Salleh, A.B., Basri, M., 2004. Microbial degradation of hydrocarbons in soil by microbial consortium. Intl. Biodeter. Biodegrad. 54, 61–67.
Gibson, D.T., Subramanian, V., 1984. Microbial degradation of aromatic hydrocarbons. In: Gibson, D.T. (Ed.), Microbial Degradation of Organic Compounds McGraw–Hill, New York, pp. 181–252.
Goodin, J.D., Webber, M.D., 1995. Persistence and fate of anthracene and Benzo[*a*]pyrene in municipal sludge treated soil. J. Environ. Qual. 24, 271–278.
Goodwin, N.S., Park, P.J.D., Rawlinson, A.P., 1983. Crude oil microbial degradation under simulated and natural conditions. In: Bjorøy, M. (Ed.), Advances in Organic Geochemistry 1981 John Wiley & Sons Inc., New York, pp. 650–658.
Greenwood, P.F., Wibrow, S., George, S.J., Tibbett, M., 2008. Sequential hydrocarbon biodegradation in a soil from arid coastal australia, treated with oil under laboratory-controlled conditions. Org. Geochem. 39, 1336–1346.
Guiliano, M., Boukir, A., Doumenq, P., Mille, G., 2000. Supercritical fluid extraction of BAL 150 crude oil asphaltenes. Energy Fuels 14, 89–94.
Hao, R., Lu, A., 2009. Biodegradation of heavy oils by halophilic bacterium. Prog. Nat. Sci. 19, 997–1001.
Harvey, R.G., 1997. Polycyclic Aromatic Hydrocarbons. John Wiley & Sons Inc.-VCH, New York.
Head, I.M., Jones, D.M., Röling, W.F.M., 2006. Marine microorganisms make a meal of oil. Nat. Rev. Microbiol. 4, 173–182.
Hernáez, M.J., Reineke, W., Santero, E., 1999. Genetic analysis of microbial degradation of tetralin by a *Sphingomonas* strain. Appl. Environ. Microbiol. 65 (4), 1806–1810.

Holba, A.G., Dzou, I.L., Hickey, J.J., Franks, S.G., May, S.J., Lenney, T., 1996. Reservoir geochemistry of South Pass 61 field, Gulf of Mexico: compositional heterogeneities reflecting filling history and microbial degradation. Org. Geochem. 24, 1179–1198.

Hsu, C.S., Robinson, P.R., 2006. Practical Advances in Petroleum Processing, vols. 1 and 2. Springer, New York.

Huang, L., Ma, T., Li, D., Liang, F., Liu, R., Li, G., 2008. Optimization of nutrient component for diesel oil degradation by *Rhodococcus erythropolis*. Mar. Pollut. Bull. 56, 1714–1718.

Jones, D.M., Douglas, A.G., Parkes, R.J., Taylor, J., Giger, W., Schaffner, G., 1983. The recognition of biodegraded petroleum-derived aromatic hydrocarbons in recent marine sediments. Mar. Poll. Bull. 14 (103), 108.

Kanaly, R.A., Harayama, S., 2000. Biodegradation of high-molecular weight polycyclic aromatic hydrocarbons by bacteria. J. Bacteriol. 182, 2059–2067.

Kim, I.S., Park, J., Kim, K.W., 2001. Enhanced biodegradation of polycyclic aromatic hydrocarbons using nonionic surfactants in soil slurry. Appl. Geochem. 16, 1419–1428.

Kim, J.-S., Crowley, D.E., 2007. Microbial diversity in natural asphalts of the rancho La Brea tar pits. Appl. Environ. Microbiol. 73, 4579–4591.

Lacotte, D.J., Mille, G., Acquaviva, M., Bertrand, J.C., 1996. Arabian light 150 asphaltene biotransformation with n-Alkanes as co-substrate. Chemosphere 32, 1755–1761.

Lavania, M., Cheema, S., Sarma, P.M., Mandal, A.K., Lal, B., 2012. Biodegradation of Asphalt by *Garciaella petrolearia* TERIG02 for viscosity reduction of heavy oil. Biodegradation 23, 15–24.

Leahy, J.G., Colwell, R.R., 1990. Microbial degradation of hydrocarbons in the environment. Microbiol. Rev. 54, 305–315.

Mielke, J.E., 1990. Oil in the ocean: the short- and long-term impacts of a spill. Congressional Research Service, 90-356 SPR, July 24, p. 11.

Miiller, D.E., Holba, A.G., Huges, W.B., 1987. Effects of microbial degradation on crude oils. In: Meyer, R.F. (Ed.), Exploration for Heavy Crude Oil and Natural Bitumen AAPG Studies in Geology #25. American Association of Petroleum Geologists, Tulsa, Oklahoma, pp. 233–241.

Milić, J.S., Beškoski, V.P., Ilić, M.V., Ali, S.A.M., Gojgić-Cvijović, G.Đ., Vrvić, M.M., 2009. Bioremediation of soil heavily contaminated with crude oil and its products: composition of the microbial consortium. J. Serb. Chem. Soc. 74 (4), 455–460.

Miller, E.C., Miller, J.A., 1981. Search for the ultimate chemical carcinogens and their reaction with cellular macromolecules. Cancer 47, 2327–2345.

Mogollon, L., Rodriguez, R., Larrota, W., Ortiz, C., Torres, R., 1998. Biocatalytic removal of nickel and vanadium from petroporphyrins and asphaltenes. Appl. Biochem. Biotechnol. 70–72, 765–777.

Mohajeri, L., Aziz, H.A., Isa, M.H., Zahed, M.A., 2010. A statistical experiment design approach for optimizing microbial degradation of weathered crude oil in coastal sediments. Bioresource Technol. 101, 893–900.

Morais, E.B., de Tauk-Tornisielo, S.M., 2009. Biodegradation of oil refinery residues using mixed-culture of microorganisms isolated from land farming. Brazilian Arch. Biol. Technol. 52 (6), 1571–1578.

Mueller, J.G., Chapman, P.J., Blattmann, B.O., Pritchard, P.H., 1990. Isolation and characterization of a fluoranthene-utilizing strain of *Pseudomonas paucimobilis*. Appl. Environ. Microbiol. 56, 1079–1086.

Naranjo, L., Urbina, H., De Sisto, A., León, V., 2007. Isolation of autochthonous Non-White rot fungi with potential for enzymatic upgrading of venezuelan extra-heavy crude oil. Biocatal. Biotransf. 25, 1–9.

Nasrollahzadeh, H.S., Najafpour, G.D., Aghamohammadi, N., 2007. Microbial degradation of phenanthrene by mixed culture consortia in a batch bioreactor using central composite Face-Entered design. Int. J. Environ. Res. 1, 80–87.

Nester, E.W., Anderson, D.G., Roberts Jr., C.E., Pearsall, N.N., Nester, M.T., 2001. Microbiology: A Human Perspective, third ed. McGraw-Hill, New York.

Obayori, O.S., Ilori, M.O., Adebusoye, S.A., Oyetibo, G.O., Omotayo, A.E., Amund, O.O., 2009. Degradation of hydrocarbons and biosurfactant production by *Pseudomonas sp.* strain LP1. World J. Microbiol. Biotechnol. 25, 1615–1623.

Obire, O., 1990. Bacterial degradation of three different crude oils in nigeria. Nig. J. Bot. 1, 81–90.

Obire, O., 1993. The suitability of various nigerian petroleum fractions as substrate for bacterial growth. Discov. Innov. 9, 25–32.

Obire, O., Nwaubeta, O., 2001. Microbial degradation of refined petroleum hydrocarbons in soil. J. Appl. Sci. Environ. Mgt. 5 (1), 43–46.

Obire, O., Okudo, I.V., 1997. Effects of crude oil on a freshwater stream in nigeria. Discov. Innov. 9, 25–32.

Okoh, A.I., 2006. Biodegradation: an alternative in the cleanup of petroleum hydrocarbon pollutants. Biotechnol. Mol. Biol. Rev. 1 (2), 38–50.

Okoh, A.I., Ajisebutu, S., Babalola, G.O., Trejo-Hernandez, M.R., 2002. Microbial degradation of Mexican heavy crude oil (Maya) by *Pseudomonas aeruginosa*. J. Trop. Biosci. 2 (1), 12–24.

Ourisson, G., Albrecht, P., Rohmer, M., 1979. The hopanoids. Paleochemistry and biochemistry of a group of natural products. Pure Appl. Chem. 51, 709–729.

Park, K.S., Sims, R.C., DuPont, R.R., Doucette, W.J., Matthews, J.E., 1990. Fate of polynuclear aromatic hydrocarbon compounds in two soil types: influence of volatilization, abiotic loss, and biological activity. Environ. Toxicol. Chem. 9, 187–195.

Parkash, S., 2003. Refining Processes Handbook. Gulf Professional Publishing, Elsevier, Amsterdam, Netherlands.

Pathak, H., Kantharia, D., Malpani, A., Madamwar, D., 2009. Naphthalene microbial degradation using Pseudomonas Sp. HOB1: In Vitro studies and assessment of naphthalene degradation efficiency in simulated microcosms. J. Hazard. Mater. 166, 1466–1473.

Payne, J.R., McNabb Jr., G.D., 1984. Weathering of petroleum in the marine environment. Marine Technol. Soc. J. 18 (3), 24.

Pineda, F.G., Mesta-Howard, A.M., Boll-Argúello, G., Lira-Galeana, C., 2004. A microbial consortium isolated from a crude oil sample that uses asphaltenes as a carbon and energy source. Biodegradation 15, 145–151.

Prince, R., 1993. Petroleum spill bioremediation in marine environments. Crit. Rev. Microbiol. 19, 217–242.

Rabus, R., Kube, M., Heider, J., Beck, A., Heitmann, K., Widdel, F., et al., 2005. The genome sequence of an anaerobic aromatic-degrading denitrifying bacterium, Strain Ebn1. Arch. Microbiol. 183, 27–36.

Rahman, K.S.M., Thahira-Rahman, J., Lakshmanaperumalsamy, P., Banat, I.M., 2002. Towards efficient crude oil degradation by a mixed bacterial consortium. Biores. Tech. 85, 257–261.

Ramos, J.L., Duque, E., Gallegos, M.T., Godyoy, P., Ramos-Gonzalez, M.I., Rojas, A., et al., 2002. Mechanisms of solvent tolerance in Gram-negative bacteria. Ann. Rev. Microbiol. 56, 743–768.

Rosenberg, E., Rubinovitz, C., Gottlieb, A., Rosenhak, S., Ron, E.Z., 1988. Production of biodispersan by *Acinetobacter calcoaceticus* A2. Appl. Environ. Microbiol. 54, 317–322.

Salam, L.B., Obayori, O.S., Akashoro, O.S., Okogie, G.O., 2011. Microbial degradation of Bonny light crude oil by bacteria isolated from contaminated soil. Int. J. Agric. Biol. 13 (2), 245–250.

Sar, N., Rosenberg, E., 1983. Emulsifier production by *Acinetobacter calcoaeticus* strains. Curr. Microbiol. 9, 309–314.

Schreiber, A.F., Winkler, U.K., 1983. Transformation of tetralin by whole cells of *Pseudomonas stutzeri* As39. Appl. Microbiol. Biotechnol. 18 (1), 6–10.

Sen, R., 2008. Biotechnology in petroleum recovery: microbial EOR. Prog. Energy Combust. Sci. 34 (2008), 714–724.
Shabani, M., Crescente, C., Torsaeter, O., 2009. The effect of bacterial solution on the wettability index and residual oil saturation in sandstone. J. Petrol. Sci. Eng. 69 (3–4), 255–260.
Shuttleworth, K.L., Cerniglia, C.E., 1995. Environmental aspects of polynuclear aromatic hydrocarbon microbial degradation. Appl. Biochem. Biotechnol. 54, 291–302.
Sikkema, J., Bont, J.A.M., 1991. Isolation and initial characterization of bacteria growing on tetralin. Microbial. Degr. 2 (1), 15–23.
Sivaraman, C., Ganguly, A., Mutnuri, S., 2010. Microbial degradation of hydrocarbons in the presence of cyclodextrins. World J. Microbiol. Biotechnol. 26, 227–232.
Soli, G., Bens, E.M., 1972. Bacteria which attack petroleum hydrocarbons in a saline medium. Biotechnol. Bioeng. 14, 319–330.
Speight, J.G., 2014. The Chemistry and Technology of Petroleum, fifth ed. CRC Press, Taylor & Francis Group, Boca Raton, Florida.
Speight, J.G., Arjoon, K.K., 2012. Bioremediation of Petroleum and Petroleum Products. Scrivener Publishing, Salem, Massachusetts.
Speight, J.G., Ozum, B., 2002. Petroleum Refining Processes. Marcel Dekker Inc., New York.
Strawinski, R.J., Stone, R.W., 1940. The utilization of hydrocarbons by bacteria. J. Bacteriol. 40 (3), 461.
Sutherland, J.B., Rafii, F., Khan, A.A., Cerniglia, C.E., 1995. Mechanisms of polycyclic aromatic hydrocarbon degradation. In: Young, L.Y., Cerniglia, C.E. (Eds.), Microbial Transformation and Degradation of Toxic Organic Chemicals John Wiley & Sons Inc., New York, pp. 269–306.
Taghvaei Ganjali, S., Nahri Niknafs, B., Khosravi, M., et al., 2007. Photo-oxidation of crude petroleum maltenic fraction I natural simulated conditions and structural elucidation of photoproducts. Iran J. Environ. Health Sci. Eng. 4 (1), 37–42.
Taylor, T.R., Jackson, K.J., Duba, A.G., Chen, C.I., 1998. In Situ Thermally Enhanced Microbial degradation of Petroleum Fuel Hydrocarbons and Halogenated Organic Solvents. United States Patent 5,753,122. May 19.
Thouand, G., Bauda, P., Oudot, J., Kirsch, G., Sutton, C., Vidalie, J.F., 1999. Laboratory evaluation of crude oil biodegradation with commercial or natural microbial inocula. Can. J. Microbiol. 45, 106–115.
Uribe-Alvarez, C., Ayala, M., Perezgasga, L., Naranjo, L., Urbina, H., Vazquez-Duhalt, R., 2011. First evidence of mineralization of petroleum asphaltenes by a strain of *Neosartorya fischeri*. Microbial Biotechnol. 4 (5), 663–672.
Van Brummelen, T.C., Verweij, R.A., Wedzinga, S.A., Van Gestel, C.A.M., 1996. Enrichment of polycyclic aromatic hydrocarbons in forest soils near a blast furnace plant. Chemosphere 32, 293–314.
Van Der Meer, J.R., Devos, W.M., Harayama, S., Zehnder, A.J.B., 1992. Molecular mechanisms of genetic adaptation to xenobiotic compounds. Microbiol. Rev. 56 (677), 694.
Venkateswaran, K., Hoaki, T., Kato, M., Maruyama, T., 1995. Microbial degradation of resins fractionated from Arabian light crude oil. Can. J. Microbiol. 41, 418–424.
Vieira, P.A., Faria, S.R., Vieira, B., De Franca, F.P., Cardoso, V.L., 2009. Statistical analysis and optimization of nitrogen, phosphorus, and inoculum concentrations for the microbial degradation of petroleum hydrocarbons by response surface methodology. J. Microbiol. Biotechnol. 25, 427–438.
Weissenfels, W.D., Beyer, M., Klein, J., 1990. Degradation of phenanthrene, fluorene and fluoranthene by pure bacterial cultures. Appl. Microbiol. Biotechnol. 32, 479–484.
Weissenfels, W.D., Beyer, M., Klein, J., Rehm, H.J., 1991. Microbial metabolism of fluoranthene: isolation and identification of ring fission products. Appl. Microbiol. Biotechnol. 34, 528–535.

Whyte, L.G., Bourbonnière, L., Greer, C.W., 1997. Biodegradation of petroleum hydrocarbons by psychrotrophic *Pseudomonas* strains possessing both alkane (*alk*) and naphthalene (*nah*) catabolic pathways. Appl. Environ. Microbiol. 63 (9), 3719–3723.

Wild, S.R., Jones, K.C., 1993. Biological and abiotic losses of polynuclear aromatic hydrocarbons from soils freshly amended with sewage sludge. Environ. Toxicol. Chem. 12, 5–12.

Willumsen, P.A., Karlson, U., Pritchard, P.H., 1998. Response of fluoranthene-degrading bacteria to surfactants. Appl. Microbiol. Biotechnol. 50, 475–483.

Wyndham, R.C., Costerton, J.W., 1981. In vitro microbial degradation of bituminous hydrocarbons and in situ colonization of bitumen surfaces within the athabasca oil sands deposit. Appl. Environ. Microbiol. 41, 791–800.

Ye, D., Siddiqi, M.A., Maccubbin, A.E., Kumar, S., Sikka, H.C., 1996. Degradation of polynuclear aromatic hydrocarbons by *Sphingomonas paucimobilis*. Environ. Sci. Technol. 30, 136–142.

Zahed, M.A., Aziz, H.A., Isa, M.H., Mohajeri, L., 2010. Enhancement of microbial degradation of n-alkanes from crude oil contaminated seawater. Int. J. Environ. Res. 4 (4), 655–664.

Zander, M., 1983. Physical and chemical properties of polycyclic aromatic hydrocarbons. In: Bjørseth, A. (Ed.), Handbook of Polycyclic Aromatic Hydrocarbons Marcel Dekker, Inc., New York, pp. 1–26.

Zekri, A., Ghannam, M., Almehaideb, R., 2003. Carbonate rocks wettability changes induced by microbial solution. Paper No. SPE 80527. Proceedings. SPE Asia Pacific Oil and Gas Conference and Exhibition. Jakarta, Indonesia. September 9–11. Society of Petroleum Engineers, Richardson, Texas.

Zengler, K., Richnow, H.H., Rossello-Mora, R., Michaelis, W., Widdel, F., 1999. Methane formation from long-chain alkanes by anaerobic microorganisms. Nature 401, 266–269.

Zrafi-Nouira, I., Guermazi, S., Chouari, R., Safi, N.M.D., Pelletier, E., Backhrouf, A., et al., 2009. Molecular diversity analysis and bacterial population dynamics of an adapted seawater microbiota during the degradation of tunisian zarzantine oil. Biodegradation 20, 467–486.

CHAPTER 8

Nonthermal Methods of Recovery

8.1 INTRODUCTION

The vast heavy oil reserves available in various parts of the world are becoming increasingly important as a secure future energy source (Safinya, 2008). In fact, the heavy oil resource is trillions of barrels, but the cumulative recovery totals to date are on the order of billions of barrels. Whether the potential and promise of heavy oil is realized depends on the evolution of recovery technologies that are appropriate for the wide range of reservoir and oil-phase conditions. Such technologies also need to be comparatively benign from an environmental aspect.

Heavy oil production has been increasing in recent years and is expected to increase in the future because of the expected supply shortfall in conventional oil and an abundance of relatively large and known heavy oil reservoirs. Heavy oil is commercially produced by primary recovery, water injection, and thermal-enhanced oil recovery (TEOR) methods (Mai et al., 2006). A majority of the recent primary production is in Venezuela's Orinoco belt and long horizontal wells have made these projects possible.

However, heavy oil has low mobility (see chapter: "General Methods of Oil Recovery") because of its high viscosity, even though the relative permeability may be close to unity. As a result, the mobility ratio would be unfavorable in any fluid displacement process. Use of a chemical (assuming its loss to the rock matrix is minimal) can help in two ways: first, the oil–water interfacial tension will decrease, resulting in a lower residual oil saturation despite an inefficient displacement, and second, the low interfacial tension would promote the formation of an emulsion which, in turn, would make the effective mobility ratio more favorable (Browne et al., 1996). The emulsion would lower the mobility of the displacing fluid through drop entrainment and entrapment.

Steam injection is the most widely applied enhanced oil recovery method. The majority of the steam injection projects are in California, Canada, Indonesia, and Venezuela. Recovery can approach 20% for cyclic

Introduction to Enhanced Recovery Methods for Heavy Oil and Tar Sands.
DOI: http://dx.doi.org/10.1016/B978-0-12-849906-1.00008-4

steaming and over 50% for continuous steam injection (Kumar, 2006). Novel methods such as solvent injection and hybrid methods are being tested for heavy oil recovery, where steam may not be the best option.

However, heavy oil in a reservoir is often very viscous (unless the reservoir temperature is high) and does not flow easily, therefore traditional methods of oil recovery, such as primary and secondary methods, are not usually applicable to heavy oil or, if they are applicable, it is often decided to bypass such methods and proceed to enhanced methods of recovery. The route to use is, of course, reservoir specific and what is applied to one heavy oil reservoir cannot necessarily be applied to another heavy oil reservoir.

The amount of oil that is recoverable is determined by a number of factors including the permeability of the rocks, the strength of natural drives (the gas present, pressure from adjacent water or gravity), and the viscosity of the oil. When the reservoir rocks are "tight" such as shale, oil generally cannot flow through but when they are permeable such as in sandstone, oil flows freely. The flow of oil is often helped by natural pressures surrounding the reservoir rocks, including natural gas that may be dissolved in the oil, natural gas present above the oil, water below the oil, and the strength of gravity. Heavy oils tend to span a large range of viscosity from manageable liquids to poorly mobile materials.

In terms of heavy oil recovery, it is more traditional to apply thermal oil recovery techniques of which steam injection is the most popular. Thus, heavy oil recovery efforts include thermal methods [steam floods, cyclic steam stimulation, steam-assisted gravity draining (SAGD)] as well as nonthermal methods [cold flow with sand production, cyclic solvent process, vapor-assisted petroleum extraction (VAPEX)] and, in some cases, a new interest has been initiated using in situ combustion processes (Table 8.1) (see chapter: "Nonthermal Methods of Recovery"). Significant improvements to the effectiveness of these methods can be achieved by developing a basic understanding of the complex displacement mechanisms and by developing new techniques for in situ characterization of fluid and reservoir characteristics. Development of optimal strategies for recovering these reserves requires the development of state-of-the-art reservoir flow simulator(s) that incorporate techniques for coupling geomechanics and fluid flow besides accurately representing thermal, phase equilibria, and mass transfer effects. Improved recovery efficiency can also be achieved by various combinations of thermal and nonthermal processes. In addition, recent advances in drilling and production from

Table 8.1 General Section of Recovery Processes for Heavy Oil and Tar Sand Bitumen

Method	Description
Vapex	Uses solvent rather than steam in SAGD-type wells
Hybrid	Solvent plus steam in SAGD, CSS, and steamflood wells
In situ combustion with vertical and horizontal wells	Uses heavy oil in reservoir and injected air
Gasification of heavy ends	Used for steam generation and hydrogen production
Downhole heating with electricity	Resistance, induction, radio-frequency (RF)

unconsolidated sands will allow the development of heavy oil recovery strategies that would not have been possible three decades ago.

Thus, the quest to produce heavy oil is a global issue. However, in many cases, the resource does not resemble the conditions that are thought to be typical for heavy oil: viscous oil held in relatively permeable, shallow sands and the fields of interest have evolved to include fractured carbonates, offshore settings, and deeper, more geologically heterogeneous heavy-oil resources. Such new settings introduce new challenges, which include highly permeable pathways and limits on reservoir access, in addition to those of large oil-phase viscosity and low reservoir energy.

Water flooding of heavy oil is summarily dismissed because of adverse mobility ratios. However, water floods have been used with heavy oil to enhance formation pressures and help displace the heavy crude (NPC, 2007; Mai and Kantzas, 2009). Heavy oil viscosity decreases rapidly with increasing temperature, therefore the presence of internal heat is an asset for production. In fact, reservoir temperature (the mobility of the heavy oil is increased at high reservoir temperature) and the temperature of the water may be deciding factors when considering heavy oil recovery by water flooding. In fact, water flooding can be used in some heavy oil reservoirs to maintain pressure during *cold production*. The term *cold production* is a nonthermal heavy oil production without sand and refers to the use of operating techniques and specialized pumping equipment to aggressively produce heavy oil reservoirs without applying heat (Tchambak et al., 2012).

The cold production process may often be confused with the cold heavy oil production with sand (CHOPS) process which encourages the associated production of large quantities of the unconsolidated noncemented reservoir sand, which, in turn, results in significantly higher oil production (such as in the CHOPS process). In contrast, conventional practices of primary heavy oil production discourage sand production and result in minimized initial unit operating costs. This practice, however, may, prevent many wells from achieving their maximum oil production rate and reserve potential.

However, on occasion, oil production and recovery by the cold production process cannot occur without the production of sand. Typically, heavy oil flows more efficiently when sand is produced from unconsolidated reservoirs. Sand production is thought to be a function of: (1) the absence of clays and cementation materials; (2) the viscosity of the oil; (3) the producing water cut and gas–oil ratio (GOR); and (4) the rate of pressure drawdown. In cold production, it is believed that a foamy fluid phase is formed when flow rates of a liquid containing dissolved gas are high and the flowing pressure at the well is below the bubble point. This high slurry rate (including oil, water, gas, and sand) could be achieved by a larger pressure gradient near the wellbore using a progressive cavity pump.

The presence of clay stabilizes the sand grains and reduces sand movement. Higher viscosity oil increases the frictional drag between the oil and the sand grains which promotes sand movement. High water or gas production inhibits sand production because gas/water is produced instead of an oil/sand mixture. Increasing the drawdown rate also promotes sand movement because of the increase in the velocity of the fluid into the wellbore and hence increased frictional drag on the sand grains.

8.2 PRIMARY RECOVERY METHODS

Primary recovery (using the inherent or natural energy of the reservoir; see chapter: "General Methods of Oil Recovery") of heavy oil from a reservoir is not the usual approach to recovery. Primary recovery techniques rely entirely on natural forces within the reservoir and, although primary recovery is widely applicable to the recovery of conventional crude oil, it is less applicable to heavy oil. In high temperature reservoirs that keep the heavy oil sufficiently fluid, primary recovery may undoubtedly be applicable. For example, production mechanisms of the heavy-oil solution-gas-drive process are not completely elucidated and, consequently, it is difficult to optimize primary recovery.

Given the option, operators will try to produce as much heavy oil as possible by primary recovery methods (cold production), but this is dependent upon the fluidity of the oil which, in turn, is dependent upon the reservoir temperature. The term cold production refers to the use of operating techniques and specialized pumping equipment to aggressively produce heavy oil reservoirs without applying heat. This encourages the associated production of large quantities of the unconsolidated uncemented reservoir sand which in turn results in significantly higher oil production.

Typical recovery factors (percentage of the oil in a reservoir that can be recovered) for cold production range from 1% to 10% of the original oil in place (OOIP) (Curtis et al., 2002). In addition, again depending on the character of the reservoir *and* the properties of the oil in the reservoir, primary production with artificial lift, for example, injection of a light oil, or diluent, to reduce viscosity, may find use. Many fields produce most efficiently with horizontal production wells.

In some cases, exploiting foamy oil behavior (see chapter: "Heavy Oil and Tar Sand Bitumen") and/or encouraging the production of sand along with the oil turn out to be the preferred production strategy. Choosing the optimal cold production strategy requires an understanding of rock and fluid properties (Chen, 2006). Foaming can complicate the recovery process even more. When heavy oil contains associated gas, foaming can occur at any point in the reservoir, wellbore, flow lines, or production equipment at the surface. When gas reaches the bubble point and comes out of suspension as the pressure and temperature of the fluids change, the character of the oil changes from the known to the unknown. However, while foaming can be a problem, it can also work in the operator's favor. Since any foam that forms in the reservoir increases pressure, it can serve as a temporary gas drive, pushing oil toward the wellbore. The issue is to understand the fluid properties from which it might be possible to predict when and where foaming is likely to occur and how the phenomenon can be used beneficially.

The term cold production refers to the use of operating techniques and specialized pumping equipment to aggressively produce heavy oil reservoirs without applying heat. This encourages the associated production of large quantities of the unconsolidated uncemented reservoir sand which in turn results in significantly higher oil production. Primary production in some heavy-oil reservoirs may be larger than that estimated by conventional calculations. The main driving force behind primary recovery is pressure depletion through solution gas drive.

The most preferential reservoirs for cold production of heavy oil vary and are very dependent upon reservoir character and heavy oil properties

(Chugh et al., 2000). For example, when the viscosity of the heavy oil is high (on the order of 2000–30,000 cP), the higher the oil viscosity, the greater the drag force on a sand particle. Higher oil viscosity is desired for higher sand production and also in foamy solution drive and higher interfacial tensions result in higher degrees of critical gas saturation and supersaturation, which should be beneficial to oil recovery. The interfacial tension between gas and oil should increase with a decrease in oil API.

Also, the more unconsolidated the formation (ie, the less cementitious the formation) and less clay cement, the more likely the sand production. In addition, as water production increases the cleanup capacity decreases. For most heavy oil reservoirs, the connate water saturations are approximately 10–50%, with most being in the range 10–30%. The water cut is normally quite low at the beginning of primary production. A high water cut could be caused by water flooding or coning of the bottom aquifer or edge water support. The bottom aquifer plays two roles in primary production: initially it provides pressure support to the reservoir and drives heavy oil upwards but once the water cone reaches the wellbore, the higher water cut impedes the oil production.

The higher the initial reservoir pressure and the better the initial drawdowns, the more the well will clean up (ie, will produce sand with oil). In general, once a field is pressure depleted, infill wells will not clean up. Most heavy oil reservoirs in Alberta and Saskatchewan are shallow and the depths are less than 3000 feet with an average depth on the order of 1300–2000 feet. If the pressure gradient is less than the hydrostatic head, the current reservoir pressure can indicate the magnitude of pressure depletion and the maximum pressure drawdown that can be achieved by progressive cavity pumps, which can significantly affect the sand production and foamy oil behavior (Chugh et al., 2000).

A cold production oil reservoir is a *solution gas drive reservoir* if the major reservoir energy for primary depletion is supplied by the release of gas from the oil and the expansion of the in-place fluids as reservoir pressure declines. The fraction of OOIP that can be recovered by solution gas drive decreases with increasing oil viscosity. For heavy oil reservoirs, the expected recovery factor by solution gas drive is typically about 5%. A number of heavy oil reservoirs under solution gas drive, however, have obtained anomalous primary performance results: low production GORs, high oil production rates, and recovery of unexpectedly large amounts of oil. This unusual behavior has been attributed to (1) the expansion of gas bubbles giving the oil a foamy character as the bubbles are trapped by the oil and recovery is then enhanced by solution—ultimate oil recovery with primary techniques can be as high as 20% for some heavy foamy oil reservoirs and (2) internal erosion in unconsolidated sand reservoirs, which can create a network of high-permeability channels (*wormholes*) which can enhance drainage by a factor of 10 or more (Chen, 2006).

On the other hand, there are several very large (>100,000 barrels per day) projects for heavy oil (approximately 12° API) recovery in the heavy oil belt (Faja 3 Del Orinoco) in Venezuela, where the main recovery method to date has been primary recovery. Heavy oil production from this belt is expected to last for 35 years at 600,000 barrels per day (Meyer and Attanasi, 2003). In this formation, the reservoir (deposit) temperature is sufficiently high and the viscosity is sufficiently low that, with the existing solution gas, the heavy oil can flow at reservoir (deposit) temperatures. Horizontal wells up to 1650 feet long allow the heavy oil to be produced at economic rates while maintaining sufficiently low drawdown pressures to prevent extensive sand production. Also, the horizontal legs are placed precisely in the target sands using logging-while-drilling and measurement-while-drilling equipment, enabling more cost-effective placement of the wells. Finally, in some locations, sand production from the unconsolidated formation is minimized using slotted liners and other sand-control methods. The foamy nature of the oil has yielded initial rates of over 1000 barrels per day, and this is not a common recovery for this gravity oil. Heavy oil production from this belt is expected to last for 35 years at a production rate of 600,000 barrels per day (Curtis et al., 2002; Meyer and Attanasi, 2003).

However, primary production in some heavy-oil reservoirs is larger than that estimated by conventional calculations (Kovscek, 2002). Conventionally, the main driving force behind primary recovery is pressure depletion through solution gas drive. Solution gas drive is the mechanism, whereby the lowering of reservoir pressure through production in an undersaturated reservoir reaches the bubble point where gas starts to evolve from solution. The evolved gas does not begin to flow until the critical gas saturation has been reached. Once the critical gas saturation point is reached, there is an increase in rate of pressure drop due to the production of the gas-phase. It has been noted that the oil at the wellhead of these heavy-oil reservoirs resembles a form of foam, hence the term *foamy oil*. A key to developing an accurate mechanistic understanding of heavy-oil solution gas drive is to delineate bubble growth, interaction, and gas flow experimentally.

The main issue for cold production is the low recovery factor for primary production. Fields are not being developed with future, secondary processes in mind. For example, wells, cement, and completions are not designed for high temperatures encountered in steam injection and other thermal recovery processes. Horizontal and fishbone wells should be drilled in the optimum location with regard to permeability, porosity, oil composition, and distances above water or below gas, and the length of the laterals.

Drilling, measurement-while-drilling, and logging-while-drilling technologies are key enablers for this. In horizontal wells and multilateral wells, being able to monitor, understand, control, and ensure the flow from different sections of the well will improve production and reduce unwanted water and/or natural gas production. In the Orinoco field, natural gas production is an issue because of interference with progressive cavity pumps' ability to lift the heavy oil.

8.3 SECONDARY RECOVERY METHODS

Even in conventional oil reservoirs, a lot of oil can be left behind after primary production since the natural reservoir pressure has dwindled to the point where it cannot force the oil to the surface. Thus, as fluid withdrawal continues from the reservoir, the pressure within the reservoir gradually decreases, and the amount of gas in solution decreases. As a result, the flow rate of fluid into the wellbore decreases, and less gas is liberated. The fluid may not reach the surface, so a pump (artificial lift) must be installed in the wellbore to continue producing the crude oil. If this is allowed to continue, the flow rate of the crude oil becomes so small, and the cost of lifting the oil to the surface becomes so great, that the well costs more to operate than the revenues that can be gained from selling the crude oil (after discounting the price for operating costs, taxes, insurance, and return on capital). The well's economic limit has then been reached and it is abandoned. At this stage, there is insufficient underground pressure to force the oil to the surface. If economical, as it often is, the remaining oil in the well is extracted using secondary oil recovery methods. However, in order to maintain well production, the flow of oil should not be allowed to decrease to the minimum. Operators usually apply secondary recovery at a point before reservoir energy is depleted.

When dealing with heavy oil reservoirs 90% or more of the OOIP can be left in the reservoir after attempts at primary or cold production (Curtis et al., 2002).

Secondary oil recovery uses various techniques to aid in recovering oil from depleted or low-pressure reservoirs. Pumps, such as horsehead pumps and electrical submersible pumps (ESPs), are used to bring the oil to the surface. Other secondary recovery techniques increase the reservoir's pressure by water injection (water flooding), natural gas injection (gas flooding), and gas lift, which injects air, carbon dioxide, or some other gas into the reservoir.

Together, primary and secondary recovery methods generally allow 25–35% of the reservoir's oil to be recovered.

8.3.1 Water Flooding

Water flooding is a form of oil recovery wherein the energy required to move the oil from the reservoir rock into a producing well is supplied from the surface by means of water injection and the induced pressure from the presence of additional water. Water injection is used to prevent low pressure in the reservoir—the water replaces the oil which has been taken, keeping the production rate and the pressure the same over the long term. Water floods are essentially artificial water drives and, at one time, were considered to be a form of enhanced recovery.

In conventional reservoirs, water floods are the most efficient of the various drive mechanisms, recovering 40–80% of the OOIP. Artificial water drives (water floods) typical average about 50% of the OOIP. If primary production (natural flow plus artificial lift) would have recovered 25% of the oil then water flooding typically recovers an additional 25%.

In a completely developed oil field, the wells may be drilled anywhere from 200 to 2000 feet (60–600 m) from one another, depending on the nature of the reservoir. If water is pumped into alternate wells in such a field, the pressure in the reservoir as a whole can be maintained or even increased. In this way the rate of production of the crude oil also can be increased; in addition, the water physically displaces the oil, thus increasing the recovery efficiency. In some reservoirs with a high degree of uniformity and little clay content, water flooding may increase the recovery efficiency to as much as 60% or more of the OOIP. Water flooding was first introduced in the Pennsylvania oil fields, more or less accidentally, in the late 19th century, and it has since spread throughout the world.

Water flood has been conducted successfully in a few high viscosity reservoirs in the past, and several projects are currently ongoing and planned around the world. Incremental recovery of approximately 2–20% of the OOIP has been reported (Kumar, 2006).

While the above numbers are not typical it should be noted that the ratio of oil recover by water flood oil to oil recovered by primary methods varies greatly with depth for heavy oil as well as for conventional oil. Shallow reservoirs tend to have both low pressures and small amounts of dissolved gas. This low available drive energy often translates into low primary recovery, sometimes less than 5%. These reservoirs often recover many times more secondary oil than primary oil. For example, if only 5% is recoverable primary and 50% is recovered by water flooding, the ratio of water flood oil to primary is 9:1.

Conversely, deeper oil reservoirs tend to have higher pressures and more dissolved gas and consequently better recoveries on primary production. In a 10,000 foot oil reservoir with a gas cap and high solution gas, 40% primary recovery is possible. This obviously leaves only 10% likely to be recovered by water flooding, giving a secondary to primary ratio of 1:4. This reduced susceptibility to water flooding as gas content increases all the way to 100% gas reservoirs, which are never good water flood candidates. Primary recovery from pure gas reservoirs often exceeds 90%. If one is unfortunate enough to have a natural water drive recovery seldom exceeds 50%. Clearly water drives, either natural or artificial are not advantageous in true gas reservoirs. Any and every source of bulk water can be, and has been, used for injection. Some aspects to consider when selecting an injection source are as follows:

Produced water is often used as an injection fluid. This reduces the potential of causing formation damage due to incompatible fluids, although the risk of scaling or corrosion in injection flow lines or tubing remains. Also, the produced water, being contaminated with hydrocarbons and solids, must be disposed of in some manner, and disposal to sea or river will require a certain level of clean-up of the water stream first. However, the processing required to render produced water fit for reinjection may be equally costly. As the volumes of water being produced are never sufficient to replace all the production volumes (oil and gas, in addition to water), additional "make-up" water must be provided. Mixing waters from different sources exacerbates the risk of scaling.

Sea water is obviously the most convenient source for offshore production facilities (Speight, 2015), and it may be pumped inshore for use in land fields. Where possible, the water intake is placed at sufficient depth to reduce the concentration of algae; however, filtering, deoxygenation, and biociding is generally required. *Aquifer water* from water-bearing formations other than the oil reservoir, but in the same structure, has the advantage of purity where available. *River water* will always require filtering and biociding before injection.

The filters must clean the water and remove any impurities, such as shells and algae. Typical filtration is to 2 μm. The filters are so fine so as not to block the pores of the reservoir. Sand filters are the easiest to use, because there is an automatic system with Delta P which cleans the filter with a backwash when the sand filter is dirty. The sand filter has different beds with various sizes of sand granules. The sea water traverses the first, finest, layer of sand down to the coarsest and to clean the filter, the process is inverted. After the water is filtered it continues on to fill the deoxygenation tower—oxygen must be removed from the water because it promotes corrosion and growth of certain bacteria. Bacterial growth in the reservoir can produce toxic hydrogen sulfide, a source of serious production problems, and block the pores in the rock.

A *deoxygenation tower* brings the injection water into contact with a dry gas stream (gas is always readily available in the oil field). The filtered water drops into the deoxygenation tower, splashing onto a series of trays, causing dissolved oxygen to be lost to the gas stream. An alternative method, also used as a backup to deoxygenation towers, is to add an oxygen scavenging agent such as sodium bisulfite.

The high pressure, high flow *water injection pumps* are placed near to the deoxygenation tower and boosting pumps. They fill the bottom of the reservoir with the filtered water to push the oil towards the wells like a piston. The result of the injection is not quick, it needs time.

Water flooding can also be used in some heavy oil reservoirs to maintain pressure during cold production. The Captain Field in the North Sea uses horizontal wells with specially designed ESPs for the heavy oil production, and horizontal wells for water injection (Etebar, 1995). The horizontal injectors provide more uniformly distributed pressures and a more efficient line-water drive. Since water viscosity (~1 cP) is much lower than the heavy oil (80–100 cP), care must be taken to avoid water fingering from the injecting wells to the producing wells.

The Captain Field in the North Sea uses horizontal wells for the heavy oil production, and horizontal wells for water injection (Etebar, 1995). The horizontal injectors provide more uniformly distributed pressures and a more efficient line-water drive. However, since the viscosity of water is much lower than the heavy oil, care must be taken to avoid water fingering from the injecting wells to the producing wells.

In cold and/or offshore environments, water flooding, and perhaps polymer-augmented water flooding, may present the most attractive recovery option following primary recovery. Steam injection is relatively mature, but cost-effective mobility and profile control by use of aqueous-phase surfactants, gels, or advanced well completions remains an open question. In situ combustion achieved by air injection is technically and economically feasible, especially for deeper, thinner, higher-pressured reservoirs. Combustion is difficult to describe and control. Nevertheless, it is attractive for in situ upgrading and sulfur removal.

8.3.2 Gas Injection

Gas injection (*also called reinjection* or *gas repressurization*) is the reinjection of natural gas into an underground reservoir, typically one already containing both natural gas and crude oil, in order to increase the pressure within the reservoir and thus induce the flow of crude oil or else sequester gas that cannot be exported. This is not to be confused with gas lift, where gas is injected into the annulus of the well rather than the reservoir. After the crude has been pumped out, the natural gas is once again recovered. Since many of the wells found around the world contain heavy crude, this process increases their production.

The basic difference between light conventional crude oil and heavy crude oil is its viscosity and pumpability—the lighter the crude the easier it is to pump. Recovery of hydrocarbons in a well is generally limited to 50% (heavy crudes) and 75–80% (light crudes). Recycling of natural gas or other inert gases causes the pressure to rise in the well, thus causing more gas molecules to dissolve in the oil, lowering its viscosity and thereby increasing the well's output. Air is not suitable for repressuring wells because it tends to cause deterioration of the oil, thus carbon dioxide or natural gas is used to repressure the well. The term *gas-reinjection* is also sometimes referred to as repressuring or repressurizing—the term being used only to imply that the pressure inside the well is being increased to aid recovery.

Injection or reinjection of carbon dioxide also takes place in order to reduce the emission of CO_2 into the atmosphere, a form of carbon sequestration. This has been mooted as a major weapon in the future fight against climate change, allowing mass storage of CO_2 over a geological timescale.

8.3.3 Cold Production

The term cold production refers to the use of operating techniques and specialized pumping equipment to aggressively produce heavy oil reservoirs without applying heat (Chugh et al., 2000). This encourages the associated production of large quantities of the unconsolidated uncemented reservoir sand which in turn results in significantly higher oil production. In contrast, conventional practices of primary heavy oil production discourage sand production and result in minimized initial unit operating costs. This practice, however, may prevent many wells from achieving their maximum oil production rate and reserve potential.

The basis of cold production is that the oil production and recovery improve when sand production occurs naturally. Sand production is thought to be a function of: (1) the absence of clays and cementation materials; (2) the viscosity of the oil; (3) the producing water cut and GOR; and (4) the rate of pressure drawdown (Chugh et al., 2000). Field production data indicate that heavy oil flows more efficiently when sand is produced from unconsolidated reservoirs. In the Elk Point and Lindberg reservoirs, sand production from wells occurs regularly (Loughead, 1992; McCaffrey and Bowman, 1991).

The presence of clay stabilizes the sand grains and reduces sand movement. Higher viscosity oil increases the frictional drag between the oil and the sand grains which promotes sand movement. High water or gas production inhibits sand production because gas/water is produced instead of an oil/sand mixture. Increasing the drawdown rate also promotes sand movement because of the increase in the velocity of the fluid into the wellbore and hence increased frictional drag on the sand grains. It has been reported that gross near-wellbore failure of the formation due to sand production results in excellent productivity.

The produced sand creates a modified wellbore geometry that could have several configurations including piping tubes (wormholes), dilated zones, sheared zones, or possibly cavities. Porosity in the dilated zones may increase leading to large increases in reservoir permeability. In addition, the flow of sand with the oil has the potential to reduce the frictional drag forces on the oil and result in increased productivity in the porous region. Furthermore, fines migration, which occurs during oil production, can block pore throats and reduce the number of flow paths available for the oil. Producing sand helps to eliminate many of these bottlenecks and the dilation of the sand also creates larger pore throats that are more difficult to block.

CHOPS is a heavy oil production process that involves the deliberate production of sand with the heavy oil and which is widely used as a production approach in unconsolidated sandstones. The main conditions for a successful CHOPS process are: (1) continuous sand failure due to the presence of unconsolidated sands; (2) active foamy oil mechanism due to the presence of gas in solution; (3) no free water zones in the reservoir; and (4) the use of PC pumps.

The process results in the development of high-permeability channels (*wormholes*) in the adjacent low cohesive strength sands, facilitating the flow of oil foam that is caused by solution gas drive. The key benefits of the process are improved reservoir access, order-of-magnitude higher oil production rates (as compared to primary recovery), and lower production costs. The outstanding technical issues involve sand handling problems, field development strategies, wormhole plugging for water shut-off, low ultimate recovery, and sand disposal. Originally, cold production mechanisms were thought to apply only to vertical wells with high-capacity pumps. It is now believed that these mechanisms may also apply to horizontal wells and lighter (heavy) oils (McLean and Kilpatrick, 1997; Zaki et al., 2002).

The CHOPS process involves the intentional coproduction of sand with oil, as it has become apparent that the exclusion of sand results in uneconomic production rates. Thus, instead of blocking sand ingress by screens or gravel packs, sand is encouraged to enter the wellbore by aggressive perforation and swabbing strategies. Vertical or slightly inclined wells (vertical to 45°) are operated with rotary progressive cavity pumps (rather than reciprocating pumps) and old fields are converting to higher-capacity progressive cavity pumps, giving production boosts to old wells. Productivity increases over conventional production and a CHOPS process can recover as much as 12% to perhaps as much as 25% of the OOIP, rather than the 0–5% typical of primary production without sand in such cases. Finally, because massive sand production creates a large disturbed zone, the reservoir may be positively affected for later implementation of thermal processes.

The CHOPS process increases productivity for the following reasons: (1) if the sand can move or is unconsolidated, the basic permeability to fluids is enhanced; (2) as more sand is produced, a growing zone of greater permeability is generated, similar to a large-radius well which gives better production; (3) gas coming out of solution in heavy oil does not generate a continuous gas phase; rather, bubbles flow with the fluid and do not coalesce, but expand down-gradient, generating an "internal" gas drive, referred to as *foamy flow*. This also helps to locally destabilize the sand, sustaining the process; (4) continuous sand production means that asphaltene or fines plugging of the near-wellbore environment potentially do not occur, so there is no possibility of an effect to impair productivity; and

(5) as sand is removed, the overburden weight acts to shear and destabilize the sand, helping to drive sand and oil toward the wellbore. Typically, a well placed on CHOPS production will initially produce a high percentage of sand, greater than 20% by volume of liquids. However, this generally drops after some weeks or months. The huge volumes of sand are disposed of by slurry fracture injection or salt cavern placement or by sand placement in a landfill in an environmentally acceptable manner.

CHOPS is used for thin subsurface oil sands (typically 1–7 m thick) in Canada, provided the oil sand is unconsolidated and provided the heavy oil contains sufficient solution gas to power the production process. To have any natural gas in solution, the oil sand must be at least a few hundred meters deep. For example, there are a large number of CHOPS wells located near Lloydminster, Alberta. In fact, at the time of writing, CHOPS is the only commercial method for exploiting thin oil sands. CHOPS wells (by definition) require sand production. Foamy oil production may occur without sand production in other areas, such as in the Faja belt, Venezuela. Alternatively, oil may be produced with sand, but without solution gas in still other areas.

CHOPS wells are vertical or slightly deviated wells; they are cased and perforated, and a downhole pump is deployed to create an aggressive pressure differential between formation and wellbore pressures. This causes natural gas to break out of solution from the heavy oil, resulting in "foamy oil." Gas bubbles evolving at the wormhole–sand interface destabilize sand grains and the expanding gas helps move the mixture through the wormholes. Gravity drive on the unconsolidated sands also provides energy for production. At the start-up of production, up to 10% sand by volume is produced along with oil, water, and gas. Sand production eventually falls to under 2% during the well lifetime.

The surface footprint for CHOPS wells is small, only requiring space for the wellhead, a storage tank, and a small project center. Any produced gas is used on site to power equipment or to heat the storage tank. Because a large volume of sand is produced, pipelines cannot be used for transportation. Instead, trucks are required to move oil, water, and sand for processing or disposal.

The main area for the use of CHOPS technology has been the Heavy Oil Belt that straddles the Alberta-Saskatchewan border, but there are also a number of other heavy oil reservoirs to which CHOPS may be applied. The major deposits suitable for the process are typically high porosity unconsolidated sandstone. In addition, the Alberta Carbonate Triangle, where carbonate reservoir rocks (limestones and dolomites) also contain heavy oil reservoirs, is certainly worthy of consideration for future sources of heavy oil.

A major breakthrough would be a secondary recovery method to tap the remaining (approximately 90%, v/v) of the OOIP. Two possible secondary recovery methods are in situ combustion and solvent flood. In addition, a primary production method that has a higher recovery factor would also have a significant impact. A high recovery factor and oil production without sand might replace trucking with pipelines, thus reducing carbon dioxide emissions as well as manpower costs, and allowing year-around production.

8.3.4 Pressure Pulse Technology

Pressure pulse technology (PPT) is a technology which can be used to enhance the recovery rate of nonaqueous phase liquid and to reduce solids clogging in wells, permeable reactive barriers, and fractured media (CRA, 2003). This technology is based on the discovery that large amplitude pressure pulses that are dominated by low-frequency wave energy generate enhanced flow rates in porous media. For example, in preliminary experiments in heavy oil reservoirs in Alberta, PPT has reduced the rate of depletion, increased the oil recovery ratio, and prolonged the life of wells.

The technology uses steady, nonseismic pulse vibrations (eg, 15 pulses per minute) that generate a low velocity wave effect to encourage flow of oils and small solid particles. It is effective in geologic formations exhibiting elastic properties, such as unconsolidated sediments and sedimentary rocks. It must be applied in a downhole manner in order to be effective. It has been used by the oil industry to improve oil recovery from otherwise exhausted reserves for many years. Also, it has been found that very large amplitude pressure pulses applied for 5–30 hours to a blocked producing well can reestablish economic production in a CHOPS well for many months, even years. Pulsing has been applied in injector wells for improving the efficiency of water flood patterns and has shown indications of increased oil production and decreased water cut. Additional potential applications include improving the effectiveness of matrix acidizing and diversion.

The mechanism by which PPT works is to generate a porosity dilation wave (a fluid displacement wave similar to a tidal wave); this generates pore-scale dilation and contraction so that oil and water flow into and out of pores, leading to periodic fluid accelerations in the pore throats. As the porosity dilation wave moves through the porous medium at a velocity of about 50–100 feet/seconds, the small expansion and contraction of the pores with the passage of each packet of wave energy helps unblock pore throats, increases the velocity of liquid flow, overcomes part of the effects of capillary blockage, and reduces some of the negative effects of instability due to viscous fingering, coning, and permeability streak channeling.

Although a very new concept dating only since 1999, PPT promises to be a major adjunct to a number of oil production processes, particularly all pressure-driven processes, where it will both accelerate flow rates and increase the oil recovery factor.

8.4 ENHANCED RECOVERY METHODS

Tertiary recovery (enhanced recovery) begins when secondary oil recovery is not enough to continue adequate production, but only when the oil can still be extracted profitably. This depends on the cost of the extraction method and the current price of crude oil. When prices are high, previously unprofitable wells are brought back into production and when they are low, production is curtailed. Tertiary recovery allows another 5–15% of the reservoir's oil to be recovered.

Tertiary oil recovery reduces the viscosity of the oil to increase oil production. TEOR methods are tertiary recovery techniques that heat the oil and make it easier to extract. Steam injection is the most common form of TEOR, and is often done with a cogeneration plant. In this type of cogeneration plant, a gas turbine is used to generate electricity and the waste heat is used to produce steam, which is then injected into the reservoir. In situ combustion is another form of TEOR, but instead of steam, some of the oil is burned to heat the surrounding oil. Occasionally, detergents are also used to decrease oil viscosity as a tertiary oil recovery method. Another method to reduce viscosity is carbon dioxide flooding.

However, there is a renewed interest in chemically enhanced oil recovery (CEOR; Figs. 8.1 and 8.2) because of diminished reserves and advances in surfactant and polymer technology. Greater understanding of the chemical reactions involved has led to good results in the field (Krumrine and Falcone, 1987). Combinations of chemicals may be applied as premixed slugs or in sequence.

The choice of the method and the expected recovery depends on many considerations, economic as well as technological. Some methods

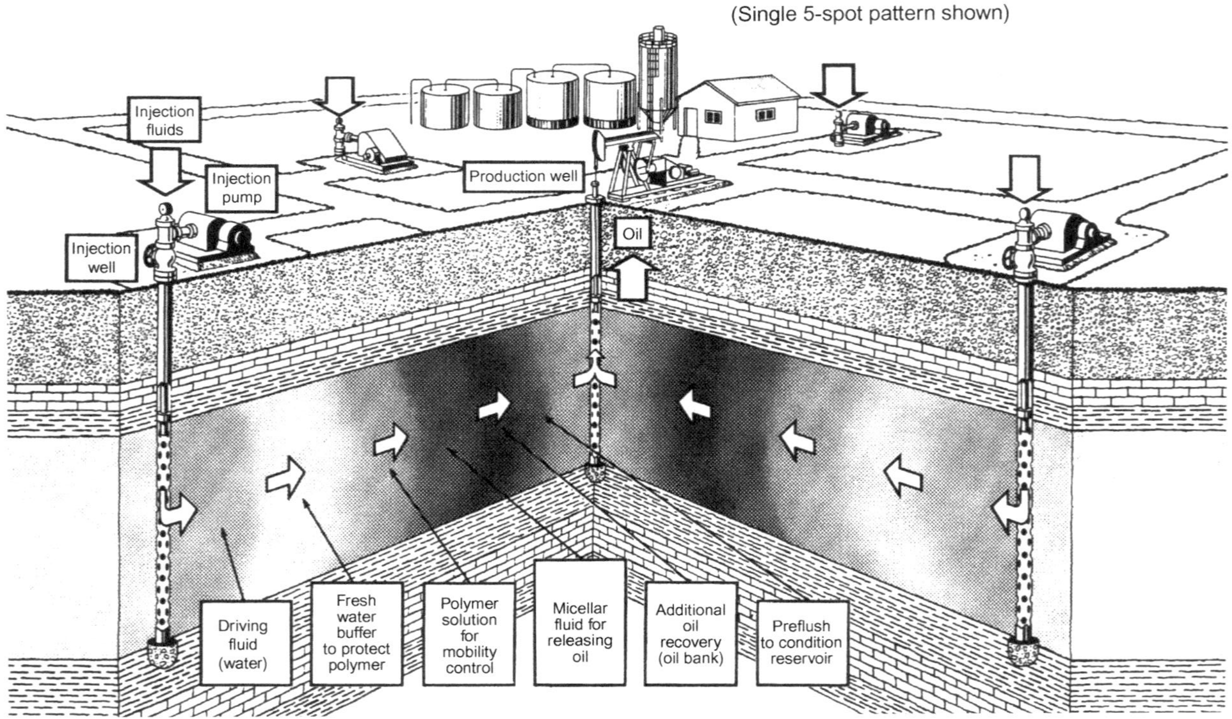

Figure 8.1 Schematic for chemical-enhanced recovery processes.

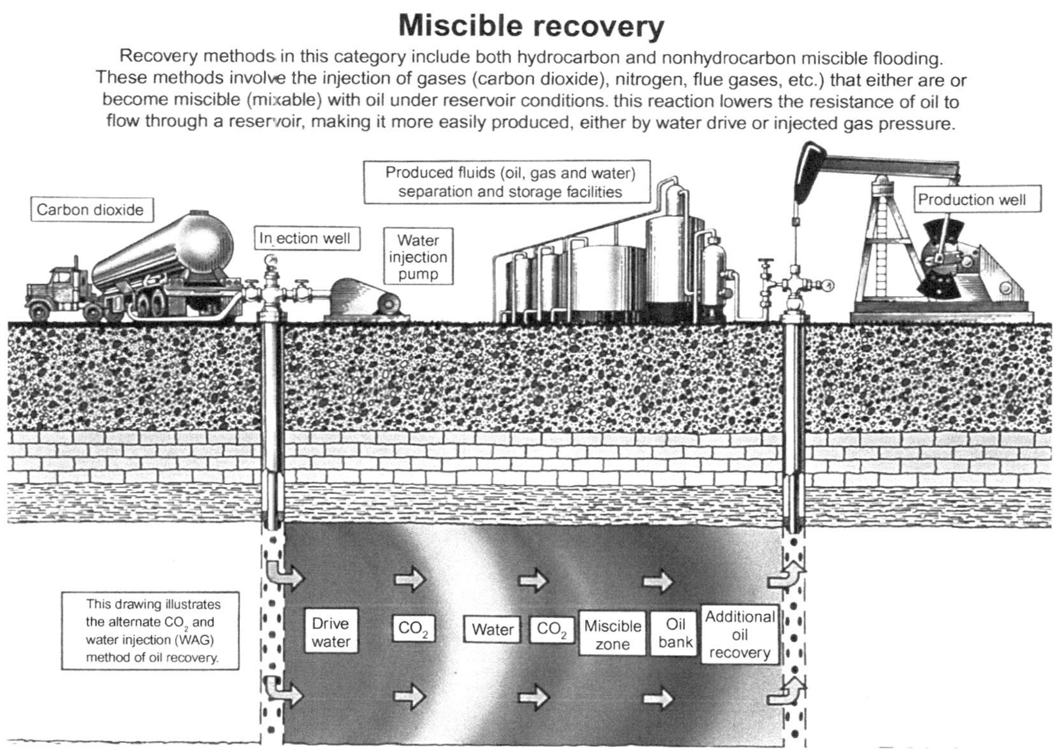

Figure 8.2 Schematic for miscible-enhanced recovery processes.

are commercially successful, while others remain largely of academic interest. Only a few recovery methods have been commercially successful, such as steam injection-based processes in heavy oils (if the reservoir offers favorable conditions for such applications) and miscible carbon dioxide for light oil reservoirs (Thomas, 2008). Methods for improving oil recovery, in particular those concerned with lowering the interstitial oil saturation, have received a great deal of attention both in the laboratory and in the field. From the vast amount of literature on the subject, one gets the impression that it is relatively simple to increase oil recovery beyond secondary (assuming that the reservoir lends itself to primary and secondary recovery) but this is not the case (Thomas, 2008).

8.4.1 Alkaline Flooding

Alkaline flooding (*caustic flooding*) is an enhanced oil recovery technique in which an alkaline chemical, such as sodium hydroxide, sodium orthosilicate, or sodium carbonate, is injected during water flooding or during polymer flooding operations. It is the intent of the very complex process to form a surfactant in situ, as a result of reactions between the injected alkali and the acid components of the crude oil. Also, some of the alkali may be consumed as a result of reactions with clay minerals.

Other mechanisms that may enhance recovery are changing rock from oil-wet to water-wet, which lower interfacial tension, and emulsification, which lowers viscosity. Alkalis such as sodium hydroxide or potassium hydroxide are used. However, these alkalis can react strongly with minerals in the connate water and with the reservoir rocks to the detriment of the process. In addition, alkaline flooding is not always suitable for application to carbonate reservoirs because of the abundance of calcium: the mixture between the alkaline chemical and the calcium ions can produce hydroxide precipitation that may damage the formation and have a detrimental effect on the process.

Alkaline flooding is one of several CEOR methods which are an advancement on conventional secondary water flooding operations. CEOR techniques produce lower amounts of greenhouse gases than TEOR approaches. A modification to the process is the addition of surfactant and polymer to the alkali, giving rise to an alkaline-surfactant polymer EOR method, essentially a less costly form of micellar-polymer flooding. The alkaline agents aid in the displacement of crude oil by raising the pH of the injected flood water. The alkali reacts with acidic components in crude oil forming an in situ surfactant at the oil–brine interface. This mixture then mobilizes the crude oil and removes it from the pore spaces in the reservoir.

In the process, the chemicals are added to water before injection into a reservoir to generate fluid properties that are more favorable for oil production. CEOR methods include polymer flooding, surfactant flooding, and alkaline flooding processes, the most widely applicable of which is surfactant flooding. Surfactant flooding consists of the injection of a suitable surfactant (which lowers the oil–water interfacial tension sufficiently, and adsorbs little on the rock matrix) to drive the oil. The reduction in interfacial tension and the resulting reduction in residual oil saturation are the expected effects. In situ formation of an emulsion would also help in improving the effective mobility ratio. The main problem with surfactants is loss to the rock matrix through several surfactant–rock interactions.

In terms of surfactant loss as a result of adsorption and reaction with minerals, such losses increase as the clay content increases. Gravity segregation of the surfactant is also a significant factor, given the slow injection rates and large areas involved in the field. Mixing of the surfactant with water, especially where the process is initiated after a water flood, can dilute it to a point where the surfactant is not effective.

Alkaline flooding has been used only in reservoirs containing specific types of high-acid crude oils. Although surfactant flooding is expensive, it has been used in a few large-scale projects and has been demonstrated to have excellent potential for improving the recovery of low-to-moderate viscosity oils. CEOR is commercially available under limited conditions, determined by reservoir characteristics, including depth, salinity, and pH. The high cost of chemicals and reservoir characterization studies need to be reduced, to allow expanded use of CEOR methods before full commercialization can take place.

The addition of silicates is an enhancement to alkaline flooding insofar as the silicates play two major functions: (1) as a buffer, maintaining a constant high pH level to produce a minimum interfacial tension and (2) by improving surfactant efficiency through the removal of hardness ions from reservoir brines, thus reducing adsorption of surfactants on rock surfaces.

8.4.2 Carbon Dioxide Flooding

Carbon dioxide (CO_2) flooding is a process whereby carbon dioxide is injected into an oil reservoir in order to increase output when extracting oil. Ideally, carbon dioxide flooding is particularly effective in reservoirs deeper than 2000 feet, where carbon dioxide will be in a supercritical state, with the oil gravity on the order of greater than 22°–25°. Carbon dioxide flooding is not affected by the lithology of the reservoir area but simply by the reservoir characteristics.

Carbon dioxide flooding is commonly used to recover oil from reservoirs in which the initial pressure has been depleted through primary production and possibly water flooding (Orr et al., 1982). Water is injected into the reservoir until pressure is restored to a desired level, and then CO_2 is introduced into the reservoir through these same injection wells. As the carbon dioxide is forced into the reservoir a zone of miscible carbon dioxide and light hydrocarbons forms a front that is soluble with the oil, making it easier to move toward production wells.

If a well has been produced before and has been designated suitable for carbon dioxide flooding, the first action is to restore the pressure within the reservoir to one suitable for production. This is done by injecting water (with the production well shut off) which will restore pressure within the reservoir to a suitable pressure for carbon dioxide flooding. Once the reservoir is at this pressure, the next step is to inject the carbon into the same injection wells used to restore pressure. The carbon dioxide gas is forced into the reservoir and is required to come into contact with the oil. This creates a miscible zone that can be moved more easily to the production well. The carbon dioxide injection may be alternated with more water injection and the water acts to sweep the oil towards the production zone.

The initial carbon dioxide slug is typically followed by alternate water and carbon dioxide injection—the water serving to improve sweep efficiency and to minimize the amount of carbon dioxide required for the flood. Production is from an oil bank that forms ahead of the miscible front. As reservoir fluids are produced through production wells, the carbon dioxide reverts to a gaseous state and provides a "gas lift" similar to that of original reservoir natural gas pressure.

Until recently, most of the carbon dioxide used for enhanced oil recovery has come from naturally occurring reservoirs. But new technologies are being developed to produce carbon dioxide from industrial applications, such as natural gas processing, fertilizer, ethanol, and hydrogen plants, in locations where naturally occurring reservoirs are not available. One demonstration at the Dakota Gasification Company's plant in Beulah, North Dakota is producing carbon dioxide and delivering it by a new 204-mile pipeline to the Weyburn oil field in Saskatchewan, Canada. EnCana, the field's operator, is injecting the carbon dioxide to extend the field's productive life, hoping to add another 25 years and as much as 130 million barrels of oil that might otherwise have been abandoned.

Additional work has examined potential improvements in carbon dioxide-enhanced oil recovery technologies beyond the state-of-the-art that can further increase this potential. This work evaluating the potential of improvements in oil recovery efficiency for carbon dioxide-enhanced oil recovery illustrates that the wide-scale implementation of next generation carbon dioxide-enhanced oil recovery technology advances have the potential to increase domestic oil recovery efficiency from about one-third to over 60%.

The presence of an oil-bearing transition zone beneath the traditionally defined base (oil–water contact) of an oil reservoir is well established. What is now clear, is that, under certain geologic and hydrodynamic conditions, an additional residual oil zone exists below this transition zone, and this resource could add another 100 billion barrels of oil resource in-place in the United States, and an estimated 20 billion barrels could be recoverable with state-of-the-art carbon dioxide-enhanced oil recovery technologies.

Large volumes of technically recoverable domestic oil resources remain undeveloped and are yet to be discovered in the United States, and this potential associated with carbon dioxide-enhanced oil recovery represents just a portion, albeit large, of this potential. Undeveloped domestic oil resources still in the ground (in-place) total 1124 billion barrels. Of this large in-place resource, 430 billon barrels are estimated to be technically recoverable. This resource includes undiscovered oil, "stranded" light oil amenable to carbon dioxide-enhanced oil recovery technologies, unconventional oil (deep heavy oil and tar sands), and new petroleum concepts (residual oil in reservoir transition zones).

8.4.3 Cyclic Carbon Dioxide Stimulation

Cyclic carbon dioxide stimulation, also known as the *huff-and-puff method*, is a single-well operation, which is developing as a method of rapidly producing oil. Similar to the cyclic steam process, carbon dioxide is injected into an oil reservoir after which the well is shut in for a time (providing for a *soak period*) and the well is then opened, allowing the oil and fluids to be produced.

The primary objective of miscible carbon dioxide-enhanced oil recovery is to remobilize and dramatically reduce the after water flooding residual oil saturation in the reservoir's pore space. The dissolving of the carbon dioxide in the oil reduces the oil's viscosity and causes it to swell, allowing the oil to flow more easily toward the well. The process can also be used in heavy oil reservoirs by high-pressure injection of carbon dioxide to facilitate miscibility between the oil and carbon dioxide, and in cases where thermal methods are not feasible.

The process is a single-well operation that is developing as a method of rapidly producing oil. It is similar in operation to the conventional cyclic or huff-and-puff steam injection process. In the process, carbon dioxide is injected into a well drilled into an oil reservoir, the well is shut-in for a time providing for a *soak period*, then is opened, allowing the oil and fluids to be produced. In this process the production of additional oil is accomplished by some or all of the following mechanisms:

1. Carbon dioxide dissolves in the oil, reducing its viscosity and allowing the oil to flow more easily toward a production well.
2. There is an increased oil-phase saturation due to carbon dioxide dissolving in the oil and causing it to swell.
3. Solution-gas drive is achieved by the evolution of carbon dioxide and natural gas from the oil phase at the lower pressures occurring during production.
4. The hydrocarbons are extracted by the supercritical carbon dioxide.

This process is also applicable to viscous (heavy) oil reservoirs that have a high oil saturation and temperatures or pressures that preclude miscibility between oil and carbon dioxide. The most important operating parameters are (1) the volume of and carbon dioxide injected per cycle; (2) the number of cycles; and (3) the degree of back pressure during production. This process can be repeated several times, but efficiency decreases with the number of cycles. Cyclic and carbon dioxide stimulation can be useful in recovering heavy oil in cases where thermal methods are not feasible.

Thus, miscible carbon dioxide-enhanced oil recovery is a multiple contact process, involving the injected carbon dioxide and the reservoir's oil. During this multiple contact process, carbon dioxide will vaporize the lighter oil fractions into the injected carbon dioxide phase and carbon dioxide will condense into the reservoir's oil phase. This leads to two reservoir fluids that become miscible (mixing in all parts), with favorable properties of low viscosity, a mobile fluid and low interfacial tension.

When insufficient reservoir pressure is available or the reservoir's oil composition is less favorable (heavier), the injected carbon dioxide will not become miscible with the reservoir's oil. As such, another oil displacement mechanism, immiscible carbon dioxide flooding, occurs. The main mechanisms involved in immiscible carbon dioxide flooding are: (1) oil phase swelling, as the oil becomes carbon dioxide saturated with carbon dioxide; (2) viscosity reduction of the swollen oil and carbon dioxide mixture; (3) extraction of lighter hydrocarbon into the carbon dioxide phase; and (4) fluid drive plus pressure. This combination of mechanisms enables a portion of the reservoir's remaining oil to be mobilized and produced. In general, immiscible carbon dioxide-enhanced oil recovery is less efficient than miscible carbon dioxide-enhanced oil recovery in recovering the oil remaining in the reservoir.

8.4.4 Nitrogen Flooding

Nitrogen flooding can be used to recover light oil that is capable of absorbing added gas under reservoir conditions, is low in methane, and at least 5000 feet deep to withstand the high injection pressure necessary for the oil to mix with the nitrogen without fracturing the producing formation. When nitrogen is injected into a reservoir, it forms a miscible front by vaporizing lighter oil components. As the front moves away from the injection wells its leading edge goes into solution, or becomes miscible, with the reservoir oil. Continued injection moves the bank of displaced oil toward production wells. Water slugs are injected alternately with the nitrogen to increase the sweep efficiency and oil recovery. Nitrogen can be manufactured on site at relatively low cost by extraction from air by cryogenic separation, and being totally inert it is noncorrosive.

In *nitrogen-carbon dioxide flooding*, because of its lower cost, the nitrogen can be used in a carbon dioxide flood to displace the carbon dioxide slug and the heavy oil bank.

8.4.5 Polymer Flooding

Polymer flooding is an enhanced oil recovery method that uses polymer solutions to increase oil recovery by decreasing the water– oil mobility ratio by increasing the viscosity of the displacing water. The process involves the addition of a very small amount of a water-soluble, high molecular weight polymer to the displacing water in order to increase the apparent viscosity of water, and thus lower the mobility ratio. However, because the viscosity of heavy oils is high, there is a limit to the polymer concentration that can be used, both for reasons of cost and mechanical considerations of injection pressure.

Polymer flooding is used under certain reservoir conditions that lower the efficiency of a regular water flood, such as fractures or high-permeability regions that channel or redirect the flow of injected water, or heavy oil that is resistant to flow. Adding a water-soluble polymer to the water flood allows the water to move through more of the reservoir rock, resulting in a larger percentage of oil recovery. Polymer gel is also used to shut off high-permeability zones. In the process, the volumetric sweep is improved and the oil is more effectively produced. Often injectivity will be one of the critical factors. The polymer solution should therefore be a non-Newtonian and shear thinning fluid, that is, the viscosity decreases with increasing shear rate.

Polymer flooding is often an accompaniment to water flooding and is the most commonly used chemical enhancement process since it is easy to apply and requires relatively small investment. Adding high molecular weight polymers increases the viscosity of water and, with some polymers, reduces the aqueous-phase permeability without changing the relative permeability to oil, which can significantly improve water flood volumetric sweep efficiency. Cross-linking, or gelling, polymers in situ with metallic ions can augment performance in a sweep profile control, helping to plug high conductivity zones or minor fractures that degrade sweep efficiency. The main drawbacks to the use of polymers are their high cost and the low injection rate caused by high viscosity (which impacts economic rate of return), degradation at higher temperatures, intolerance to high salinity, polymer deterioration from shear stress imparted by pumping, flow through tubulars and perforations, and long-term instability in the reservoir environment.

In the process, a water-soluble polymer is added into the flood water. This increases the viscosity of water. There are three potential ways in which a polymer flooding makes the oil recovery process more efficient: (1) through the effects of polymers on fractional flow; (2) by decreasing the water–oil mobility ratio; and (3) by diverting injected water from zones that have been swept. The most important preconditions for polymer flooding are reservoir temperature and the chemical properties of reservoir water. At high temperature or with high salinity in reservoir water, the polymer cannot be kept stable and the polymer will lose most of its viscosity.

Polymer flooding is usually not recommended for heavy oils and the more viscous oils (Taber et al., 1997). Nevertheless, with the need for increasing sustainable oil supplies, polymer flooding tends to be implemented for heavy oils with a much higher viscosity than the limited one in screen criteria (Wassmuth et al., 2007; Zhang et al., 2010). Polymer flooding can be used to enhance the sweep efficiency by way of three mechanisms: (1) increasing water viscosity; (2) decreasing water permeability through swept areas; and (3) contacting a large volume of the reservoir. The addition of nanoparticles can improve the pseudo-plasticity behavior of polymer solutions with low shear rate values. Nanosilica minerals are frequently used in improved processes with aqueous solution. In fact, silica nanoparticles and other types of nanoparticles (clay minerals and titanium oxide nanoparticles) have shown great potential for enhanced oil recovery of both light and heavy oil (Cheraghian et al., 2014; Cheraghian and Khalili Nezhad, 2015).

8.4.6 Micellar-Polymer Flooding

Micellar-polymer flooding (also known as detergent flooding or microemulsion flooding) uses low concentrations of surfactants in water to reduce the interfacial tension between oil and water. A surfactant flood must be designed for a specific crude oil in a specific reservoir taking into account such factors as salinity, temperature, pressure, and clay content. Generally, multiple slugs are used and the alkalinity range is subject to adsorption and retention through ionic exchange with reservoir rocks.

The process uses the injection of a micellar slug into a reservoir. The micellar slug contains a mixture of a surfactant, cosurfactant, alcohol, brine, and oil that moves through the oil-bearing formation, and which acts to release oil from the pores of the reservoir rock much as a dishwashing detergent releases grease from dishes so that it can be flushed away by flowing water. As the micellar solution moves through the oil-bearing formation in the reservoir, it releases much of the oil trapped in the rock. To further enhance production, polymer-thickened water for mobility control (as described in the polymer flooding process) is injected behind the micellar slug. Here again, a buffer of fresh water is normally injected following the polymer and ahead of the drive water to prevent contamination of the chemical solutions. This method has one of the highest recovery efficiencies of the current enhanced oil recovery methods but it is also one of the most costly to implement.

8.4.7 Solvent-Based Processes

Solvent-based methods have been developed to move heavy oil. A diluent such as naphtha or light oil may be injected near the pump to reduce the viscosity of the heavy oil and allow it to be more easily pumped. Alternatively, diluent may be added at the surface to facilitate pipeline transport.

VAPEX is a nonthermal solvent-based relatively cold (40°C), low-pressure process in which two parallel horizontal wells are drilled with about a 15-feet vertical separation (Yazdani and Maini, 2008). The VAPEX technology is similar to the SAGD technology but instead of steam, solvent is injected into the oil sands resulting in significant viscosity reduction. The injection of vaporized solvents, such as ethane or propane, helps to create a vapor-chamber through which the oil flows due to gravity drainage. The process can be applied in paired horizontal wells, single horizontal wells, or a combination of vertical and horizontal wells. The key benefits are significantly lower energy costs, potential for in situ upgrading, and application to thin reservoirs (Butler and Mokrys, 1991, 1997; Butler and Jiang, 2000). The vapor travels to the oil face where it condenses into a liquid and the solvent mixed with the oil flows to the lower well (gravity drainage) and is pumped to the surface.

The physics of the VAPEX process are essentially the same as for the SAGD process (see chapter: "Nonthermal Methods of Recovery") and the configuration of wells is generally similar. However, the selection of well spacing in the VAPEX process is as a function of oil mobility at reservoir conditions. For reservoirs where the heavy oil or bitumen mobility is low, the well spacing should be small. The lateral spacing between injector and producer has a negligible effect on the ultimate oil recovery and just causes a long delay in the start of well communication. Moreover, the length of horizontal well is important. However, long horizontal wells lead to an economical production rate because more solvent can spread into porous media and affect a bigger area of untouched heavy oil.

The key benefits are claimed to be (1) significantly lower energy costs; (2) the potential for in situ upgrading; and (3) application to thin reservoirs, with bottom water or reactive mineralogy. In addition to concern over the cost of the solvent there are the usual concerns about the interaction of the solvent with the reservoir minerals—clay is known to adsorb organic solvents very strongly—and the integrity of the reservoir formations and associated strata—a minor fault can cause loss of the solvent as well as environmental havoc. There is also the concern over the deposition of asphaltic material and its effect on reservoir permeability. While the function of the solvents might be to extract soluble components of heavy oil, initial contact between the solvent and the oil at a low solvent-to-oil ratio will cause solubilization of the asphaltic constituents (Mitchell and Speight, 1973) with later deposition of these constituents as the solvent-to-oil ratio increases in the later stages of the process.

Because of the slow diffusion of gases and liquids into viscous oils, this approach, used alone, perhaps will be suited only for less viscous oils, although preliminary tests indicate that there are micromechanisms that act so that the VAPEX dilution process is not diffusion rate limited and the process may be suitable for the highly viscous heavy oil and tar sand bitumen (Yang and Gu, 2005a, 2005b). Nevertheless, VAPEX can undoubtedly be used in conjunction with SAGD methods. As with SAGD and inert gas injection (IGI), a key factor is the generation of a three-phase system with a continuous gas phase so that as much of the oil as possible can be contacted by the gaseous phases, generating the thin oil film drainage mechanism. As with IGI, vertical permeability barriers are a problem, and must be overcome through hydraulic fracturing to create vertical permeable channels, or undercut by a the lateral growth of the chamber beyond the lateral extent of the limited barrier, or baffle.

Another solvent process, the N-Solv process (N-Solv, 2015), uses warm solvent vapor to extract bitumen in situ (N-Solv, 2015). In the process, a warm solvent vapor is pumped into the injection well of a tar sand deposit and, after reaching the extraction chamber, condenses and dissolves the bitumen, draining downward to the production well as a solution of both solvent and oil. Naturally occurring water and noncondensable gases are also carried along and out of the reservoir. Coke-forming asphaltene constituents remain in the deposit. At the surface, the partially upgraded oil mixed with solvent is separated into three components: deasphalted oil, solvent, and naturally occurring water. The deasphalted oil is sent to refineries for further processing and the solvent is pumped back into the separator and repurified before being recycled back into the reservoir. It is estimated that a solvent quantity equivalent to 20–30% (v/v) of the extracted deasphalted oil must be added to the system as make-up. As the chamber grows, the make-up solvent is continually replenished and at the end of the life of a well, the solvent can be recovered and reused on another well or marketed.

8.4.8 Microbial-Enhanced Oil Recovery

Microbial-enhanced oil recovery (MEOR) processes involve the use of reservoir microorganisms or specially selected natural bacterial to produce specific metabolic events that lead to enhanced oil recovery (Tables 8.2 and 8.3). The processes that facilitate oil production are complex and may involve multiple biochemical processes. Microbial biomass or biopolymers may plug high-permeability zones and lead to a redirection of the water

Table 8.2 Types of Microbial Processes for Oil Recovery

Process	Production Problem	Types of Activity or Product Needed
Wellbore clean-up (improve oil drainage into wellbore)	Paraffin and scale deposits	Emulsifiers, biosurfactants, solvents, acids, hydrocarbon degradation
Well stimulation (stimulate release of oil entrapped by capillaries and brine)	Formation damage, pore damage	Gas, acids, solvents, biosurfactants
	High water production	Biomass and polymer production
Enhanced water flood (reduce permeability variation and block water channels)	Poor displacement efficiency	Biosurfactants, solvents, polymers
	Poor sweep efficiency	Biomass and polymer production
	Scouring	Nitrate reduction

flood, produce surfactants which lead to increased mobilization of residual oil, increase gas pressure by the production of carbon dioxide, or reduce the oil viscosity due to digestion of large molecules.

In MEOR processes, microbial technology is exploited in oil reservoirs to improve recovery (Banat, 1995; Clark et al., 1981; Stosur, 1991). From a scientific perspective, MEOR processes are somewhat akin to in situ bioremediation processes (see chapter: "Microbial Enhanced Oil Recovery") (Speight and Arjoon, 2012). Injected nutrients, together with indigenous or added microbes, promote in situ microbial growth and/or generation of products which mobilize additional oil and move it to producing wells through reservoir repressurization, interfacial tension/oil viscosity reduction, and selective plugging of the most permeable zones (Bryant et al., 1989; Bryant and Lindsey, 1996). Alternatively, the oil-mobilizing microbial products may be produced by fermentation and injected into the reservoir.

Typically, nutrients, such as sugars, nitrates, or phosphates, are regularly injected to stimulate the growth of the microbes, which are indigenous to some reservoirs, and aid their performance. The microbes then generate surfactants and carbon dioxide that help to displace the oil in a similar way to other displacement methods. Since growth occurs at exponential rates, the process quickly generates considerable surfactant in a cost-effective manner. Studies have shown that several microbially produced biosurfactants compare favorably with chemically synthesized surfactants.

Table 8.3 Microbial Products and Their Contribution to Enhanced Oil Recovery

Microbial	Effect
Acids	Modification of reservoir rock
	Improvement of porosity and permeability
	Reaction with calcareous carbon dioxide production
Biomass	Selective or nonselective plugging
	Emulsification through adherence to hydrocarbons
	Modification of solid surfaces, eg, wetting
	Degradation and alteration of oil
	Reduction of oil viscosity and oil pour point
	Desulfurization of oil
Gases (CO_2, $CH_4{\cdot}H_2$)	Reservoir repressurization
	Oil swelling
	Viscosity reduction
	Increase of permeability due to solubilization of carbonate rocks by CO_2
Solvents	Oil dissolution
Surface-active agent	Lowering interfacial tension
Polymers	Mobility control
	Selective or nonselective plugging

This technology requires consideration of the physicochemical properties of the reservoir in terms of salinity, pH, temperature, pressure, and nutrient availability (Khire and Khan, 1994a, 1994b). Only bacteria are considered promising candidates for MEOR. Molds, yeasts, algae, and protozoa are not suitable due to their size or inability to grow under the conditions present in reservoirs. Many petroleum reservoirs have high concentrations of sodium chloride (Jenneman, 1989) and require the use of bacteria which can tolerate these conditions (Shennan and Levi, 1987). Bacteria producing biosurfactants and polymers can grow at sodium concentrations up to 8% and selectively plug sandstone to create a biowall to recover additional oil (Raiders et al., 1989).

Organisms that participate in oil recovery produce a variety of fermentation products, for example, carbon dioxide, methane, hydrogen, biosurfactants, and polysaccharides from crude oil, pure hydrocarbons, and a variety of nonhydrocarbon substrates. Organic acids produced through fermentation readily dissolve carbonates and can greatly enhance permeability in limestone reservoirs, and attempts have been made to promote anaerobic production.

The MEOR process may modify the immediate reservoir environment in a number of ways that could also damage the production hardware or the formation itself. Certain sulfate reducers can produce H_2S, which can corrode pipeline and other components of the recovery equipment. Thus, despite numerous MEOR tests, considerable uncertainty remains regarding process performance. Ensuring success requires an ability to manipulate environmental conditions to promote growth and/or product formation by the participating microorganisms. Exerting such control over the microbial system in the subsurface is itself a serious challenge. In addition, conditions vary from reservoir to reservoir, which calls for reservoir-specific customization of the MEOR process, and this alone has the potential to undermine microbial process economic viability.

MEOR differs from CEOR in the method by which the enhancing products are introduced into the reservoir. *Cyclic microbial recovery* requires the injection of a solution of microorganisms and nutrients down a well into an oil reservoir. This injection can usually be performed in a matter of hours, depending on the depth and permeability of the oil-bearing formation. Once injection is accomplished, the injection well is shut in for days to weeks. During this time, known as an incubation or soak period, the microorganisms feed on the nutrients provided and multiply in number. These microorganisms produce substances metabolically that affect the oil in-place in ways that facilitate its flow, making it easier to produce. Depending on the microorganisms used, these products may be acids, surfactants, and certain gases, most notably hydrogen and carbon dioxide. At the end of this period, the well is opened, and the oil and products resulting from this process are produced. The method eliminates the need for continual injection, but after the production phase is completed a new supply of microorganisms and nutrients must be injected if the process is to be repeated.

Oil recovery by *microbial flooding* also involves the use of microorganisms but in this case, the reservoir is usually conditioned by a water flush after which a solution of microorganisms and nutrients is injected into the formation. As this solution is pushed through the reservoir by water drive, gases and surfactants are formed, the oil is mobilized, and pumped through the well. However, even though microbes produce the necessary chemical reactions in situ, whereas surface injected chemicals may tend to follow areas of higher permeability, resulting in decreased sweep efficiency, there is a need for caution and astute observation of the effects of the microorganisms on the reservoir chemistry.

The mechanisms by which MEOR processes work can be quite complex and may involve multiple biochemical processes. In selective plugging approaches, microbial cell mass or biopolymers plug high-permeability zones and lead to a redirection of the water flood. In other processes, biosurfactants are produced in situ which leads to increased mobilization of residual oil. In still other processes, microbial production of carbon dioxide and organic solvents reduces the oil viscosity as the primary mechanism for enhanced oil recovery.

In a MEOR process, conditions for microbial metabolism are supported via injection of nutrients. In some processes, this involves injecting a fermentable carbohydrate into the reservoir. Some reservoirs also require inorganic nutrients as substrates for cellular growth or for serving as alternative electron acceptors in place of oxygen or carbohydrates. The stimulation of oil production by in situ bacterial fermentation can proceed by one or a combination of the following mechanisms: (1) improvement of the relative mobility of oil to water by biosurfactants and biopolymers; (2) partial repressurization of the reservoir by methane and carbon dioxide; (3) reduction of oil viscosity through the dissolution of organic solvents in the oil phase; (4) increase of reservoir permeability and widening of the fissures and channels through the etching of carbonaceous rocks in limestone reservoirs by organic acids produced by anaerobic bacteria; (5) cleaning the wellbore region through the acids and gas from *in* situ fermentation in which the gas pushes oil from dead space and dislodge debris that plugs the pores—the average pore size is increased and, as a result, the capillary pressure near the wellbore is made more favorable for the flow of oil; and (6) selective plugging of highly permeable zones by injecting slime-forming bacteria followed by sucrose solution which initiates the production of extracellular slimes and aerial sweep efficiency is improved.

The target for enhanced oil recovery processes is the quantity of unrecoverable oil in known reservoirs and bitumen in known deposits. One of the major attributes of MEOR technologies is its low cost but there must be the recognition that MEOR is a single process. Furthermore, reports on the deleterious activities of microorganisms in the oil field contribute to the skepticism of employing technologies using microorganisms. It is also clear that scientific knowledge of the fundamentals of microbiology must be coupled with an understanding of the geological and engineering aspects of oil production in order to develop MEOR technology.

However, MEOR has the advantage that microbes do not consume large amounts of energy and they are independent of the price of crude oil, compared to other processes. Also, with increasing subsurface depth, temperature appears to be the principal factor limiting microbial life, besides availability of suitable nutrients. They are also susceptible to salinity which limits the use of microbes. In addition, the microbes also have the potential to assist in partial upgrading of the heavy oil or bitumen (Speight and Arjoon, 2012; El-Gendy and Speight, 2015). This, along with recent developments in *upgrading* of heavy oil and bitumen (Speight, 2014), indicates that the near future could see a reduction of the differential cost of upgrading heavy oil. These processes are based on a better understanding of the issues of asphaltene solubility effects at high temperatures, incorporation of a catalyst that is chemically precipitated internally during the upgrading, and improving hydrogen addition or carbon rejection.

8.5 TAR SAND MINING

Tar sands are sedimentary rocks containing bitumen, a heavy hydrocarbonaceous compound. Tar sands deposits may be divided into two major types (1) a breached reservoir where erosion has removed the capping layers from a reservoir of relatively heavy petroleum, allowing the more volatile petroleum hydrocarbons to escape; and (2) a deposit formed when liquid petroleum has seeped into a near-surface reservoir from which the more volatile petroleum hydrocarbons have escaped. In either type of deposit, the lighter, more volatile hydrocarbons have escaped to the environment, leaving the heavier, less volatile constituents in place. The material left in place is altered by contact with air, bacteria, and groundwater. Because of the very viscous nature of the bitumen in tar sands, tar sands cannot be processed by normal petroleum production techniques. Also, tar sands deposits are not uniform. Differences in the permeability and porosity of the reservoir rock and varying degrees of alteration by contact with air, bacteria, and groundwater mean that there is a large degree of uncertainty in the estimates of the bitumen content of a given tar sands deposit.

In the tar sand mining process, the mined tar sand is transported to a processing plant where the bitumen is extracted (separated) and disposal of the waste sand. Such a procedure is often referred to as *oil mining*. This is the term applied to the surface or subsurface excavation of petroleum-bearing formations for subsequent removal of the heavy oil or bitumen by washing, flotation, or retorting treatments. Oil mining also includes recovery of heavy oil by drainage from reservoir beds to mine shafts or other openings driven into the rock, or by drainage from the reservoir rock into mine openings driven outside the tar sand but connected with it by bore holes or mine wells.

Oil mining is the term applied to the surface or subsurface excavation of petroleum-bearing formations for subsequent removal of the heavy oil or bitumen by washing, flotation, or retorting treatments. Oil mining also includes recovery of heavy oil by drainage from reservoir beds to mine shafts or other openings driven into the rock, or by drainage from the reservoir rock into mine openings driven outside the tar sand but connected with it by bore holes or mine wells. The method is not new and mining of petroleum and bitumen occurred in the Sinai Peninsula, the Euphrates valley, and in Persia prior to 5000 BC. In addition, subsurface oil mining was used in the Pechelbronn oil field in Alsace, France, as early as 1735. This early mining involved the sinking of shafts to the reservoir rock, only 100–200 feet (30–60 m) below the surface and the excavation of the tar sand in short drifts driven from the shafts. These tar sands were hoisted to the surface and washed with boiling water to release the bitumen. The drifts were extended as far as natural ventilation permitted. When these limits were reached, the pillars were removed and the openings filled with waste. This type of mining continued at Pechelbronn until 1866, when it was found that oil could be recovered from deeper, and more prolific, sands by letting it drain in place through mine openings with no removal of sand to the surface for treatment. Nevertheless, mining for petroleum is a new challenge facing the petroleum industry.

The oil mining method of recovery has received considerable attention since it was chosen as the technique of preference for the only two commercial bitumen recovery plants in operation in North America. In situ processes have been tested many times in the United States, Canada, and other parts of the world and are ready for commercialization. There are also conceptual schemes that are a combination of both mining (aboveground recovery) and in situ (nonmining recovery) methods. Oil mining methods should be applied in reservoirs that have significant residual oil saturation and have reservoir or fluid properties that make production by conventional methods inefficient or impossible. The high well density in improved oil mining usually compensates for the inefficient production caused by reservoir heterogeneity. However, close well spacing can also magnify the deleterious effects of reservoir heterogeneity. If a high-permeability streak exists with a lateral extent that is less than the interwell spacing of conventional wells but is comparable to that of improved oil mining, the channeling is more unfavorable for the improved oil mining method.

Engineering a successful oil mining project must address a number of items because there must be sufficient recoverable resources, the project must be conducted safely, and the project should be engineered to maximize recovery within economic limits. The use of a reliable screening technique is necessary to locate viable candidates. Once the candidate is defined, this should be followed by an exhaustive literature search covering the local geology, drilling, production, completion, and secondary and tertiary recovery operations.

The reservoir properties, which can affect the efficiency of heavy oil or bitumen production by mining technology, can be grouped into three classes: (1) *primary properties*, that is, those properties that have an influence on the fluid flow and fluid storage properties and include rock and fluid properties, such as porosity, permeability, wettability, crude oil viscosity, and pour point; (2) *secondary properties*, that is, those properties that significantly influence the primary properties, including pore size distribution, clay type, and content; and (3) *tertiary properties*, that is, those other properties that mainly influence oil production operation (fracture breakdown pressure, hardness, and thermal properties) and the mining operations (eg, temperature, subsidence potential, and fault distribution).

There are also important rock mechanical parameters of the formation in which a tunnel is to be mined and from where all oil mining operations will be conducted. These properties are mostly related to the mining aspects of the operations, and not all are of equal importance in their influence on the mining technology. Their relative importance also depends on the individual reservoir. Many of the candidate reservoirs for application of improved oil mining are those with high oil saturation resulting from the adverse effects of reservoir heterogeneity. Faulting, fracturing, and barriers to fluid flow are features that cause production of shallow reservoirs by conventional methods to be inefficient. Production of heterogeneous reservoirs by underground oil production methods requires consideration of the manner in which fractures alter the flow of fluids.

In a highly fractured formation with low matrix permeability, the fluid conductivity of the fracture system may be many times that of the matrix rock. In a highly fractured reservoir with low matrix permeability and reasonably high porosity, the fracture system provides the highest permeability to the flow of oil but the matrix rock contains the greater volume of the oil in place. The rate of the flow of oil from the matrix rock into the fracture system, the extent and continuity of the fracture system, and the degree to which the production wells effectively intersect the fracture system determine the production rate. Special consideration must be given to these factors in predicting production rates in fractured reservoirs. Under favorable circumstances, higher production rates may be achieved in fractured reservoirs by improving mining methods than in less heterogeneous reservoirs. Other reservoirs that are good candidates for oil mining are those that are shallow but have high oil saturation, have a nearby formation that is competent enough to support the mine, and cannot be efficiently produced by conventional methods.

8.5.1 Mining Operations

To utilize a tar sands resource in a mining operation, the bitumen must be recovered from its natural setting, extracted from the inorganic matrix (largely sand and silt) in which it occurs, and upgraded to produce a synthetic crude oil suitable as a feedstock for a conventional refinery. In general, it takes about 2.0 tonnes (2.2 tons) of surface-mined Athabasca tar sands to produce 159 L or 1 barrel (42 gal) of synthetic crude oil. Nonmining operations recover the bitumen already free of the matrix (sand and clay) in which it originally occurred. Preparation may require removal of bitumen or vaporized bitumen from steam, other gases, water, or solvents. Depending on the end product required, upgrading may not be required.

Surface mining is the mining method that is currently being used by Suncor Energy and Syncrude Canada Limited to recover oil sand from the ground. Surface mining can be used in mineable oil sand areas which lie under 75 m (250 feet) or less of overburden material. Only 7% of the Athabasca Oil Sands deposit can be mined using the surface mining technique, as the other 93% of the deposit has more than 75 m of overburden. This other 93% will have to be mined using different mining techniques.

The first step in surface mining is the removal of muskeg and overburden. Muskeg is a water-soaked area of decaying plant material that is 1–3 m thick and lies on top of the overburden material. Before the muskeg can be removed it must be drained of its water content. The process can take up to 3 years to complete. Once the muskeg has been drained and removed, the overburden must also be removed. Overburden is a layer of clay, sand, and silt that lies directly above the oil sands deposit. Overburden is used to build dams and dykes around the mine and will eventually be used for land reclamation projects. When all of the overburden is removed, the oil sand is exposed and is ready to be mined.

There are two methods of mining currently in use in the Athabasca Oil Sands. Suncor Energy uses the truck-and-shovel method of mining whereas Syncrude uses the truck-and-shovel method of mining, as well as draglines and bucket-wheel reclaimers. These enormous draglines and bucket-wheels are being phased out and soon will be completely replaced with large trucks and shovels. The shovel scoops up the oil sand and dumps it into a heavy hauler truck. The heavy hauler truck takes the oil sand to a conveyor belt that transports the oil sand from the mine to the extraction plant. Presently, there are extensive conveyor belt systems that transport the mined oil sand from the recovery site to the extraction plant. With the development of new technologies these conveyors are being phased out and replaced with hydrotransport technology. Hydrotransport is a combination of ore transport and preliminary extraction. After the bituminous sands have been recovered using the truck-and-shovel method, it is mixed with water and caustic soda to form a slurry and is pumped along a pipeline to the extraction plant. The extraction process thus begins with the mixing of the water and agitation needed to initiate bitumen separation from the sand and clay.

Mine spoils need to be disposed of in a manner that assures physical stabilization. This means appropriate slope stability for the pile against not only gravity but also earthquake forces. Since return of the spoils to the mine excavations is seldom economical, the spoil pile must be designed as a permanent structure whose outline blends into the landscape. Straight, even lines in the pile must be avoided.

Even though estimates of the recoverable oil from the Athabasca deposits are only of the order of 27×109 bbl of synthetic crude oil (representing <10% of the total in-place material), this is, for the Canadian scenario, approximately six times the estimated volume of recoverable conventional crude oil. In addition, the comparative infancy of the development of the alternative options almost ensured the adoption of the mining option for the first two (and even later) commercial ventures.

Underground mining options have also been proposed but for the moment have not been developed because of the fear of collapse of the formation onto any operation/equipment. This particular option should not, however, be rejected out-of-hand because a novel aspect or the requirements of the developer (which remove the accompanying dangers) may make such an option acceptable. Currently, bitumen is recovered commercially from tar and deposits by a mining technique. This produces tar sand that is sent to the processing plant for separation of the bitumen from the sand prior to upgrading.

The oil mining method of recovery has received considerable attention since it was chosen as the technique of preference for the only two commercial bitumen recovery plants in operation in North America. In situ processes have been tested many times in the United States, Canada, and other parts of the world and are ready for commercialization. There are also conceptual schemes that are a combination of both mining (above-ground recovery) and in situ (nonmining recovery) methods.

The bitumen occurring in tar sand (oil sand) deposits poses a major recovery problem. The material is notoriously immobile at formation temperatures and must therefore require some stimulation (usually by thermal means) in order to ensure recovery. Alternately, proposals have been noted which advocate bitumen recovery by solvent flooding or by the use of emulsifiers. There is no doubt that with time one or more of these functions may come to fruition, but for the present the two commercial operations rely on the mining technique.

The main challenges for the surface mining process are (1) to minimize the environmental impact; (2) to reclaim the land; and (3) to restore the vegetation, especially restoration of any forested areas.

Typically, for every cubic yard of synthetic crude oil produced at the plant, six cubic yards of sand and 1.5 cubic yards of fine tailings must be transported. Stockpiled overburden, sand, and tailing ponds can occupy a significant area of the mine, and may have to be moved to access oil sands beneath them. In addition, tar sand bitumen (as processed in Alberta) contains approximately 4.5–5% (w/w) sulfur, which is removed in the upgrading process and which, when the market cannot absorb the sulfur, is stockpiled in large mounds on site. Furthermore, large quantities of water have to be captured during spring flood and stored in reservoirs for use during the low water months of the year. Also, large quantities of high-sulfur and high-metals coke produced during the upgrading process must be stored (it is often not suitable for use as fuel for the processing plant) and eventually disposed.

8.5.2 Bitumen Recovery

The *hot-water process* (Clark, 1944) is, to date, the only successful commercial process to be applied to bitumen recovery from mined tar sands in North America (Clark, 1944; Carrigy, 1963a, 1963b; Fear and Innes, 1967; Speight and Moschopedis, 1978). Many process options have been tested with varying degrees of success and one of these options may even supersede the hot-water process.

The process utilizes the linear and the nonlinear variation of bitumen density and water density, respectively, with temperature so that the bitumen that is heavier than water at room temperature becomes lighter than water at 80°C (180°F). Surface-active materials in the tar sand also contribute to the process. The essentials of the hot-water process involve conditioning, separation, and scavenging.

Thus, in the hot water extraction process, the tar-sand feed is introduced into a *conditioning* drum. In this step the oil sand is heated, mixed with water, and agglomeration of the oil particles begins. The conditioning is carried out in a slowly rotating drum that contains a steam-sparging system for temperature control as well as mixing devices to assist in lump size reduction and a size ejector at the outlet end. The oil sand lumps are reduced in size by ablation and mixing action. The conditioned *pulp* has the following characteristics: (1) solids 60–85% and (2) pH 7.5–8.5.

In the conditioning step, also referred to as mixing or pulping, tar-sand feed is heated and mixed with water to form a pulp of 60% by weight to 85% by weight solids at 80–90°C (175–196°F). First the lumps of tar sand as-mined are reduced in size by ablation, that is, successive layers of lump are warmed and sloughed off revealing cooler layers. The conditioned pulp is screened through a double-layer vibrating screen. Water is then added to the screened material (to achieve more beneficial pumping conditions) and the pulp enters the separation cell through a central feed well and distributor. The bulk of the sand settles in the cell and is removed from the bottom as tailing, but the majority of the bitumen floats to the surface and is removed as froth. A middlings stream (mostly of water with suspended fines and some bitumen) is withdrawn from approximately midway up the side of the cell wall. Part of the middlings is recycled to dilute the conditioning drum effluent for pumping. Clays do not settle readily and generally accumulate in the middlings layer.

The combined froth from the separation cell and scavenging operation contains an average of about 10% by weight mineral material and up to 40% by weight water. The dewatering and demineralizing is accomplished in two stages of centrifuging; in the first stage the coarser mineral material is removed but much of the water remains. The feed then passes through a filter to remove any additional large-size mineral matter that would plug up the nozzles of the second stage centrifuges.

Bituminous froth from the hot-water process may be mixed with a hydrocarbon diluent, for example, coker naphtha, and centrifuged. The Suncor process employs a two-stage centrifuging operation and each stage consists of multiple centrifuges of conventional design installed in parallel. The bitumen product contains 1% by weight to 2% by weight mineral (dry bitumen basis) and 5% by weight to 15% by weight water (wet diluted basis). Syncrude also utilizes a centrifuge system with naphtha diluent.

Environmental regulations in Canada or the United States will not allow the discharge of tailings streams into (1) the river; (2) on to the surface; or (3) on to any area where contamination of groundwater domains or the river may occur. The tailings streams are essentially high in clays and contain some bitumen, hence the current need for tailings ponds, where some settling of the clay occurs. In addition, an approach to acceptable reclamation of the tailings ponds will have to be accommodated at the time of site abandonment.

The bitumen isolated by the hot-water separation process is then sent for upgrading to a synthetic crude oil.

8.5.3 Other Processes

It is conceivable that the problems related to bitumen mining and bitumen recovery for the tar sand may be alleviated somewhat by the development of process options that require considerably less water in the sand/bitumen separation step. Such an option would allow a more gradual removal of the tailings ponds.

The proposed *cold-water process* for bitumen separation from mined tar sand has also been recommended (Miller and Misra, 1982). The process uses a combination of cold water and solvent and the first step usually involves disintegration of the tar sand charge that is mixed with water, diluent, and reagents. The diluent may be a petroleum distillate fraction such as kerosene and is added in approximately a 1:1 weight ratio to the bitumen in the feed. The pH is maintained at 9–9.5 by the addition of wetting agents and approximately 0.77 kg of soda ash per ton of tar sand. The effluent is mixed with more water, and in a raked classifier the sand is settled from the bulk of the remaining mixture. The water and oil over flow the classifier and are passed to thickeners where the oil is concentrated. Clay in the tar-sand feed has a distinct effect on the process; it forms emulsions that are hard to break and are washed with the underflow from the thickeners.

The *sand-reduction process* is a cold-water process without solvent. In the first step, the tar-sand feedstock is mixed with water at approximately 20°C (68°F) in a screw conveyor in a ratio of 0.75–3 ton/ton of tar sand (the lower range is preferred). The mixed pulp from the screw conveyor is discharged into a rotary-drum screen, which is submerged in a water-filled settling vessel. The bitumen forms agglomerates that are retained by an 840-μm (20-mesh) screen. These agglomerates settle and are withdrawn as oil product. The sand readily passes through the 840-μm (20-mesh) screen and is withdrawn as waste stream. The process is called sand reduction because its objective is the removal of sand from the tar sand to provide a feed suitable for a fluid coking process; c. 80% of sand is removed. Nominal composition of the oil product is 58% by weight bitumen, 27% by weight mineral matter, and 15% by weight water.

The *spherical agglomeration process* resembles the sand-reduction process. Water is added to tar sands and the mixture is ball-milled. The bitumen forms dense agglomerates of 75–87% by weight bitumen, 12–25% by weight sand, and 1–5% by weight water.

An *oleophilic sieve process* (Kruyer, 1982, 1983) offers the potential for reducing tailings pond size because of a reduction in the water requirements. The process is based on the concept that when a mixture of an oil phase and an aqueous phase is passed through a sieve made from oleophilic materials, the aqueous phase and any hydrophilic solids pass through the sieve but the oil adheres to the sieve surface on contact. The sieve is in the form of a moving conveyor, the oil is captured in a recovery zone, and recovery efficiency is high.

An anhydrous *solvent extraction process* for bitumen recovery has been attempted and usually involves the use of a low-boiling hydrocarbon. The process generally involves up to four steps. In the mixer step, fresh tar sand is mixed with recycled solvent that contains some bitumen and small amounts of water and mineral. Solvent-to-bitumen weight ratio is adjusted to approximately 0.5. The drain step consists of a three-stage counter-current wash. Settling and draining time is approximately 30 minutes for each stage. After each extraction step, a bed of sand is formed and the extract is drained through the bed until the interstitial pore volume of the bed is emptied. From time to time, the bed is plugged with fine mineral or emulsion. In these cases, the drainage rate is essentially zero and the particular extraction stage is ineffective. The last two steps of the process are devoted to solvent recovery. Stripping of the solvent from the bitumen is straightforward. The solvent recovery from the solids holds the key to the economic success of an anhydrous process.

Another above-ground method of separating bitumen from mined tar sand involves *direct heating of the tar sand* without previous separation of the bitumen. Thus, the bitumen is not recovered as such but is an upgraded overhead product. In the process, the sand is crushed and introduced into a vessel, where it is contacted with either hot (spent) sand or with hot product gases that furnish part of the heat required for cracking and volatilization. The volatile products are passed out of the vessel and are separated into gases and (condensed) liquids. The coke that is formed as a result of the thermal decomposition of the bitumen remains on the sand, which is then transferred to a vessel for coke removal by burning in air. The hot flue gases can be used either to heat incoming tar sand or as refinery fuel. As expected, processes of this type yield an upgraded product but require various arrangements of pneumatic and mechanical equipment for solids movement around the refinery.

One approach toward recovering a significant portion of heavy oil from a reservoir is to use a combination of petroleum recovery and mining technologies. Through the use of mining technology, access is developed below the petroleum reservoir within or beneath a permeability barrier. Underground development consists of providing room for subsurface drilling and petroleum production operations as well as a life-supporting atmosphere and safe working conditions. Wells are drilled at inclinations from the horizontal to the vertical upward into the reservoir. Wells produce as a result of a combination of pressure depletion and gravity drainage.

In *improved mining*, directional (horizontal or slant) wells are drilled into the reservoir from a mine in an underlying formation to drain oil by pressured depletion and gravity drainage. In the process of gravity drainage extraction of liquid crude oil, the wells are completed so that only the forces acting within the reservoir are used. The forces acting on the reservoir are left intact, perhaps maintained or increased. A large number of closely spaced wells can be drilled into a reservoir from an underlying tunnel more economically than the same number of wells from the surface. In addition, only one pumping system is required in underground drainage, whereas at the surface each well must have a pumping system. The objective of using a large number of wells is to produce each well slowly so that the gas–oil and water–oil interfaces move toward each other efficiently. By maintaining the reservoir pressures because of forces acting on the reservoir, it is then assured that the oil production is provided by the internal forces due to gravity (the buoyancy effect) and capillary effects.

The recovery efficiency is improved by applying an enhanced oil recovery method. The production drain holes should be surveyed while drilling, and the drilling should be accomplished using mud-operated drills with a bent sub for control of direction. The drain holes must be controlled so that a network of uniformly spaced holes conforming to the data output from the computer modeling can be drilled in the production zone. The drill string should be equipped with check valves to minimize the backflow of mud during the installation of additional drill pipe. The return mud line should be equipped with a blow line to safely vent to the surface any formation gas encountered. All drilling should be accomplished working through a blowout preventer or diverter. Drill cuttings should be contained in a closed system and not allowed to encumber the mine atmosphere.

Large vertical shafts sunk from the surface are generally the means through which underground openings can be excavated. These shafts are one means of access to offer an outlet for removal of excavated rock, provide sufficient opening for equipment, provide ventilation, and allow the removal of oil and gas products during later production. These requirements plus geological conditions and oil reservoir dimensions determine the shaft size. It is expected that an access shaft will range from 8 to 20 feet in diameter.

Completing wells from a level of drifts beneath the reservoir is the most economical application of gravity drainage. In this case, only one pumping system is required rather installing a pump in each well as is necessary in wells drilled from the surface. When wells are drilled from beneath the producing formation in an oil–water system, the developer has two options for completion. The casing may be set totally through the formation and the region opposite the oil saturation perforated to permit production. If desired, the easing may pass only through the water-saturated zone and then a jet slotting process can be used to wash out or drill slots in the oil-saturated zone to increase the productivity of the individual well. Other shaft configurations include those for mining in weak rock and shafts for pumping drainage.

Excavation of shafts and tunnels during the mining process results in waste rock that must be stored or disposed of at the ground surface. They cannot be dumped in a heap near the mine shaft as an eyesore for future generations and a source of air and water pollution for untold years. Air, water, and esthetics must be conserved not only for the present but also against reasonable future contingencies, either natural or man-made.

8.6 NONMINING METHODS

The successful recovery technique that is applied to one deposit/resource is not necessarily the technique that will guarantee success for another deposit. There are sufficient differences between the tar sand deposits of the United States and Canada that general applicability is not guaranteed. Hence, caution is advised when applying the knowledge gained from one resource to the issues of another resource. Although the principles may at first sight appear to be the same, the technology must be adaptable.

In principle, the *nonmining recovery of bitumen from tar sand deposits* is an enhanced recovery technique and requires the injection of a fluid into the formation through an injection wall. This leads to the in situ displacement of the bitumen from the recovery and bitumen production at the surface through an egress (production well). There are, however, several serious constraints that are particularly important and relate to bulk properties of the tar sand and the bitumen. In fact, both must be considered *in toto* in the context of bitumen recovery by nonmining techniques. For example, such processes need a relatively thick layer of overburden to contain the driver substance within the formation between injection and production wells.

One of the major deficiencies in applying mining techniques to bitumen recovery from tar sand deposits is (next to the immediate capital costs) the associated environmental problems. Moreover, in most of the known deposits, the vast majority of the bitumen lies in formations in which the overburden–pay zone ratio is too high. Therefore, it is not surprising that over the last two decades a considerable number of pilot plants have been applied to the recovery of bitumen by nonmining techniques from tar sand deposits where the local terrain and character of the tar sand may not always favor a mining option.

There are, however, several serious constraints that are particularly important and relate to bulk properties of the tar sand and the bitumen. In fact, both must be considered in the context of bitumen recovery by nonmining techniques. For example, the Canadian deposits are unconsolidated sands with a porosity ranging up to about 45% whereas other deposits may range from predominantly low-porosity, low-permeability consolidated sand to, in a few instances, unconsolidated sands. In addition the bitumen properties are not conducive to fluid flow under deposit conditions. Nevertheless, where the general nature of the deposits prohibits the application of a mining technique, a nonmining method may be the only feasible bitumen recovery option.

Another general constraint to bitumen recovery by nonmining methods is the relatively low injectivity of tar sand formations. Thus it is usually necessary to inject displacement or recovery finds at a pressure such that fracturing (parting) is achieved. Such a technique therefore changes the reservoir profile and introduces a series of channels through which fluids can flow from the injection well to the production well. On the other hand, the technique may be disadvantageous insofar as the fracture occurs along the path of least resistance, giving undesirable (ie, inefficient) flow characteristics within the reservoir between the injection and production wells, leaving a large part of the reservoir relatively untouched by the displacement or recovery fluids.

Many innovative concepts in heavy oil production have been developed and major new technologies have positively affected the heavy oil industry in the last 10–20 years. The oil industry has pioneered drilling shallow (500–3500 feet deep) horizontal wells at cost-per-meter values that are now only 1.2–1.3 times those of vertical wells. These wells, in the shallowest cases (150–1500 feet deep) are often drilled using masts inclined to reduce curvature build rates required to *turn the corner* from vertical to horizontal. In "reduced footprint" developments, where there are a number of producing wells and injecting wells on the same pad (savings in roads, services,...), many horizontal wells (2–6) may be drilled from a small area, no larger than a hectare.

Coiled tubing drilling and work-over have been introduced and perfected in the last decade, further reducing costs of horizontal well drilling. Good seismic control and cuttings analysis allows precise steering in thin zones (20) to place the well in the optimum position in the reservoir. In the production phase, the long drainage length of a single well, as much as 1200 m in many cases, allows much more effective production, also giving higher production percentages of OOIP when used in gravity drainage technologies.

Long horizontal wells with several multilateral branches have been used widely in the development of the heavy oils of Venezuela, where production rates as high as 2000–2500 bbl/day in some wells have been achieved through the use of aggregate horizontal lengths as large as 10,000 m in oil of 1200–5000 cP viscosity. Unfortunately, this technology can only achieve 8–15% recovery, and only from the best high-permeability zones. Thus, for the more efficient development of these resources, other technologies will necessarily be implemented in the future.

Horizontal wells using cold primary production in heavy oils have been widely used in Canada since the late 1980s, but the successes (usually well published) have been substantially offset by the failures (rarely published). Furthermore, low recovery factors (seldom close to 10%), early water breakthrough (usually impossible to plug when it happens), short well life (30–40% decline per year) and other factors such as expensive work-over if sand plugging takes place have combined to make this a technology that has little attraction in Canada.

IGI is a technology for conventional oils in reservoirs where good vertical permeability exists, or where it can be created through propped hydraulic fracturing. It is generally viewed as a *top-down* process with nitrogen or methane injection, through vertical wells at the top of the reservoirs, creating a gas–oil interface that is slowly displaced toward long horizontal production wells. As with all gravity drainage processes, it is essential to balance the injection and production volumes precisely so that the system does not become pressure driven, but remains in the gravity-dominated flow regime.

High recovery ratios are achieved because, in the absence of elevated pressure gradients, a thin oil film is maintained between the gas and water phases. The film is maintained because the sum of the oil–water and gas–oil surface tensions is always less than the water–gas surface tension. Thus, the thin film configuration is thermodynamically stable, and allows the oil to drain to values far lower than the *residual oil saturation* value.

The interfaces in IGI are gravity-stabilized because of the difference in phase densities, so that at slow drainage rates the interfaces remain approximately horizontal without viscous fingering. In one configuration, the horizontal wells are produced under a back-pressure equal to the pressure in the underlying water phase, so no water coning can occur. During production, if the water cut increases, the production rate is reduced so that the interface becomes stable. Alternatively, if gas is injected too quickly, gas coning can develop, and if this is observed, the gas injection rate must be reduced to sustain stability. The process is continued until the oil zone is "pinched down" to the horizontal well, achieving the high recovery ratios possible with gravity drainage methods. These principles are fundamental to all gravity-dominated processes and failure to adhere to them will drive the system into conditions of instability (such as coning and fingering).

In reservoirs with excellent vertical permeability, the bottom water zone can also be injected with water to cause the oil–water interface to rise slowly toward the production well. First implemented in Canada, IGI is used extensively in carbonate pinnacle reefs with excellent vertical permeability, and recovery ratios exceeding 80% are systematically achieved. As with all gravity drainage processes, it is generally necessary to place the horizontal wells as low in the structure as possible.

Hybrid approaches that involve the simultaneous use of several technologies are evolving and will see greater applications in the future. Some of the evolving options that will be tried at the field scale in the next decade include:

1. A mixture of steam and miscible and noncondensable hydrocarbons is being field tested as a hybrid SAGD-VAPEX approach, with apparent good success and reduction of steam–oil ratios.
2. Single horizontal laterally offset wells can be operated as moderate pressure cyclic steam stimulation wells in combination with SAGD pairs to widen the steam chamber and reduce steam–oil ratios by about 20%.
3. Simultaneous CHOPS and SAGD, with CHOPS used in offset wells until steam breakthrough occurs. Then the CHOPS wells are converted to slow gas and hot water (or steam) injection wells to control the process. The high-permeability zones generated by CHOPS should accelerate the SAGD recovery process.
4. Incorporating PPT along with CHOPS has already been field tested with economic success, and PPT has potential applications in other hybrid approaches.
5. PPT may aid in partially stabilizing water flood through reducing the viscous fingering and coning intensity.

In addition to hybrid approaches, the new production technologies, along with older, pressure-driven technologies, will be used in successive phases to extract more oil from reservoirs, even from reservoirs that have been abandoned after primary exploitation. Old reservoirs can be redeveloped with horizontal wells, even linking up the wells to bypassed oil because of the physics of oil film spreading between water and gas phases. These staged approaches hold the promise of significantly increasing recoverable reserves worldwide, not just in heavy oil cases.

Finally, recent developments in *upgrading* of heavy oil and bitumen (Speight, 2013, 2014) indicate that the near future could see a reduction of the differential cost of upgrading heavy oil. These processes are based on a better understanding of the issues of asphaltene solubility effects at high temperatures, incorporation of a catalyst that is chemically precipitated internally during the upgrading, and improving hydrogen addition or carbon rejection.

REFERENCES

Banat, I.M., 1995. Biosurfactant production and possible uses in microbial enhanced oil recovery and oil pollution remediation. Biores. Technol. 51, 1–12.

Browne, G.E., Hass, G.R., Sell, R.D., 1996. Downhole emulsification: viscosity reduction increases production. J. Can. Petrol. Technol. 35, 25–31.

Bryant, R.S., Donaldson, E.C., Yen, T.F., Chilingarian, G.V., 1989. Microbial enhanced oil recovery. In: Donaldson, E.C., Chilingarian, G.V., Yen, T.F. (Eds.), Enhanced Oil Recovery II: Processes and Operations Elsevier, Amsterdam, pp. 423–450.

Bryant, R.S., Lindsey, R.P., 1996. World-wide applications of microbial technology for improving oil recovery Proceedings of the SPE Symposium on Improved Oil Recovery. Society of Petroleum Engineers, Richardson, TX.27–134

Butler, R.M., Jiang, Q., 2000. Improved recovery of heavy oil by VAPEX with widely spaced horizontal injectors and producers. J. Can. Petrol. Technol. 39 (1), 48–56.

Butler, R.M., Mokrys, I.J., 1991. A new process (VAPEX) for recovering heavy oils using hot water and hydrocarbon vapor. J. Can. Petrol. Technol. 30 (1), 97–106.

Butler, R.M., Mokrys, I.J., 1997. Process and apparatus for the recovery of hydrocarbons from a hydrocarbon deposit. United States Patent 5,607,016. March 4.

Carrigy, M.A., 1963a. The oil sands of Alberta. Bulletin No. 14. Alberta Research Council, Edmonton, Alberta, Canada.

Carrigy, M.A., 1963b. The oil sands of Alberta. Information Series No. 45. Alberta Research Council, Edmonton, Alberta, Canada.

Chen, X., 2006. Heavy Oils, Part 1. SIAM News, Society of Industrial and Applied Mathematics, Philadelphia, Pennsylvania. 39(3), April.

Cheraghian, G., Khalili Nezhad, S.S., 2015. Effect of nanoclay on heavy oil recovery during polymer flooding. Petrol. Sci. Technol. 33, 9.

Cheraghian, G., Khalili Nezhad, S., Kamari, M., Hemmati, M., Masihi, M., Bazgir, S., 2014. Effect of nanoclay on improved rheology properties of polyacrylamide solutions used in enhanced oil recovery. J. Petrol. Explor. Prod. Tech. 5 (2), 189.

Chugh, S., Baker, R., Telesford, A., Zhang, E., 2000. Mainstream options for heavy oil: Part I—cold production. J. Can. Petrol. Technol. 39 (4), 31–39.

Clark, J.B., Munnecke, D.M., Jenneman, G.E., 1981. In situ microbial enhancement of oil production. Dev. Ind. Microbiol. 15, 695–701.

Clark, K.A., 1944. Hot-water separation of alberta bituminous sand. Trans. Can. Inst. Min. Met. 47, 257–274.

CRA, 2003. Pressure Pulse Technology. Innovative Technology Group, Conestoga-Rovers and Associates, Niagara Falls, New York. Volume 3, Number 1 (January). <http://www.craworld.com/en/newsevents/resources/inn_2003_jan.pdf> (accessed 1.06.14.).

Curtis, C., Kopper, R., Decoster, E., Guzmán-Garcia, A., Huggins, C., Knauer, L., et al., 2002. Heavy-oil reservoirs. Oilfield Rev. 14 (3), 42–46.

El-Gendy, N.S., Speight, J.G., 2015. Handbook of Refinery Desulfurization. CRC Press, Taylor & Francis Group, Boca Raton, FL.

Etebar, S., 1995. Captain Innovative Development Approach, SPE 30369 (September).

Fear, J.V.D., Innes, E.D., 1967. Canada's first commercial tar sand development. Proceedings. 7th World Petroleum Congress. Elsevier, Amsterdam, vol. 3, pp. 633–650.

Jenneman, G.E., 1989. The potential for in-situ microbial applications. Dev. Petrol. Sci. 22, 37–74.

Khire, J.M., Khan, M.I., 1994a. Microbially enhanced oil recovery (MEOR). Part 1. Importance and mechanisms of microbial enhanced oil recovery. Enzyme Microb. Technol. 16, 170–172.

Khire, J.M., Khan, M.I., 1994b. Microbially enhanced oil recovery (MEOR). Part 2. Microbes and the subsurface environment for microbial enhanced oil recovery. Enzyme Microb. Technol. 16, 258–259.

Kovscek, A.R., 2002. Heavy and thermal oil recovery production mechanisms. Quarterly Technical Progress Report. Reporting Period: April 1 through June 30, 2002. DOE Contract Number: DE-FC26-00BC15311. July.

Krumrine, P.H., Falcone Jr., J.S., 1987. Beyond alkaline flooding: design of complete chemical systems. SPE 16280 1987.

Kruyer, J., 1982. Oleophilic separation of tar sands. Oil–water mixtures and minerals. Proceedings. Second International Conference on Heavy Crude and Tar Sands, Caracas, Venezuela, February 7–17.

Kruyer, J., 1983. Oleophilic separation of tar sands. Oil–water mixtures and minerals. Preprint No. 3d. Summer National Meeting. American Institute of Chemical Engineers, Denver, Colorado, August 28–31.
Kumar, M., 2006. Heavy oil recovery—recent developments and challenges. Los Angeles Monthly Petroleum Technology Forum. L.A. Basin Section, SPE. <http://www.laspe.org/petrotech/petrooct10906.html> (accessed 1.06.14.).
Loughead, D.J., 1992. Lloydminster heavy oil production—why so unusual? Ninth Annual Heavy Oil and Oil Sands Technology Symposium, Calgary, Alberta, Canada, March 11.
Mai, A., Bryan, J., Goodarzi, N., Kantzas, A., 2006. Insights into non-thermal recovery of heavy oil. Paper No. 2006-553. Proceedings. 1st World Heavy Oil Conference. Petroleum Society of Canada, Calgary, Alberta, Canada. Also: Journal of Canadian Petroleum Technology, 2009, 48(3), 1–9.
Mai, A., Kantzas, A., 2009. Heavy oil waterflooding: effects of flow rate and oil viscosity. J. Can. Petrol. Technol. 48 (3), 42–51.
McCaffrey, W.J., Bowman, R.D., 1991. Recent successes in primary bitumen production. HOOS Technical Symposium Challenges and Innovation. Annual Technical Meeting, Petroleum Society of the Canadian Institute of Mining.
McLean, J.D., Kilpatrick, P.K., 1997. Effects of asphaltene solvency on stability of water-in-crude-oil emulsions. J. Colloid Interface Sci. 189, 242–253.
Meyer R.F., Attanasi E.D., 2003. Fact Sheet 70-03, Heavy Oil and Natural Bitumen—Strategic Petroleum Resources. United States Geological Survey, Reston, Virginia. <http://pubs.usgs.gov/fs/fs070-03/> (accessed 31.03.14.).
Miller, J.C., Misra, M., 1982. Hot water process development for utah tar. Sands Fuel Proc. Technol. 6, 27–59.
Mitchell, D.L., Speight, J.G., 1973. The solubility of asphaltenes in hydrocarbon solvents. Fuel 52, 149.
NPC, 2007. Heavy oil working document. Topic Paper #22. NPC Global Oil & Gas Study, National Petroleum Council, Washington, DC, July 18.
N-Solv, 2015. <https://www.nsolv.ca/technology/#sthash.xeWLBbxB.dpuf> (accessed 24.06.15.).
Orr Jr., F.M., Heller, J.P., Taber, J.J., 1982. Carbon dioxide flooding for enhanced oil recovery: promise and problems. J. Am. Oil Chem. Soc. 59 (10).
Raiders, R.A., Knapp, R.M., McInerney, M.J., 1989. Microbial selective plugging and enhanced oil recovery. J. Ind. Microbiol. 4, 215–230.
Safinya, K., 2008. Heavy oil recovery—the road ahead. Alberta Oil 4 (1), 14–19.
Shennan, J.L., Levi, J.D., 1987. In situ microbial enhanced oil recovery. In: Kosaric, N., Cairns, W.L., Gray, N.C.C. (Eds.), Biosurfactants and Biotechnology Marcel Dekker, New York, pp. 163–180.
Speight, J.G., 2013. Heavy and Extra Heavy Oil Upgrading Technologies. Gulf Professional Publishing, Elsevier, Oxford.
Speight, J.G., 2014. The Chemistry and Technology of Petroleum, fifth ed. CRC-Taylor and Francis Group, Boca Raton, FL.
Speight, J.G., 2015. Handbook of Offshore Oil and Gas Operations. Gulf Professional Publishing, Elsevier, Oxford.
Speight, J.G., Arjoon, K.K., 2012. Bioremediation of Petroleum and Petroleum Products. Scrivener Publishing, Salem, MA.
Speight, J.G., Moschopedis, S.E., 1978. Factors affecting bitumen recovery by the hot water process. Fuel Proc. Technol. 1, 261.
Stosur, G.J., 1991. Unconventional EOR concepts. Crit. Rep. Appl. Chem. 33, 341–373.
Taber, J.J., Martin, F.D., Seright, R.R., 1997. EOR screening criteria revisited. SPE Res. Eng. 12, 189–206.

Tchambak, E., Oyeneyin, B., Oluyemi, G., 2012. Heavy oil recovery: a cold process using CO_2-EOR technique. Adv. Mater. Res. 367, 421–429.
Thomas, S., 2008. Enhanced oil recovery: an overview. Oil Gas Sci. Technol.—Revue Institut Français du Pétrole 63 (1), 9–19.
Wassmuth, F., Green, K., Hodgins, L., Turta, A., 2007. Polymer flood technology for heavy oil recovery. Paper No. 2007. Proceedings. Canadian International Petroleum Conference, Calgary, Alberta, June 12–14. Society of Petroleum Engineers, Richardson, TX.
Yang, C., Gu, Y., 2005a. A novel experimental technique for studying solvent mass transfer and oil swelling effect in a vapor extraction (VAPEX) process. Paper No. 2005-099. Proceedings. 56th Annual Technical Meeting. The Canadian International Petroleum Conference, Calgary, June 7–9.
Yang, C., Gu, Y., 2005b. Effects of solvent-heavy oil interfacial tension on gravity drainage in the VAPEX process. Paper No. SPE 97906. Society of Petroleum Engineers International Thermal Operations and Heavy Oil Symposium, Calgary, Alberta, Canada, November 1–3.
Yazdani, A., Maini, B.B., 2008. Modeling the VAPEX process in a very large physical model. Energy Fuels 22, 535–544.
Zaki, N.N., Poindexter, M.K., Kilpatrick, P.K., 2002. Factors contributing to petroleum foaming. 2. Synthetic crude oil systems. Energy Fuels 16, 711–717.
Zhang, Y.P., Huang, S., Luo, P., 2010. Coupling immiscible CO_2 technology and polymer injection to maximize EOR performance for heavy oils. J. Can. Petrol. Technol. 49, 25.

CHAPTER 9

Thermal Methods of Recovery

9.1 INTRODUCTION

Heavy oil is characterized by high viscosity (ie, resistance to flow measured in centipoises or cp) and high densities compared to conventional oil (IEA, 2005). Although there are many definitions for heavy oil, the most appropriate definition is drawn from United States Federal Energy Administration (FE-76-4), which relates to tar sand

> *...the several rock types that contain an extremely viscous hydrocarbon which is not recoverable in its natural state by conventional oil well production methods including currently used enhanced recovery techniques. The hydrocarbon-bearing rocks are variously known as bitumen-rocks oil, impregnated rocks, oil sands, and rock asphalt.*

By inference, petroleum and heavy oil are recoverable by well production methods and currently used enhanced recovery techniques. For convenience, it is assumed that before depletion of the reservoir energy, conventional crude oil is produced by primary and secondary techniques, whereas heavy oil requires tertiary (enhanced) oil recovery techniques. While this is an oversimplification, it may be used as a general guide. The term *natural state* cannot be defined out of context; in the context of Federal Energy Administration (FEA) Ruling 1976-4, the term is defined in regards to the composition of the heavy oil or bitumen. The final determinant of whether or not a reservoir is a tar sand deposit is the character of the viscous phase (bitumen) and the method that is required for recovery. Using this definition removes reliance upon a single property, such as viscosity or API gravity (subject to experimental differences in the analytical method as well as the interpretation of the user of the data), which varies from country to country. While a single property might serve to generally define heavy oil, it is by no means accurate and should not be taken as definitive.

In some cases, the majority of subsurface heavy oil will not flow towards a wellbore in sufficient quantity to be economically viable. Thermal recovery involves heating up the reservoir, thereby lowering the

Introduction to Enhanced Recovery Methods for Heavy Oil and Tar Sands.
DOI: http://dx.doi.org/10.1016/B978-0-12-849906-1.00009-6

heavy oil's viscosity and enabling the oil to flow to the wellbore. The API gravity of heavy oil is usually less than 20°, depending upon the reservoir, and viscosity is very high at reservoir temperature. Heavy oil viscosity decreases rapidly with increasing temperatures; therefore, external heat may be required for production. High-temperature steam is commonly used to deliver heat to the formation. The SOR or fuel oil ratio is an important measure of the energy required to produce heavy oil; an SOR of 3 means that three barrels of water (converted to steam) are needed to produce one barrel of oil.

At sufficiently low viscosity, the oil saturation upstream of the displacement front is essentially irreducible, leading to what is known as piston-like displacement. For a crude of high viscosity, such as heavy oil, the drop in oil saturation at the displacement front is much smaller, and a substantial amount of mobile crude is bypassed by the leading edge of the displacement front. Because of its high viscosity, the crude has a much higher flow resistance than does the water. The water simply moves faster than the crude, leaving some mobile crude behind. For the same volume of water injected, then, the oil displacement front would be closer to a producer in a heavy-oil reservoir. This means that for the more viscous crudes, at the same water injection rate and injected volume, water will break through at the producer earlier; more water must be handled earlier in the life of the project; and less oil will be produced. Oil will continue to be produced after water breakthrough but the average oil cut (the fraction of oil in the produced stream) will be lower. With high-viscosity crudes, many pore volumes of injected water are required to reduce the remaining oil to near its irreducible value.

Not only is the displacement efficiency poor, but the volumetric sweep efficiency is decreased at the higher crude viscosity. This decrease in volumetric sweep efficiency is the result of two different mechanisms. First, during the displacement of a viscous crude by water, the macroscopic displacement front is never planar but varies according to the velocity of the flow lines. Since the flow resistance of a flow channel is least where the displacement front is most advanced, the flow velocities in those channels are the highest. The result is that the displacement front arrives at the producer soonest along the flow line of highest velocity and that breakthrough occurs earlier (for a given injection rate) for the more viscous oil. An example is the effect of increasing oil viscosity (which is proportional to mobility ratio) on the sweep efficiency at the time of water breakthrough, which applies to a homogeneous repeated five-spot well pattern.

Layers of different permeability and gravity tongues are examples of other conditions leading to macroscopic bypassing.

The second mechanism reducing the volumetric sweep efficiency is associated with small-scale viscous instability which in the displacement front will develop in the most homogeneous porous media; they arise from small perturbations in an otherwise uniform flow pattern. These perturbations may be caused, for example, by small differences in the shape of grains or pores. The distance between the viscous fingers that are most likely to form decreases with increasing crude viscosity, and the fingers are likely to be formed at relatively low rates. In addition, some of the oil between adjacent fingers may be bypassed and trapped more easily when, because of high viscosities, the fingers are close together, that is, the displacement efficiency is lower when the oil viscosity is higher. Thus, the presence of heterogeneities in any reservoir not only ensures the presence of microscopic viscous fingering but also results in bypassing and trapping of oil on quite a large scale and the effect being more pronounced the more viscous the oil.

The other way to alter crude viscosity chemically is by thermal decomposition (visbreaking, thermal cracking, or pyrolysis). The crude itself is destroyed by pyrolysis, the products of which are small amounts of solid carbonaceous residue (popularly but erroneously called coke) and gas and a relatively large fraction of a less-viscous liquid. Visbreaking is a mild form of pyrolysis that decreases viscosity without significantly changing other crude oil properties of interest. Some crude alteration occurs in all thermal recovery processes, but a significant amount of conversion by pyrolysis occurs only at high temperatures.

Thermal methods are the main means of recovering heavy oils, typically the heavy oils with an API gravity less than 20°, which typically are not amenable to recovery by primary production or waterflooding so initial oil saturation is typically high at the start of a thermal recovery project. The principle of thermal recovery is simple: (1) increase the temperature dramatically and reduce the viscosity of the oil; and (2) improve the mobility ratio. The two primary methods of heating reservoir oil are injection of fluid heated at the surface or production of heat directly within the reservoir by burning some of the oil in place. The recovery mechanism of the steam flooding process is complex—in addition to viscosity reduction, another important criterion (not always recognized and fully appreciated) is steam distillation of the lower-boiling constituents of the heavy oil.

In fact, thermal processes have been the only practical means of improving heavy oil recovery performance of reservoirs containing viscous crudes. The use of heat in wellbores to decrease the crude viscosity and increase crude production rates has been accepted as viable for many decades. Cyclic steam injection has been commercially successful since it was discovered in the early 1960s. Hot water, hot gas, steam, and various kinds of combustion drives have gained varying degrees of acceptance. Because of its success, relative simplicity, and low cost, the use of heat to improve oil recovery has gained wide acceptance in the oil industry.

Although the development of thermal recovery processes was prompted initially by difficulties in displacing and producing heavy crudes, the drive processes technically can be applied also to low-viscosity crudes. Steam- and combustion-drive processes have recovered low-viscosity crudes in several fields. Even though the amount of oil recovered may be higher, these processes seldom can compete commercially with cold-water flooding. Hot-water flooding and thermal stimulation methods are somewhat ineffective for low-viscosity crudes. However, it is an oversimplification to state that thermal recovery processes are successful because they tend to reduce the viscosity of the crude oil; there are several important factors related to oil chemistry and oil–rock interactions that play a role in heavy oil recovery (Prats, 1986; Lake and Walsh, 2004). For example, the role of steam-distillation and other physical effects in steam-based processes have long been ignored or not understood. It is only when there is a thorough knowledge of such phenomena that heavy oil recovery will be fully understood.

Generally, the thermal recovery processes used today fall into two classes: (1) processes in which a hot fluid such as steam is injected into the reservoir; and (2) processes in which heat is generated within the reservoir itself, such as the combustion processes (Fig. 9.1). The former are usually the steam-based processes, whereas the latter are processes such as in situ combustion or fire-flooding. Thermal recovery processes also can be classified as thermal drives or stimulation treatments. A thermal drive not only reduces the viscosity but also provides a force to increase the flow rates of the oil to the production well. In thermal stimulation treatments, only the reservoir near production wells is heated.

In thermal processes, fluid is injected continuously into a number of injection wells to displace oil and obtain production from other wells. The pressure required to maintain the fluid injection also increases the driving forces in the reservoir, increasing the flow of crude oil. Driving

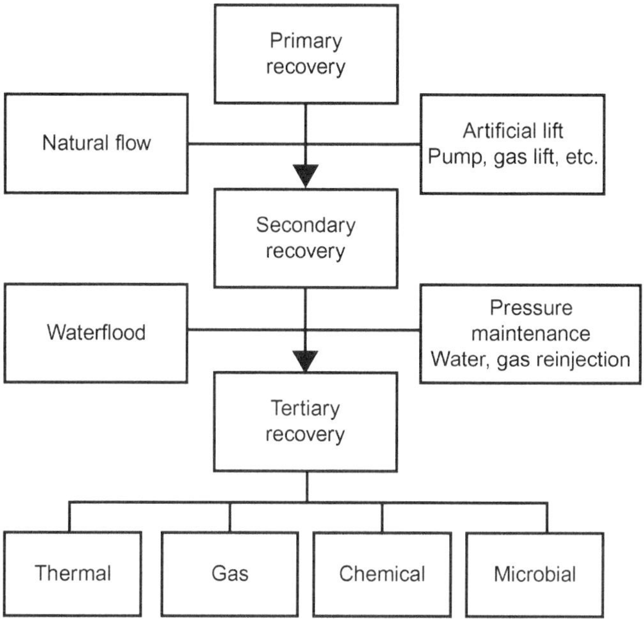

Figure 9.1 Methods for oil recovery.

forces present in the reservoir, such as gravity, solution gas, and natural water drive, effect the improved recovery rates once the flow resistance is reduced. Stimulation treatments also can be combined with thermal drives, in which case the driving forces are both natural and imposed.

However, the successful recovery technique that is applied to one heavy oil reservoir is not necessarily the technique that will guarantee success for another reservoir. General applicability of the techniques is not guaranteed. Caution is advised when applying the knowledge gained from one resource to the issues of another resource. Although the principles may at first sight appear to be the same, the technology must be adaptable.

In fact, each production method must be tailored for the particular resource and its fluid properties. A method that works in one situation may fail utterly in a different one. It is essential that the properties of the resource (in this case, of the reservoir and the oil) be fully understood before selecting a production scheme. Essential properties include the geological setting; the depth, areal extent, and thickness of the resource; oil composition, density, viscosity, and gas content; the presence of bottom water or top gas zones; petrophysical and geomechanical properties, such as porosity, permeability, and rock strength; the presence of shale layers;

vertical and horizontal permeability; and the variation of these properties across the reservoir.

In fact, geomechanics (measuring and understanding the formation and overburden mechanical properties under drilling and production conditions) is an important aspect of prerecovery operations that is a necessity for optimal process performance. Borehole stability is always an issue when drilling horizontal wells in unconsolidated formation (eg, for steam-assisted gravity drainage (SAGD) processes). In addition, fracturing the formation may be desirable or undesirable depending on the process. For example, fracturing the overburden and allowing steam to escape would adversely affect the SAGD, cyclic steam stimulation, and steam-flood processes.

Thermal processes have been the only practical means of improving the oil recovery from heavy crude oil reservoirs, and the use of heat in wellbores to increase production rates for heavy crude oils has been accepted as viable for several decades. For example, cyclic steam injection has been commercially successful since it was first applied in the early 1960s. Processes using hot water, hot gas, steam, and various kinds of combustion drives have gained varying degrees of acceptance. Even though the application of thermal recovery processes was initiated because of the difficulties encountered in recovering heavy crude oil (specifically, the difficulties in encouraging flow to a production well), the drive processes can also be applied to recovery of the residual oil in energy-depleted reservoirs that hold considerable amounts of conventional (low-viscosity) crude oil.

Thermal decomposition (visbreaking or thermal cracking) can also be used to decrease the viscosity of heavy crude oil. The intramolecular balance of the crude oil is changed with the resulting deposition of a carbonaceous residue that separates from the oil, leaving a lighter (easier-flowing) oil product phase. Visbreaking is a surface process, and its chemistry offers opportunities for thermal recovery of heavy oil. The partial upgrade of oil in situ with the deposition of the carbonaceous residue, which contains most of the nitrogen and metals originally in the oil, and the production of partially upgraded oil at the surface is to be preferred. However, the disadvantage is that the deposition of a carbonaceous solid has the potential to result in blockage of the pores and flow channels within the reservoir. Such interference with the recovery process is to be avoided.

All thermal recovery processes tend to reduce the reservoir flow resistance by reducing the viscosity of the crude. The thermal recovery processes used today fall into two classes: those in which a hot fluid is injected into the reservoir and those in which heat is generated within the reservoir itself. The latter are known as in situ processes, an example of which is in situ combustion or fire flooding. Processes combining injection and in situ generation of heat have been tested but currently are not practiced to any great extent. Thermal recovery processes also can be classified as thermal drives or stimulation treatments. In thermal drives, fluid is injected continuously into a number of injection wells to displace oil and obtain production from other wells. The pressure required to maintain the fluid injection also increases the driving forces in the reservoir, increasing the flow of crude. Thus, a thermal drive not only reduces the flow resistance but also provides a force that increases flow rates. In thermal stimulation treatments, only the reservoir near production wells is heated. Driving forces present in the reservoir—such as gravity, solution gas, and natural water drive—effect the improved recovery rates once the flow resistance is reduced. Stimulation treatments also can be combined with thermal drives, in which case the driving forces are both natural and imposed. In thermal stimulation treatments, the reduction in flow resistance also may result from the removal of organic or other solids from openings in the casing, the liner, the screen, and even from the pores of the reservoir rock.

It is the purpose of this chapter to present the various types of thermal recovery processes in common use. But first, the next section points out the need for thermal recovery processes by discussing the pertinent features of a conventional waterflood applied to a reservoir containing a very viscous crude. Typically, it is generally the case that some degree of crude oil alteration occurs in all thermal recovery processes, but a significant amount of conversion by thermal changes occurs only at the high temperatures encountered during combustion processes.

9.2 THERMAL STIMULATION

By definition, stimulation processes decrease resistance to flow and, thus, allow the driving forces present in the reservoir to increase crude production. In thermal stimulation, the reduction in flow resistance is achieved by heating the wellbore and the reservoir near it. One mechanism that is always in force in thermal stimulation is a reduction in the viscosity of the crude and of the water; reducing the viscosities tends to reduce the flow resistance. A second mechanism is wellbore clean-up, in which the following might occur: organic solids near a wellbore may be melted or dissolved; clays may be stabilized; the absolute permeability may be increased by the high temperatures; or fines that could be inhibiting flow in gravel packs may be flushed away. Wellbore clean-up usually has a rather minor effect after the first stimulation cycle.

Thermal stimulation currently is the only effective treatment for viscous oil reservoirs with poor lateral continuity. Because the effects are confined to the neighborhood of the wellbore, thermal stimulation improves oil production rates rather quickly.

9.2.1 Wellbore Heating

In wellbore heating, the wellbore is heated at the producing interval. Generally, production and heating are performed concurrently, but sometimes they are alternated. The reservoir is heated by thermal conduction countercurrent to the transfer of heat out of the reservoir by the produced fluids. The wellbore normally is heated either by using a gas-fired downhole burner or a downhole electric heater hanging on an electric power cable or by circulating fluids heated at the surface. Heating by passing electric currents through a reservoir also has been proposed.

Wellbore heating largely has been supplanted by cyclic steam injection, which is generally much more effective. As is the case with many recovery processes, the mechanisms involved in the production of oil during cyclic steam injection are diverse and complex. There is no doubt that a reduction in the viscosity of the crude in the heated zone near the well greatly affects the production response.

9.2.2 Downhole Heating

Electrical heating consists of providing electrical currents to generate heat and increase the temperature near the wellbore. Initially, most downhole heating devices were designed for combustion or downhole steam generation (Jamaluddin et al., 1998 and references cited therein). Currently, electric downhole heating systems are finding common use to lower the viscosity of heavy oil in the reservoirs to increase the flow rates of crude and improve recovery. Generally, the systems are very efficient, reliable, and easily controlled solutions to allow production of heavy, high viscosity oils. Various methods of electric heating have been used for more than 25 years to control not only heavy viscous oil but also wax and hydrates. Currently, two kinds types of downhole heaters are in use by the oil industry: (1) inductive heaters; and (2) resistive heaters. In the latter case, the heat transfer is by conduction and it requires an extended time period to heat the reservoir. The key parameters to be understood in the downhole heating process are the variation in the viscosity of the heavy oil with temperature and the rate of heat provided by the heater associated with the generated temperature gradient in the volume around the well (Rodriguez et al., 2008).

The Orinoco belt (Faja) in Venezuela contains one of the world's largest resources of heavy and extra heavy oil. Because of the production decline of conventional light crude oil, projects must focus on increasing the recovery of heavy and extra heavy oils by use of thermal and nonthermal methods. Steam-based thermal-recovery processes are more efficient in low-pressure reservoirs; however, because of their depth, the initial pressures of the reservoirs in the Faja are relatively high, ranging from 600 to 1500 psi with viscosities typically greater than 2000 cP. For these reasons, it is important to decrease the pressure of the reservoirs with primary-production techniques to facilitate the economical implementation of steam-injection-based methods. In such cases, the initial production of heavy and viscous oils can be accelerated by use of downhole heaters that, by providing energy to the vicinity of the well, decrease oil viscosity and increase the oil production rate. A consequential advantage of using downhole heaters as a prelude to a steam-injection process is that they accelerate early production and reservoir-pressure depletion.

As an example of heavy oil recovery, the method has been applied to recovery of Venezuelan crude oil (API gravity on the order of 8° to 9°) with a viscosity of 18,000 cP at the reservoir temperature of 47°C (117°F). The heating varied over the horizontal length from the toe to the heel with more heat (higher watt density) delivered to the toe to equalize the wellbore pressure drop in the long horizontal producing zone. The temperature increased to 76°C (168°F) at the toe. In another example, for a California crude oil (API in the range 8.5° to 13°), heating the oil to temperatures above 93°C (200°F) substantially reduced the viscosity, resulting in increased production (Parman and Ojeda, 2013).

9.2.3 Hot-Fluid Injection

Hot-fluid injection processes are those processes in which preheated fluids are injected into a relatively cold reservoir. The injected fluids are usually heated at the surface, although wellbore heaters (downhole heaters) are seeing a wider use. Fluids range from water (both liquid and vapor) and air to others, such as natural gas, carbon dioxide, exhaust gases, and even solvents. It should be noted here that if water vapor is injected into the reservoir, the process might be more correctly designated as a steam-based process.

In this process, the water is filtered, treated to control corrosion and scale, heated, and if necessary, treated to minimize the swelling of clays in the reservoir. The primary role of the heated water is to reduce the oil viscosity and, thereby, improve the displacing efficiency over that obtainable from a conventional waterflood. The design and operation of hot-water drives have many elements in common with conventional waterflooding (see chapter: Microbial Enhanced Oil Recovery).

The current common practice is to inject steam into a formation for a few weeks, wait a few days to let the heat "soak in" and allow the steam to condense, and then put the well on production. This process is known by a number of names: steam soak, steam stimulation, huff-and-puff, and cyclic steam injection. In the process, a single well for both injection and production and steam is used for injection for several days or weeks. Then the well is shut in for several days to a month or more, the soak period. After the soak period, the well is produced for up to 6 months, after which the process is repeated. The steam heats the rock and fluids surrounding the wellbore and also provides some drive pressure—by the time production resumes, the steam has condensed and oil and water are produced. Other fluids can be used instead of steam, but none have been found to be as effective. Hot water introduces a larger volume of water per unit of heat injected, and its use results in higher water saturations. This can affect the producing oil cut adversely.

The advantage of the *huff and puff* process is the relatively short time required for steam injection time so that the well is producing most of the time. The disadvantage is that only the reservoir near the wellbore is stimulated. As a result, this method therefore often has to be followed by continuous steam injection to drive oil toward a separate production well. Reservoir pressure, dictated by depth, imposes a limit on steam-flooding: higher pressures require more fuel to generate higher temperature on the surface to produce saturated steam needed for efficient steam flood. The higher temperature also leads to greater heat losses.

Cyclic steam injection is popular because the production response is obtained earlier and the amount of recovered oil per amount of steam injected is often higher than in thermal drives. Also, relatively small steam boilers can be used, and they can be moved from well to well. Wells can be steam-soaked several times, the main requirements being natural driving forces, such as solution gas or gravity drainage, and sufficient oil near the wells.

Generally, the injected fluids are heated at the surface, although wellbore heaters have been used on occasion. Fluids range from the common ones such as water (both liquid and vapor) and air, to others such as natural gas, carbon dioxide, exhaust gases, and even solvents. The choice is controlled by cost, expected effect on crude production response, and availability of fluids. Sometimes, the choice of injection fluid is governed by the nature of the reservoir. Hot gas or even steam (but not hot water) may be considered for injection into the crest of a reservoir with a small, low-pressure gas cap,making sure to maintain gravitational stability by not exceeding the critical rate. Injection of hot water near zones swept by an encroaching natural water drive may prevent oil resaturation of the zones already swept. But injection of steam under similar circumstances should be considered carefully, as the cooling by any encroaching water could have a significantly adverse effect on the rate of growth of the steam zone. Also, steam or fresh water should not be injected, without due consideration of the possible consequences, into reservoirs containing significant amounts of swelling clays.

The effective mobility ratio associated with hot-fluid injection is very unfavorable for noncondensable gases, less so for hot water, and the least unfavorable (or even favorable) for condensable gases, such as steam. The effectiveness of steam displacement processes is related to the fact that the ratio of the pressure gradients upstream and downstream of the condensation front, evaluated at the condensation front, is affected greatly by the condensation mechanism itself. Since thermal drives involve moving hot fluids from one well to another, they require reservoir continuity over distances somewhat greater than the well spacing.

All processes in which hot fluid is injected through the wellhead suffer from heat losses from the injection wellbore to the formations overlying the reservoir (the overburden). Such heat losses can be a significant fraction of the injected heat when the wells are deep or poorly insulated and the injection rates are low. Under such conditions the temperature of an injected noncondensable fluid entering the formation may be significantly lower than at the wellhead. When the injected fluid is condensable, as in the case of steam, the heat losses cause some of the vapors to be condensed. However, the temperature remains approximately constant as long as there is vapor present, unless there is a large friction pressure drop or a large gravity head—two effects that generally tend to offset each other and are negligible in most current operations.

Surface lines generally are insulated, and injection wells can be completed in such a way as to reduce heat losses. Because of the thermal stresses on the casing, it is usual to use new wells, cemented to surface, for injection. Existing wells in good condition often are used for production in drive projects.

In its simplest form, a hot-water flood (hot-water drive) involves the flow of only two phases: water and oil (Dietz, 1967). Hot-water flooding is basically a displacement process in which oil is displaced immiscibly by both hot and cold water. Except for temperature effects and the fact that they generally are applied to relatively viscous crude oils, hot-water floods have many elements in common with conventional waterfloods.

Because of the pervasive presence of water in all petroleum reservoirs, displacement by hot water must occur to some extent in all thermal recovery processes. It is known to contribute to the displacement of oil in the zones downstream of both steam drives and combustion drives. Thus, many elements of the discussion on hot-water drives presented in this chapter are applicable to appropriate regions in other processes.

In the hot-water flood, the leading edge of the injected hot water loses heat so rapidly that it quickly reaches the initial reservoir temperature, and the oil mobility is that of the unheated oil at the leading edge of the displacement front. On the other hand, the viscosity of the injected hot water is lower than in conventional waterfloods. Thus, the mobility ratio of the oil ahead of the displacement front and the injected water near the injection well is less favorable in hot-water floods than in conventional waterfloods (which should result in somewhat earlier water breakthrough than hot-water floods). On the other hand, the mobility ratio of the fluids in the heated zones is more favorable in the hot-water floods than in conventional waterfloods. This results in better displacement efficiency from the heated zone and would improve the ultimate recovery even where residual oil saturation does not decrease with increasing temperature.

Typically, in hot-water flooding, the water is filtered, treated to control corrosion and scale, heated, and if necessary, treated to minimize the swelling of clays in the reservoir. The complex mineralogy of clay minerals (Grim, 1968) must be taken into account wherever clay occurs in a reservoir.

Thermal expansion of oil also contributes to the improved displacement efficiency of thermal projects. Gravity separation usually is accentuated by the increased density difference between water and oil as temperatures increase. In addition, most of the reservoir heat is in the zones from which most of the oil already has been displaced. Indeed, all thermal drives are characterized by the presence of large amounts of heat in oil-depleted portions of the reservoir; this has prompted a number of modifications aimed at scavenging or recycling the heat to improve the efficiency of the process (Lake and Walsh, 2004; see chapter: Occurrence and Reservoir Evaluation).

The amount of oil displaced in a hot-water drive is invariably larger than the amount produced and the oil that is displaced but not produced is stored in unswept portions of the reservoir. In the case of viscous crude oils especially, the mobility ratio between the advancing oil and any gas or water in the reservoir is very favorable. This means that crude oil tends to fill regions of the reservoir initially filled with mobile gas and water before it is produced. Where an oil bank forms, consideration of these effects allows an estimation of the recovery history from estimates of the oil displacement history.

Although there are no simple methods for predicting oil recovery from hot-water floods, an approach is suggested that (1) is based on conventional waterflood technology; (2) appears to have some of the elements necessary to describe hot-water floods; and (3) considers only the effects of permeability variations and mobility ratio. It is by no means a proven method and is offered merely as an example of how it may be possible to adapt the existing technology from a related field when there is no other method for making the desired estimates of oil recovery.

In addition to cost, the choice of the fluid is controlled by the expected effect of the fluid on the production response of the crude oil as well as on availability of fluids. For example, seawater may be used for injection into an undersea reservoir where the cost of water delivery to the platform would be prohibitive. On the other hand, the choice of injection fluid is governed by the nature of the reservoir, with each reservoir being considered a specific entity subject to the character of the oil and the mineralogy of the reservoir rock (Fig. 9.2). For example, steam or hot water should not be injected without first considering the possible consequences of its effects on reservoirs containing significant amounts of swelling clays.

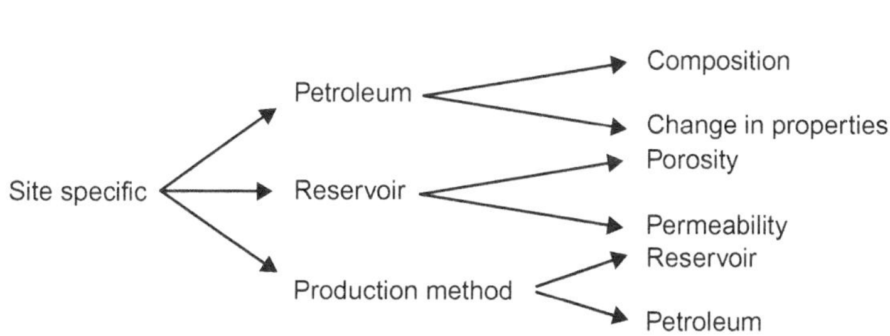

Figure 9.2 Parameters that constitute to site specificity.

Finally, thermal recovery processes may be used for either stimulation or flooding, and the choice is usually dictated by the properties of the oil and the reservoir. In reservoirs that are small or that have relatively poor continuity, it may not be feasible to drill enough wells to ensure adequate communication to implement a flooding operation. It may be necessary to consider a stimulation treatment to increase both recovery rates and ultimate economic recovery. Cyclic steam injection has proved to be a successful stimulation technique for heavy oil as long as there is the means to produce the crude oil once its viscosity is reduced. Combustion stimulation has been found to be successful in burning solid organic particles, in stabilizing clays, and in increasing absolute permeability near the treated well. These effects are particularly attractive in low-permeability reservoirs.

9.3 STEAM-BASED PROCESSES

The steam-based processes are the most advanced of all enhanced oil recovery methods in terms of field experience and thus have the least uncertainty in estimating performance (EPRI, 1999; Curtis et al., 2002; Advanced Resources International, 2005). However, as with all enhanced oil recovery processes, a good reservoir description is a necessity (see chapter: Occurrence and Reservoir Evaluation; Lake et al., 1992). Steam processes are most often applied in reservoirs containing viscous oils and tars, usually in place of rather than following secondary or primary methods. Commercial application of steam processes has been underway since the early 1960s. In contrast, although in situ combustion has been field tested under a wide variety of reservoir conditions, few projects have proved to be economical and advanced to commercial scale.

Once-through boilers customarily are used to produce steam and varying amounts of hot water. The hot water usually is injected into the reservoir along with the steam, although sometimes the amount of hot water injected is reduced by separating the vapor and liquid phases upstream of the wellhead. The separated hot water usually is recycled through the boiler feed system (if the water is sufficiently clean) or used to preheat the feed water to the boiler. Because heating promotes deposition of dissolved solids, it is especially important in steam injection operations to treat the boiler feed-water properly. Precautions for reducing heat losses in surface and injection lines are similar to those to be taken in hot-water flooding. Because of the condensation resulting from heat losses, the water content of the steam entering a formation is generally higher than that at the wellhead. Injected or condensed fresh (hot) water may create problems in reservoirs containing swelling clays.

In addition to possessing the displacement mechanisms and thermal phenomena present in hot-water floods (which lead to improvements in displacement efficiency resulting from a decrease in viscosity, a reduction in capillary forces, and thermal expansion of the crude), steam drives also (1) cause boiling and steam stripping of the light components in the crude; (2) promote the formation of a low-viscosity oil bank near the condensation front; (3) yield unusually low residual oil saturations; and (4) improve the effective mobility ratio of the displacing process.

In the steam-stripping process, steam removes relatively light components from the crude, much as a dry current of air passing over a pool of water lowers the vapor pressure and picks up moisture. Steam stripping removes a larger fraction of the crude than would be suggested by its boiling point distribution (eg, the boiling point distribution of a particular crude might indicate that 20% of the crude boils at 400°F, but when the vapor pressure is maintained at a low level by the flowing vapor, more than 20% of the crude would be removed). As the steam condenses, most of the stripped components in the steam also condense, and these light ends condensing at the steam condensation front help to improve the displacement efficiency. The mixing of the condensing light ends with crude reduces the viscosity of the crude contacted by a subsequent advance of the condensation front. Eventually, the oil bypassed by the advancing condensation front is so light and distillable that it shrinks significantly as steam continues to pass by. Such shrinkage of the bypassed oil can result in very low ultimate values for the residual oil in steam drives (These effects would be more pronounced with crudes containing a high percentage of light fractions but would occur to some extent with any crude). In addition, the condensation phenomenon itself, quite apart from any distillation effects, results in more favorable effective mobility ratios than would have been expected on the basis of viscosity ratios alone.

In *steam stimulation* processes, heat and drive energy are supplied in the form of steam injected through wells into the heavy oil reservoir. In most instances, the injection pressure must exceed the formation fracture pressure in order to force the steam into the reservoir and into contact with the oil. When sufficient heating has been achieved, the injection wells are closed for a soak period of variable length and then allowed to produce, first applying the pressure created by the injection and then using pumps as the wells cool and production declines.

The concept behind the steam-based processes is generally viscosity reduction so that the heavy oil can flow to the production well. Other effects such as emulsion drive, thermal expansion, solution gas drive, and steam distillation are also believed to occur. The extent of each is unknown and oil dependent. In the steam-stripping processes, steam removes relatively low-boiling components from the heavy oil. In fact, steam stripping removes a larger fraction of the crude oil than would be suggested by the boiling point distribution because of *steam distillation* (Northrop and Venkatesan, 1993). There is no chemical change to the constituents of the oil, although there may be favorable compositional changes to the oil insofar as lighter fractions are recovered and heavier materials remain in the reservoir. As the steam condenses, most of the stripped or steam-distilled components in the steam also condense and form a solvent front that will assist in displacing heavy oil (as a solvent/oil mix) towards the production well. These effects help to improve the displacement efficiency.

Whether or not steam distillation occurs and the role it pays in oil recovery depends on the character of the heavy oil as well as the downhole conditions. Steam distillation can indeed augment the process when the steam-distilled material moves with the steam front and acts as a solvent for oil ahead of the steam front (Prats, 1986).

When dealing with heavy oil, it can be reasonably assumed that the steam-distilled material is composed of aromatics and naphthenic constituents which are excellent solvents for oil (Mitchell and Speight, 1973). On the other hand, if the steam-distilled constituents have a predominantly paraffinic character (which is not considered likely for heavy oil but very likely for residual oil in an energy-depleted conventional oil reservoir), this will cause deposition of asphaltic material ahead of the steam front. The ever-present disadvantage is that any deposited material will block the reservoir flow channels, thereby restricting oil flow to the production well. The distinct advantage is that if flow to the recovery well is not restricted, there is the potential for recovery of partially upgraded heavy oil.

Another effect is that the oil bypassed by the advancing condensation front is so volatile that it shrinks significantly as steam continues to pass by, resulting in low recovery of the heavy oil. These effects would be more pronounced with the residual oil from conventional crude oil recovery, which contains relatively high levels of volatile constituents. Overall, the condensation steam-stripping and steam-distillation effects result in more favorable recovery than is expected on the basis of viscosity reduction alone.

Steam can also be injected into one or more wells with production coming from other wells (*steam drive*). This technique is very effective in heavy oil formations but has found little success during application to heavy oil reservoirs because of the difficulty in connecting injection and production wells. However, once the flow path has been heated, the steam pressure is cycled, alternately moving steam up into the oil zone, and the oil is allowed to drain down into the heated flow channel to be swept to the production wells.

If the viscous bitumen in a tar-sand formation can be made mobile by admixture of either a hydrocarbon diluent or an emulsifying fluid, a relatively low temperature secondary recovery process is possible (*emulsion steam drive*). If the formation is impermeable, communication problems exist between injection and production wells. However, it is possible to apply a solution or dilution process along a narrow fracture plane between injection and production wells.

All of the processes in which hot fluid is injected through the wellhead suffer from heat losses from the injection wellbore to the overburden formations. These heat losses can be a significant fraction of the injected heat when the wells are deep or poorly insulated and the injection rates are low. Under such conditions, the temperature of an injected noncondensable fluid entering the formation may be significantly lower than at the wellhead. When the injected fluid is condensable, as in the case of steam, the heat losses cause condensation of the vapor. If the steam is unable to maintain the heat and enters the reservoir as hot water, the oil in the reservoir actually comes into contact with hot water rather than steam. Whether or not the reservoir is injected with steam is subject to the depth of the reservoir (ie, the distance the surface-generated steam has to travel to enter the reservoir). Cooling of the steam by any encroaching water could have a significantly adverse effect not only on the rate of growth of the steam zone but also on the initiation of the steam zone. However, there are several schools of thought that contend that the majority of steam-based processes are, in fact, hot-water-based processes. To overcome the steam-or-hot-water issue, surface lines generally are insulated, and injection wells are completed in a manner that reduces heat losses.

9.3.1 Steam Drive Injection

Steam drive injection (steam injection) has been commercially applied since the early 1960s. Steam drives differ markedly in performance from hot-water drives, the difference being solely due to the presence and effects of the condensing vapor. The presence of the gas phase causes light components in the crude oil to be distilled and carried along as hydrocarbon components in the gas phase. Where the steam condenses, the condensable hydrocarbon components do likewise, thus reducing the viscosity of the crude oil at the condensation front. Moreover, the condensing steam makes the displacement process more efficient and improves the sweep efficiency. The net effect is that recovery from steam drives is significantly higher than from hot-water drives.

The process occurs in two steps: (1) steam stimulation of production wells (ie, direct steam stimulation); and (2) steam drive by steam injection to increase production from other wells (ie, indirect steam stimulation; see chapter: General Methods of Oil Recovery). When there is some natural reservoir energy, steam stimulation normally precedes steam drive. In steam stimulation, heat is applied to the reservoir by the injection of high-quality steam into the production well.

Huff and puff employs a pattern of vertical wells. Typically, pressurized steam is pumped down a well for weeks and sometimes months, thoroughly heating the reservoir near the well. The process is then halted, usually for several weeks; this allows the heavy oil to become separated from the reservoir sand. Then, the heavy oil is artificially lifted from the same well. This technology has been successfully used for many years but is fast being replaced in Canada by the SAGD method. It is still considered applicable in heavy oil regions where SAGD is not suitable.

9.3.2 Cyclic Steam Injection

Cyclic steam injection (see chapter: General Methods of Oil Recovery) is the alternating injection of steam and production of oil with condensed steam from the same well or wells. This process is predominantly a vertical well process, with each well alternately injecting steam and producing heavy oil and steam condensate. In practice, steam is injected into the formation at greater than fracturing pressure. This is followed by a soak period after which production is commenced. The heat injected warms the heavy oil and lowers its viscosity. A heated zone is created through which the warmed heavy oil can flow back into the well. This is a well-developed process; the major limitation is that less than 30% (usually less than 20%) of the initial oil in place can be recovered.

Cyclic steam stimulation is sometimes applied to heavy oil reservoirs to boost recovery during the primary production phase. During this time it assists natural reservoir energy by thinning the oil so it will more easily move through the formation to the injection/production wells. The minimum depth for applying cyclic steam stimulation is on the order of 1000 feet; depth varies depending upon the type and structure of the overlying formations.

In the cyclic steam stimulation, a predetermined amount of high-pressure, high temperature steam (estimated to be at approximately 350°C, 650°F) is injected into wells that have been drilled or converted for injection purposes into the reservoir or deposit. The pressure of the steam fractures the oil sand, while the heat of the steam melts the heavy oil or bitumen. The wells are then shut in to allow the steam to heat (soak period) the producing formation around the well. As the steam soaks into the deposit, the heated material flows to a producing well and is pumped to the surface. After a sufficient time has elapsed to allow adequate heating, the injection wells are returned to production until the heat is dissipated with the produced fluids. This cycle of soak-and-produce, or (huff-and-puff) may be repeated until the response becomes marginal because of declining natural reservoir pressure and increased water production. At this time a continuous steam flood is usually initiated to continue the heating and thinning of the oil and to replace declining reservoir pressure so that production may continue. When the steam flooding is started, some of the original injection wells will be converted for use as production wells, along with the others drilled or designated for that purpose. This process can be repeated several times in a formation, and it can take between 120 days and 2 years to complete a steam stimulation cycle.

Cyclic steam stimulation works best when there are thick pay zones (>10 m) with high porosity sands (>30%). Shale layers that reduce vertical permeability are not a problem for vertical wells that penetrate thick pay zones. However, good horizontal permeability (>1 Darcy) is important for production. Recently, cyclic steam stimulation has been applied to wells with multilateral horizontal legs.

There are three phases in cyclic steam stimulation. First, high-temperature, high-pressure steam is injected for up to one month. Second, the formation is allowed to soak for 1 or 2 weeks to allow the heat to diffuse and lower the heavy oil viscosity. Third, heavy oil is pumped out of the well until production falls to uneconomic rates, which may take up to 1 year. Then the cycle is repeated, as many as 15 times, until production can no longer be recovered. Artificial lift is required to bring the heavy oil to surface.

Typical recovery factors for cyclic steam stimulation are 20% to 35% with SORs of 3 to 5.22. Steam floods may follow cyclic steam stimulation.

While cyclic steam stimulation produces the heavy oil around a single wellbore, steamflooding recovers the heavy oil between wells. For example, a five-spot pattern with four producing wells surrounding a central steam injection well is a common configuration. The well spacing can be less than two acres for a field in steam flood. The steam heats the oil to lower its viscosity and provides pressure to drive the heavy oil toward the producing wells. In most steamflood operations, all of the wells are steam-stimulated at the beginning of the flood. In a sense, cyclic steam stimulation is always the beginning phase of a steamflood. In some cases, even the steamflood injection wells are put on production for one or two cyclic steam stimulation cycles to help increase initial project production and pay out the high steamflood capital and operating costs.

Cyclic steam stimulation and steam floods are used in California, western Canada, Indonesia, Oman, and China. California's Kern River production rose from less than 20,000 barrels per day in the late 1950s before cyclic steam stimulation to over 120,000 barrels per day by 1980 after the introduction of cyclic steam stimulation. The Duri field in Indonesia is the world's largest steamflood and produces 230,000 barrels per day with an estimated ultimate recovery factor of 70% in some locations.

Cyclic steam injection is used extensively in heavy-oil reservoirs, tar sand deposits, and in some cases, to improve injectivity prior to steam flooding or in situ combustion operations. The technique has also been applied to the California tar sand deposits (Bott, 1967) and in some heavy oil reservoirs north of the Orinoco deposits (Franco, 1976; Ballard et al., 1976). The steamflooding technique has been applied, with some degree of success, to the Utah tar sands (Watts et al., 1982) and has been proposed for the San Miguel (Texas) tar sands (Hertzberg et al., 1983).

Technical challenges for cyclic steam stimulation and steamflooding are primarily related to reducing the cost of steam, which is generated in most locations using natural gas. The economics may be improved by also generating and selling electricity and by using waste heat for cogeneration. Alternative fuels (coal, heavy ends, coke) are discussed separately below; they could also reduce the cost of steam generation. Monitoring and controlling the steam front could also reduce costs by redirecting steam to zones where the heavy oil has not been produced. Steam could be shut off from zones that have been successfully swept and directed toward unswept regions.

Gravity override is a natural occurrence in every steamflood. The steam breaks through to the producers, at which time the process turns into a gravity drainage process. The steam chest at the top of the formation expands downward, and the heated heavy oil drains by gravity to the producing wells. Although the geometry configuration is different to that of the SAGD process, the basic effects are the same.

The measurement of the produced fluids (oil, water, and natural gas) at the surface for each well can be used to optimize production by adjusting artificial lift rates and steam injection rates. Downhole fluid-flow measurements could be used to identify which zones are producing oil, water, or gas in a producing well. Monitoring may involve drilling observation wells where permanent sensors may be deployed or where logging can be periodically performed. Downhole temperature and pressure sensors may use fiber-optic or wireline technology. Water and steam saturation outside the casing of an observation well can be measured with nuclear spectroscopy logs. Time-lapse, cross-well electrical imaging can be used to identify bypassed heavy oil zones between closely spaced (500 m) observation wells. Cross-well seismic and surface seismic measurements might be used to locate steam fronts. High-resolution imaging of the formation and the fluid saturations before completing the wells and during production, however, is an open technical challenge.

Technologies must be reliable and have long operating periods between service periods. High-temperature- (up to 300°C) and corrosion-resistant equipment, including pumps (artificial lift), cements, completions, liners, packers, valves, electronics, and sensors, are needed. Thermal expansion of the formation can also cause the casing to fail.

Most cyclic steam stimulation and steamflood wells have been vertical wells. More recently, vertical wells with multilateral branches and horizontal wells are being tried. The advantage is a reduced footprint while tapping large subsurface regions. Optimal control and configuration of these wells for cyclic steam stimulation and steam flood recovery processes are still being developed.

Cyclic steam injection also is used as a precursor to steam drive technology. In reservoirs containing heavy crude oil, the resistance to flow between the wells may be sufficiently high that steam injection rates are severely limited, making steam drive technically inefficient. Cyclic steam injection reduces the flow resistance near wells, where the resistance is most pronounced; this alone improves the injection rate attainable during steam drives by reducing the resistance to flow between wells. Repeated cyclic steam injection reduces the flow resistance still farther from the wells and may lead to connecting the heated zones of adjacent wells and further improving the operability of the steam drive.

A drawback to cyclic steam injection is that the ultimate recovery may be low relative to the total oil in-place in the reservoir. Ultimate recoveries from steam drives are generally much larger than those from cyclic steam injection. Thus, cyclic steam injection followed by a steam drive is an attractive combination; crude production is accelerated quickly, and the ultimate recovery is quite high.

9.3.3 Steam Drive

Steam drive involves the injection of steam through an injection well into a reservoir and the production of the mobilized heavy oil and steam condensate from a production well. Steam drive is usually a logical follow-up to cyclic steam injection. Steam drive requires sufficient effective permeability to allow injection of the steam at rates sufficient to raise the reservoir temperature to move the heavy oil to the production well.

Two expected problems inherent in the steam drive process are steam override and reservoir plugging. Any in situ thermal process tends to override (migrate to the top of the effected interval) because of differential density of the hot and cold fluids. These problems can be partially mitigated by rapid injection of steam at the bottom or below the target interval through a high-permeability water zone or fracture. Each of these options will raise the temperature of the entire reservoir by conduction and, to a lesser degree, by convection, and the effectiveness of the following injection of steam into the target interval will be enhanced.

For a successful steam drive project, the porosity of the reservoir rock should be at least 20%, the permeability should be at least 100 mD, and the heavy oil saturation should be at least 40%. The reservoir oil content should be at least 800 bbl per acre-foot. The depth of the reservoir should be less than 3000 feet, and the thickness should be at least 30 feet. Other preferential parameters have also been noted on the basis of success with several heavy oil reservoirs.

Finally, injection of hot noncondensable gases has been attempted only infrequently. The low viscosity of noncondensing gases results in poor mobility ratios and, consequently, poor displacement efficiencies (Fig. 2.2). Stripping of light ends also may be important in hot-gas drives. Here condensation of the stripped components results from cooling of the gas stream (which is analogous to meteorological precipitation) rather than, as in the case of steam, from the condensation of the carrier vapor phase. Because of the low density of gases, high volumetric gas injection rates (referred to standard conditions) are required to provide heat injection rates comparable with those obtainable with water or steam. At per-well injection rates normally encountered in field operations, wellbore heat losses reduce the temperature of hot dry gases significantly.

9.3.4 Steam-Assisted Gravity Drainage

SAGD was initially developed to recover bitumen from the Canadian oil sands (Dusseault et al., 1998). The key element of SAGD is that the two wells need to be parallel and horizontal. It is only in the last 10 to 15 years that directional drilling technology has been able to achieve these two characteristics with any degree of certainty.

In the simplest form, the SAGD technology requires the drilling of two parallel horizontal wells through the oil-bearing formation. Into the upper well, steam is injected creating a high-temperature steam chamber. The increased heat loosens the thick crude oil causing it to flow downward in the reservoir to the second horizontal well. This second well is located parallel to and below the steam injection well. This heated, thinner oil is then pumped to the surface via the second horizontal, or production well. Water is injected into the bitumen-drained area to maintain the stability of the deposit.

Several variations of this process have been developed—one variation uses a single horizontal well, with steam injection through a central pipe and production along the annulus while another variation involves steam injection through existing vertical wells and production through an underlying horizontal well. The key benefits of the SAGD process are an improved steam–oil ratio and high ultimate recovery (on the order of 60% to 70%). The outstanding technical issues relate to low initial oil rate, artificial lifting of bitumen to the surface, horizontal well operation, and the extrapolation of the process to reservoirs having low permeability, low pressure, or bottom water. Steam, perhaps along with a mixture of hydrocarbons that dissolve into the oil and help reduce its viscosity, is injected into the upper well. The heat reduces the oil viscosity to values as low as 1 to 10 cP (depending on temperature and initial conditions) and develops a "steam chamber" that grows vertically and laterally. The steam and gases rise because of their low density, and the oil and condensed water are removed through the lower well. The gases produced during SAGD tend to be methane, with some carbon dioxide and traces of hydrogen sulfide.

Even though the injection and production wells can be very close (between 5 and 7 m), the mechanism causes the steam-saturated zone (known as the steam chamber) to rise to the top of the reservoir, expand gradually sideways, and eventually allow drainage from a very large area. The method is claimed to significantly improve heavy oil recovery by between 50% and 60% of the original oil in place (OOIP) and is therefore more efficient than most other thermal recovery methods.

To a small degree, the noncondensable gases tend to remain high in the structure, filling the void space and even acting as a partial "insulating blanket" that helps to reduce vertical heat losses as the chamber grows laterally. At the pore scale, and at larger scales as well, flow is through countercurrent, gravity-driven flow, and a thin and continuous oil film is sustained, giving high recoveries estimated to be as large as 70% to 80% in suitable reservoirs.

Operating the production and injection wells at approximately the same pressure as the reservoir eliminates viscous fingering and coning processes and also suppresses water influx or oil loss through permeable streaks. This keeps the steam chamber interface relatively sharp, and reduces heat losses considerably. Injection pressures are much lower than the fracture gradient, which means that the chances of breaking into a thief zone (an instability problem that plagues all high-pressure steam injection processes, such as cyclic steam soak) are essentially zero.

The SAGD process, like all gravity-driven processes, is extremely stable because the process zone grows only by gravity segregation, and there are no pressure-driven instabilities such as channeling, coning, and fracturing. It is vital in the SAGD process to maintain a volume balance, replacing each unit volume withdrawn with a unit volume injected to maintain the processes in the gravity-dominated domain. If bottom-water influx develops, this indicates that the pressure in the water is larger than the pressure in the steam chamber, and steps must be taken to balance the pressures. Because it is not possible to reduce the pressure in the water zone, the pressure in the steam chamber and production well region must be increased. This can be achieved by increasing the operating pressure of the steam chamber through the injection rate of steam or through reduction of the production rate from the lower well. After some time, the pressures become more balanced and the water influx ceases. Thereafter, maintaining the volume balance carefully is necessary.

Clearly, a low pressure gradient between the bottom water and the production well must be sustained. If pressure starts to build up in the steam chamber zone, then loss of hot water can take place as well. In such cases, the steam chamber pressure must be reduced and perhaps the production rate must be increased slightly to balance the pressures. In all these cases, the system tends to return to a stable configuration because of the density differences between the phases.

SAGD seems to be relatively insensitive to shale streaks and similar horizontal barriers, even up to several meters thick (3 to 6 feet), that otherwise would restrict vertical flow rates. This occurs because as the rock is heated, differential thermal expansion causes the shale to be placed under a tensile stress, and vertical fractures are created, which serve as conduits for steam (up) and liquids (down). As high temperatures hit the shale, the kinetic energy in the water increases, and adsorbed water on clay particles is liberated. Thus, instead of thermal expansion, dehydration (loss of water) occurs, and this leads to volumetric shrinkage of the shale barriers. As the shale shrink, the lateral stress (fracture gradient) drops until the pore pressure exceeds the lateral stress, which causes vertical fractures to open. The combined processes of gravity segregation and shale thermal fracturing make SAGD so efficient that recovery ratios of 60% to 70% are probably achievable even in cases where there are many thin shale streaks. However, there are limits on the thickness of the shale bed that can be traversed in a reasonable time.

Heat losses and deceleration of lateral growth mean that there is an economic limit to the lateral growth of the steam chamber. This limit is thought to be a chamber width of four times (4×) the vertical zone thickness. For thinner zones, horizontal well pairs would therefore have to be placed close together, increasing costs as well as providing lower total resources per well pair. Consequently, the zone thickness limit (net pay thickness) must be defined for all reservoirs.

The cost of heat is a major economic constraint on all thermal processes. Currently, steam is generated with natural gas, and when the cost of natural gas rises, operating costs rise considerably. SAGD is about twice as thermally efficient as cyclic steam stimulation for similar cases, with steam–oil ratios that are now approaching two (instead of four for cyclic steam soak). If high recovery ratios are possible, SAGD will likely displace pressure-driven thermal process in all cases where the reservoir is reasonably thick and favorable to the process; not all reservoirs are favorable to the SAGD process. Additionally, because of the lower pressures associated with SAGD in comparison to high-pressure processes such as cyclic steam soak and steam drive, greater wellbore stability reduces substantially the number of sheared wells that are common in cyclic steam soak projects.

Finally, because of the lower pressures associated with SAGD, in comparison to high pressure processes such as cyclic steam soak and steam drive, greater wellbore stability should be another asset, reducing substantially the number of sheared wells that are common in cyclic steam soak projects. However, SAGD is not suitable for all reservoirs. It performs best in clean, continuous sands, and it requires continuous vertical permeability. It did not work well on the Clearwater formation at Cold Lake (Alberta, Canada), for example, in part because the Clearwater formation is neither as permeable nor as vertically continuous as the McMurray (Hart Energy, 2006).

Recently, one of the new improved techniques in SAGD is *FAST-SAGD*, which combines conventional SAGD wells with cyclic steam stimulation wells (offset wells) (Shin and Polikar, 2004, 2005). These offset wells are drilled at the same depth as that of SAGD production well. After short period of time of operation of the SAGD, which are started to operate similarly to the conventional SAGD, steam is injected through cyclic steam stimulation wells with high injection rate under high operation pressure to make heat easily transfer to reservoir.

However, in the FAST-SAGD operation, steam has a tendency to bypass other wells during the injection period due to the operation of the cyclic steam stimulation wells. Hence live steam will be produced at producer, which has an adverse effect on the thermal efficiency of the process thereby causing the production rate to decrease significantly. In order to offset this condition, there are several constrains that need to be considered: (1) the location of the offset wells should be calculated carefully so that the steam chambers are operated at the expected time to get the effective thermal process; (2) the cyclic steam stimulation wells should be planned with consideration of reservoir parameters—permeability, oil saturation, formation heat conduction, and heterogeneity (such as the presence of a shale barrier or clay lenses) in order to achieve high efficiency.

As a result of recent innovations a collection of hybrid variants are available for thermal recovery processes (Table 9.1). Recognizing that there is no one process that will satisfy the requirements for heavy oil and bitumen recovery, a hybrid process (as the name suggests) is based on a recovery concept (that has been tested and authenticated) and consists of an added option to assure increased recovery. For example, the hybrid SAGD process (HSAGD process) uses a similar well configuration to the FAST-SAGD method (fracture-assisted steam technology (FAST) coupled with the SAGD process). However, the wells are operated very differently in terms of the operating conditions. In Fast-SAGD process, the SAGD wells are operated first and cyclic steam stimulation wells (offset wells) start later and require higher injection pressure and injection rate.

Table 9.1 Hybrid variants proposed for thermal recovery processes

Steam-assisted gravity drainage (SAGD)
FAST-SAGD
ES-SAGD (expanding solvent-SAGD)
HASD (horizontal alternate steam drive)
SAGP (steam and gas push)
SAGOGD (steam-assisted gas oil gravity drainage)
In situ combustion
THAI (toe-heel air injection)
Basal combustion
Top-down combustion
Solvent-based injection
VAPEX (vapor extraction)
LASER (liquid addition to steam for enhancing recovery)
SAS (steam-alternating solvent process)
Carbon dioxide-based VAPEX process

Therefore steam is easy bypass to other wells but the HSAGD process can improve this phenomenon. In the HSAGD process, all cyclic steam stimulation wells are placed in a staggered pattern and the wells are operated at the same pressure and placed in operation earlier than SAGD wells.

Another concept related to the VAPEX process but used in a different manner can be used in conjunction with a steam-based process, such as SAGD. In the N-Solv process (http://www.n-solv.com/process.htm) (Canadian Patent Applications, 2299790, 2351148, 2374115), heated solvent vapor is injected at moderate pressures into the gravity drainage chamber. The vapor flows from the injection well to the colder perimeter of the chamber, where it condenses. This delivers heat and fresh solvent directly to the bitumen extraction interface. The extraction conditions are mild compared to in situ steam processes, so the valuable components in the bitumen are preferentially extracted, and the problematic high-molecular-weight coke-forming species (asphaltenes) are left behind. The condensed solvent and oil then drain by gravity to the bottom of the drainage chamber and are recovered via the production well.

9.4 IN SITU COMBUSTION PROCESSES

The other significant thermal recovery process, in situ combustion (fire flooding) attempts to recover oil by igniting a portion of the in-place crude by injecting air or oxygen or by chemical or electrical means. The process has always appeared to be attractive—especially for the recovery of bitumen from tar sand deposits and the potential for in situ upgrading of heavy oil and bitumen. Thus, in addition to providing the heat to mobilize the oil, in situ combustion of heavy oil can provide some in situ upgrading through the use of minerals or additives (Dabbous and Fulton, 1972; Fassihi et al., 1984a, b; He, 2004; He et al., 2005; Shallcross et al., 1991; Strycker et al., 1999; Castanier and Kovscek, 2005). During in situ combustion of heavy oils, temperatures of up to 700°C (1290°F) can be observed at the combustion front, which is sufficient to promote some upgrading (depending upon the residence time of the heavy oil or bitumen in the hot zone).

In situ combustion is normally applied to reservoirs containing low-gravity oil but has been tested over perhaps the widest spectrum of conditions of any enhanced oil recovery process. In the process, heat is generated within the reservoir by injecting air and burning part of the crude oil. This reduces the oil viscosity and partially vaporizes the oil in place, and the oil is driven out of the reservoir by a combination of steam, hot water, and gas drive. Forward combustion involves movement of the hot front in the same direction as the injected air; reverse combustion involves movement of the hot front opposite to the direction of the injected air.

During *in situ combustion* or *fire flooding*, energy is generated in the formation by igniting bitumen in the formation and sustaining it in a state of combustion or partial combustion. The high temperatures generated decrease the viscosity of the oil and make it more mobile. Some cracking of the bitumen also occurs and an upgraded product rather than bitumen itself is the fluid recovered from the production wells.

Heavy oil and bitumen upgrading is of major economic importance. Numerous field observations have shown upgrading of 2° to 6° API for heavy oils undergoing combustion (Ramey et al., 1992). For the most past, in situ combustion processes are not new; work on various aspects of the processes has continued since at least 1923 (Howard, 1923; Wolcott, 1923; Kuhn and Koch, 1953; Grant and Szasz, 1954). In fact, in situ combustion has been applied in over a hundred fields (Farouq Ali, 1972; Chu, 1977, 1982; Brigham et al., 1980). The South Belridge project, which began commercial operations in 1964, is of special significance since it was a commercial success and analyses of the field data yielded a number of useful, new concepts and correlations. Certain ideas carried over from waterflooding were found to be inappropriate to the in situ combustion process (Gates and Ramey, 1958, 1980; Ramey et al., 1992).

The main parameters required in the design of an in situ combustion project are, in addition to operating costs: (1) the fuel concentration per unit reservoir volume burned; (2) the composition of the fuel; (3) the amount of air required to burn the fuel; (4) the volume of reservoir swept by the combustion zone; (5) the required air injection rates and pressures; and (6) the oil production rate (Alexander et al., 1962). Injection of air alone is known as dry underground combustion, in situ combustion, or fire flooding (Kuhn and Koch, 1953). However, there are several variants of the in situ combustion process.

In the process, either dry air or air mixed with water can be injected into the reservoir. This process is can be started by lowering a heater or ignitor into an injection well. Air is then injected down the well, and the heater is operated until ignition is accomplished. After heating the surrounding rock, the heater is withdrawn, but air injection is continued to maintain the advancing combustion front. Water is sometimes injected simultaneously or alternately with air creating steam, which contributes to better heat utilization and reduced air requirements.

In terms of the chemistry of the process, at temperatures in excess of approximately 345°C (650°F), reactions between oxygen and the organic fuel result in the production of carbon dioxide, carbon monoxide, and water as the principal reaction products. However, if the temperature is less than approximately 345°C (650°F), water and oxygenated organic compounds (such as phenol derivatives) are the principal reaction products, along with carboxylic acids, aldehydes, ketones, alcohols, which may be the result of low-temperature oxidation (Burger and Sahuquet, 1972). Moreover, in many cases, it is often assumed that the minerals and the water in the reservoir are inert. However, the presence of pyrite (FeS_2) increases air requirements and naturally-occurring carbonates (MCO_3, M is a divalent metal) and sulfates (MSO_4, M is a divalent metal) decompose at high temperatures, as well as react with acid products formed by low-temperature oxidation (Poettmann et al., 1967).

The combustion zone is burned out as the combustion front advances and any water formed or injected will turn to steam in this zone due to residual heat. This steam flows on into the unburned area of the formation, helping to heat it. The high temperature just ahead of the combustion zone causes lower-boiling constituents of the oil or products to vaporize, leaving a heavy residual coke or carbon deposit as fuel for the advancing combustion front. In the vaporizing that contains combustion products, vaporized low-boiling hydrocarbons, and steam, cooling (owing to the distance of the zone from the combustion front) causes light hydrocarbons to condense and steam to revert back to hot water. This action displaces oil and, with combustion gases, aids in driving the oil to production wells. The oil bank (an accumulation of displaced oil) cools as it moves toward production wells, and temperatures will drop to near the initial reservoir temperature. When the oil bank reaches the production wells, the oil, water, and gases will be brought to the surface and separated—the oil to be sold and the water and gases sometimes reinjected. The process is terminated by stopping air injection when predesignated areas are burned out or the burning front reaches production wells.

Ideally, the fire propagates uniformly from the air injection well to the producing well, moving oil and combustion gases ahead of the front and any coke remaining behind the moved oil provides the fuel. Temperatures in the thin combustion zone may reach several hundred degrees centigrade, sufficient to crack the heavy oil into lower-boiling products. The oil is subjected to a combination of miscible displacement by the condensed light hydrocarbons, hot water drive, vaporization, and steam and gas drive. As the temperature in the volume element exceeds about 345°C (650°F), the oil will more than likely undergo thermal cracking to form a volatile fraction and a low-volatility, coke-like residue. The volatile products are carried in the gas stream, while the coke-like residue is burned as fuel in the combustion zone. The heat generated at the combustion zone is transported ahead of the front by conduction through the formation matrix and by convection of the vapors and liquids (Wu and Fulton, 1971).

The thickness of the combustion zone is variable, and the temperature is usually on the order of 345°C to 650°C (650°F to 1200°F) but may be difficult to control (this is the issue with many combustion-related recovery processes). As the combustion front moves forward, a zone of clean sand is left behind where only air flows. As a result of distillation and thermal cracking, the quality of the produced oil is improved. For example, in South Belridge, the produced oil gravity was as high as 18° API, compared to 12.9° API for the original oil.

Furthermore, in the South Belridge work, it became evident that reservoir lithology is an important parameter in fuel deposition (Gates and Ramey, 1958), and the amount of fuel deposited increases with the addition of clay to the sample of oil and sand (Bousaid and Ramey, 1968). This is in agreement chemically with the known catalyst activity of minerals in refining processes and the ability of clay to adsorb polar constituents such as the carbenes and carboids that are the thermal precursors to coke (Speight, 2014 and references cited therein).

In situ combustion is normally applied to reservoirs containing low-gravity oil but has been tested over perhaps the widest spectrum of conditions of any enhanced oil recovery process. In the process (see chapter: General Methods of Oil Recovery), heat is generated within the reservoir by injecting air and burning part of the crude oil. This reduces the oil viscosity and partially vaporizes the oil in place. The oil is driven out of the reservoir by a combination of steam, hot water, and gas drive.

In *forward combustion*, the combustion front moves in the same direction as the air flow but *reverse combustion* occurs when the combustion front moves in a direction opposite to the flow of the injected air. Reverse combustion is achieved by igniting the crude near a production well while temporarily injecting air into it. Upon resumption of the normal air injection program, the combustion front will move toward the injection wells.

The fire front can be difficult to control, and it may propagate in a haphazard manner, resulting in premature breakthrough to a producing well. There is a danger of a ruptured well with hot gases escaping to the surface. The produced fluid may contain an oil–water emulsion that is difficult to break, and contrary to expectations, it may also contain heavy metal compounds that are difficult to remove in the refinery.

The wellbore near the pay zone, or for that matter any part of the injection well that might come in contact with free oxygen and fuel (crude oil), should be designed for high thermal stresses. Crude oil is likely to enter the wellbore by gravity drainage where the air enters the formation preferentially over a short segment of a large open interval that has adequate vertical permeability. This crude inflow may be increased as the reservoir temperature near the wellbore increases as a result of the heat generated either by the ignition system used in the wellbore or by the combustion process itself (including reverse combustion following spontaneous ignition a short distance into the reservoir). When designing injection wells, precautions should be taken against any likelihood of combustion in the wellbore.

Wet and partially quenched combustion, also known by the acronym COFCAW (combination of forward combustion and waterflooding) (Dietz Weijdema, 1968a; Parrish and Craig, 1969), uses water injection during the combustion process to recuperate the heat from the burned zone and adjacent strata. In this process, the ratio of injected water to air is used to control the rate of advance of the combustion front, the size of the steam zone, and the temperature distribution.

The appeal of an in situ combustion process is the potential for partial upgrading of the oil in the reservoir, providing the undesirable constituents of the oil remain in the reservoir. In addition, the process has the rapid kinetics of a thermal process, and there is no need to generate energy at the surface.

During in situ combustion or fire flooding, energy is generated in the formation by igniting heavy oil in the formation and sustaining it in a state of combustion or partial combustion. The high temperatures generated decrease the viscosity of the oil and make it more mobile. Some cracking of the heavy oil occurs, and an upgraded product rather than heavy oil itself is the fluid recovered from the production wells.

The relatively small portion of the oil that remains after the displacement mechanisms have acted becomes the fuel for the in situ combustion process. Production is obtained from wells offsetting the injection locations. In some applications, the efficiency of the total in situ combustion operation can be improved by alternating water and air injection. The injected water tends to improve the utilization of heat by transferring heat from the rock behind the combustion zone to the rock immediately ahead of the combustion zone.

The performance of in situ combustion is predominantly determined by four factors: (1) the quantity of oil that initially resides in the rock to be burned; (2) the quantity of air required to burn the portion of the oil that fuels the process; (3) the distance to which vigorous combustion can be sustained against heat losses; and (4) the mobility of the air or combustion product gases.

In many field projects, the high gas mobility has limited recovery through its adverse effect on the sweep efficiency of the burning front. Because of the density contrast between air and reservoir liquids, the burning front tends to override the reservoir liquids. To date, combustion has been most effective for the recovery of viscous oils in moderately thick reservoirs in which reservoir dip and continuity provide effective gravity drainage or operational factors permit close well spacing.

The use of combustion to stimulate oil production is regarded as attractive for deep reservoirs. In contrast to steam injection, it usually involves no loss of heat. The duration of the combustion may be less than 30 days or as much as 90 days, depending on requirements. In addition, backflow of the oil through the hot zone must be prevented or coking will occur.

The process (1) has the potential to be thermally more efficient than steam-based processes; (2) has no depth restriction, other than advised caution in shallow-depth reservoirs/deposits; and (3) is well suited to relatively thin (less than 25 feet) reservoir/deposit sands. However, the process is complicated insofar as it is difficult to predict or control, and operational problems from the high temperature include cement failures, sanding/erosion, corrosion at both injection and production well because of oxygen and moisture, and high gas production rate.

9.4.1 Forward Combustion

The most common form of in situ combustion is *dry forward combustion*, which involves movement of the hot front in the same direction as the injected air while reverse combustion involves movement of the hot front opposite to the direction of the injected air. In forward combustion, the hydrocarbon products released from the zone of combustion move into a relatively cold portion of the formation. Thus, there is a definite upper limit of the viscosity of the liquids that can be recovered by a forward-combustion process. On the other hand, since the air passes through the hot formation before reaching the combustion zone, burning is complete; the formation is left completely cleaned of hydrocarbons.

In this process, air is injected into a heavy oil reservoir, the crude is ignited in situ, and the resulting combustion front moves away from the injection well. The heat generated at the combustion front propagates through the reservoir, reduces the oil viscosity, and thereby increases the oil production rate and recovery. The propagation of a combustion front in a reservoir is the most rapid method of thermal recovery. The combustion front can move more rapidly than heat can be moved by conduction and convection in a reservoir, and the convective heat wave velocity for the case of air injection is about one quarter that of the combustion front (Martin et al., 1958; Ramey, 1971).

In forward combustion, the hydrocarbon products released from the zone of combustion move into a relatively cold portion of the formation. Thus, there is a definite upper limit of the viscosity of the liquids that can be recovered by a forward combustion process. On the other hand, since the air passes through the hot formation before reaching the combustion zone, burning is complete; the formation is left completely cleaned of hydrocarbons.

Generally, forward combustion is referred to as *dry forward combustion*, and the effects of any reservoir water are, for unknown reasons, ignored. Temperature levels in dry forward combustion, which affect the displacement, distillation, stripping, cracking, and formation of solid fuel downstream of the combustion front, are affected by the amount of fuel burned per unit volume of reservoir rock. At high temperatures (approximately 815°C, 1500°F), the combustion zone is very thin, whereas at lower temperatures (approximately 345°C, 650°F), a smoldering reaction with the bypassed air may occur over distances of several feet and generate heat and ultimately cause spontaneous ignition. At intermediate temperatures (approximately 600°C, 1200°F), the combustion reaction proceeds slowly enough to allow significant leakage of free oxygen in the direction of flow, thus increasing the thickness of the reaction zones.

A characteristic of the dry forward combustion process is that the temperature of the burned zone remains quite high because the heat capacity of the air injected is too low to transfer a significant amount of heat. For this reason, water sometimes is used during or after the combustion process to help transfer the heat from the burned zone to downstream areas.

Forward combustion is particularly applicable to reservoirs containing mobile heavy oil and/or with a high effective permeability. Even though lower effective reservoir permeability is required for air injection compared with steam injection, the reservoir ahead of the combustion front is subject to plugging as the vaporized fluids cool and condenses. Consequently, a relatively high permeability (400 to 1000 mD) and relatively low bitumen saturation (45% to 65% of pore volume) are most favorable for this process. The combustion process yields a partially upgraded product because the temperature gradient ahead of the combustion front mobilizes the lighter hydrocarbon components that move toward the cooler portion of the reservoir and mix with unheated bitumen. This mixture is eventually produced through a production well. The heavier components (eg, coke) are left on the sand grains and are consumed as fuel for the combustion. Under certain operating conditions a significant cost saving is attained by injecting oxygen or oxygen-enriched air rather than atmospheric air because of reduced compression costs and a lower produced gas/oil ratio.

Another form of in situ combustion is the *wet combustion* method, in which air and water are injected concurrently or alternately. The purpose of injecting water is to recuperate and transport heat from the burned zone to the colder regions downstream of the combustion front. This method may be considered for thin reservoirs, where heat loss to adjacent formations is significant (Dietz and Weijdema, 1963; Parrish and Craig, 1969; Dietz, 1970; Beckers and Harmsen, 1970; Burger and Sahuquet, 1973).

The addition of water during the combustion process means that heat is transferred more effectively than with air alone. In addition, the steam zone ahead of the combustion front is larger, and the reservoir is swept more efficiently than with air alone. The improved displacement from the steam zone results in lower fuel availability and consumption in the combustion zone, so a greater volume of the reservoir is burned for a given volume of air injected.

Water must be injected in the wet combustion process, but it may be difficult to inject both air and water simultaneously at the desired rates in low-permeability reservoirs. In such cases, the water and air can be injected alternately, and the duration of the air and water injection periods can be controlled to achieve the desired average water/air ratio, which is essential to obtain the desired combustion front velocity and temperature level. At a low water/air ratio, the water that reaches the combustion front has been converted to steam, whereas at a high water/air ratio, the water that reaches the combustion front is, for the most part, in the liquid phase.

One of the benefits of the combustion process is the production of a partially upgraded product. The temperature gradient ahead of the combustion front either causes the lower-molecular-weight (more volatile) constituents to distill and move toward the cooler portion of the reservoir and mix with unheated oil or some of the higher-molecular-weight (less volatile) constituents crack (thermally decompose) and the volatile products also move toward the cooler portion of the reservoir and mix with unheated oil. The cracking process also produces a carbonaceous residue that deposits on the reservoir rock and is consumed as fuel during the combustion process.

The operative term for any in situ combustion project is *process control* or *flame front control*, which related to the amount of air and the amount and type of fuel in the subterranean formation(s). The fuel burned in a reservoir also is known as the *fuel content*, the *fuel consumption*, or the *fuel availability*. In terms of the fuel itself, the fuel that is burned during the forward in situ combustion process is not the crude oil in the reservoir but it is, for the most part, the carbonaceous residue that was produced by thermal cracking and distillation (near the combustion front) of the residual crude oil—in the current context, the fuel will be part of the heavy oil or tar sand bitumen. The amount of fuel present per unit bulk volume of reservoir is an extremely important parameter in combustion operations and is a determinant in the amount of air required to burn a unit bulk volume of reservoir. In wet combustion operations, however, the addition of sufficient water may reduce the temperature to the point that the fuel will not burn completely. Where the combustion of all the fuel is unnecessary, incomplete combustion may be an advantage since it reduces the amount of air required to burn through a unit bulk volume of reservoir. The process also can be carried out under partially quenched conditions so that a mid-boiling fraction of the crude oil is burned or oxidized at low temperatures. On the other hand, the reverse combustion process also consumes a mid-boiling fraction of the crude oil.

Air consumption is affected not only by reactions with organic fuel but also by reactions between oxygen and some minerals (such as pyrite—iron sulfide—FeS_2) in the reservoir. However, when minerals are involved in the process and react, special attention must be paid to the amount of fuel burned.

In addition to the fuel available and/or consumed, the parameters of interest are (1) the amount of air required to burn a unit bulk volume of reservoir rock; and (2) the amount of crude available for displacement from the burned zone. The general assumption is that any unburned crude oil is displaced, this is not typically the case and there is usually a (partially-combusted or thermally-generated) carbonaceous residue, similar to coke or semicoke. In such cases, the estimate of fuel available—the carbonaceous residue plus the fuel burned—cannot, in principle, be assumed to be the fuel burned. However, where combustion occurs at high temperatures, the carbonaceous residue may be negligible. In fact, there is a reasonable correlation between fuel burned and air consumed when the combustion occurs at high temperatures. At lower temperatures some fraction of the consumed oxygen generally is reacted with the crude without generating carbon oxide gases or water.

9.4.2 Reverse Combustion

A third variation of the in situ combustion process is the *reverse combustion* method. In this technique, the combustion zone is initiated at a production well. The reverse combustion front travels countercurrent to the air towards the injection well where air is injected. The oil flows towards the production well, through the combustion zone. Since no oil bank is formed, the total flow resistance decreases with time, and thus this method is particularly suitable for reservoirs containing very viscous crude oils. One disadvantage of this method is the likelihood of spontaneous ignition. Spontaneous ignition would result in oxygen being consumed near the injector, and the process would change to forward combustion (Dietz and Weijdema, 1968b). Another disadvantage of reverse combustion is the inherent instability of the process, which results in narrow combustion channels being formed and therefore an inefficient burn (Gunn and Krantz, 1980; Johnson et al., 1980). Historically, reverse combustion has been difficult to maintain because the oxygen is depleted not far from the injection well (Dietz and Weijdema, 1968b). Furthermore, sustained air injection into an unheated reservoir generally leads to spontaneous ignition near the injection well (Elkins et al., 1974; Burger, 1976; Tadema and Weijdema, 1970).

Reverse combustion is particularly applicable to reservoirs with lower effective permeability, in contrast with forward combustion. The method is more effective because the lower permeability would cause the reservoir to be plugged by the mobilized fluids ahead of a forward combustion front. In the reverse combustion process, the vaporized and mobilized fluids move through the heated portion of the reservoir behind the combustion front. The reverse combustion partially cracks the oil, consumes a portion of the oil as fuel, and deposits residual coke on the sand grains, leaving 40% to 60% of the oil as recoverable oil. This coke deposition serves as a cementing material, reducing movement and production of sand.

As the combustion front reaches the heavy oil, a significant amount of cracking occurs, and a relatively large amount of solid fuel (compared to the amount formed during forward combustion) is deposited on the reservoir rock. However, recovery of the partially upgraded product is lower than in forward combustion because some is burned in the process, and a high degree of *equivalent oil saturation* occurs in the burned zone (including unburned solid fuel). On the other hand, the API gravity of the recovered product is increased significantly by extensive cracking because the products flow through the hot burned zone and are subject to secondary, tertiary, or higher level cracking reactions.

The addition of water or steam to an in situ combustion process can result in a significant increase in the overall efficiency of the process. Modes of application include injection of alternate slugs of air (oxygen) and water or coinjection of air (oxygen) and steam. Again, the combination of air (oxygen) injection and steam or water injection increases injectivity costs that may be justified by increased bitumen recovery. Two major benefits may be derived. Heat transfer in the reservoir is improved because the steam and condensate have greater heat-carrying capacity than combustion gases and gaseous hydrocarbons. Sweep efficiency may also be improved because of the more favorable mobility ratio of steam–oil compared with gas–oil.

Process efficiency is affected by reservoir heterogeneity that reduces horizontal sweep. The underburden and overburden must provide effective seals to avoid loss of injected air and produced oil. Process efficiency is enhanced by the presence of some interstitial water saturation. The water is vaporized by the combustion and enhances the heat transfer by convection. The combustion processes are subject to override because of differences in the densities of injected and reservoir fluids. Production wells should be monitored for and equipped to cool excessively high temperatures (>1095°C, >2000°F) that may damage downhole production tools and tubulars.

Applying a preheating phase before the recovery phase may significantly enhance the steam or combustion extraction processes. Preheating can be particularly beneficial if the saturation of the more viscous oil (API gravity = 12° or lower) is sufficiently great to lower the effective permeability to the point that production is precluded by reservoir plugging. Preheating increases the mobility of the oil by raising its temperature and lowering its viscosity, and the outcome is a lower required pressure to inject steam or air to recover the oil.

Using combustion to stimulate oil production is regarded as attractive for deep reservoirs. In contrast to steam injection, it usually involves no loss of heat. The duration of the combustion may be less than 30 days or approximately 90 days, depending upon requirements. In addition, backflow of the oil through the hot zone must be prevented or coking occurs.

When it is determined that the reservoir should be preheated, there are several methods by which this can be accomplished. Conducting a reverse combustion phase in a zone of relatively high effective permeability and low oil saturation is one method. Steam or hot gases may be rapidly injected into a high-permeability zone in the lower portion of the reservoir. In the *FAST* process, steam is injected rapidly into an induced horizontal fracture near the bottom of the reservoir to preheat the reservoir. This process has been applied successfully in three pilot projects in southwest Texas. Shell has accomplished the same preheating goal by injecting steam into a high-permeability bottom water zone in the Peace River (Alberta) field. Electrical heating of the reservoir by radio-frequency waves may also be an effective method.

A variation of the combustion process involves use of a heat-up phase, a blow-down (production) phase, and then a displacement phase using a fire-water flood (a *combination of forward combustion and waterflood*, COFCAW). In the process, water is injected simultaneously or alternately with air into the formation and is actually wet forward combustion, which was developed to use the great amount of heat that would otherwise be lost in the formation. The injected water recovers the heat from behind the burning front and transfers it to the oil bank ahead. Because of this additional energy, the oil displacement is more efficient and requires less air. In spite of these advantages, a wet combustion process cannot avoid liquid-blocking problems and use of wet combustion is limited by the oil viscosity.

In any field in which primary recovery operations are followed by secondary or enhanced recovery operations, there is a change in product quality. Product oils recovered by the thermal stimulation of heavy oil reservoirs show some improvement in properties over those of the heavy oil in place. Although this improvement in properties may not appear to be too drastic, nevertheless it usually is sufficient to have major advantages for refinery operators. Any incremental increase in the units of the hydrogen/carbon ratio can save amounts of costly hydrogen during upgrading. The same principles are also operative for reductions in the nitrogen, sulfur, and metals content. This removal of nitrogen, sulfur, and metals from the products also improves catalyst life and activity when the product oil is refined. In short, in situ recovery processes may have the added benefit of *leaving* some of the more obnoxious constituents (from the processing objective) in the ground.

A combustion displacement process may be more attractive than a steam drive process, assuming that a downhole steam generator is also not available to the reservoir. Conditions that might preclude a steam drive and be in favor of a combustion process include (1) high sustained injection pressures—above 1500 psi; (2) excessive heat losses from the injection well in reservoirs more than 4000 feet deep; (3) a lack of a supply of fresh water or treatment costs that make the use of steam prohibitively expensive; (4) serious clay swelling problems; and (5) thin or low-porosity sands where heat management using wet combustion processes tends to make them more efficient than steam drives (Wilson and Root, 1966).

Wet combustion would be considered instead of dry combustion where there is ample available water and where water/air injectivity is favorable. Wet combustion would not be used where there is little likelihood that the water would move through the burned zone to recuperate heat effectively, as in gravity-dominated operations (Koch, 1956; Gates and Sklar, 1971).

9.4.3 The THAI Process

In situ combustion may make a comeback with a relatively recent concept, or at least a derivative thereof. THAI (toe-to-heel air injection) (Fig. 9.3) is based on the geometry of horizontal wells that may solve the problems

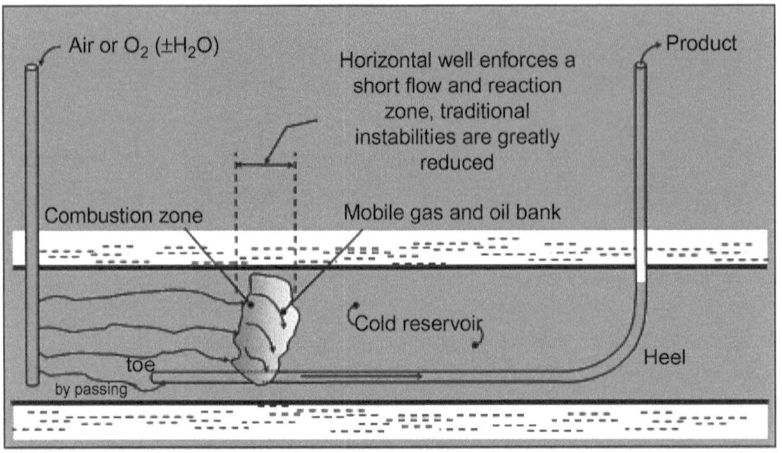

Figure 9.3 The THAI process.

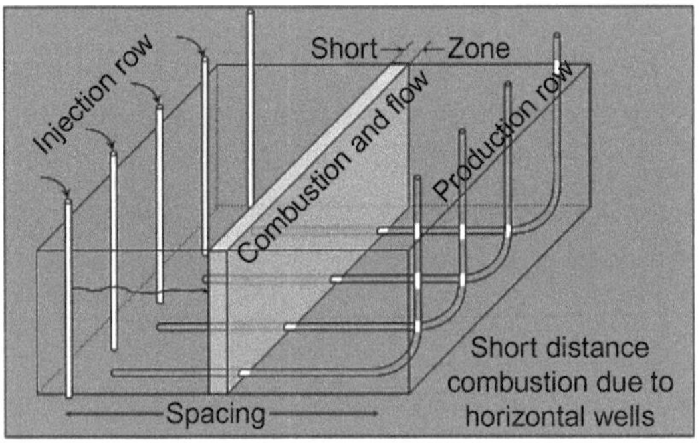

Figure 9.4 Configuration of the THAI process.

that have plagued conventional in situ combustion. The well geometry (Fig. 9.4) enforces a short flow path so that any instability issues that are associated with conventional combustion are reduced or even eliminated.

In the process, a vertical air injection well is combined with a horizontal production well and offers a way to control the forefront and produce a product that is partially upgraded and ready for the refinery. In the process, a combustion front is created where part of the oil in the reservoir is burned, generating heat that reduces the viscosity of the oil, allowing it to flow by gravity to the horizontal production well. The combustion front sweeps the oil from the toe to the heel of the horizontal producing well, recovering an estimated 80% of the OOIP while partially upgrading the crude oil in situ.

The THAI process has the potential to operate in reservoirs that are lower in pressure, of a lower quality, thinner, and deeper than required for the SAGD process. In addition, the THAI process integrates existing proven technologies and provides the opportunity to create a step change in the development of heavy oil resources globally. The horizontal well arrangement provides a unique gravity and pressure drawdown geometry. Another key feature is that oil recovery occurs via a short displacement mechanism, which requires oil to move downwards (with the help of gravity) typically just 15 to 30 feet, as opposed to the lateral movement of several hundred feet in the usual combustion processes.

The THAI process begins with preheating both wellbores using steam to initiate oil mobility and clear pore space between the injector and the toe of the producing well. After ignition (autoignition with some oils), the energy to sustain in situ combustion comes from the burning of the coke that is continuously laid down within the reservoir. Product sulfur is reduced, as are heavy metals, which are left as an inert residue on the reservoir rock. No water or gas fuel is required during production, and the produced water can be treated to usable industrial quality.

A further benefit of the THAI process is that it performs in situ upgrading through thermal cracking of the heavy oil. Laboratory tests achieved upgrading by up to 10° API. On this basis, a 10° API oil could be expected to yield an 18° to 20° API oil at the surface. This is a very desirable feature of any recovery process since every increase of 1° API can mean refinery savings of several dollars per barrel.

The reservoir matrix is an important aspect of the THAI process; some clay content is required to help thermal cracking of the oil ahead of the combustion front. The process can break down intermittent shale breaks that could otherwise obstruct permeability. The THAI process can also cope with a bottom water zone by effectively steaming it out. Tests also indicate that the THAI process can handle a gas cap; gravity and pressure draws down all the gasses ahead of the combustion front.

Ahead of the combustion front (typically around 600°C, 1112°F) is the coking zone, in front of which is a 10 to 15-feet-wide mobile oil zone through which drainage takes place into the horizontal well. These zones move through the reservoir at about 1 to 3 feet per day, depending on the air injection rate. The temperature drops to between 200°C and 350°C at the front of the mobile oil zone, with a corresponding reduction in the rate of drainage. Ahead of the mobile oil zone is the cold immobile virgin oil layer, through which there is no communication for gas. This characteristic of the process geometry means that the only way out is down into the open section of the horizontal well. The horizontal well trajectory is thus a built-in self-controlling guidance system for fluid flow. This makes the process much more controllable than conventional in situ recovery systems, in which fluids are less controlled and can move and penetrate anywhere in the reservoir.

In summary, the THAI process has many potential technical benefits including: (1) higher source recovery that is estimated at 70% to 80% of the oil in place, and it is potentially feasible for use over a broader range of reservoirs, including reservoirs having low pressure, thin reservoirs, and previously steamed reservoirs; (2) well geometry that enforces a short flow path so that the instabilities associated with the longer flow path in conventional combustion methods are reduced or even eliminated; and (3) a lower environmental impact insofar as there is negligible fresh water use, fewer greenhouse gas emissions, a smaller surface footprint, and easier reclamation.

As an extension of the THAI process, the CAPRI process involves a layer of refinery-type catalyst along the outside of the horizontal producer well; it is therefore the catalytic variant of the THAI process. It uses an annular sheath of solid catalyst surrounding the horizontal producer well in the bottom of the oil layer. The thermally cracked oil produced drains into the horizontal producer well, first passing through the layer of catalyst where the high pressure and temperature in the reservoir enable thermal cracking and hydroconversion reactions to take place so that only light, converted oil is produced at the surface.

In summary, the THAI technology offers many potential advantages over the SAGD technology, including higher resource recovery of the OOIP, lower production and capital costs, minimal usage of natural gas and fresh water, a partially upgraded crude oil product, reduced diluent requirements for transportation, and significantly lower greenhouse gas emissions. The THAI process also has the potential to operate in reservoirs that are lower in pressure, contain more shale, are lower in quality, and are thinner and deeper than SAGD. This type of technology could be utilized in deep heavy oil resources both onshore and offshore.

9.5 COMPARISON OF RECOVERY METHODS

Thermal recovery processes are used for either stimulation or flooding. The choice frequently is governed by the properties of the reservoir.

In reservoirs that are small or that have relatively poor continuity, it may not be feasible to drill enough wells to implement a flooding operation. Indeed, if it is prohibitively costly to drill additional wells to ensure adequate communication over close spacing, there may be no choice but to consider a stimulation treatment to increase both recovery rates and ultimate economic recovery. Where the crude viscosity is high, cyclic steam injection has proved to be a successful stimulation technique as long as natural driving forces are available to produce the crude once its viscosity is reduced. Combustion stimulation has been found to be successful in burning solid organic particles, in stabilizing clays, and in increasing absolute permeability near the treated well. These effects are particularly attractive in low-permeability reservoirs.

Cyclic steam injection also is used as a precursor to steam drives. In reservoirs containing very viscous crudes, the flow resistance between wells may be so great that steam injection rates are severely limited, making steam drives both technically inefficient and uneconomic. Cyclic steam injection reduces the flow resistance near wells, where the resistance is most pronounced; this alone improves the injection rate attainable during steam drives by reducing the resistance to flow between wells. Repeated cyclic steam injection reduces the flow resistance still farther from the wells and may lead to connecting the heated zones of adjacent wells and further improving the operability of steam drives.

In addition to sometimes being a necessary precursor to a steam drive, cyclic steam injection also is preferred for economic reasons. Because it is a stimulation treatment, it accelerates oil production. There is no long delay in obtaining a production response, as is the case for steam drives. Furthermore, since steam injection into any well lasts only a few weeks and the resulting production lasts several months, portable steam generators have been developed. A single portable steam generator can be moved from well to well to apply the cyclic steam injection process to several wells at a reasonable capital cost. A drawback to cyclic steam injection, however, is that the ultimate recovery may be low relative to the total oil in-place in the reservoir. Ultimate recoveries from steam drives are generally much larger than those from cyclic steam injection. Thus, cyclic steam injection followed by a steam drive is an attractive combination in that crude production is accelerated quickly and the ultimate recovery is quite high.

Combustion displacement processes may be more attractive than steam drives. Conditions that might call for combustion rather than steam injection include: (1) high sustained injection pressures (above 1500 psi); (2) excessive heat losses from the injection well (in reservoirs more than 4000 ft deep); (3) a lack of a supply of fresh water or treatment costs that make the use of steam prohibitively expensive; (4) serious clay swelling problems due to fresh condensate; (5) undesirable or prohibited use of fuel to fire steam generators; and (6) thin or low-porosity sands, where the heat recuperation in wet combustion processes tends to make them more efficient than steam drives. Wilson and Root have discussed the comparative costs of reservoir heating using steam and air injection. The successful development of downhole steam generators would affect conditions 1 and 2 and probably the comparative costs of the processes.

Wet combustion would be considered instead of dry combustion where there is ample available water and where water/air injectivity is favorable. Wet combustion would not be used where there is little likelihood that the water would move through the burned zone to recuperate heat effectively, as in gravity-dominated processes.

The effective mobility ratio of steam drives is more favorable than that of combustion processes. Also, for combustion processes there is a rather high reported incidence of well failure associated with high temperatures, corrosion, and erosion. These factors prompt us to recommend that *when the economics are the same (laying aside considerations of risk)*, steam injection is to be preferred to a combustion drive. However, each process has its limitations, and sometimes the conditions may point to only one thermal process. That process may well be a combustion one.

Calculation of the production schedules, that is, the amount of oil estimated to be produced as a function of time (for any producing mechanism), is one of the main goals of reservoir engineering. The approach followed in this monograph regarding the estimation of recovery resulting from the several thermal recovery processes is to present the calculation methods that are applicable to simple geometries and that can be used to explain the various mechanisms pertinent to field projects. The geometry most often considered is the uniform parallel (linear) flow system, and emphasis is given to those methods for which the use of hand-held calculators is sufficient. In connection with the use of hot water and steam, buoyancy and permeability layering effects often prevent the flow from being uniform over a vertical section of the reservoir. The same is true of combustion. In most cases, there are no simple methods for predicting the crude recovery under such conditions. Indeed, numerical simulation may be the only way to estimate crude production response where there is buoyancy or lack of vertical uniformity. Although, in terms of accuracy, numerical simulation methods of predicting recovery are far superior to desk-top methods, the latter (ie, desktop methods) can be very useful because of their availability and ease of use, remembering that methods developed for residual oil recovery are not immediately applicable to heavy oil recovery and to bitumen recovery.

Finally, it is likely that one or more hybrid approaches that involve the simultaneous use of several technologies will see increased future application. Each will be chosen according to the efficiency of the recovery phase and the ability to encourage increased production.

As an addendum to this chapter, mention must be made to the proposed recovery process that involves microwave heating and the potential for in situ upgrading. Microwave technology has been identified as having potential application to tar sand recovery (Bosisio et al., 1977; Mutyala et al., 2010). The ability of microwaves to heat the deposit selectively and volumetrically leads to the possibility for the extraction of a product in high yield, and that the oil can be upgraded in the process (Bosisio et al., 1977). The potential for the success of microwave heating for heavy oil or bitumen extraction leads to the hypothesis that microwave heating can remove the hydrocarbonaceous products (bitumen or a bituminous product) from tar sand deposits by selectively heating the water phases that are present interstitially, and within the hydrophilic clay minerals. Preliminary analysis has shown the product quality to be favorable but a rigorous assessment of the concept is still necessary. Whilst the preliminary analysis has shown that the oil quality can be favorable, a rigorous assessment is needed to understand the requirements for further upgrading of the product. Furthermore, it is possible that the clay in the sand also may play a role as a catalyst.

REFERENCES

Advanced Resources International, 2005. Basin Oriented Strategies for CO2 Enhanced Oil Recovery: Onshore California Oil Basins. Prepared for the US Department of Energy, Washington, DC.

Alexander, J.D., Martin, W.L., Dew, J.N., 1962. Factors affecting fuel availability and composition during in-situ combustion. J. Petrol. Technol. 4, 1154–1164.

Ballard, J.R., Lanfranchi, E.E., Vanags, P.A., 1976. Proceedings. Twenty-seventh Annual Meeting Petroleum Society, Canadian Institute of Mining. Calgary, Alberta, Canada, June.

Beckers, H.L., Harmsen, G.J., 1970. The effect of water injection on sustained combustion in a porous medium. Soc. Petrol. Eng. J. June, 145–163.

Bosisio, R., Cambon, J., Chavarie, C., Klvana, D., 1977. Experimental results on the heating of athabasca tar sand samples with microwave power. J. Microwave Power 12, 301–307.

Bott, R.C., 1967. Cyclic steam project in a virgin tar reservoir. J. Petrol. Technol. 19, 585–591.
Bousaid, I.S., Ramey Jr., H.J., 1968. Oxidation of crude oil in porous media. Soc. Petrol. Eng. J. June, 137–148.
Brigham, W.E., Satman, A., Solitaire, M.Y., 1980. Recovery correlations for in-situ combustion field projects and application to combustion pilots. J. Petrol. Technol. December, 2132–2138.
Burger, J.G., Sahuquet, B.C., 1972. Chemical aspects of in-situ combustion: heat of combustion and kinetics. Soc. Petrol. Eng. J. October, 410–422.
Burger, J.G., Sahuquet, B.C., 1973. Laboratory research on wet combustion. J. Petrol. Technol. October, 1137–1146.
Burger, J.G., 1976. Spontaneous ignition in oil reservoirs. Soc. Petrol. Eng. J. April, 73–81.
Castanier, L.M., Kovscek, A.R., 2005. Heavy-Oil upgrading in situ via solvent injection and combustion: a "New" method. European Association of Geoscientists and Engineers 67th Conference and Exhibition. Madrid, Spain, June 13–16.
Chu, C., 1977. A study of fireflood field projects. J. Petrol. Technol. February, 171–179.
Chu, C., 1982. State-of-the-Art review of fireflood field projects. J. Petrol. Technol. January, 19–36.
Curtis, C., Kopper, R., Decoster, E., Guzmán-Garcia, A., Huggins, C., Knauer, L., et al., 2002. Heavy oil reservoirs. Oilfield Rev. Autumn, 30–51.
Dabbous, M.K., Fulton, P.F., 1972. Low temperature oxidation kinetics and effects on the in-situ combustion process. Paper No. SPE 4143. Society of Petorleum Engineers-American Institute of Mechanical Engineers 47th Annual Fall Meeting. San Antonio, Texas, Oct. 8–11.
Dietz, D.N., 1967. Hot water drive. Proceedings. Volume 3, 451–457. Seventh World Petroleum Congress. Mexico City, Mexico.
Dietz, D.N., 1970. Wet underground combustion, state of the art. J. Petrol. Technol. May, 605–617.
Dietz, D.N., Weijdema, J., 1963. Wet and partially quenched combustion. J. Petrol. Technol. April, 411–415.
Dietz, D.N., Weijdema, J., 1968a. Wet and partially quenched combustion. J. Petrol. Technol. 20, 411–413.
Dietz, D.N., Weijdema, J., 1968b. Reverse combustion seldom feasible. Producers Monthly 32 (5), 10.
Dusseault, M.B., Geilikman, M.B., Spanos, T.J.T., 1998. J. Petrol. Technol. 50 (9), 92–94.
Elkins, L.F., Skov, A.M., Martin, P.J., Lutton, D.R., 1974. Experimental fireflood – carlyle field, kansas. Paper No. SPE 5014. Society of Petroleum Engineers Annual Meeting, Houston, Texas, October 6–9.
EPRI, 1999. Enhanced oil recovery scoping study. Report TR-113836. Electric Power Research Institute, Palo Alto, California.
Farouq Ali, S.M., 1972. A current appraisal of in-situ combustion field tests. J. Petrol. Technol. April, 477–486.
Fassihi, M.R., Brigham, W., 1984b. Reaction kinetics of in-situ combustion. Soc. Petrol. Eng. J. Sept, 399–416.
Fassihi, M.R., Brigham, W.E., Ramey, H.H., 1984a. Reaction kinetics of in-situ combustion: Part 1– observations. Soc. Petrol. Eng. J. August, 399–416.
Franco, A., 1976. Oil Gas J. 74 (14), 132.
Gates, C.F., Sklar, I., 1971. Combustion as a primary recovery process – midway sunset field. J. Petrol. Technol. Pet. 23, 981–986.
Gates, G.F., Ramey Jr., H.J., 1958. Field results of south belridge thermal recovery experiment. Trans. Am. Inst. Mech. Eng. 213, 236–244.

Gates, G.F., Ramey Jr., H.J., 1980. Method of engineering in-situ combustion oil-recovery projects. J. Petrol. Technol. February, 285–294.

Grant, B.F., Szasz, S.E., 1954. Development of an underground heat wave for oil recovery. J. Petrol. Technol. May, 22–23.

Grim, R.E., 1968. Clay Mineralogy. McGraw Hill, New York.

Gunn, R.D., Krantz, W.B., 1980. Reverse combustion instabilities in tar sands and coal. Soc. Petrol. Eng. J. August, 267–277.

Hart Energy, 2006. Heavy oil: unleashing the potential. Suppl. E&P Oil Gas Investor http://www.hartenergy.com.

He, B., 2004. The effect of metallic salt additives on in-situ combustion performance. M.S. Report. Stanford University, Stanford, California. <http://ekofisk.stanford.edu/pereports/web/default.htm>.

He, B., Chen, Q., Castanier, L.M., Kovscek, A.R., 2005. Improved in-situ combustion performance with metallic salt additives. Paper No. SPE 93901. Proceedings of the Society of Petroleum Engineers Western Regional Meeting. Irvine, California, March 30–April 1.

Hertzberg, R., Hojabri, F., Ellefson, L., 1983. Preprint No. 35e. Summer National Meeting of the American Institute of Chemical Engineers. Denver, Colorado, August 28–31.

Howard, F.A., 1923. Method of Operating Oil Wells. U.S. Patent No. 1,473,348.

IEA, 2005. Resources to reserves—oil and gas technologies for the energy markets of the future. International Energy Agency, Paris. <www.iea.org/Textbase/publications/free_new_Desc.asp?PUBS_ID=1568> (accessed 31.03.15.).

Jamaluddin, A.K.M., Mehta, S.A., Moore, R.G., 1998. Downhole heating device to remediate near-wellbore formation damage related to clay swelling and fluid blocking. Paper No. 98–73. Proceedings. Annual Technical Meeting, Petroleum Society of the Canadian Institute of Mining, Calgary, Alberta, Canada. June 8–10. Canadian Institute of Mining, Westmount, Quebec, Canada.

Johnson, L.A., Fahy, L.J., Romanowski, L.J., Barbour, R.V., Thomas, K.P., 1980. An echoing hl-Situ combustion oil recovery project in a utah tar sand. J. Petrol. Technol. February, 295–305.

Koch, R.L., 1956. Practical use of combustion drive at west newport field. Petrol. Eng. January, 72.

Kuhn, C.S., Koch, R.L., 1953. In-Situ combustion – newest method of increasing oil recovery. Oil Gas J. 52 (14), 92.

Lake, L.W., Schmidt, R.L., Venuto, P.B., 1992. A niche for enhanced oil recovery in the 1990s. Oilfield Rev. January, 55–61.

Lake, L.W., Walsh, M.P., 2004. Primary Hydrocarbon Recovery. Elsevier, Amsterdam, Netherlands.

Martin, W.L., Alexander, J.D., Dew, J.N., 1958. Process variables of in-situ Combustion. Trans. AIME 213, 28–35.

Mitchell, D.L., Speight, J.G., 1973. The solubility of asphaltenes in hydrocarbon solvents. Fuel 52, 149.

Mutyala, S., Fairbridge, C., Paré, J.R.J., Bélanger, J.M.R., Ng, S., Hawkins, R., 2010. Microwave applications to oil sands and petroleum: a review. Fuel Process. Technol. 91, 127–135.

Northrop, P.S., Venkatesan, V.N., 1993. Analytical steam distillation model for thermal enhanced oil recovery processes. Ind. Eng. Chem. Res. 32 (9), 2039–2046.

Parman, D., Ojeda, S., 2013. Use of electric downhole heaters to improve production and recovery of heavy, viscous oil in california and venezuela. Paper No. SPE 167347. Proceedings. SPE Kuwait Oil and Gas Show and Conference, Kuwait City, Kuwait. October 8–10. Society of Petroleum Engineers, Richardson, Texas.

Parrish, D.R., Craig Jr., F.F., 1969. Laboratory study of a combination of forward combustion and waterflooding – the COFCAW process. J. Petrol. Technol. 21, 753–761.
Poettmann, F.H., Schilson, R.E., Surkalo, H., 1967. Philosophy and technology of in-situ combustion in light oil reservoirs. Proceedings. 7th World Petroleum Congress, 3: 487.
Prats, M., 1986. Thermal Recovery. Society of Petroleum Engineers, Richardson, Texas.
Ramey Jr., H.J., 1971. In-Situ combustion. Proceedings. World Petroleum Congress, Moscow, pp. 253–262.
Ramey Jr., H.J., Stamp, V.V., Pebdani, F.N., 1992. Case history of south belridge, california, in-situ combustion oil recovery. Paper No. SPE 24200. Proceedings. 9th Society of Petroleum Engineers/US Department of Energy Enhanced Oil Recovery Symposium. Tulsa, Oklahoma. April 21–24.
Rodriguez, R., Bashbush, J.L., Rincón, A., 2008. Feasibility of using electrical downhole heaters in faja heavy-oil reservoirs. Paper No. SPE 117682. Proceedings. SPE International Thermal Operations and Heavy Oil Symposium, Calgary, Alberta, Canada. Society of Petroleum Engineers, Richardson, Texas.
Shallcross, D.C., Rios, C.F., De Los, Castanier, L.M., 1991. Modifying in-situ combustion performance by the use of water-soluble additives. SPERE 6 (3), 287–294.
Shin, H., Polikar, M., 2004. Review of reservoir parameter to optimize SAGD and FAST-SAGD operating conditions. Paper No. 2004-211. Proceedings. Canadian International Petroleum Conference (CIPC), Petroleum Society, Canadian Institute Of Mining, Metallurgy & Petroleum, Calgary, Alberta, Canada. June.
Shin, H., Polikar, M., 2005. Optimizing the SAGD process in three major canadian oil-sand areas. Paper No. SPE 95754. Proceedings. SPE Annual Technical Conference and Exhibition, Dallas, Texas. October.
Speight, J.G., 2014. The Chemistry and Technology of Petroleum, fifth ed. CRC Press, Taylor and Francis Group, Boca Raton, Florida.
Strycker, A., Sarathi, P., Wang, S., 1999. Evaluation of in situ combustion for schrader bluff. Topical Report, National Petroleum Technology Office, United States Department of Energy, Washington, DC. http://www.osti.gov/bridge/.
Tadema, H.J., Weijdema, J., 1970. Spontaneous ignition in oil sands. Oil Gas J. December, 77–80.
Watts, K.C., Hutchinson, H.L., Johnson, L.A., Barbour, R.V., Thomas, K.P., 1982. Proceedings. 54th Annual Fall Meeting, Society of Petroleum Engineers, American Institute of Mechanical Engineers. New Orleans. September 26–29.
Wilson, L.A., Root, P.J., 1966. Cost comparison of reservoir heating using steam or air. J. Petrol. Technol. 18, 233–239.
Wolcott, E.R., 1923. Method of Increasing the Yield of Oil Wells. U.S. Patent No. 1,457,479.
Wu, C.H., Fulton, P.F., 1971. Experimental simulation of the zones preceding the combustion front of an in-situ combustion process. Soc. Petrol. Eng. J. March, 38–46.

CHAPTER 10

Upgrading During Recovery

10.1 INTRODUCTION

In spite of the recent discoveries of reservoirs of conventional oil, heavy oil and bitumen still represent a large amount of the worldwide oil reserves which will be essential if future supply to meet the global energy demand for fuels and petroleum derivatives is to be met. The strategy to recover heavy oil and tar sand bitumen, in addition to the properties of the heavy oil and bitumen, depends on the properties of the reservoir or deposit. Once recovered, the heavy oil and bitumen must be transported to the refinery. This requires lowering the viscosity to improve the flow characteristics.

Upgrading heavy crude oil and bitumen as a means of viscosity reduction and quality improvement is of major economic importance (Hedrick et al., 2006; Rana et al., 2007; Shah et al., 2010; Speight, 2013c). Heavy crude oils exist in large quantities in the western hemisphere, but are difficult to produce and transport because of their high viscosity, which is influenced by their composition (Bukka et al., 1991; Speight, 2000, 2014a). Some crude oils, but most heavy oil and bitumen, contain compounds, such as sulfur and/or heavy metals, that cause additional refining problems and costs. In situ upgrading could be a very beneficial process for leaving the unwanted elements in the reservoir and increasing API gravity.

Fluids produced from a well are seldom pure crude oil. In fact, a variety of materials may be produced by oil wells in addition to liquid and gaseous hydrocarbons. The natural gas itself may contain as impurities one or more nonhydrocarbon substances. The most abundant of these impurities is hydrogen sulfide, which imparts a noticeable odor to the gas. A small amount of this compound is considered advantageous as it gives an indication of leaks and where they occur. A larger amount, however, makes the gas obnoxious and difficult to market. Such gas is referred to as *sour gas* (see chapter: "Heavy Oil and Tar Sand Bitumen") and much of it is used in the manufacture of carbon black. A few natural gases contain helium, and this element does in fact occur in commercial quantities in certain gas

Introduction to Enhanced Recovery Methods for Heavy Oil and Tar Sands.
DOI: http://dx.doi.org/10.1016/B978-0-12-849906-1.00010-2

fields; nitrogen and carbon dioxide are also found in some natural gases. Gas is usually separated at as high a pressure as possible, reducing compression costs when the gas is to be used for gas lift or delivered to a pipeline. Lighter hydrocarbons and hydrogen sulfide are removed as necessary to obtain a crude oil of suitable vapor pressure for transport yet retaining most of the natural gasoline constituents.

By far the most abundant extraneous material is water. Many wells, especially during their declining years, produce vast quantities of salt water, and disposing of it is both a serious and an expensive problem. Furthermore, the brine may be corrosive, which necessitates frequent replacement of casing, pipe, and valves, or it may be saturated so that the salts tend to precipitate upon reaching the surface. In either case the water produced with the oil is a source of continuing trouble. Finally, if the reservoir rock is an incoherent sand or poorly cemented sandstone, large quantities of sand are produced along with the oil and gas. On its way to the surface, the sand has been known to scour its way completely through pipes and fittings.

It must also be remembered that in any field where primary production is followed by a secondary or enhanced production method, there will be noticeable differences in properties between the fluids produced. In fact, during primary production, there are reports of the oil showing a gradual increase in density, viscosity, and other properties over time, indicating fractionation during flow through the formation (Reichert et al., 1989). There are also the effects of reverse gas deasphalting to be considered in terms of the change of composition of recovered oil with time (Speight, 2014a). The differences in elemental composition may not reflect these differences to any great extent (Zou et al., 1989), but more significant differences will be evident from an inspection of the physical properties. One issue that arises from the physical property data is that such oils may be outside the range of acceptability for refining techniques other than thermal options. In addition, overloading of thermal process units will increase as the proportion of the heavy oil in the refinery feedstock increases. Obviously there is a need for more and more refineries to accept larger proportions of heavy crude oils as the refinery feedstock and to have the capability to process such materials.

In summary, the technologies applied to oil recovery involve different concepts, some of which can cause changes to the oil during production. Technologies such as alkaline flooding, microemulsion (micellar/emulsion) flooding, polymer augmented water flooding, and carbon dioxide miscible/immiscible flooding do not require or cause any change to the oil. The steaming technologies may cause some steam distillation that can augment the process when the steam distilled material moves with the steam front and acts as a solvent for oil ahead of the steam front (Pratts, 1986). Again, there is no chemical change to the oil although there may be favorable compositional changes to the oil insofar as lighter fractions are recovered and heavier materials remain in the reservoir (Richardson et al., 1992; Sharpe et al., 1995).

Known technologies for refinery upgrading of heavy oil and bitumen can be broadly divided into *carbon rejection* and *hydrogen addition* processes. *Carbon rejection* redistributes hydrogen among the various components, resulting in fractions with increased H/C atomic ratios and fractions with lower H–C atomic ratios. On the other hand, *hydrogen addition* processes involve the reaction of heavy crude oils with an external source of hydrogen and result in an overall increase in H–C ratio. Within these broad ranges, the more common upgrading technologies can be subdivided as follows:

1. *Carbon rejection*
   Visbreaking and coking, which could be achieved during the thermal recovery processes.
2. *Hydrogen addition*
   Hydrovisbreaking, which could be achieved during the steam-based thermal recovery processes.
3. *Separation processes*
   Distillation and deasphalting, which could be achieved (steam distillation) during the steam-based thermal processes and during the various solvent processes.

In addition, the occurrence of steam distillation can produce volatile constituents that serve as a solvent and the combustion processes might be adjusted to produce a small amount of the heavy oil or bitumen as a solvent-type product which will assist in flushing the remaining heavy oil or bitumen out of the reservoir or deposit. More options are now being sought in order to increase process efficiency in terms of the yield of the produced oil or oil product.

An additional issue that no doubt must be overcome is the control of the complex chemistry of the reactions in underground formations. For example, the carbon-rejection processes are thermal processes through which the high molecular weight constituents in the feedstock disproportionate, generating distillates with a high H–C ratio and releasing a highly aromatic residue (typically, coke). This is a free-radical process and involves the homolysis of the carbon–carbon and carbon–heteroatom bonds, followed by β-scission reactions through which, as the reaction progresses, increasingly lower-boiling hydrocarbon products are formed resulting in the production of distillates and gas. However, the aromatic radicals produced by dealkylation tend to react intramolecularly and intermolecularly, giving rise to highly condensed polynuclear structures which become ever less soluble in the reaction mixture and lead to the formation of mesophase and, hence, coke above a certain level. This, in itself is not disastrous but can have serious adverse (disastrous) effects if the coke blocks the flow channels within the reservoir or deposit. The residue's tendency to form coke is linked to the degree of polycondensation of the high molecular weight aromatic structures and is quantified by the formation of thermal coke as determined by application to the feedstock (heavy oil or bitumen) of the standard test method for determination of the carbon residue (ASTM D189, ASTM D524, ASTM D4530; see chapter: "Evaluation of Reservoir Fluids") (Speight, 2014a, 2015).

In terms of recovery operations, thermal methods have been the most successful because of the wide experience acquired during the long period of their utilization. Among the thermal methods, cyclic steam stimulation (see chapter: "Nonthermal Methods of Recovery") is especially promising because it has fast payout, although the recovery factors achieved can be considered to be low in relation to other thermal methods. Chemical flooding (see chapter: "Microbial Enhanced Oil Recovery") brings a large technical efficiency, but aspects related to its technological complexity and its high level of expertise and experience needed to successfully implement it in the oilfield should be taken into account to preserve the economic requirements. On the other hand, partial upgrading during the recovery operation or partial upgrading at the surface after recovery has received considerable attention. Dilution (see chapter: "Microbial Enhanced Oil Recovery") and heating (see chapter: "Nonthermal Methods of Recovery") have been the most widely applied methods in pipeline transportation of heavy oil and bitumen, even with the disadvantage of the expensive cost of heat and the scarcity of diluents and light oils with which the heavy oil and/or bitumen could be mixed (Martínez-Palou et al., 2011). Emulsion formation is another option that is of value in certain cases, as long as the emulsions can be broken at the receiving terminal (Zhang et al., 1991; Salager et al., 2001; Langevin et al., 2004).

In general, thermal processes are not very selective towards the production of distillates since, as the severity of the process increases, the gas yield increases, and problems are encountered in relation to the stability of the reaction products (Speight, 2014a). The quality of the distillates is poor because thermal cracking alone is not capable of removing, in any significant way, the heteroatoms present in the heavy feedstocks. Moreover, the products are typically rich in olefins and dienes and, hence, must be stabilized through hydrotreating. In the hydrogen-addition processes, conversion of the heavy feedstocks and the distillates is achieved through the combined reaction effects of cracking and catalytic hydrogenation of the reactive fragments. In this way, it is possible to control the propagation of the reactions more effectively, especially with regard to reducing the problem of coke formation.

In summary, the principal types of reaction that operate in the thermal (or combustion) processes are, therefore, the dealkylation of aromatic structures, the dehydrogenation of naphthene structures, and condensation of the various (alicyclic and aromatic) systems to form polynuclear aromatic systems. All of these reactions are favored by high temperatures, that is, temperatures generally in excess of 450°C (840°F) (Speight and Ozum, 2002; Parkash, 2003; Hsu and Robinson, 2006; Gary et al., 2007; Speight, 2014a). Such temperatures are not typically attained in steam-based processes but can be anticipated to be operative in combustion-based processes (see chapter: "Nonthermal Methods of Recovery").

A final (nonupgrading and ex situ) option that is already in common practice is to use conventional crude oil (which is readily available in the general area where heavy crude oil and/or bitumen is recovered) to dilute or convert the heavier material in order to produce an acceptable material to transport by pipeline (Yaghi and Al-Bemani, 2002). The dilution option is workable but does not represent any form of upgrading of the heavy oil—the operative word is *dilution*—but there is always the risk of incompatibility and the formation of a separate phase if the properties of the components of the blend are not well understood and offers a limitation due to the potential for incompatibility. If the heavy oil (or bitumen) and the conventional crude oil have limited compatibility this would limit the amount of dilution but since this option does not involve upgrading, it will not be discussed further here and is discussed elsewhere (Speight, 2014a, 2015).

Thus, it is the purpose of this chapter to present an outline of the processes that might be available for in situ upgrading.

10.2 UPGRADING IN SITU

The purpose of converting and upgrading heavy oil and tar sand bitumen during the recovery process is to transform a substrate consisting of high molecular weight constituents, which are viscous and rich in toxic elements and metals, into a recovered product that is less viscous and, while still not fully comparable to conventional crude oil, can be further upgraded to yield saleable products in a conventional refinery (Speight and Ozum, 2002; Parkash, 2003; Hsu and Robinson, 2006; Gary et al., 2007; Speight, 2014a). In such an upgrading scheme, any coke formed remains underground and does not have to be disposed at the surface. This transformation can be achieved directly through thermal or hydrocracking processes, or indirectly through transformation of the feedstock (a quantity of material for feeding the reactor) into syngas (ie, a mixture of carbon monoxide, CO, and hydrogen, H_2) through gasification, and the subsequent production of paraffin through Fischer-Tropsch (FT) synthesis (Chadeesingh, 2011; Speight, 2013a, 2014b; Luque and Speight, 2015).

By all definitions, the quality of heavy oil and bitumen from tar sand deposits is poor as a refinery feedstock. In order for heavy oil bitumen to be processed in refineries, it would be preferable if the heavy oil and especially the tar sand bitumen could receive some degree of upgrading during the recovery process. Both materials are relatively carbon-rich and hydrogen-poor and upgrading requires that some of the carbon be removed (as coke) while adding additional hydrogen to make more valuable hydrocarbon products and it is the most complex area of operation in a tar sand recovery and processing plant. Thus, upgrading is using three main processes: coking removes carbon and breaks large bitumen molecules into smaller parts; distillation sorts mixtures of hydrocarbon molecules into their components; catalytic conversions help transform the heavy oil and bitumen feedstocks into more valuable forms and hydrotreating is used to help remove sulfur and nitrogen and add hydrogen to the reacting feedstock (Moschopedis et al., 1998; Speight, 1984, 1986, 2013c, 2014a).

The new technologies such as steam-assisted gravity draining (SAGD), vapor-assisted petroleum extraction (VAPEX), and the older concept of in situ combustion (ISC; Moore et al., 1995) have the potential to recover heavy oil and/or tar sand bitumen with altered properties, that is, partial upgrading during the recovery operation (Duerkson, 1993). For the SAGD process, in situ conversion of bitumen may require specialized equipment to be installed downhole in each injection well to generate the high steam temperatures required to initiate conversion. In addition, the potential for combustion-assisted gravity drainage (CAGD)—an integrated horizontal well air injection process for recovery and upgrading of heavy oil and bitumen from tar sands—would involve short-distance air injection and direct mobilized oil production, which are the main features of this process that lead to stable sweep and high oil recovery (Ranhema and Daulat, 2010). These characteristics identify the CAGD process as a high-potential oil recovery method either in primary production or as a follow-up process in reservoirs that have been partially depleted. The CAGD process combines the advantages of both gravity drainage and conventional ISC. A combustion chamber develops in a wide area in the reservoir around the horizontal injector and consists of flue gases, injected air, and mobilized oil.

The fuel actually burned in forward ISC is not always the heavy oil the reservoir or the bitumen in the deposit. Typically, it is predominantly the carbon-rich residue that is the product from thermal cracking and distillation of the heavy oil or bitumen near to the combustion front (see chapter: "Nonthermal Methods of Recovery"). The amount of fuel present per unit bulk volume of reservoir is an important parameter in the combustion process and determines the air required to burn a unit bulk volume of reservoir. In wet combustion operations, caution is advised since the addition of water may reduce the temperature to the point that the fuel will not burn completely. Where the combustion of all of the fuel is unnecessary, incomplete combustion may be an advantage since it reduces the amount of air required to burn through a unit bulk volume of reservoir or deposit.

However, whether or not upgrading will be planned to occur at the recovery stage, the changed properties of the recovered product can have a major impact on processing options for an associated refinery. However, in terms of recovery during upgrading there can be hurdles to overcome. For example, the application of low-boiling hydrocarbon solvents to reduce or eliminate natural gas for steam generation has received significant recent interest but these low-boiling hydrocarbons also have a natural tendency to cause asphaltene constituents to separate from the oil, thereby offering the promise of in situ upgrading (leaving the high coke formers below ground), but plugging of the underground transport channels by the separated asphaltene constituents is also a possibility.

As in any field in which primary recovery operations are followed by secondary or enhanced recovery operations and there is a change in product quality, such is also the case for tar sand recovery operations. Thus, product oils recovered by the thermal stimulation of tar sand deposits show some improvement in properties over those of the bitumen in-place. Although this improvement in properties may not appear to be too drastic, nevertheless it usually is sufficient to have major advantages for refinery operators. Any incremental increase in the units of the hydrogen–carbon ratio can save amounts of costly hydrogen during upgrading. The same principles are also operative for reductions in the nitrogen, sulfur, and oxygen contents. This latter occurrence also improves catalyst life and activity, as well as reduces the metals content. In short, in situ recovery processes (although less efficient in terms of bitumen recovery relative to mining operations) may have the added benefit of *leaving* some of the more obnoxious constituents (from the processing objective) in the ground (Kuhlman, 2000).

Recent developments in *upgrading* of heavy oil (Ancheyta and Speight, 2007; Speight, 2014a) indicate that the near future could see a reduction of the differential cost of upgrading heavy oil. These processes are based on a better understanding of asphaltene solubility effects at high temperatures, the incorporation of a catalyst that is chemically precipitated internally during the upgrading, and the improvement of hydrogen addition or carbon rejection.

In situ upgrading can reduce the viscosity of heavy oil by cracking long hydrocarbon chains and can improve oil quality by reducing or removing asphaltene constituents and resin constituents. Asphaltenes may contain iron, nickel, and vanadium, which are damaging to refineries. Excess carbon, in the form of coke, may be left in the reservoir. The upgraded oil flows more readily into the wellbore (increasing recovery factor), is easier to lift to the surface, and may eliminate the need for a diluent for pipeline transportation. Furthermore, in situ upgrading might eliminate the need for surface upgrading facilities, thus reducing capital investments. In a conventional thermal process (eg, SAGD), the heavy oil is heated in situ, but it may cool after being produced to surface. It then has to be reheated for upgrading. Therefore, in situ upgrading may be more energy efficient as well.

There are three main approaches for heating the reservoir: steam injection; ISC; and electric heating. Steam injection pressures are limited because most heavy oil deposits are relatively shallow. The maximum steam temperature is limited by the ideal gas law. For example, at a 1000 m depth, the formation pressure is approximately 10 MPa, which permits a steam temperature of approximately only 300°C. This is too low to provide significant upgrading on a short timescale. ISC, which involves injection of an oxidizing gas, such as air or oxygen-enriched air, to generate heat by burning a portion of the oil, is capable of much higher temperatures (approximately 700°C), which should allow significant upgrading. Electric heating [resistance, induction, or radio frequency (RF)] should also be able to achieve the high temperatures required for in situ upgrading (Mut, 2005).

Heavy oil constituents can be cracked into lighter hydrocarbon molecules at high enough temperatures and pressures. In the pyrolysis of heavy oil, carbon–carbon bonds in the hydrocarbon chain are broken by heat; essentially the vibrational energy exceeds the chemical energy in the carbon–carbon bonds. Pyrolysis occurs in the absence of oxygen or a catalyst, but steam may be present. For example, steam cracking and thermal cracking are done in refineries at temperatures at or above 800°C. Such high temperatures are difficult to achieve in the reservoir. Pyrolysis can still occur at lower temperatures, but at much, much slower rates. For example, heavy oil produced under primary and fireflood conditions shows a gradual increase in density, viscosity, and other properties over time (Reichert et al., 1989).

Adding a catalyst (such as iron, which would be reminiscent of the older *red mud process* for upgrading coal to coal liquids; Speight, 2013a) to a thermal process may enhance in situ upgrading, even at the lower temperatures for steam injection (Jiang et al., 2005). Laboratory experiments combining ISC with a catalyst in a horizontal producing well produced significantly upgraded oil. Thermal cracking occurred in the combustion zone, and additional upgrading was achieved by catalytic cracking in the production well (Xia et al., 2002). The downhole catalytic upgrading produced light oil, characterized by a low viscosity, that was readily converted into gasoline and diesel fractions, with a higher conversion in a refinery fluid catalytic cracking (FCC) unit than that obtained with normal virgin bitumen vacuum gas oil (Greaves and Xia, 2004; Shah et al., 2010).

In the process, the level of in situ upgrading of heavy oil (Wolf Lake, Alberta, Canada) was achieved by a downhole catalytic process—a catalytic variant of the Toe-Heel Air Injection (THAI) process (Chapter 9). The data provided an indication of the potential for downhole upgrading in the field. The product was a light oil having a low viscosity consistent with facile conversion (on an FCC basis) into gasoline and diesel fractions than that obtained with normal virgin bitumen vacuum gas oil. In addition, the process offers significant potential benefits regarding in situ removal (and retention) of heavy metals, and reduction of sulfur in the product oil.

Thus, in situ conversion, or underground refining, is a promising technology to tap the extensive reservoirs of heavy oil and deposits of bitumen. The new technology (Gregoli, 1985; Gregoli and Rimmer, 2000; Gregoli et al., 2000) features the injection of high-temperature, high-quality steam and hot hydrogen into a formation containing heavy hydrocarbons to initiate conversion of the heavy hydrocarbons into lighter hydrocarbons. In effect, the heavy hydrocarbons undergo partial underground refining that converts them into a synthetic crude oil (*syncrude*). The heavier portion of the syncrude is treated to provide the fuel and hydrogen required by the process, and the lighter portion is marketed as a conventional crude oil. The potential advantages of an in situ process for bitumen and heavy oil include: (1) leaving the carbon forming precursors in the ground; (2) leaving the heavy metals in the ground; (3) reducing sand handling; and (4) bringing a partially upgraded product to the surface. The extent of the upgrading can, hopefully, be adjusted by adjusting the exposure of the bitumen of heavy oil to the underground thermal effects. The concept of downhole catalytic upgrading of heavy oil using ISC is not new as evidenced by interest in the concept and work performed over the past four decades (Reichert et al., 1989; Mamora et al., 1993; Weissman and Kessler, 1996; Weissman, 1997; Moore et al., 1999; Fan and Liu, 2002; Motaghi et al., 2010). The downhole catalytic upgrading process requires suitable placement of a catalyst bed in the reservoir, the flow of oil through the catalytic bed at a specified temperature and pressure, and finally drainage of the upgraded oil into the production well. In order to accomplish in situ upgrading there are two requirements: (1) addition of hydrogen or (2) deposition of thermal coke or a coke-like sediment that removes much of the asphaltene and resin constituents from the oil. Since hydrogen addition must be used during surface upgrading in order to stabilize the upgraded bitumen and product saleable products, this means that the cost of partial upgrading can be offset by the reduced hydrogen requirements in a surface upgrader.

The most promising enhanced recovery methods for partial upgrading during recovery are the combustion methods. The concept of any combustion technology requires that the oil be partially combusted and that thermal decomposition occur to other parts of the oil. This is sufficient to cause irreversible chemical and physical changes to the oil to the extent that the product is markedly different to the oil in-place. Recognition of this phenomenon is essential before combustion technologies are applied to oil recovery.

Although this improvement in properties may not appear to be too drastic, nevertheless it usually is sufficient to have major advantages for refinery operators. Any incremental increase in the units of the hydrogen–carbon ratio can save amounts of costly hydrogen during upgrading. The same principles are also operative for reductions in the nitrogen, sulfur, and oxygen contents. This latter occurrence also improves catalyst life and activity as well as reduces the metals content.

10.2.1 Solvent-Based Processes

The application of light hydrocarbon solvents to reduce or eliminate natural gas for steam generation has received significant recent interest. These light hydrocarbons also have a natural tendency to cause asphaltene constituents to separate thereby offering the promise of some in situ upgrading. VAPEX is the most advanced process in this area. Physical and chemical separations into fractions might lead to segregated and more targeted process steps, including more efficiently targeted hydrogen addition. There may be some overlap here with demetallization.

An extension of solvent recovery is the combined use of solvents and thermal stimulation to achieve some degree of in situ upgrading. Several companies or joint ventures are known to be piloting variations of this approach. In all cases, a minor factor is a degree of upgrading that may occur in new recovery methods but a potentially major factor is the likely conversion of bitumen-based residues in future for energy, power, and hydrogen at production or upgrading stages and the possible application of mild in situ field upgrading to reduce dependence on diluent for transport to distant refineries. The potential move to *less severe* primary upgrading will place more emphasis on *conversion* at the secondary stage as well as heteroatom removal. The desire to reduce overall hydrogen consumption will place emphasis on lower light by-product production and targeted hydrogen addition to synthetic crude oil fractions.

The main solvent process for quick and convenient surface upgrading using a solvent is *dilution*. By this means, heavy oil can be diluted sufficiently so that the blend meets the specifications for shipping by pipeline or other means of transportation where a specification must be met.

Dilution of heavy oil has been studied using various solvents in terms of aromaticity, chemical nature, and viscosity (Motaghi et al., 2010; Speight, 2014a). Dilution with low viscosity hydrocarbons, such as light crude oil or naphtha, has shown that the viscosity reduction efficiency is controlled by the sole viscosity of the diluent and not by its aromaticity. However, while blending diluents of different chemical nature and polarity can enhance viscosity reduction there is always the risk (especially with paraffin-based solvents) that separation of asphaltene constituents will occur (Mitchell and Speight, 1973; Speight, 1979, 2007).

Solvent deasphalting processes allow removal of sulfur and nitrogen compounds as well as metallic constituents by balancing yield with the desired feedstock properties (Ditman, 1973). In the process, the feedstock is mixed with dilution solvent from the solvent accumulator and then cooled to the desired temperature before entering the extraction tower. Because of its high viscosity, the charge oil can neither be cooled easily to the required temperature nor will it mix readily with solvent in the extraction tower. By adding a relatively small portion of solvent upstream of the charge cooler (insufficient to cause phase separation), the viscosity problem is avoided.

The *choice of solvent* is vital to the flexibility and performance of the unit. The solvent must be suitable not only for the extraction of the desired oil fraction, but also for control of the yield and/or quality of the deasphalted oil at temperatures which are within the operating limits (Mokrys and Butler, 1993).

10.2.2 Bulk Thermal Processes

Downhole upgrading offers the prospect of both improving recovery, and decreasing environmental impact. However, use of catalysts to enhance downhole upgrading is limited by the need for one that can survive the extreme coking conditions arising from the cracking of heavy oil. The potential of hydrogen donors to improve upgrading and enhance catalyst lifetime should also be considered (Dim et al., 2015).

The mobilization of heavy oil in the reservoir or bitumen in the deposit by partial combustion is not a new idea, and still has many hurdles to overcome before it can be considered close to being commercial. However, the product oil is likely less viscous. In short, in situ recovery processes (although less efficient in terms of bitumen recovery relative to mining operations) may have the added benefit of *leaving* some of the more obnoxious constituents (from the processing objective) in the ground (Kuhlman, 2000).

In the *modified in situ extraction* processes, combinations of in situ and mining techniques are used to access the reservoir. A portion of the reservoir rock must be removed to enable application of the in situ extraction technology. The most common method is to enter the reservoir through a large-diameter vertical shaft, excavate horizontal drifts from the bottom of the shaft, and drill injection and production wells horizontally from the drifts. Thermal extraction processes are then applied through the wells. When the horizontal wells are drilled at or near the base of the tar sand reservoir, the injected heat rises from the injection wells through the reservoir, and drainage of produced fluids to the production wells is assisted by gravity.

Generally, as opposed to heavy oil recovery, bitumen recovery requires a higher degree of thermal stimulation because bitumen, in its immobile state, is extremely difficult to move to a production well. Extreme processes are required, usually in the form of a degree of thermal conversion that produces free-flowing product oil that will flow to the well and reduce the resistance of the bitumen to flow.

Other techniques involve methods such as the establishment of hydrovisbreaking in situ, such as injection of superheated steam and hot hydrogen into the formation (Fig. 10.1). The injected fluids initiate hydrovisbreaking, producing a partially upgraded lighter product (that may have solvent properties) and driving the oil in the formation to the producing wells. In some process schemes, specialized downhole equipment is used to generate superheated steam and heat the injected hydrogen. However, control of ISC processes has been difficult and the THAI process offers some relief to the potential for an out-of-control combustion reaction. Not that the THAI process should be construed as an easy task.

Thermal cracking processes offer attractive methods of feedstock conversion at low operating pressure without requiring expensive catalysts. Currently, the most widely operated residuum conversion processes are

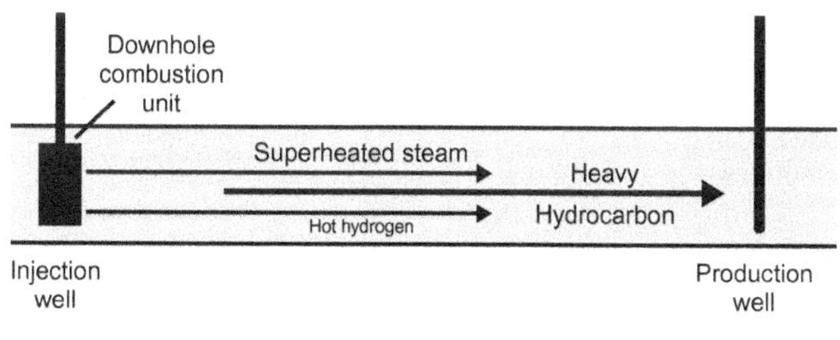

Figure 10.1 Schematic of in situ upgrading by hydrovisbreaking.

visbreaking, delayed coking, and fluid coking, which are still attractive processes for refineries from an economic point of view (Dickenson et al., 1997). The visbreaking process offers wide potential because of the mild nature of the process (Kuhlman, 2000; Joshi et al., 2008; Radovanović and Speight, 2011; Speight, 2012).

Visbreaking (*viscosity reduction, viscosity breaking*) is a mild form of thermal cracking insofar as the thermal reactions are not allowed to proceed to completion and are interrupted by quenching. Process conditions range from 455°C to 510°C (850–950°F) at a short residence time and from 50 to 300 psi at the heating coil outlet. It is the short residence time that brings to visbreaking the concept of being a mild thermal reaction in contrast to, for example, the delayed coking process where residence times are much longer and the thermal reactions are allowed to proceed to completion. Liquid-phase cracking takes place under these low-severity conditions to produce some naphtha, as well as material in the kerosene and gas oil boiling range. The gas oil may be used as additional feed for catalytic cracking units, or as heating oil.

Hydrovisbreaking, a noncatalytic process, is conducted under similar conditions to visbreaking and involves treatment with hydrogen under mild conditions (RAROP, 1991, p. 57). The presence of hydrogen leads to more stable products (lower flocculation threshold) than can be obtained with straight visbreaking, which means that higher conversions can be achieved, producing a lower viscosity product.

Visbreaking and variants, and the recently demonstrated ORMAT process are examples of bulk thermal processes that convert residues without progressing all the way to solid coke. These processes have significant potential to be integrated with deasphalting to produce varying yields of residua to meet future alternative energy and hydrogen production needs.

A recent variant of the visbreaking process is the *Aquaconversion process* (Marzin et al., 1998a, 1998b) (Fig. 10.2), which is a catalytic visbreaking process which operates in the presence of steam. The novel catalytic system allows hydrogen from the water to be transferred to the residua when operated at the conditions normally used for the visbreaking process. Similar operating conditions (pressure and temperature) are used and with the hydrogen incorporation into the product, a higher conversion level can be reached, thus enabling higher API and viscosity improvements to be achieved while maintaining product stability.

The typical visbreaking technology is limited in conversion level because of the stability of the resulting product, because one process

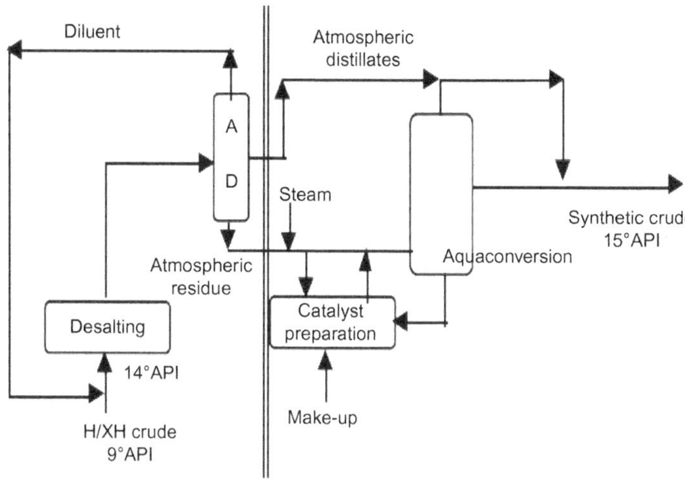

Figure 10.2 The Aquaconversion process.

requirement is that the product has to be stable. Standard visbreaking allows only an approximate 2°–6° API upgrading of the heavy crude and only a limited viscosity reduction, which does not ensure its transport without external diluent. The Aquaconversion process pushes this maximum conversion level within the stability specification by adding a homogeneous catalyst in the presence of steam.

Other variants of visbreaking that may find use at the wellhead could the Tervahl T process and the Tervahl H process. In the *Tervahl T process* (RAROP, 1991, p. 25; Hedrick et al., 2006), the feedstock is heated to the desired temperature using the coil heater and heat is recovered in the stabilization section and held for a specified residence time in the soaking drum. The soaking drum effluent is quenched and sent to a conventional stabilizer or fractionator where the products are separated into the desired streams. The gas produced from the process is used for fuel. In the *Tervahl H process*, the feedstock and hydrogen-rich stream are heated using heat recovery techniques and a fired heater and held in the soak drum as in the Tervahl T process. The gas and oil from the soaking drum effluent are mixed with recycled hydrogen and separated in the hot separator where the gas is cooled by being passed through a separator and recycled to the heater and soaking drum effluent. The liquids from the hot and cold separator are sent to the stabilizer section where purge gas and synthetic crude are separated. The gas is used as fuel and the synthetic crude can now be transported or stored.

ISC has long been used as an enhanced oil recovery method. For heavy oils, numerous field observations have shown upgrading of 2°–6° API for heavy oils undergoing combustion (Ramey et al., 1992). During ISC of heavy oils, temperatures of up to 700°C can be observed at the combustion front. Some processes offer the potential for partial upgrading during recovery and many typically involve in situ hydrogenation of the oil or bitumen (Hamrick and Rose, 1977; Gregoli, 1985; Hewgill and Kalfayan, 1992; Gregoli et al., 2000; Graue, 2001).

ISC involves an injection of an oxidizing gas (air or oxygen-enriched air) to generate heat by burning a portion of the oil. Most of the oil is driven toward the producers by a combination of gas drive (from the combustion gases) and steam and water drive. This process is also called fire flooding to describe the movement of the burning front inside the reservoir. Based on the respective directions of front propagation and air flow, the process can be forward, when the combustion front advances in the same direction as the air flow, or reverse, when the front moves against the air flow.

Forward combustion can be further characterized as "dry" when only air or enriched air is injected or "wet" when air and water are coinjected. In the process, air is injected in the target formation for a short time, usually a few days to a few weeks and the oil in the formation is ignited. Ignition can be induced using downhole gas burners, electrical heaters, and/or injection of pyrophoric agents (not recommended) or steam (Rodríguez et al., 2008). In some cases, auto-ignition occurs when the reservoir temperature is fairly high and the oil reasonably reactive. This often happens for California oils.

After ignition, the combustion front is propagated by a continuous flow of air. As the front progresses into the reservoir, several zones can be found between the injector and the producer as a result of heat, mass transport, and the chemical reactions occurring in the process. The burned zone is the volume already burned. This zone is filled with air and may contain small amounts of residual unburned organic solids. As it has been subjected to high temperatures, mineral alterations are possible. Because of the continuous air flow from the injector to the burned zone, temperature increases from the injected air temperature at the injector to the near combustion front temperature near the combustion front. There is no oil left in this zone.

The combustion front is the highest temperature zone. It is very thin, often no more than several inches thick. It is in that region that oxygen combines with the fuel and high-temperature oxidation occurs. The products of the burning reactions are water and carbon oxides. The fuel is often misnamed coke. In fact, it is not pure carbon but a hydrocarbon with H–C atomic ratios ranging from about 1 to 2.0. This fuel is formed in the thermal cracking zone just ahead of the front and is the product of cracking and pyrolysis, which is deposited on the rock matrix. The amount of fuel burned is an important parameter because it determines how much air must be injected to burn a certain volume of reservoir.

Chemical reactions are of two main categories: (1) oxidation, which occurs in the presence of oxygen and (2) pyrolysis, which is caused mainly by elevated temperatures. In general at low temperature, oxygen combines with the oil to form oxidized hydrocarbons, such as peroxides, alcohols, or ketones. This generally increases the oil viscosity but could increase oil reactivity at higher temperature. When oxygen contacts the oil at higher temperature, combustion occurs resulting in the production of water and carbon oxides. Of all the reactions that can occur during ISC, only low temperature oxidation can increase the viscosity of the oil. If the fireflood is conducted properly, low temperature oxidations are minimized because most of the oxygen injected is consumed at the burning front.

Distillation allows transport and production of the light fractions of the oil leaving behind the heavy ends. These heavy ends often contain the majority of the undesirable compounds, which may contain sulfur or metals.

Another possible in situ upgrading technique involves a combination of solvent injection and combustion. Cyclic oil recovery has numerous advantages both technically and economically. It can also be easily optimized in a given oil reservoir. Cyclic injection of solvents, either gas or liquid, followed by ISC of a small part of the reservoir to increases the temperature near the well but also cleans the wellbore region of all the residues left by the solvents. Alternate slugs of solvent and air would be injected and production would occur after each solvent slug injection and after each combustion period. The process could be repeated until an economic limit is reached. One important point to note is that both solvent injection and ISC have been proven to be effective in a variety of reservoirs; however, the combination of the two methods has never been tried.

The most significant effect will be the precipitation and/or deposition of heavy hydrocarbons, such as asphaltenes or paraffins. The produced oil is expected to be slightly upgraded by the solvent cycle. Unlike the classic well to well ISC, we would only try to improve near wellbore conditions by burning the solid residues left after the solvent cycle. The benefits of using combustion at this stage are expected to include: (1) productivity improvement through removal of the heavy ends left from the solvent cycle; (2) possible deactivation of the clays near the wellbore due to the high temperature of the combustion; and (3) reduced viscosity of the oil due to temperature increase.

Downhole upgrading of virgin Athabasca tar sand bitumen has been investigated in a series of experiments using the THAI process, which uses combinations of vertical injection wells and horizontal producer wells, arranged in a direct, or staggered line drive. Downhole upgrading of the bitumen was significant, with the API gravity of the produced oil increasing by an average of 8° API, compared to the original bitumen. The produced oil viscosity was also dramatically reduced and saturates, aromatics, resins, and asphaltenes (SARA) analysis was used to assess the quality of the produced oil, showing that the saturate fraction of the bitumen was increased from approximately 16% by weight to 72% by weight.

The THAI process could well have a wider range of application than SAGD, but in any case, a detailed knowledge of the reservoir is essential (Speight, 2013b). SAGD generally works best in relatively thick (40 m) homogeneous pay zones. It is possible that the THAI process will be effective down to about 6 m thickness, as is common in many Saskatchewan heavy oil pools (Greaves and Xia, 2004).

The CAPRI process involves the addition of a gravel-packed catalyst, as used in a conventional refinery, between the tubing and the horizontal wellbore. Test results have shown this technique to add 6–8 API points on top of the THAI in situ upgrades. Based on these data, the combination could deliver in situ upgrading to above the 22° API requirement for produced fluids that can be transported by pipeline without diluent which also represents a major saving in surface upgrading and refining costs.

The benefits of the introduction of hydrogen during in situ retorting offer much promise. The possible application of such methods for selective separation of the metal constituents is an obvious benefit. For example, partial oxidation in the presence of steam may produce hydrogen for immediate pick up, and result in integrated recovery and significant upgrading.

10.2.3 The Future of In Situ Upgrading

Finally, recent developments in *upgrading* of heavy oil (Ancheyta and Speight, 2007) indicate that the near future could see a reduction of the differential cost of upgrading heavy oil as refineries evolve beyond this century (Speight, 2011). These processes are based on a better understanding of asphaltene solubility effects at high temperatures, the incorporation of a catalyst that is chemically precipitated internally during the upgrading, and the improvement of hydrogen addition or carbon rejection.

In situ upgrading can reduce the viscosity of heavy oil by cracking long hydrocarbon chains and can improve oil quality by reducing or removing asphaltene constituents and resin constituents. Asphaltene constituents may contain iron, nickel, and vanadium, which are damaging to refineries. Excess carbon, in the form of coke, may be left in the reservoir.

The upgraded oil flows more readily into the wellbore (increasing recovery factor), is easier to lift to surface, and may eliminate the need for a diluent for pipeline transportation. Furthermore, in situ upgrading might eliminate the need for surface upgrading facilities, thus reducing capital investments. In a conventional thermal process (eg, SAGD), the heavy oil is heated in situ, but it may cool after being produced to surface. It then has to be reheated for upgrading. In situ upgrading may be more energy efficient as well.

However, the increased mobilization of heavy oil in the reservoir by partial upgrading is not a new idea and still has many hurdles to overcome before it can be considered close to being commercial. The product will be less viscous than the heavy oil in place but some property changes, such as high olefin content from cracking, are not necessarily positive.

There are three main approaches for heating the reservoir: (1) steam distillation; (2) mild thermal cracking—visbreaking; and (3) partial combustion.

10.2.3.1 Steam Distillation

Steam injection pressures are limited during recovery of heavy oil because most heavy oil deposits are relatively shallow. The maximum steam temperature is limited by the ideal gas law. For example, at a 1000m depth, the formation pressure is approximately 1450psi, which permits a steam temperature of approximately 300°C (570°F), which is too low to provide significant upgrading on a short timescale. ISC is capable of much higher temperatures (approximately 700°C, 1290°F), which has the potential for measurable upgrading. Electric heating (resistance, induction, or RF) should also be able to achieve the high temperatures required for in situ upgrading (Mut, 2005). But the influence of the use of steam, such as steam distillation then becomes an option (Lim et al., 1993; Monin and Audibert, 1994).

The principal mechanisms responsible for oil recovery during heavy oil recovery by steam-based processes are: (1) thermal expansion of the oil; (2) viscosity reduction; and (3) steam distillation. Steam distillation, once an ignored phenomenon, deserves recognition as the main mechanism that reduces the residual oil saturation behind the hot water front during steam flood, particularly for light and medium oils. Furthermore, steam distillation can also play an important role during heavy oil recovery by steam flooding.

The vaporizing effect of hydrocarbons is induced by increasing the system temperature, and it is reflected by the increase in the system pressure. Steam will evolve from the aqueous phase and strip some hydrocarbon components from the oleic phase. Since the pressure of the system is increased; the apparent bubble point of the water is consequently increased, which causes the higher molecular weight, boiling point hydrocarbons to vaporize at the elevated saturation temperatures causing displacement of the oil by steam displacement *and* steam distillation (Volek and Pryor, 1972; Wu and Brown, 1975; Sarathi et al., 1988; Sharpe et al., 1995).

To reduce viscosity of heavy crude oil, solvents frequently are used for dilution—this is one of the most efficient methods of pipeline transportation of heavy oil. Solvents also are injected into the reservoir for well cleaning, stimulation, fracturing, and, less frequently, for miscible displacement. Thus, one benefit resulting from the distillation of heavy oil during a steam flood is that the lower-boiling constituents can act as a solvent to lower the viscosity of the original oil. In the steam distillation process, the vaporized distillate mixes with the original oil ahead of the steam condensation zone. Depending on the quantity of light hydrocarbon available to mix with the oil, viscosity can be substantially decreased—provided the distillate has a sufficient amount of aromatic constituents or naphthenic constituents to maintain the asphaltene constituents in solution (Mitchell and Speight, 1973; Shu, 1984).

It has also been shown that that, during steam–propane distillation of oil, the steam/propane vapor strips the more volatile components from the heavy oil and encourages a decrease of the boiling point of these components. The steam, enriched with the propane and light components, flows through the steam zone to the condensation front where both steam and light hydrocarbons condense. The condensed hydrocarbons are miscible with the oil, reducing overall viscosity of liquids hydrocarbon. The steam/propane distillation begins when the total vapor pressure (steam and propane) in the presence of two immiscible liquids (water and oil exerting its own vapor pressure at the temperature of the system) equals the total pressure on the system. It is also possible that the role of propane is to reduce the boiling point of the lower molecular weight components. As a result, heavy oil will begin distilling at temperatures much lower than the normal boiling points of the constituents and, as a consequence, the distillation yield increases.

As promising as this may seem, the steam distillation yields are mainly dependent on the oil composition and may not correlate with crude API gravity. Changes in steam saturation pressure and temperature have insignificant effect on the yields; however, superheated steam significantly increases the yields for some crude oils.

10.2.3.2 Mild Thermal Cracking

Heavy oil constituents can be cracked into lighter hydrocarbon molecules at high enough temperatures and pressures. In the pyrolysis of heavy oil, carbon–carbon bonds in the hydrocarbon chain are broken by heat; essentially the vibrational energy exceeds the chemical energy in the carbon–carbon bonds. Pyrolysis occurs in the absence of oxygen or a catalyst, but steam may be present. For example, steam cracking and thermal cracking are done in refineries at temperatures at or above 800°C (1470°F). Such high temperatures are difficult to achieve in the reservoir. Pyrolysis can still occur at lower temperatures, but at much, much slower rates.

For example, heavy oil produced under primary and fireflood conditions shows a gradual increase in density, viscosity, and other properties over time. This indicates fractionation during flow through the formation and there is also a decrease in viscosity, density, and other properties as a result of the products of thermal cracking of the oil and the formation of smaller straight-chain alkanes and small aromatics molecules (Reichert et al., 1989).

Thus, in situ upgrading is also an additional benefit to the combustion recovery process. As the combustion front moves through the reservoir, the lower-boiling constituents of the heavy oil are vaporized, thermal cracking occurs, and the coke product is consumed as fuel. The resulting, produced oil has a much lower viscosity than the original bitumen.

It is also conceivable that under the appropriate conditions (as yet undefined) electric heating (resistance, induction, or RF) should also be able to achieve the high temperatures required for in situ upgrading (Cunha, 2005).

10.2.3.3 Partial Combustion

The mobilization of heavy oil in the reservoir by partial combustion is not a new idea, and still has many hurdles to overcome before it can be considered close to being commercial. However, the product oil is likely less viscous.

Forward ISC by itself is already an effective in situ upgrading method with improvements in API gravity by as much as 6°. Other work (Xia and Greaves, 2006; Greaves and Xia, 2004, 2008; Greaves et al., 2005) has followed up on this potential but direct application of these laboratory results to the field is difficult, such technology deserves further research.

However, while ISC is a relatively inexpensive process, it has major drawbacks. The high temperatures in the presence of oxygen, which are encountered when the process is applied, cause coke formation and the production of olefins and oxygenated compounds such as phenols and ketones, which in turn cause major problems when the produced liquids are processed in refinery units. Commonly, the processing of products from thermal cracking is restricted to delayed or fluid coking because the hydrocarbon is degraded to a degree that precludes processing by other methods.

One concept that relates to upgrading during in situ recovery utilizes a unique combination of operations to achieve hydrovisbreaking in formations in which heavy oil and commonly encountered levels of formation permeability combine to limit fluid mobility. The process uses either (1) a continuous operation or (2) a cyclic operation. In the continuous operation, one or more vertical injection boreholes and one or more vertical production boreholes are used in which multiple, uncased, horizontal boreholes may extend from the vertical boreholes. In the cyclic operation, both injection and production occur in the same vertical boreholes in which multiple, uncased, horizontal boreholes may extend from the vertical boreholes. A mixture of reducing gases, oxidizing gases, and steam are fed to downhole combustion devices located in the injection boreholes. Combustion of the reducing gas-oxidizing gas mixture is carried out to produce superheated steam and hot reducing gases for injection into the formation to convert and upgrade the heavy crude or bitumen into lighter hydrocarbons. Communication between the injection and production boreholes in the continuous operation and fluid mobility within the formation in the cyclic operation is induced by fracturing, multiple horizontal boreholes extending from vertical boreholes, or other related methods. In the continuous mode, the injected steam and reducing gases drive upgraded hydrocarbons and virgin hydrocarbons to the production boreholes for recovery. In the cyclic operation, wellhead pressure is reduced after a period of injection causing injected fluids, upgraded hydrocarbons, and virgin hydrocarbons in the vicinity of the boreholes to be produced. Injection and production are then repeated for additional cycles. In both operations, the hydrocarbons produced are collected at the surface for further processing.

The THAI process potentially allows the inclusion of a catalytic upgrading stage since it provides favorable operating temperatures at the production well as the combustion zone is "anchored" to the horizontal well (Xia and Greaves, 2006; Xia et al., 2002). The cracking reactions happening in the mobile oil zone create precursor conditions for the CAPRI process. The reactants are usually water (steam) and combustion gases, which pass through the mobile oil zone and are in contact with the catalyst layer around the horizonta1 production well. The catalyst reacts with partially upgraded THAI oil and further upgrades it.

Thus, adding a catalyst (such as iron) to a thermal process may enhance in situ upgrading, even at the lower temperatures for steam injection (Jiang et al., 2005). Laboratory experiments combining ISC with a catalyst in a horizontal producing well produced significantly upgraded oil. Thermal cracking occurred in the combustion zone, and additional upgrading was achieved by catalytic cracking in the production well (Xia et al., 2002). The downhole catalytic upgrading produced light oil, characterized by a low viscosity, that was readily converted into gasoline and diesel fractions, with a higher conversion on an FCC basis than that obtained with normal virgin bitumen vacuum gas oil (Greaves and Xia, 2004; Greaves et al., 2005).

Thus, one form of in situ heavy oil upgrading involves the injection of a catalyst—hence the CAPRI process, involves the addition of gravel-packed catalyst, as used in a conventional refinery, between the tubing and the horizontal wellbore. Test results have shown the technique to add 6–8 API points on top of the THAI in situ upgrades.

10.2.3.4 Solvent Deasphalting

The application of light hydrocarbon solvents to reduce or eliminate natural gas for steam generation has received significant recent interest. These lower-boiling hydrocarbons also have a natural tendency to cause asphaltene constituents to separate (Mitchell and Speight, 1973; Speight, 2014a) thereby offering the promise of some in situ upgrading.

An alternate to catalytic upgrading is in situ deasphalting. Here a solvent, such as propane, is injected to drop out some of the asphaltene constituents and lower the viscosity of the heavy oil. This could be a stand-alone process or an additional benefit to a solvent-based recovery process such as VAPEX.

During a solvent-based heavy oil recovery process, such as vapor extraction (VAPEX), a condensable solvent is injected into a heavy oil reservoir. Solvent dissolution into heavy oil and possible asphaltene precipitation drastically reduce its viscosity so that the diluted heavy oil can flow towards a production well. In the past, several physical modeling studies have shown that the produced heavy oil has much lower amounts of heavy components than the original heavy oil. This phenomenon is often referred to as in situ upgrading.

Typically, after a solvent is made in contact with heavy oil at a relatively high pressure for a sufficiently long time, the solvent-heavy oil system at equilibrium state can be roughly divided into three different layers. The top layer is a solvent-enriched liquid phase, the middle layer comprises heavy oil with the dissolved solvent, and the bottom layer mainly consists of heavy components. The solvent-heavy oil mixtures in these three layers show rather different chemical and physical properties, such as solvent concentration, carbon number distribution, and viscosity. The top layer has the highest concentrations of solvent and light components and the lowest viscosity of heavy oil even after its dissolved solvent is flashed off. The heavy oil in the middle layer has similar carbon number distribution to the original heavy oil. The bottom layer has the lowest solvent concentration and the highest concentration of heavy components.

The heavy oil in the bottom layer after its dissolved solvent is flashed off has much higher viscosity than the original heavy oil. These experimental results indicate that in a solvent-based heavy oil recovery process, the solvent-heavy oil mixtures in the top and middle layers can be recovered because of their lower viscosity, whereas the heavy oil in the bottom layer may be left behind in the heavy oil reservoir because of its higher viscosity. In this way, the produced heavy oil is in situ upgraded during the solvent-based heavy oil recovery process.

The application of low-boiling hydrocarbon solvents has received significant recent interest. These light hydrocarbons have a natural tendency to separate asphaltene constituents and offers the promise of some in situ upgrading. However, there has been little work done to determine reservoir damage or deposit damage caused by asphaltene separation and, in addition, the site specificity of each reservoir and each deposit needs to be considered before any company charges ahead into a *Light Brigade at Balaclava* situation. An extension of solvent recovery is the combined use of solvents and thermal stimulation to achieve some degree of in situ upgrading. The potential move to *less severe* primary upgrading will place more emphasis on *conversion* at the secondary stage as well as heteroatom removal.

Another solvent-based process, the N-Solv process, uses warm solvent vapor to extract bitumen in situ (N-Solve, 2015). In the process, a warm solvent vapor is pumped into the injection well of a tar sand deposit and, after reaching the extraction chamber, condenses and dissolves the bitumen, draining downward to the production well as a solution of both solvent and oil. Naturally occurring water and noncondensable gases are also carried along and out of the reservoir. Coke-forming asphaltene constituents remain in the deposit. At the surface, the partially upgraded oil mixed with solvent is separated into three components: deasphalted oil, solvent, and naturally occurring water. The upgraded bitumen, which appears as deasphalted oil, is sent to refineries for further processing and the solvent is pumped back into the separator and repurified before being recycled back into the reservoir. It is estimated that a solvent quantity equivalent to 20–30% (v/v) of the extracted deasphalted oil must be added to the system as make-up. As the chamber grows, the make-up solvent is continually replenished and at the end of the life of a well, the solvent can be recovered and reused on another well or marketed.

10.2.3.5 Microbial-Enhanced Oil Recovery

Microbial-enhanced oil recovery (MEOR) is another process that has the potential for in situ upgrading through bioconversion bacterial upgrading of the heavy feedstock (Clark et al., 1981; Saxman and Crull, 1984; Shennan and Levi, 1987; Bryant et al., 1989; Jenneman, 1989; Raiders et al., 1989; Khire and Khan, 1994a, 1994b; Banat, 1995; Bryant and Lindsey, 1996; Bailey et al., 2001; Kirkwood et al., 2004; El-Gendy and Speight, 2015). MEOR is a general term for any enhanced petroleum recovery technique utilizing single-celled microorganisms or their by-products.

The process depends on microbes that can convert the bitumen to lower viscosity oil or methane. These microbes may be naturally occurring or injected into the reservoir. As additional nutrients are normally required for this process to take place at reasonable rates, these must be injected. However, since, microbial recovery methods are relatively slow processes there is likelihood (unfortunately) that such a process would be applied solely as a final *clean-up* process.

There are several processes involving MEOR (Saxman and Crull, 1984): (1) the processes that are in situ insofar as the processes take place within the petroleum formation and (2) the processes that are ex situ and are used either to manufacture production chemicals or to treat produced production chemicals or to treat produced hydrocarbons. Both types of process have the advantageous of desulfurization and denitrogenation of the produced hydrocarbons (in situ or ex situ) which removes any environmentally disadvantageous sulfur and nitrogen compounds.

Finally, there are also proposals that involve the use microwave heating for recovery of products from tar sand formations and the potential for in situ upgrading. Microwave technology has been identified as having potential application to tar sand recovery (Bosisio et al., 1977; Mutyala et al., 2010). The ability of microwaves to heat the deposit selectively and volumetrically leads to the possibility for the extraction of a product in high yield, and that the oil can be upgraded in the process (Bosisio et al., 1977). The potential for the success of microwave heating for heavy oil or bitumen extraction leads to the hypothesis that microwave heating can remove the hydrocarbonaceous products (bitumen) from tar sand deposits by selectively heating the water phases that are present interstitially, and within the hydrophilic clay minerals (Robinson et al., 2010). Preliminary analysis has shown the product quality to be favorable, a rigorous assessment of the process is still necessary. Preliminary analysis has shown showed that the product quality was favorable but rigorous assessment of the concept is needed to understand the requirements for further upgrading of the product. Furthermore, it is possible that the clay in the sand also played a role as a catalysts.

10.3 PARTIAL UPGRADING AT THE SURFACE

The influx of the heavy feedstocks (heavy oil, extra heavy oil, and tar sand bitumen) into the refinery system can offset the shortages of conventional crude oil but there is also a need for increased refining capacity as well as the need for lighter crude oil feedstocks. While new residue processing capacity needs be added to existing refineries (Speight, 2011), there is the need for simple primary upgrading systems to make the oil acceptable to a pipeline—many heavy oils are too viscous to transport by pipeline and fall outside of the specification required by pipeline owners (Wagner and Kresnyak, 2015).

10.3.1 General Aspects of Upgrading

The influx of heavy oils into the refinery system can offset the shortages of conventional crude oil but there is also a need for increased refining capacity. New residue processing capacity needs be added to existing refineries, or it could be built in separate, stand-alone upgrading facilities. If the oil is too viscous to transport by pipeline, and/or there is the need for heat or energy at the production site, heavy oil upgrading in the field is attractive, and may avoid extensive modifications of existing refineries. Traditional residue processing such as coking or hydrocracking are very expensive processes, and require large scale to be viable. Thus, petroleum refining is now in a significant transition period as the industry moves further into the 21st century and the demand for petroleum and petroleum products has shown a sharp growth in recent decades.

In order to satisfy the changing pattern of product demand, significant investments in refining conversion processes will be necessary to profitably utilize heavy feedstocks. The most efficient and economical solution to this problem will depend to a large extent on individual refinery situations. However, the most promising technologies will likely involve the conversion of vacuum residua. But technologies are needed that will take the feedstock beyond current limits and, at the same time, reduce the amount of coke and other nonessential products. Such a goal may require the use of two or more technologies in series rather than an attempt to develop a whole new one-stop conversion technology, which is in keeping with current refinery operations (Speight and Ozum, 2002; Parkash, 2003; Hsu and Robinson, 2006; Gary et al., 2007; Speight, 2014a).

A refinery is an integrated collection of unit processes (Speight and Ozum, 2002; Parkash, 2003; Hsu and Robinson, 2006; Gary et al., 2007; Speight, 2014a) and can have any one, or a conjunction, of several configurations but the refinery of the future will, of necessity be required to be a *conversion refinery*. A conversion refinery incorporates all the basic building blocks found in both the *topping refinery* and the *hydroskimming refinery*, but it also features gas oil conversion plants, such as catalytic cracking and hydrocracking units, olefin conversion plants, such as alkylation or polymerization units, and, frequently, coking units for sharply reducing or eliminating the production of residual fuels. Conversion refineries currently produce as much as two-thirds of their output as unleaded gasoline, with the balance distributed between liquefied petroleum gas, high-quality jet fuel, low-sulfur diesel fuel, and coke. Many such refineries also incorporate solvent extraction processes for manufacturing lubricants and petrochemical units with which to recover high-purity propylene, benzene, toluene, and xylenes for further processing into polymers.

The manner in which refineries convert heavy oil into low-boiling high-value products has become a major focus of operations with new concepts evolving into new processes (Khan and Patmore, 1998; Speight, 2000, 2007). Even though they may not be classed as conversion processes *per se*, pretreatment processes for removing asphaltene constituents, metals, sulfur, and nitrogen constituents are also important and can play an important role.

New processes for the conversion of residua and heavy oils will probably be used perhaps not only in place of but also in conjunction with visbreaking and coking options with some degree of hydroprocessing as a primary conversion step. In addition, other processes may replace or, more likely, augment the deasphalting units in many refineries. An exception, which may become the rule, is the upgrading of bitumen from tar sands (Speight, 1990, 2000, 2007).

The Aquaconversion technology offers a novel way of upgrading heavy crude oil. This is a catalytic hydrovisbreaking process which operates in the presence of steam. Visbreaking technology is limited in conversion level because of the stability of the resulting product. Because one process requirement is that the syncrude has to be stable, standard visbreaking allows only a 2° API upgrading of the heavy crude and only a limited viscosity reduction, which does not ensure its transport without external diluent. The process pushes this maximum conversion level within the stability specification by adding a homogeneous catalyst in the presence of steam (Marzin et al., 1998a, 1998b).

This novel catalytic system allows hydrogen from the water to be transferred to the residue when operated at the conditions normally used for the visbreaking process. Similar operating conditions (pressure and temperature) are used. Hydrogen incorporation is much lower than that obtained when using a deep hydroconversion process under high hydrogen partial pressure. Nevertheless, it is high enough to saturate the reactive species formed within the thermal process that would normally lead to the formation of coke precursors and coke. Because of hydrogen incorporation, a higher conversion level can be reached, thus enabling higher API and viscosity improvements to be achieved while maintaining product stability.

The important aspect of the Aquaconversion technology is that it does not produce any solid by-product, such as coke, nor does it require any hydrogen source or high-pressure equipment. In addition, the Aquaconversion process can be implanted in the production area, and thus the need for external diluent and its transport over large distances is eliminated. Light distillates from the raw crude can be used as diluent for both the production and desalting processes.

Conceivably, heavy oil could be upgraded at the wellhead by a primary upgrading (in fact a partial upgrading) step, such as Aquaconversion. Thus, upgrading heavy oil in the future will adapt to the nature of the recovered oil by an as-yet-to-be-specified upgrading step at the recovery stage. However, technologies that are adequate for this task will need to be identified and the level of upgrading will also need to be determined. The uses of visbreaking, hydrovisbreaking, or deasphalting are options of such an upgrading step. Indeed, partial upgrading as part of the recovery process may be a determinant of the price differentials between the recovered oil and a benchmark crude oil.

Visbreaking and variants, and the recently demonstrated ORMAT process, which is a combination of visbreaking and deasphalting offer milder thermal and physical processes that can play one of two possible roles: as field upgraders to reduce or eliminate the need for diluent for transport, or as primary processes in future upgraders where some residue is removed and consumed in hydrogen production.

On the other hand there are a number of technologies and technology concepts that have been described since the 1980s that are believed to be suitable for upgrading in the refinery (Table 8.1). These technologies are in addition to the more standard and recognizable processes such as coking technologies and hydrocracking technologies (Moschopedis et al., 1998; Speight and Ozum, 2002). Coking technologies can be carbon-wasteful in terms of the amount of coke produced (when the goal is to produce liquid fuels) and hydrocracking technologies do not fare much better since heavy oil feedstocks can have a serious effect on catalyst activity and life. In addition, blending heavy oil with lighter oil may not always be an option for introduction of heavy oil to the refinery. Shortages of lighter oil(s) or incompatibility of the heavy oil with the heavy oil may be limitations to the blending option.

However, it is not abundantly clear if the newer process concepts will change not only the quantity of the product but also, sometimes inadvertently, change the quality of the product. While an upward trend in product yield is usually looked upon as a positive change, a change in product quality is not always a positive change and may lead to a reevaluation of the steps needed to produce a specification product for sales. Furthermore, ancillary technologies that are linked to upgrading process selection (eg, residue gasification) will also provide the key to efficient heavy oil upgrading.

While not truly a full upgrading system, there some preferences for partial upgrading of heavy crude oil, extra heavy oil, or tar sand bitumen at the wellhead (Hedrick et al., 2006). This would involve upgrading the material at the wellhead and leaving much of the material behind as coke, and then pipeline the upgraded material out as synthetic crude. The products of the coking operation, and in some cases some of the residue, are hydrotreated. The hydrotreated materials are recombined with the fractionated light materials to form synthetic crude that is then transported to market in a pipeline. A portion of the crude may or may not be bypassed around the processing units. There are several examples of this type of processing in the current Canadian oil sands operations seen around Fort McMurray in Alberta, Canada. This option is made attractive by the presence of abundant natural gas in the area as well a local electrical power source. The current operations leave the coke produced by the various operators as back fill into the open pit mines producing the oil sands.

The option of partial upgrading at the surface involves recovery of the heavy crude and sufficient upgrading to make the crude transportable by pipeline. As a result of such upgrading, the properties of the heavy crude would be modified to meet the specification required by the pipeline company. This could either be though dilution with a suitable (presumably aromatic solvent that would prevent the asphaltene constituents from separating as a separate phase) or through partial thermal upgrading.

Traditional heavy feedstock processing, such as coking or hydrocracking, are very expensive processes, and require a large scale to be viable. Thus, adaptation of such processes to the wellhead is not always viable. Thus, the manner in which refineries convert heavy oil into low-boiling high-value products has become a major focus of operations with new concepts evolving into new processes (Speight, 2011). Even though they may not be classed as conversion processes *per se*, pretreatment processes for removing asphaltene constituents, metals, sulfur, and nitrogen constituents are also important and can play an important role in selecting a suitable wellhead processing option.

Conceivably, heavy oil or bitumen could be upgraded at the wellhead and sent through a pipeline to a refinery for further upgrading. However, this is not to be construed that heavy oil upgrading will *always* involve a coking step as the primary upgrading step. Other options, including some presented elsewhere could well become predominant methods for upgrading in the future.

Partial upgrading requires less processing than full upgrading, and it is cheaper, and may be well suited for locations adjacent to the upstream-primarily in situ facilities. The partial upgrading process will target the production of a partially processed product, or partial upgraded heavy oil or bitumen to meet pipeline specifications for API gravity, sulfur content, and viscosity. The primary objective is to upgrade heavy oil and the bitumen to an easily transportable product without the addition of diluent, either naphtha or a conventional crude oil, either of which may initiated asphaltene separation. Another option is to cause asphaltene separation followed by gasification of the asphaltene fraction followed by the FT synthesis (Chadeesingh, 2011; Speight, 2013a, 2014b).

There is (or will be) an obvious future need for partial upgrading during or immediately after recovery. On the other hand, hydrogen addition must be used during upgrading in order to stabilize the upgraded heavy oil—which could means that the cost of partial upgrading is not much reduced as compared to full upgrading. Therefore, the only choice currently is no upgrading or full upgrading. Other goals could be to achieve breakthroughs in upgrading technologies, such as nonthermal coking methods, that would use far less energy, or such as gasification at 800°C (1470°F), which is far lower than the current commercial temperatures. The technology where changes do occur involves combustion of the oil *in situ*. The concept of any combustion technology requires that the oil be partially combusted and that thermal decomposition occurs to other parts of the oil. This is sufficient to cause irreversible chemical and physical changes to the oil to the extent that the product is markedly different to the oil in-place. Recognition of this phenomenon is essential before combustion technologies are applied to oil recovery.

Although this improvement in properties may not appear to be too drastic, nevertheless it usually is sufficient to have major advantages for refinery operators. Any incremental increase in the units of hydrogen–carbon ratio can save amounts of costly hydrogen during upgrading. The same principles are also operative for reductions in the nitrogen, sulfur, and oxygen contents. This latter occurrence also improves catalyst life and activity as well as reduces the metals content.

In short, in situ recovery processes (although less efficient in terms of bitumen recovery relative to mining operations) may have the added benefit of *leaving* some of the more obnoxious constituents (from the processing objective) in the ground. Processes that offer the potential for partial upgrading during recovery are varied but usually follow a surface process. Not that this be construed as an easy task, there are many disadvantages that arise from attempting in situ upgrading.

Finally, there is not a single in situ recovery process that will be applicable to all reservoirs and no single recovery process will be able to access all the heavy oil in a given reservoir. To achieve maximum recovery it will be necessary to apply a combination of different processes. For example a steam-based recovery process followed by ISC, followed by in situ upgrading, followed by bioconversion of the residual hydrocarbons. This type of sequential recovery will require careful planning to ensure that the optimum sequence and timing is applied.

Then to achieve partial upgrading during recovery requires a further sequential operation before a transportable produced. The big breakthrough will be in the area of in situ upgrading through innovations in downhole conversion equipment and downhole catalysts. Innovations in downhole equipment that can withstand the high temperature for the thermal recovery methods and downhole catalysts that will allow optimization of the in situ upgrading and recovery of the partially upgraded product. In order to accomplish this, a multistep system may be required to achieve the necessary aims of heavy oil and bitumen recovery with partial upgrading. What this might be is currently unknown but there are possibilities.

10.3.2 Biodesulfurization

Another option for partial upgrading at the surface is biodesulfurization as might be made to occur during MEOR and continued at the surface. The occurrence of sulfur in ring systems and its removal by biological means could well produce a bevy of hydrocarbon liquids that are suitable for recovery and conventional refining.

In fact, the application of biotechnology to petroleum refining (biorefining) is a possible alternative to some of the current refining processes. The major potential applications of biorefining are biodesulfurization, biodenitrogenation, biodemetallization, and biotransformation of heavy crude oils into lighter crude oils (Le Borgne and Quintero, 2003; El-Gendy and Speight, 2015). In the biodesulfurization process, organic sulfur is removed using enzyme-catalyzed reactions (*biocatalyzed desulfurization*) and has the capability of producing low sulfur products (hydrocarbons). Certain microbial biocatalysts have been identified that can transform sulfur compounds found in fuels, including ones that selectively remove sulfur from dibenzothiophene derivatives (Abbad-Andaloussi et al., 2003; Mužic and Sertić-Bionda, 2013).

Biodesulfurization can be considered as a potential alternative to the conventional deep hydrodesulfurization processes used in refineries. In this process, microorganisms, their enzymes or cellular extracts as catalysts remove organosulfur from petroleum fractions without degrading the carbon skeleton of the organosulfur compounds and the use of this pathway has been proposed for the desulfurization of petroleum in production fields and also refineries (Soleimani et al., 2007). In addition, biodesulfurization offers specificity of enzymes, relatively lower capital and operating costs, mild processing conditions and reduces the need for hydrogen. Hydrodesulfurization is not equally effective in desulfurizing all classes of sulfur compounds present in fossil fuels. The biodesulfurization process, on the other hand, is effective regardless of the position of alkyl substitutions (Pacheco, 1999). However, the hydrodesulfurization process conditions are sufficient not only to desulfurize sensitive (labile) organosulfur compounds, but also to (1) remove nitrogen and metals from organic compounds, (2) induce saturation of at least some carbon–carbon double bonds, (3) remove substances having an unpleasant smell or color, (4) clarify the product by drying it, and (5) improve the cracking characteristics of the material. Therefore, with respect to these advantages, placing the biodesulfurization unit at the surface of an oil recovery operation as a complementary upgrading technology to achieve low-sulfur liquids that are suitable for pipeline transportation should also be considered (Monticello et al., 1996; Pacheco, 1999; Soleimani et al., 2007).

For example, there is a method for reducing the viscosity of viscous petroleum liquids, such as heavy crude oil and bitumen (Monticello et al., 1996). The method is appropriate for use with viscous petroleum liquids that contain heterocyclic sulfur, the physicochemical properties of the heterocycles contributing significantly to the viscosity of the liquid. The method involves contacting the viscous petroleum liquid with a biocatalyst that converts sulfur-bearing heterocycles into molecules that lack physicochemical properties conducive to viscosity. The biocatalyst works in a sulfur-specific manner, such that the sulfur-bearing heterocycle is altered at the sulfur heteroatom thereof. Through biocatalysis, carbon–sulfur bonds are cleaved and/or polar substituents, such as hydroxyl groups, are joined to the sulfur heteroatom, the hydrocarbon framework of the sulfur-bearing heterocycle, or both. Preferred biocatalysts for viscosity reduction include preparations of *Rhodococcus rhodochrous* ATCC No. 53968 microorganisms and enzymes obtained the microbe.

10.4 THE FUTURE

In summary, no one in situ recovery process with some degree of upgrading will be applicable to all reservoirs and no single recovery process will be able to access all of the heavy oil and extra heavy oil in the reservoir, or bitumen in the deposit. To achieve maximum recovery along with a degree of upgrading it will be necessary to apply a combination of different processes, that is, piggy-back one process with another process or with other processes. For example, a steam-based recovery process followed by ISC of the residual material in the reservoir (deposit) which would be accompanied by in situ upgrading. Alternatively, bioconversion might be the process of choice for partial upgrading of the reservoir (deposit) material. Other options can be listed recognizing that any type of sequential recovery will require careful planning to ensure that the optimum sequence or processes is applied and product quality is produced.

In summary, in situ upgrading of heavy oil, extra heavy oil, and tar sand bitumen upgrading will become the technology or *must do* process in the near future. Success has already been achieved and noted in the literature. For example, one process (or, more correctly, several processes) involves the injection of a catalyst. Since this process also involves contacting and mixing of the catalyst with the bitumen, it would be applied as an intermediate or supplemental process.

An alternate to catalytic upgrading is in situ deasphalting. Here a solvent, such as propane, is injected to drop out some of the asphaltenes and lower the bitumen viscosity. This could be a stand-alone process or an additional benefit to a solvent-based recovery process such as VAPEX. In situ upgrading is also an additional benefit to the combustion recovery process. As the combustion front moves through the reservoir, the lighter ends of the bitumen are distilled off, thermal cracking takes place, and the coke product is consumed as fuel. The resulting, produced oil, has a much lower viscosity than the original in situ bitumen.

Another mechanism for in situ upgrading is bioconversion or bacterial upgrading. The process depends on microbes that can convert the bitumen to lower viscosity oil or methane. These microbes may be naturally occurring or injected into the reservoir. As additional nutrients are normally required for this process to take place at reasonable rates, these must be injected.

REFERENCES

Abbad-Andaloussi, S., Warzywoda, M., Monot, F., 2003. Microbial desulfurization of diesel oils by selected bacterial strains. Révue Institut Français de Pétrole, Oil Gas Sci.Technol. 58 (4), 505–513.

Ancheyta, J., Speight, J.G., 2007. Hydroprocessing of Heavy Oils and Residua. CRC-Taylor & Francis Group, Boca Raton, FL.

Bailey, A.S., Kennedy, T.M., Schneider, D.R., 2001. Microbial enhanced oil recovery: diverse successful applications of biotechnology in the oil field. Paper No. SPE 72129. Proceedings. Asia Pacific Improved Oil Recovery Conference, Kuala Lumpur, Malaysia, October 6–9.Society of Petroleum Engineers, Richardson, TX.

Banat, I.M., 1995. Biosurfactant production and possible uses in microbial enhanced oil recovery and oil pollution remediation. Biores.Technol. 51, 1–12.

Bosisio, R., Cambon, J., Chavarie, C., Klvana, D., 1977. Experimental results on the heating of athabasca tar sand samples with microwave power. J. Microwave Power 12, 301–307.

Bryant, R.S., Donaldson, E.C., Yen, T.F., Chilingarian, G.V., 1989. Microbial enhanced oil recovery. In: Donaldson, E.C., Chilingarian, G.V., Yen, T.F. (Eds.), Enhanced Oil Recovery II: Processes and Operations Elsevier, Amsterdam, pp. 423–450.

Bryant, R.S., Lindsey, R.P., 1996. World-wide applications of microbial technology for improving oil recovery Proceedings of the SPE Symposium on Improved Oil Recovery. Society of Petroleum Engineers, Richardson, TX.27–134

Bukka, K., Miller, J.D., Oblad, A.G., 1991. Fractionation and characterization of utah tar sand bitumen: influence of chemical composition on bitumen viscosity. Energy Fuels 5, 333–340.

Chadeesingh, R., 2011. The fischer-tropsch process. In: Speight, J.G. (Ed.), The Biofuels Handbook The Royal Society of Chemistry, London. Part 3, (Chapter 5).

Clark, J.B., Munnecke, D.M., Jenneman, G.E., 1981. In situ microbial enhancement of oil production. Dev. Ind. Microbiol. 15, 695–701.

Cunha, L.B., 2005. Recent in situ oil recovery techniques for heavy and extra heavy oil reserves. SPE 94986.

Dickenson, R.L., Biasca, F.E., Schulman, B.L., Johnson, H.E., 1997. Refiner options for converting and utilizing heavy fuel oil. Hydrocarb. Process. 76 (2), 57–62.

Dim, P., Hart, A., Wood, J., Macnaughtan, W., Rigby, S.P., 2015. Characterization of pore coking in catalyst for thermal down-hole upgrading of heavy oil. Chem. Eng. Sci. 131, 138–145.

Ditman, J.G., 1973. Solvent deasphalting. Hydrocarb. Process. 52 (5), 110.

Duerkson, C.H., 1993. Gravity drainage processes for recovery and in situ upgrading of heavy oil and bitumen. Proceedings. 16th Rocky Mountain Fuel Society Symposium, Salt Lake City, UT.

El-Gendy, N.S., Speight, J.G., 2015. Handbook of Refinery Desulfurization. CRC Press, Taylor & Francis Group, Boca Raton, FL.

Fan, H., Liu, Y., 2002. Downhole catalysts upgrades heavy oil. Oil Gas J. 100 (11), 60–62.

Gary, J.G., Handwerk, G.E., Kaiser, M.J., 2007. Petroleum Refining: Technology and Economics, fifth ed. CRC Press, Taylor & Francis Group, Boca Raton, FL.

Graue, D.J., 2001. Upgrading and recovering heavy crude oils and natural bitumens by in situ hydrovisbreaking. United States Patent 6,328,104. December 11.

Greaves, M., Xia, T.X., 2004. Downhole catalytic process for upgrading heavy oil: produced oil properties and composition. J. Can. Petrol. Technol. 43 (9), 25–30.

Greaves, M., Xia, T.X., 2008. Stability of THAI process—theoretical and experimental observations. J. Can. Petrol. Technol. 47, 65.

Greaves, M., Xia, T.X., Ayasse, C., 2005. Underground upgrading of heavy oil using THAI—Toe-to-Heel air injection. SPE 97728.

Gregoli, A.A., 1985. Method of in situ hydrogenation of carbonaceous material. United States Patent 4,501,445. February 26.

Gregoli, A.A., Rimmer, D.P., 2000. Production of synthetic crude oil from heavy hydrocarbons recovered by in situ hydrovisbreaking. United States Patent 6,016.868. January 25.

Gregoli, A.A., Rimmer, D.P., Graue, D.J., 2000. Upgrading and recovery of heavy crude oils and natural bitumen by in situ hydrovisbreaking. United States Patent 6,016,867. January 25.

Hamrick, J.T., Rose, L.C., 1977. In situ hydrogenation of hydrocarbons in underground formations. United States Patent 4,050,525. September 27.

Hedrick, B.W., Seibert, K.D., Crewe, C., 2006. A new approach to heavy oil and bitumen upgrading. Report No. AM-06-29. UOP LLC, Des Plaines, IL.

Hewgill, G.S., Kalfayan, L.J., 1992. Enhanced oil recovery technique using hydrogen precursors. United States Patent 5,105,887. April 21.

Hsu, C.S., Robinson, P.R., 2006. Practical Advances in Petroleum Processing, vols. 1 and 2. Springer, New York.

Jenneman, G.E., 1989. The potential for in situ microbial applications. Dev. Petrol. Sci. 22, 37–74.

Jiang, S., Liu, X., Liu, Y., Zhong, L., 2005. In situ upgrading heavy oil by aquathermolytic treatment under steam injection conditions. Paper No. SPE 91973. Proceedings. International Symposium on Oilfield Chemistry, Woodlands, TX, February 2004. Society of Petroleum Engineers, Richardson, TX.

Joshi, J.B., Pandit, A.B., Kataria, K.L., Kulkarni, R.P., Sawarkar, A.N., Tandon, D., et al., 2008. Petroleum residue upgradation via visbreaking: a review. Ind. Eng. Chem. Res. 47, 8960–8988.

Khan, M.R., Patmore, D.J., 1998.. In: Speight, J.G. (Ed.), In Petroleum Chemistry and Refining Taylor & Francis, Washington, DC. (Chapter 6).

Khire, J.M., Khan, M.I., 1994a. Microbially enhanced oil recovery (MEOR). Part 1. Importance and mechanisms of microbial enhanced oil recovery. Enzyme Microb. Technol. 16, 170–172.

Khire, J.M., Khan, M.I., 1994b. Microbially enhanced oil recovery (MEOR). Part 2. Microbes and the subsurface environment for microbial enhanced oil recovery. Enzyme Microb. Technol. 16, 258–259.

Kirkwood, K.M., Foght, J.M., Gray, M.R., 2004. Prospects for biological upgrading of heavy oils and asphaltenes In: Vazquez-Duhalt, R. Quintero-Ramirez, R. (Eds.), Studies in Surface Science and Catalysis, vol. 151 Elsevier Science, Amsterdam, pp. 113–143.

Kuhlman, M., 2000. The benefits of in situ upgrading reactions to the integrated operations of the orinoco heavy-oil fields and downstream facilities. Paper No. SPE 62560. Proceedings. SPE/AAPG Western Regional Meeting, Long Beach, CA, June 19–22. Society of Petroleum Engineers, Richardson, TX.

Langevin, D., Poteau, S., Hénaut, I., Argillier, J.F., 2004. Crude oil emulsion properties and their application to heavy oil transportation. Oil Gas Sci. Technol. 59 (511), 2004.

Le Borgne, S., Quintero, R., 2003. Biotechnological processes for the refining of petroleum. Fuel Process. Technol. 81, 155–169.

Lim, K.T., Ramey Jr., H.J., Brigham, W.E., 1993. Steam distillation effects and oil quality changes during steam injection. Report No. Report DOE/BC/14600-18. United States Department of Energy, Washington, DC, January.

Luque, R., Speight, J.G. (Eds.), 2015. Gasification for Synthetic Fuel Production: Fundamentals, Processes, and Applications Woodhead Publishing, Elsevier, Cambridge.

Mamora, D.D., Ramey Jr., H.J., Brigham, W.E., Castanier, L.M., 1993. Kinetics of in situ combustion. Report No. DOE.BC/14600-51. US Department of Energy, Washington, DC, July.

Martínez-Palou, R., Mosqueira, M.L., Zapata-Rendón, B., Mar-Juárez, E., Bernal-Huicochea, Clavel-López, J.C., Aburto, J., 2011. Transportation of heavy and extra-heavy crude oil by pipeline: a review. J. Petrol. Sci. Eng. 75, 274–282.

Marzin, R., Pereira, P., McGrath, M.J., Feintuch, H.M., Thompson, G., 1998a. Oil Gas J. 97 (44), 79.

Marzin, R., Pereira, P., Zacarias, L., Rivas, L., McGrath, M., Thompson, G.J., 1998b. Resid conversion through the aquaconversion technology—an economical and environmental solution. Paper No. 1998.086. Society of Petroleum Engineers, Richardson, TX.

Mitchell, D.L., Speight, J.G., 1973. The solubility of asphaltenes in hydrocarbon solvents. Fuel 52, 149.

Mokrys, I.J., Butler, R.M., 1993. In situ upgrading of heavy oils and bitumen by propane deasphalting: the VAPEX process. Paper No. SPE 25452. Proceedings. Production Operations Symposium, Oklahoma City, Oklahoma, March 12–13. Society of Petroleum Engineers, Richardson, TX.

Monin, J.C., Audibert, A., 1994. Alteration of heavy crudes under high pressure steam conditions Preprints. Am. Chem. Soc. Div. Pet. Chem. 29, 4.

Monticello, D.J., Haney, I.I.I., William, M., 1996. Biocatalytic process for reduction of petroleum viscosity. United States Patent 5,529,930. June 25.

Moore, R.G., Laureshen, C.J., Belgrave, J.D.M., Ursenbach, M.G., Mehta, S.A., 1995. In situ combustion in canadian heavy oil reservoirs. Fuel 74, 1169–1175.

Moore, R.G., Laureshen, C.J., Mehta, S.A., Ursenbach, M.G., Belgrave, J.D.M., Weissman, J.G., et al., 1999. A downhole catalytic upgrading process for heavy oil using in situ combustion. J. Can. Petrol. Technol. 38 (13), 96.

Moschopedis, S.E., Ozum, B., Speight, J.G., 1998. Upgrading heavy oils. Rev. Process Chem. Eng. 1, 201.

Motaghi, M., Saxena, P., Ravi, R., 2010. Partial upgrading of heavy oil reserves. Petrol. Technol. Quart. Q4, 55–64.

Mut, S., 2005. Advances in an in situ upgrading process for unconventional oils. AAPG International Conference and Exhibition, Paris, France, September 11–14.

Mutyala, S., Fairbridge, C., Paré, J.R.J., Bélanger, J.M.R., Ng, S., Hawkins, R., 2010. Microwave applications to oil sands and petroleum: a review. Fuel Process. Technol. 91, 127–135.

Mužic, M., Sertić-Bionda, K., 2013. Alternative processes for removing organic sulfur compounds from petroleum fractions. Chem. Biochem. Eng. Q. 27 (1), 101–108.

N-Solve, 2015. <https://www.nsolv.ca/technology/#sthash.xeWLBbxB.dpuf> (accessed 24.06.15.).

Pacheco, M.A., 1999. Recent advances in biodesulfurization (biodesulfurization) of diesel fuel. Proceedings. NPRA Annual Meeting, San Antonio, TX, March 21–23.

Parkash, S., 2003. Refining Processes Handbook. Gulf Professional Publishing, Elsevier, Amsterdam.

Pratts, M., 1986. Thermal Recovery, vol. 7. Society of Petroleum Engineers, New York.

Radovanović, L., Speight, J.G., 2011. Visbreaking: a technology of the future. Proceedings. First International Conference—Process Technology and Environmental Protection (PTEP 2011). University of Novi Sad, Technical Faculty Mihajlo Pupin, Zrenjanin, Republic of Serbia, December 7, pp. 335–338.

Raiders, R.A., Knapp, R.M., McInerney, M.J., 1989. Microbial selective plugging and enhanced oil recovery. J. Ind. Microbiol. 4, 215–230.
Ramey Jr., H.J., Stamp, V.V., Pebdani, F.N., 1992. Case history of South Belridge, California, in situ combustion oil recovery, SPE 24200. Proceedings. 9th SPE/DOE EOR Symposium, Tulsa, Oklahoma, April 21–24.
Rana, M.S., Sámano, V., Ancheyta, J., Diaz, J.A.I., 2007. A review of recent advances on process technologies for upgrading of heavy oils and residua. Fuel 86, 1216–1231.
Ranhema, H., Daulat, M., 2010. Combustion assisted gravity drainage (CAGD) appears promising. No. 135821. Proceedings. Canadian Unconventional Resources and International Petroleum Conference, Calgary, Alberta, Canada, October 19–21. Society of Petroleum Engineers, Richardson, TX.
RAROP, 1991. RAROP heavy oil processing handbook. Research Association for Residual Oil Processing. T. Noguchi (Chairman), Ministry of Trade and International Industry (MITI), Tokyo, Japan.
Reichert, C., Fuhr, B., Sawatzky, H., Lafleur, R., Verkoczy, B., Soveran, D., et al., 1989. Fire flood recovery process effects upon heavy oil properties. Petrol. Sci. Technol. 7, 851–878.
Richardson, W.C., Fontaine, M.F., Haynes, S., 1992. Compositional changes in distilled, steam-distilled, and steam-flooded crude oils. Paper No. SPE 24033. Western Regional Meeting, Bakersfield, CA, March 30–April 1.
Robinson, J., Binner, E., Saeid, A., Al-Harahsheh, M., Kingman, S., 2010. Microwave processing of oil sands and contribution of clay minerals. Fuel 93, 153–161.
Rodríguez, R., Bashbush, J.L., Rincón, A., 2008. Feasibility of using electrical downhole heaters in faja heavy oil reservoirs. Paper No. SPE-117682. Proceedings of International Thermal Operations and Heavy Oil Symposium, Calgary, AB, Canada, October 20–23. Society of Petroleum Engineers, Richardson, TX.
Salager, J.L., Briceno, M.I., Brancho, C.L., 2001. Heavy hydrocarbon emulsions. In: Sjöblom, J. (Ed.), Encyclopedic Handbook of Emulsion Technology Marcel Dekker, New York.
Sarathi, P.S., Roark, D., Strycker, A.R., 1988. Light oil steamflooding a laboratory study. Paper No. SPE 17447. Proceedings. 58th Annual California Regional Meeting of the SPE, Bakersfield, CA, March 23–25. Society of Petroleum Engineers, Richardson, TX, pp. 431–442.
Saxman, D.B., Crull, A., 1984. Biotechnology and petroleum production. Paper No. SPE 13146. Proceedings. Annual Technical Conference and Exhibition, Houston, September 16–19. Society of Petroleum Engineers, Richardson, TX.
Shah, A., Fishwick, R., Wood, J., Leeke, G., Rigby, S., Greaves, M., 2010. A review of novel techniques for heavy oil and bitumen extraction and upgrading. Energy Environ. Sci. 3, 700–714.
Sharpe, H.N., Richardson, W.C., Lolley, C.S., 1995. Representation of steam distillation and in situ upgrading processes in heavy oil simulation. Paper No. SPE 30301. Proceedings. International Heavy oil Symposium, Calgary, Alberta, Canada, June 19–21.
Shennan, J.L., Levi, J.D., 1987. In situ microbial enhanced oil recovery. In: Kosaric, N., Cairns, W.L., Gray, N.C.C. (Eds.), Biosurfactants and Biotechnology Marcel Dekker, New York, pp. 163–180.
Shu, W.R., 1984. A viscosity correlation for mixtures of heavy oil, bitumen, and petroleum fractions. Soc. Pet. Eng. J. 24 (3), 277–292.
Soleimani, M., Bassi, A., Margaritis, A., 2007. Biodesulfurization of refractory organic sulfur compounds in fossil fuels. Biotechnol. Adv. 25, 570–596.
Speight, J.G., 1979. Studies on bitumen fractionation—(a) fractionation by a cryoscopic method, (b) effect of solvent type on asphaltene solubility. Information Series No. 84, Alberta Research Council, Edmonton, Alberta, Canada.
Speight, J.G., 1984. Upgrading heavy oils and residua: the nature of the problem. Plenary Lecture. Symposium on Catalysis, Canadian Society for Chemical Engineering, Quebec City, Quebec, Canada, September 30–October 3.

Speight, J.G., 1986. Upgrading heavy feedstocks. Ann. Rev. Energy 11, 253.
Speight, J.G., 1990. Tar sands. In: Speight, J.G. (Ed.), Fuel Science and Technology Handbook Marcel Dekker, Inc., New York. (Chapters 12–16).
Speight, J.G., 2000. The Desulfurization of Heavy Oils and Residua, second ed. Marcel Dekker, Inc., New York.
Speight, J.G., 2011. The Refinery of the Future. Gulf Professional Publishing, Elsevier, Oxford.
Speight, J.G., 2012. Visbreaking: a technology of the past and the future. Scientia Iranica 19 (3), 569–573.
Speight, J.G., 2013a. The Chemistry and Technology of Coal, third ed. CRC-Taylor and Francis Group, Boca Raton, FL.
Speight, J.G., 2013b. Oil Sand Production Processes. Gulf Professional Publishing, Elsevier, Oxford.
Speight, J.G., 2013c. Heavy and Extra Heavy Oil Upgrading Technologies. Gulf Professional Publishing, Elsevier, Oxford.
Speight, J.G., 2014a. The Chemistry and Technology of Petroleum, fifth ed. CRC-Taylor and Francis Group, Boca Raton, FL.
Speight, J.G., 2014b. Gasification of Unconventional Feedstocks. Gulf Professional Publishing, Elsevier, Oxford.
Speight, J.G., 2015. Handbook of Petroleum Product Analysis, second ed. John Wiley & Sons Inc., Hoboken, NJ.
Speight, J.G., Ozum, B., 2002. Petroleum Refining Processes. Marcel Dekker, Inc., New York.
Volek, C.W., Pryor, J.A., 1972. Steam distillation drive—Brea Field, California. J. Petrol. Technol. 24 (8), 899–906.
Wagner, J., Kresnyak, S., 2015. Bitumen and partial targeting upgrading: the next step. Hydrocarb. Process. 94 (9), 73–78.
Weissman, J.G., 1997. Review of processes for downhole catalytic upgrading of heavy crude oil. Fuel Process Technol. 50 (2–3), 199.
Weissman, J.G., Kessler, R.V., 1996. Downhole heavy crude oil hydroprocessing. Appl. Catalysis General 140 (1), 16.
Wu, C.H., Brown, A., 1975. A laboratory study on steam distillation in porous media. Paper No. 5569. Proceedings. 1975 SPE Annual Meeting, Dallas, TX, September 28–October 1.
Xia, T.X., Greaves, M., 2006. In situ upgrading of athabasca tar sand bitumen using THAI. Chem. Eng. Res. Design 84 (9), 856–864.
Xia, T.X., Greaves, M., Werfilli, W.S., Rathbone, R.R., 2002. Downhole conversion of lloydminster heavy oil using THAI-CAPRI processes. SPE 78998.
Yaghi, B.M., Al-Bemani, A., 2002. Heavy crude oil viscosity reduction for pipeline transportation. Energy Sources 24, 93–102.
Zhang, J., Chen, D., Yan, D., Yang, X., Shen, C., 1991. Pipelining of heavy crude oil as oil-in-water emulsions. Paper No. SPE 21733. Proceedings. SPE Production Operations Symposium, Oklahoma City, Oklahoma, April 7–9. Society of Petroleum Engineers, Richardson TX.
Zou, J., Gray, M.R., Thiel, J., 1989. Structural changes in cold lake bitumen due to in situ combustion. AOSTRA J. Res. 5, 75.

APPENDIX

CONVERSION FACTORS

1 acre = 43,560 ft^2

1 acre foot = 7758.0 bbl

1 atmosphere = 760 mm Hg = 14.696 psia = 29.91 inch Hg

1 atmosphere = 1.0133 bars = 33.899 ft H_2O

1 barrel (oil) = 42 gal = 5.6146 ft^3

1 barrel (water) = 350 lb at 60°F

1 barrel per day = 1.84 cm^3/second

1 Btu = 778.26 ft-lb

1 centipoise × 2.42 = lb mass/(ft) (hour), viscosity

1 centipoise × 0.000672 = lb mass/(ft) (second), viscosity

1 $foot^3$ = 28,317 cm^3 = 7.4805 gal

Density of water at 60°F = 0.999 g/cm^3 = 62.367 lb/ft^3 = 8.337 Ib/gal

1 gallon = 231 $inch^3$ = 3785.4 cm^3 = 0.13368 ft^3

1 horsepower-hour = 0.7457 kilowatt hours (kWhr) = 2544.5 Btu

1 horsepower = 550 ft-lb/second = 745.7 W

1 inch = 2.54 cm

1 m = 100 cm = 1000 mm = 10 μm = 10 angstroms (Å)

1 ounce = 28.35 g

1 pound = 453.59 g = 7000 grains

1 $mile^2$ = 640 acres

SI METRIC CONVERSION FACTORS

Acre-foot × 1.233482	E + 03 = m^3
Barrels × 1.589873	E − 01 = m^3
Centipoise × 1.000000	E − 03 = Pa seconds
Darcy × 9.869233	E − 01 = μm^2
Feet × 3.048000	E − 01 = m
Pounds/acre p foot × 3.677332	E − 04 = kg/m^3
Pounds/$inch^2$ × 6.894757	E + 00 = kPa
Dyne/cm × 1.000000	E + 00 = mN/m
Parts per million × 1.000000	E + 00 = mg/kg

GLOSSARY

Abandon to cease work on a well which is nonproductive, to plug off the well with cement plugs and salvage all recoverable equipment; also used in the context of field abandonment.

Abandonment converting a drilled well to a condition that can be left indefinitely without further attention and will not damage freshwater supplies, potential petroleum reservoirs, or the environment.

Abandonment pressure this is a direct function of the economic premises and it corresponds to the static bottom pressure at which the revenues obtained from the sales of the hydrocarbons produced are equal to the well's operation costs.

Absolute permeability the ability of a rock to conduct a fluid when only one fluid is present in the pores of the rock.

Accumulation an individual body of moveable petroleum, heavy oil, or tar sand bitumen.

Acid catalyst a catalyst having acidic character; the alumina minerals are examples of such catalysts.

Acid deposition acid rain; a form of pollution depletion in which pollutants, such as nitrogen oxides and sulfur oxides, are transferred from the atmosphere to soil or water; often referred to as atmospheric self-cleaning. The pollutants usually arise from the use of fossil fuels.

Acidity the capacity of an acid to neutralize a base such as a hydroxyl ion (OH^-).

Acidizing a technique for improving the permeability of a reservoir by injecting acid.

Acid number a measure of the reactivity of petroleum with a caustic solution and given in terms of milligrams of potassium hydroxide that are neutralized by one gram of petroleum.

Acid rain the precipitation phenomenon that incorporates anthropogenic acids and other acidic chemicals from the atmosphere to the land and water (see acid deposition).

Acoustic log see Sonic log.

Acre-foot a measure of bulk rock volume where the area is one acre and the thickness is one foot.

Additions the reserve provided by the exploratory activity. It consists of the discoveries and delimitations in a field during the study period.

Additive a material added to another (usually in small amounts) in order to enhance desirable properties or to suppress undesirable properties.

Adsorption the physical/chemical phenomenon whereby a molecule or aggregate of molecules attaches itself to the rock surface; the transfer of a substance from a solution to the surface of a solid resulting in relatively high concentration of the substance at the place of contact.

After flow (Afterflow) flow from the reservoir into the wellbore that continues for a period after the well has been shut in; after-flow can complicate the analysis of a pressure transient test.

Air injection an oil recovery technique using air to force oil from the reservoir into the wellbore.

Alkaline flooding see EOR process.

Alicyclic hydrocarbon a compound containing carbon and hydrogen only which has a cyclic structure (eg, cyclohexane); also collectively called naphthenes.

Aliphatic hydrocarbon a compound containing carbon and hydrogen only which has an open-chain structure (eg, as ethane, butane, octane, butene) or a cyclic structure (eg, cyclohexane).

Alkaline a high pH usually of an aqueous solution; aqueous solutions of sodium hydroxide, sodium orthosilicate, and sodium carbonate are typical alkaline materials used in enhanced oil recovery.

Alkalinity the capacity of a base to neutralize the hydrogen ion (H^+).

Alkanes hydrocarbons that contain only single carbon–hydrogen bonds. The chemical name indicates the number of carbon atoms and ends with the suffix "ane."

Alkenes hydrocarbons that contain carbon–carbon double bonds. The chemical name indicates the number of carbon atoms and ends with the suffix "ene."

Alkyl groups a group of carbon and hydrogen atoms that branch from the main carbon chain or ring in a hydrocarbon molecule. The simplest alkyl group, a methyl group, is a carbon atom attached to three hydrogen atoms.

Alpha-scission the rupture of the aromatic carbon–aliphatic carbon bond that joins an alkyl group to an aromatic ring.

Alumina (Al_2O_3) used in separation methods as an adsorbent and in refining as a catalyst.

American Petroleum Institute (API) a trade association comprised of larger integrated oil companies that works for the common goals of the oil industry.

American Society for Testing and Materials (ASTM) the official organization in the United States for designing standard tests for petroleum and other industrial products.

Analogous reservoir a reservoir in the same geographic area that is formed by the same, or very similar geological processes as, a reservoir in question (or under study for reserves evaluation) as regards sedimentation, diagenesis, pressure, temperature, chemical and mechanical history, and structure; also has the same or similar geologic age, geologic features, and reservoir rock and fluid properties. Analogous features and characteristics can include approximate depth, pressure, temperature, reservoir-drive mechanism, original fluid content, oil gravity, reservoir size, gross thickness, pay thickness, net-to-gross ratio, lithology, heterogeneity, porosity, and permeability. The development scheme for a reservoir (eg, as reflected by well spacing) can also be important in establishing the relevance of the analogy.

Annex B an operator's development plan for an offshore installation which requires government approval before it can be implemented.

Anticline structural configuration of a package of folding rocks and in which the rocks are tilted in different directions from the crest.

API gravity a measure of the *lightness* or *heaviness* of petroleum which is related to density and specific gravity.

$°API = (141.5/sp\ gr\ @\ 60°F) - 131.5$

Apparent bulk density the density of a catalyst as measured; usually loosely compacted in a container.

Apparent viscosity the viscosity of a fluid, or several fluids flowing simultaneously, measured in a porous medium (rock), and subject to both viscosity and permeability effects; also called effective viscosity.

Appraisal well a well drilled as part of an appraisal drilling program which is carried out to determine the physical extent, reserves and likely production rate of a field.

Aquifer a subsurface rock interval that will produce water; often the underlay of a petroleum reservoir.

Areal sweep efficiency (horizontal sweep efficiency) the fraction of the flood pattern area that is effectively swept by the injected fluids.

Aromatic hydrocarbon a hydrocarbon characterized by the presence of an aromatic ring or condensed aromatic rings; benzene and substituted benzene, naphthalene and substituted naphthalene, phenanthrene and substituted phenanthrene, as well as the higher condensed ring systems; compounds that are distinct from those of aliphatic compounds or alicyclic compounds.

Aromatization the conversion of nonaromatic hydrocarbons to aromatic hydrocarbons by: (1) rearrangement of aliphatic (noncyclic) hydrocarbons into aromatic ring structures and (2) dehydrogenation of alicyclic hydrocarbons (naphthenes).

Aromatics hydrocarbons characterized by their uniform carbon ring structure and their often pleasant aroma; commercial petroleum-derived aromatics are benzene, toluene, and xylene (BTX) or benzene, toluene, ethylbenzene, and xylene (BTEX).

Artificial production system any of the techniques used to extract petroleum from the producing formation to the surface when the reservoir pressure is insufficient to raise the oil naturally to the surface.

Asphaltene constituents the fraction of conventional oil, heavy oil, or bitumen that is insoluble in *n*-pentane or *n*-heptane.

Asphaltene onset pressure the pressure at a given test temperature that first causes asphaltene to separate as a separate phase from a reservoir fluid as pressure decreases.

Asphalt a black cementitious material manufactured in refineries for petroleum residua and used primarily for road construction.

Asphaltene (asphaltenes) the brown to black powdery material produced by treatment of petroleum, petroleum residua, or bituminous materials with a low-boiling liquid hydrocarbon, for example, pentane or heptane; soluble in benzene (and other aromatic solvents), carbon disulfide, and chloroform (or other chlorinated hydrocarbon solvents).

Asphaltene association factor the number of individual asphaltene species which associate in nonpolar solvents as measured by molecular weight methods; the molecular weight of asphaltenes in toluene divided by the molecular weight in a polar nonassociating solvent, such as dichlorobenzene, pyridine, or nitrobenzene.

Assay a procedure for determining the general distillation characteristics (eg, distillation profile, and other quality information of crude oil; also tests performed on a sample of ores or minerals to determine the physical and chemical properties.

Associated gas natural gas that is in contact with and/or dissolved in the crude oil of the reservoir. It may be classified as gas cap (free gas) or gas in solution (dissolved gas).

Associated gas in solution (or dissolved gas) natural gas dissolved in the crude oil of the reservoir, under the prevailing pressure and temperature conditions.

Associated molecular weight the molecular weight of asphaltenes in an associating (nonpolar) solvent, such as toluene.

Atmospheric distillation unit a unit sued to separate petroleum constituents using a distillation column at atmospheric pressure.

Atmospheric equivalent boiling point (AEBP) a mathematical method of estimating the boiling point at atmospheric pressure of nonvolatile fractions of petroleum.

Atmospheric residuum the nonvolatile residuum obtained by distillation of a crude oil under atmospheric pressure and which boils above 350°C (660°F).

Attainment area a geographical area that meets NAAQS for criteria air pollutants (see also nonattainment area).

Attapulgus clay see Fuller's earth.

Bank a concentration of oil (oil bank) or other fluid in a reservoir that moves cohesively through the reservoir.

Barrel the unit of measurement of liquids in the petroleum industry; equivalent to 42 US standard gallons or 33.6 imperial gallons; 7.3 barrels = 1 ton: 6.29 barrels = 1 m^3.

Basement a geological term to indicate the foot or base of a sedimentary sequence composed of igneous or metamorphic rocks.

Base number the quantity of acid, expressed in milligrams of potassium hydroxide per gram of sample that is required to titrate a sample to a specified end-point.

Basic nitrogen nitrogen (in petroleum) which occurs in pyridine form.

Basic sediment and water (BS&W, BSW) the material which collects in the bottom of storage tanks, usually composed of oil, water, and foreign matter; also called bottoms, bottom settlings.

Basin receptacle in which a sedimentary column is deposited that shares a common tectonic history at various stratigraphic levels.

Battery equipment to process or store crude oil from one or more wells.

Baumé gravity the specific gravity of liquids expressed as degrees on the Baumé (°Bé) scale.

For liquids lighter than water:

$$\text{Sp gr } 60°\text{F} = 140/(130 + °\text{Bé})$$

For liquids heavier than water:

$$\text{Sp gr } 60°\text{F} = 145/(145 - °\text{Bé})$$

Bauxite mineral matter used as a treating agent; hydrated aluminum oxide formed by the chemical weathering of igneous rock.

Behind-pipe reserves reserves that are expected to be recovered from zones in existing wells, which will require additional completion work or future recompletion prior to the start of production.

Bell cap a hemispherical or triangular cover placed over the riser in a (distillation) tower to direct the vapors through the liquid layer on the tray; see Bubble cap.

Benchmarking measures data and information used as a point of reference against which industry performance is measured.

Bentonite montmorillonite (a magnesium-aluminum silicate); used as a treating agent.

Benzene a low-boiling aromatic hydrocarbon, which occurs naturally as a part of oil and natural gas activity. It's considered to be a nonthreshold carcinogen and is an occupational and public health concern.

Billion 1×10^9

Biofuel a fuel derived by a variety of processes from an organic source that is renewable, that is, capable of being replenished naturally.

Biogenic material derived from bacterial or vegetation sources.

Biological lipid any biological fluid that is miscible with a nonpolar solvent. These materials include waxes, essential oils, chlorophyll, etc.

Biological oxidation the oxidative consumption of organic matter by bacteria by which the organic matter is converted into gases.

Biomass biological organic matter; any naturally occurring carbonaceous material derived from trees, grasses, and any other vegetation; used to produce biofuels.

Bitumen portion of petroleum that exists in the reservoirs in a semisolid or solid phase. In its natural state, it generally contains sulfur, metals and other nonhydrocarbon compounds. Natural bitumen has a viscosity of more than 10,000 cP, measured at the original temperature of the reservoir, at atmospheric pressure and gas-free. It frequently requires treatment before being refined.

Bituminous containing bitumen or constituting the source of bitumen.

Bituminous rock see Bituminous sand.

Bituminous sand a formation in which the bituminous material (see Bitumen) is found as a filling in veins and fissures in fractured rock or impregnating relatively shallow sand, sandstone, and limestone strata; a sandstone reservoir that is impregnated with a heavy, viscous black petroleum-like material that cannot be retrieved through a well by conventional production techniques.

Black oil a nonscientific term used to indicate any of the dark-colored oils; a term now often applied to heavy oil.

Block a North Sea acreage subdivision measuring approximately 10 × 20 km, forming part of a quadrant. For example, Block 9/13 is the 13th block in Quadrant 9.

Blow-down condensate and gas is produced simultaneously from the outset of production.

Blow-out when well pressure exceeds the ability of the wellhead valves to control it; oil and gas "blow wild" at the surface.

Blow-out preventers (BOPs) high pressure wellhead valves, designed to shut off the uncontrolled flow of hydrocarbons.

Boiling point a characteristic physical property of a liquid at which the vapor pressure is equal to that of the atmosphere and the liquid is converted to a gas.

Boiling range the range of temperature, usually determined at atmospheric pressure in standard laboratory apparatus, over which the distillation of oil commences, proceeds, and finishes.

Borehole the hole as drilled by the drill bit.

Bottoms the liquid which collects in the bottom of a vessel (tower bottoms, tank bottoms) either during distillation; also the deposit or sediment formed during storage of petroleum or a petroleum product.

Brackish water deep groundwater that is high in dissolved salt and unsuitable for domestic or agricultural uses.

British thermal unit see Btu.

BS&W see Basic sediment and water.

BTEX benzene, toluene, ethylbenzene, and the xylene isomers.

Btu (British thermal unit) the energy required to raise the temperature of one pound of water 1°F.

Bubble point the temperature (at a given pressure) where the first bubble of vapor is formed when heating a liquid consisting of two or more components; the temperature at which incipient vaporization of a liquid in a liquid mixture occurs, corresponding with the equilibrium point of 0% vaporization or 100% condensation.

Buckley-Leverett method a theoretical method of determining frontal advance rates and saturations from a fractional flow curve.

Butane (C_4H_{10}) a gaseous straight-chain (*n*-butane) or branch-chain hydrocarbon (iso-butane) extracted from natural gas or refinery gas streams.

Butylene (Butene; C_4H_8) an olefin hydrocarbon recovered from refinery processes; either straight chain (*n*-butene) or branched chain (iso-butene).

C_1, C_2, C_3, C_4, C_5 fractions a common way of representing fractions containing a preponderance of hydrocarbons having 1, 2, 3, 4, or 5 carbon atoms, respectively, and without reference to hydrocarbon type.

Calorific equivalence of dry gas to liquid factor (CEDGLF) the factor used to relate dry gas to its liquid equivalent. It is obtained from the molar composition of the reservoir gas, considering the unit heat value of each component and the heat value of the equivalence liquid.

Capillary forces interfacial forces between immiscible fluid phases, resulting in pressure differences between the two phases. C_1, C_2, C_3, C_4, C_5 fractions: a common way of representing fractions containing a preponderance of hydrocarbons having 1, 2, 3, 4, or 5 carbon atoms, respectively, and without reference to hydrocarbon type.

Capillary number N_c, the ratio of viscous forces to capillary forces, and equal to viscosity times velocity divided by interfacial tension. Thus:

$$N_c = \mu V/\gamma$$

In the equation, μ is the dynamic viscosity of the liquid, V is a characteristic velocity, and γ is the surface or interfacial tension between the two fluid phases.

Capillary pressure a force per area unit resulting from the surface forces to the interface between two fluids; the difference in the pressure between two fluids measured at the interface between the two fluids.

Carbene the pentane- or heptane-insoluble material that is insoluble in benzene or toluene but which is soluble in carbon disulfide (or pyridine); a type of rifle used for hunting bison.

Carboid the pentane- or heptane-insoluble material that is insoluble in benzene or toluene and which is also insoluble in carbon disulfide (or pyridine).

Carbonate reservoir a reservoir composed of sedimentary rocks deposited in marine environments with clear, shallow, warm waters and are mostly of biological origin; the rocks are made up by fragments of marine organisms, skeletons, coral, algae and precipitation, and consist mostly of calcium carbonate, which is chemically active compared to the sand which makes sandstones.

Carbon capture and storage (CCS) the process of taking waste carbon dioxide and transporting it to a storage site, normally underground in a specific type of geological formation.

Carbon dioxide augmented waterflooding waterflooding by injection of brine that is fully or nearly saturated with carbon dioxide; also called carbonated waterflooding.

Carbon-forming propensity see Carbon residue.

Carbonization the conversion of an organic compound into char or coke by heat in the substantial absence of air; often used in reference to the destructive distillation (with simultaneous removal of distillate) of coal.

Carbon rejection upgrading processes in which coke is produced, for example, coking.

Carbon residue the amount of carbonaceous residue remaining after thermal decomposition of petroleum, a petroleum fraction, or a petroleum product in a limited amount of air; also called the *coke-* or *carbon-forming propensity*; often prefixed by the terms Conradson or Ramsbottom in reference to the inventor of the respective tests.

Casing metal pipe inserted into a wellbore and cemented in place to protect both subsurface formations (such as groundwater) and the wellbore; a surface casing is set first to protect groundwater and the production casing is the last one set; the production tubing (through which hydrocarbons flow to the surface) will be suspended inside the production casing.

Casinghead gasoline the liquid hydrocarbon product extracted from casinghead gas by one of three methods: compression, absorption, or refrigeration; see also Natural gasoline.

Casing string the steel tubing that lines a well after it has been drilled; formed from sections of steel tube screwed together.

Catagenesis the alteration of organic matter during the formation of petroleum that may involve temperatures in the range 50°C (120°F) to 200°C (390°F); see also Diagenesis and Metagenesis.

Catagenesis the alteration of organic matter during the formation of petroleum that may involve temperatures in the range 50°C (120°F) to 200°C (390°F); see also Diagenesis and Metagenesis.

Catalyst a chemical agent which, when added to a reaction (process) will enhance the conversion of a feedstock without being consumed in the process.

Catalytic cracking the refining process of breaking down via heat and pressure the larger, heavier and more complex hydrocarbon molecules into simpler and lighter molecules, primarily gasoline: accomplished by the use of a catalytic agent and is an effective process for increasing the yield of gasoline from crude oil.

Catalytic hydrocracking a refining process that uses hydrogen and catalysts with relatively low temperatures and high pressures for converting middle boiling or residual material to high-octane gasoline, reformer charge stock, jet fuel, and/or high grade fuel oil; the process uses one or more catalysts, depending upon product output, and can handle high sulfur feedstocks without prior desulfurization.

Catalytic hydrotreating a refining process for treating petroleum fractions from atmospheric or vacuum distillation units (eg, naphtha, middle distillates, reformer feeds, residual fuel oil, and heavy gas oil) and other petroleum (eg, cat-cracked naphtha, coker naphtha, gas oil, etc.) in the presence of catalysts and substantial quantities of hydrogen; hydrotreating includes desulfurization, removal of substances (eg, nitrogen compounds) that deactivate catalysts, conversion of olefins to paraffins to reduce gum formation in gasoline, and other processes to upgrade the quality of the fractions.

Catalytic reforming a refining process using controlled heat and pressure with catalysts to rearrange certain hydrocarbon molecules, thereby converting paraffinic and naphthenic type hydrocarbons (eg, low-octane gasoline boiling range fractions) into petrochemical feedstocks and higher octane stocks suitable for blending into finished gasoline.

Cementitious material material have the properties of cement that is able to hold mineral grains together.

Centrifugal pump a rotating pump, commonly used for large-volume oil and natural gas pipelines, that takes in fluids near the center and accelerates them as they move to the outlet on the outer rim.

CFR Code of Federal Regulations; Title 40 (40 CFR) contains the regulations for protection of the environment.

Characterization factor the UOP characterization factor K, defined as the ratio of the cube root of the molal average boiling point, T_B, in degrees Rankine ($°R = °F+460$), to the specific gravity at 60°F/60°F:

$$K = (T_B)^{1/3}/\text{sp gr}$$

The value ranges from 12.5 for paraffin stocks to 10.0 for the highly aromatic stocks; also called the Watson characterization factor.

Chemical flooding see EOR process.

Christmas tree the assembly of fittings and valves on the top of the casing which control the production rate of oil.

Chromatographic adsorption selective adsorption on materials such as activated carbon, alumina, or silica gel; liquid or gaseous mixtures of hydrocarbons are passed through the adsorbent in a stream of diluent, and certain components are preferentially adsorbed.

Clay minerals silicate minerals that also usually contain aluminum and have particle sizes are less than 0.002 μm; used in separation methods as an adsorbent and in refining as a catalyst.

Chromatographic separation the separation of different species of compounds according to their size and interaction with the rock as they flow through a porous medium; also a standard laboratory separation method.

Clastic composed of pieces of preexisting rock.

Cloud point the temperature at which paraffin wax or other solid substances begin to crystallize or separate from the solution, imparting a cloudy appearance to the oil when the oil is chilled under prescribed conditions.

Coal an organic rock.

Coalbed methane (CBM) methane (natural gas) generated during coal formation and trapped in coal seams.

Coal tar the specific name for the tar produced from coal.

Coal tar pitch the specific name for the pitch produced from coal.

COFCAW an enhanced oil recovery process that combines forward combustion and waterflooding.

Cogeneration an energy conversion method by which electrical energy is produced along with steam generated for EOR use.

Coke number used, particularly in Great Britain, to report the results of the Ramsbottom carbon residue test, which is also referred to as a coke test.

Coker the processing unit in which coking takes place.

Coking a process for the thermal conversion of petroleum in which gaseous, liquid, and solid (coke) products are formed.

Cold production the use of operating and specialized exploitation techniques in order to rapidly produce heavy oils without using thermal recovery methods.

Combustible liquid a liquid with a flash point in excess of 37.8°C (100°F) but below 93.3°C (200°F).

Combustion zone the volume of reservoir rock wherein petroleum is undergoing combustion during enhanced oil recovery.

Commercial field an oil and/or gas field judged to be capable of producing enough net income to make it worth developing.

Commercial project a project where the degree of commitment is such that the accumulation is expected to be developed and placed on production within a reasonable time frame. A reasonable time frame for the initiation of development depends on the specific circumstances but, in general, should be limited to around 5 years.

Committed project a project where firm commitments have been made for the expenditures and activities needed to bring a discovered accumulation to the production stage. Undeveloped projects are committed only when it can be clearly demonstrated that there is intent to develop them and bring them to production. Intent may be demonstrated with firm funding/financial plans, declarations of commerciality, regulatory approvals and satisfaction of other conditions that would otherwise prevent the project from being developed and brought to production. These commitments should be unconditional, except for timing that may be dependent on the development of prior committed projects. An example of this would be where production is dedicated to a long-term sales contract and will only be developed as and when the capacity is required to satisfy the contract.

Completion (of a well) the installation of permanent wellhead equipment for the production of oil and gas; the process by which a well is brought to its final classification—dry hole, producer, or injector. A dry hole is completed by plugging and abandonment. A well deemed to be producible of petroleum, or used as an injector, is completed by establishing a connection between the reservoir(s) and the surface so that fluids can be produced from, or injected into the reservoir. Various methods are utilized to establish this connection, but they commonly involve the installation of some combination of borehole equipment, casing and tubing, and surface injection or production facilities.

Completion interval the portion of the reservoir formation placed in fluid communication with the well by selectively perforating the wellbore casing.

Complex a series of fields sharing common surface facilities.

Composition the general chemical make-up of petroleum.

Compressor a device installed in the gas pipeline to raise the pressure and guarantee the fluid flow through the pipeline.

Concession a grant of access for a defined area and time period that transfers certain rights to hydrocarbons that may be discovered from the host country to an enterprise. The enterprise is generally responsible for exploration, development, production, and sale of hydrocarbons that may be discovered. Typically granted under a legislated fiscal system where the host country collects taxes, fees and sometimes royalty on profits earned.

Condensate a mixture of low-boiling hydrocarbon liquids obtained by condensation of hydrocarbon vapors: predominately butane, propane, and pentane with some heavier hydrocarbons and relatively little methane or ethane; see also Natural gas liquids.

Condensate recovery factor (CRF) it is the factor used to obtain liquid fractions recovered from natural gas in the surface distribution and transportation facilities. It is obtained from the gas and condensate handling statistics of the last annual period in the area corresponding to the field being studied.

Conductivity a measure of the ease of flow through a fracture, perforation, or pipe.

Conformance the uniformity with which a volume of the reservoir is swept by injection fluids in area and vertical directions.

Consolidated mineral matter of variable composition assembled in masses or considerable quantities in nature, as by the action of heat or water.

Contact angle the angle of intersection of the interface between two fluids at a solid surface and which is measured from the solid surface through the aqueous phase, or in an oil and gas test through the oil phase; surface roughness affects the equilibrium contact angle, so measurements are typically made on smooth, flat surfaces.

Contact angle test the test uses carefully captured and preserved samples of reservoir oil to determine the wetting preference. In the test method, a droplet of the crude oil is suspended between two parallel plates of quartz (SiO_2) or calcite ($CaCO_3$) inside a simulated formation water bath at reservoir temperature and sometimes at reservoir pressure. By periodically displacing one of the plates sideways, a contact angle is determined at the side of the droplet where water is forcing the oil from the solid. A small angle indicates water-wetting preference, while a large angle indicates oil-wetting. Angles near 90° are intermediate-wetting. Different minerals display different wetting preferences, although most are more likely to be water-wet.

Contingent project a project where development and production of recoverable quantities has not been justified, due to conditions that may or may not be fulfilled.

Contingent resource The amounts of hydrocarbons estimated at a given date, and which are potentially recoverable from known accumulations, but are not considered commercially recoverable under the economic evaluation conditions corresponding to such date.

Continuous-type reservoir or deposit an accumulation of crude oil (heavy crude oil) or tar sand bitumen that is pervasive throughout a large area and which is not significantly affected by hydrodynamic influences.

Conventional crude oil crude oil (petroleum) found in liquid form, flowing naturally or capable of being pumped without further processing or dilution.

Conventional deposit a discrete accumulation related to a localized geological structural feature and/or stratigraphic condition, typically with each accumulation bounded by a down-dip contact with an aquifer, and which is significantly affected by hydrodynamic influences, such as the buoyancy of petroleum in water.

Conventional gas natural gas occurring in a normal porous and permeable reservoir rock, either in the gaseous phase or dissolved in crude oil, and which technically can be produced by normal production practices.

Conventional limit the reservoir limit established according to the degree of knowledge of, or research into the geological, geophysical or engineering data available.

Conventional recovery primary and/or secondary recovery.

Conversion cost the cost of changing a production well to an injection well, or some other change in the function of an oil field installation.

Core a cylindrical rock sample taken from a formation when drilling for analysis in order to determine its permeability, porosity, hydrocarbon saturation, and other productivity-associated properties.

Core floods laboratory flow tests through samples (cores) of porous rock.

Coring taking rock samples from a well by means of a special tool—a *core barrel*.

Cp (centipoise, cP) a unit of viscosity.

Cracking the thermal processes by which the constituents of petroleum are converted to lower molecular weight products.

Cracking parameters the heat and pressure conditions that transform the hydrocarbons with a high molecular weight and boiling point to hydrocarbons with a lower molecular weight and boiling point.

Cracking temperature the temperature (350°C; 660°F) at which the rate of thermal decomposition of petroleum constituents becomes significant.

Craig-Geffen-Morse method a method for predicting oil recovery by water flood.

Crane barge a large barge, capable of lifting heavy equipment onto offshore platforms: also known as a *derrick barge*.

Creaming theory a statistical technique which recognizes that in any exploration province after an initial period in which the largest fields are found, success rates and average field sizes decline as more exploration wells are drilled and knowledge of the area matures.

Criteria air contaminants (CAC) emissions of various air pollutants that affect our health and contribute to air pollution problems such as smog.

Cross-linking combining of two or polymer molecules by use of a chemical that mutually bonds with a part of the chemical structure of the polymer molecules.

Crude assay a procedure for determining the general distillation characteristics (eg, distillation profile, *q.v.*) and other quality information of crude oil.

Crude distillation unit a refinery unit that separates crude oil into different products according to their individual boiling point ranges; allows for the materials to be separated without being subjected to conditions that would cause cracking or decomposition.

Crude oil see Petroleum.

Crude oil equivalent the means for converting gas volumes to the oil equivalent which is customarily done on the basis of the heating content or calorific value of the fuel. There are a number of methodologies in common use. Before aggregating, the gas volumes first must be converted to the same temperature and pressure. Common industry gas conversion factors usually range between 1.0 barrel of oil equivalent (boe) = 5.6 thousand standard cubic feet of gas (mscf) to 1.0 boe = 6.0 mscf.

Crude oil production the volume of crude oil produced from oil reservoirs during given periods of time and measured as volumes delivered from lease storage tanks (ie, the point of custody transfer) to pipelines, trucks, or other media for transport to refineries or terminals with adjustments for (1) net differences between opening and closing lease inventories and (2) basic sediment and water (BS&W).

Cryogenic plant processing plant capable of producing liquid natural gas products, including ethane, at very low operating temperatures.

Cryogenics the study, production and use of low temperatures.

Cumulative effects changes to the environment caused by an activity in combination with other past, present, and reasonably foreseeable human activities.

Cumulative production production of oil or gas to date.

Cut point the boiling-temperature division between distillation fractions of petroleum.

Cuttings rock chippings cut from the formation by the drill bit, and brought to the surface with the mud; used by geologists to obtain formation data.

Cyclic steam injection the alternating injection of steam and production of oil with condensed steam from the same well or wells.

Cyclic steam stimulation (CSS) injecting steam into a well in a heavy-oil reservoir which introduces heat and thins the oil, allowing it to flow through the same well.

Darcy (Darcys or Darcies) a measure of rock permeability—the degree to which natural gas and crude oil can move through the rocks.

Dealkylation the removal of an alkyl group (or alkyl groups) from aromatic compounds usually by thermal means.

Deasphaltened oil the fraction of petroleum after the asphaltene constituents have been removed.

Deasphaltening removal of a solid powdery asphaltene fraction from petroleum by the addition of the low-boiling liquid hydrocarbons such as *n*-pentane or *n*-heptane under ambient conditions.

Deasphalting the removal of the asphaltene fraction from petroleum by the addition of a low-boiling hydrocarbon liquid such as *n*-pentane or *n*-heptane; more correctly the removal asphalt (tacky, semisolid) from petroleum (as occurs in a refinery asphalt plant) by the addition of liquid propane or liquid butane under pressure.

Degradation the loss of desirable physical properties of EOR fluids, for example, the loss of viscosity of polymer solutions.

Delimitation the exploration activity that increases or decreases reserves by means of drilling delimiting wells.

Density the mass (or weight) of a unit volume of any substance at a specified temperature; a measurement of the heaviness of crude oil, indicating the proportion of large, carbon-rich molecules, generally measured in kilograms per cubic meter (kg/m^3) or degrees on the American Petroleum Institute (API) gravity scale; see also Specific gravity.

Derrick the tower-like structure that houses most of the drilling controls.

Desalting removal of mineral salts (mostly chlorides) from crude oils.

Desulfurization the removal of sulfur or sulfur-containing compounds from petroleum or from a petroleum product.

Deterministic estimate the method of estimation of reserves or resources is called deterministic if a single best estimate is made based on known geological, engineering, and economic data.

Developed proved area plant projection of the extension drained by the wells of a producing reservoir.

Developed proved reserves reserves that are expected to be recovered in existing wells, including reserves behind pipe, which may be recovered with the current infrastructure through additional work and with moderate investment costs. Reserves associated with secondary and/or enhanced recovery processes will be considered as developed when the infrastructure required for the process has been installed or when the costs required for such are lower. This category includes reserves in completed intervals which have been opened at the time when the estimation is made, but that have not started flowing due to market conditions, connection problems or mechanical problems, and whose rehabilitation cost is relatively low.

Developed reserves reserves that are expected to be recovered from existing wells including reserves behind pipe. Improved recovery reserves are considered developed only after the necessary equipment has been installed, or when the costs to do so are relatively minor; may be subcategorized as producing or nonproducing.

Development the activity that increases or decreases reserves by means of drilling exploitation wells.

Development well a well drilled in a proved area in order to produce hydrocarbons; a well drilled within the proved area of an oil or gas reservoir to the depth of a stratigraphic horizon known to be productive; a well drilled in a proven field for the purpose of completing the desired spacing pattern of production: a well drilled in or adjacent to a proven part of a pool to optimize petroleum production.

Dew point pressure the pressure at which the first drop of liquid is formed, when it goes from the vapor phase to the two-phase region; also, the temperature to which air must be cooled to reach saturation (assuming air pressure and moisture content are constant)—a higher dew point indicates more moisture present in the air; the frost point is the dew point when temperatures are below freezing.

Diagenesis the concurrent and consecutive chemical reactions which commence the alteration of organic matter (at temperatures up to 50°C (120°F) and ultimately result in the formation of petroleum from the marine sediment; see also Catagenesis and Metagenesis.

Diagenetic rock rock formed by conversion through pressure or chemical reaction from a rock, for example, sandstone is a diagenetic.

Differential-strain analysis measurement of thermal stress relaxation in a recently drilled well.

Dilbit tar sand bitumen diluted with a diluent, such as naphtha.

DilSynBit a blend of tar sand bitumen, condensate, and synthetic crude oil similar to medium sour crude.

Diluent low-boiling petroleum liquid (eg, naphtha) used to dilute heavy oil or tar sand bitumen so they can flow through a pipeline.

Directional well (deviated well) a well drilled at an angle from the vertical by using a slanted drilling rig or by deflecting the drill bit; directional wells are used to drill multiple wells from a common drilling pad or to reach a subsurface location beneath land where drilling cannot be done.

Discovered petroleum-initially-in-place the quantity of petroleum which is estimated, on a given date, to be contained in known accumulations, plus those quantities already produced therefrom. Discovered petroleum-initially-in-place may be subdivided into Commercial and Subcommercial categories, with the estimated potentially recoverable portion being classified as Reserves and Contingent Resources, respectively.

Discovered resource volume of hydrocarbons tested through wells drilled.

Discovery the incorporation of reserves attributable to drilling exploratory wells that test hydrocarbon-producing formations.

Discovery well an exploratory well that encounters a previously untapped oil or gas deposit.

Dispersion a measure of the convective mi fluids due to flow in a reservoir.

Displacement efficiency the ratio of the amount of oil moved from the zone swept by the reprocess to the amount of oil present in the zone prior to start of the process. Thus, the overall recovery efficiency E of any fluid displacement process is given by the product of the macroscopic, or volumetric, displacement efficiency E_v and the microscopic displacement efficiency E_d:

$$E = E_v E_d.$$

Dissolved gas–oil ratio ratio of the volume of gas dissolved in oil compared to the volume of oil containing gas. The ratio may be original (Rsi) or instantaneous (Rs).

Distillation range the difference between the temperature at the initial boiling point and at the end point, as obtained by the distillation test.

Distillation profile the distillation characteristics of petroleum or petroleum products showing the temperature and the percent distilled.

Distribution coefficient a coefficient that describes the distribution of a chemical in reservoir fluids, usually defined as the equilibrium concentrations in the aqueous phases.

Dome geological structure with a semispherical shape or relief.

Downhole a term used to describe tools, equipment, and instruments used in the wellbore, or conditions or techniques applying to the wellbore.

Downhole steam generator a generator installed downhole in an oil well to which oxygen-rich air, fuel, and water are supplied for the purposes of generating steam for it into the reservoir. Its major advantage over a surface steam generating facility is the losses to the wellbore and surrounding formation are eliminated.

Downstream sector the refining and marketing sector of the petroleum industry.

Drainage radius the distance from which fluids flow to the well, that is, the distance reached by the influence of disturbances caused by pressure drops.

Drill cuttings the small pieces of rock created as a drill bit moves through underground formations while drilling.

Drilling rig a drilling unit that is not permanently fixed to the surface or to the seabed, for example, a drillship, a semisubmersible or a jack-up unit; also means the derrick and its associated machinery.

Drilling mud fluid circulated down the drill pipe and up the annulus during drilling to remove cuttings, cool and lubricate the bit, and maintain desired pressure in the well.

Drill stem test (formation test) Conventional formation test method; the drill stem test tool consists of a packer or packers, valves or ports that may be opened and closed from the surface, and two or more pressure-recording devices—the tool is lowered on the drill string to the zone to be tested and the packer or packers are set to isolate the zone from the drilling fluid column.

Dry forward combustion a type of in-situ combustion in which the burning front moves in the same direction as the injected air. As air is continuously supplied at the injection well, the fire ignited at this location moves toward the production wells. During forward combustion, the temperature behind the burning front is high, indicating a great amount of heat stored in the formation matrix. The injected gas heats on contact with the matrix and recovers only a small amount of the heat, with considerable losses to the surrounding formations. Another drawback of dry forward combustion is the presence of a highly viscous oil zone surrounding the production well. The fluid in this zone remains at the original reservoir temperature and its forward displacement by the heated oil is normally difficult.

Dry gas natural gas containing negligible amounts of hydrocarbons heavier than methane. Dry gas is also obtained from the processing complexes.

Dry gas equivalent to liquid (DGEL) volume of crude oil that because of its heat rate is equivalent to the volume of dry gas.

Dry hole any exploratory or development well that does not find commercial quantities of hydrocarbons; a well found to be incapable of producing either oil or gas in sufficient quantities to justify completion as an oil or gas well.

Dry point the temperature at which the last drop of petroleum fluid evaporates in a distillation test.

Dykstra-Parsons coefficient an index of reservoir heterogeneity arising from permeability variation and stratification.

Economic limit the point at which the revenues obtained from the sale of hydrocarbons match the costs incurred in its exploitation.

Economic reserves the accumulated production that is obtained from a production forecast in which economic criteria are applied.

Effective permeability a relative measure of the conductivity of a porous medium for a fluid when the medium is saturated with more than one fluid. This implies that the effective permeability is a property associated with each reservoir flow, for example, gas, oil, and water. A fundamental principle is that the total of the effective permeability is less than or equal to the absolute permeability.

Effective porosity a fraction that is obtained by dividing the total volume of communicated pores and the total rock volume.

Effective viscosity see Apparent viscosity.

Electrical precipitation a process using an electrical field to improve the separation of hydrocarbon reagent dispersions; the process may be used in chemical treating processes on a wide variety of refinery stocks.

Emulsion a dispersion of very small drops of one liquid in an immiscible liquid, such as oil in water.

Emulsion breaking the settling or aggregation of colloidal-sized emulsions from suspension in a liquid medium.

Enhanced oil recovery (EOR) petroleum recovery following recovery by conventional (ie, primary and/or secondary) methods; any method that increases oil production by using techniques or materials that are not part of normal pressure maintenance or waterflooding operations; for example, natural gas can be injected into a reservoir to "enhance" or increase oil production; the process of producing liquid hydrocarbons by methods other than the conventional use of reservoir energy and reservoir repressurizing procedures with gas or water—typically conventional production methods will produce from a reservoir approximately 30% (v/v) of the initial oil-in-place and the remaining oil, approximately 70% (v/v) of the initial resource, is a large technically and financially attractive target for enhanced oil recovery methods.

Enhanced oil recovery (EOR) process a method for recovering additional oil from a petroleum reservoir beyond that economically recoverable by conventional primary and secondary recovery methods. EOR methods are usually divided into three main categories: (1) *chemical flooding:* injection of water with added chemicals into a petroleum reservoir. The chemical processes include: surfactant flooding, polymer flooding, and alkaline flooding; (2) *miscible flooding:* injection into a petroleum reservoir of a material that is miscible, or can become miscible, with the oil in the reservoir. Carbon dioxide, hydrocarbons, and nitrogen are used; and (3) *thermal recovery:* injection of steam into a petroleum reservoir, or propagation of a combustion zone through a reservoir by air or oxygen-enriched air injection. The thermal processes include: steam drive, cyclic steam injection, and in-situ combustion.

Environmental footprint the impact of a commercial operation on the environment, in terms of resource use, waste generation and changes to the physical environment.

Established reserves the portion of the discovered resource base that is estimated to be recoverable using known technology under present and anticipated economic conditions. Includes proved plus a portion of probable (usually 50%, v/v).

Estimated ultimate recovery the quantities of petroleum which are estimated, on a given date, to be potentially recoverable from an accumulation, plus those quantities already produced therefrom.

Ethylene (C_2H_4, $CH_2{=}CH_2$, sometimes but not commonly called ethene) an olefin hydrocarbon recovered from refinery processes or petrochemical processes; used as a petrochemical feedstock for numerous chemical applications and the production of consumer goods.

Evaporation a process for concentrating nonvolatile solids in a solution by boiling off the liquid portion of the waste stream.

Evaporites sedimentary formations consisting primarily of salt, anhydrite, or gypsum, as a result of evaporation in coastal waters.

Expanding clays clays that expand or swell on contact with water, for example, montmorillonite.

Exploration well drilling carried out to determine whether hydrocarbons are present in a particular area or structure; also known as a *wildcat well*.

Exploratory well a well that is drilled without detailed knowledge of the underlying rock structure in order to find hydrocarbons whose exploitation is economically profitable.

Extra heavy oil a bitumen-type material crude oil with relatively high fractions of high molecular weight components, high specific gravity (low API density), and high viscosity at reservoir conditions but which is mobile under reservoir/deposit conditions; the production of this kind of oil generally implies difficulties in extraction and high costs—thermal recovery methods are the most common form of commercially exploiting this kind of material.

Facies one or more layers of rock that differs from other layers in composition, age, or mineral content.

FAST fracture-assisted steam flood technology.

Faujasite a naturally occurring silica-alumina (SiO_2-Al_2O_3) mineral.

Fault fractured surface of geological strata along which there has been differential movement; a planar fracture or discontinuity in a volume of rock, across which there has been significant displacement as a result of rock mass movement; large faults within the crust of the Earth result from the action of plate tectonic forces, with the largest forming the boundaries between the plates, such as subduction zones or transform faults.

Feedstock petroleum as it is fed to the refinery; a refinery product that is used as the raw material for another process; the term is also generally applied to raw materials used in other industrial processes.

Field the surface area above one or more underground petroleum pools sharing the same or related infrastructure; an area consisting of a single reservoir or multiple reservoirs all grouped on, or related to, the same individual geological structural feature and/or stratigraphic condition. There may be two or more reservoirs in a field that are separated vertically by intervening impermeable rock, laterally by local geologic barriers, or both. It could be defined differently by regulatory authorities.

Field-scale the application of EOR processes to a significant portion of a field.

Fingering the formation of finger-shaped irregularities at the leading edge of a displacing fluid in a porous medium which move out ahead of the main body of fluid.

Fireflood (Fire flood) a thermal recovery method in which the oil in the reservoir is ignited, the heat vaporizes lighter hydrocarbons and water pushes the warmed oil toward a producing well; also called in-situ combustion.

First-contact miscibility see miscibility.

Fischer-Tropsch process a process for synthesizing hydrocarbons and oxygenated chemicals from a mixture of hydrogen and carbon monoxide.

Five-spot an arrangement or pattern of wells with four injection wells at the comers of a square and a producing well in the center of the square.

Flaring/venting the controlled burning (flare) or release (vent) of natural gas that cannot be processed for sale or use because of technical or economic reasons.

Flocculation threshold the point at which constituents of a solution (eg, asphaltene constituents or coke precursors) will separate from the solution as a separate (solid) phase.

Floc point the temperature at which wax or solids separate as a definite floc.

Flood, flooding the process of displacing petroleum from a reservoir by the injection of fluids.

Flow line pipe, usually buried, through which oil or gas travels from the well to a processing facility.

Flow test an operation on a well designed to demonstrate the existence of moveable petroleum in a reservoir by establishing flow to the surface and/or to provide an indication of the potential productivity of that reservoir. Some flow tests, such as drill stem tests (DSTs), are performed in the open hole. A DST is used to obtain reservoir fluid samples, static bottomhole pressure measurements, indications of productivity and short-term flow and pressure buildup tests to estimate permeability and damage extent. Other flow tests, such as single-point tests and multipoint tests, are performed after the well has been cased. Single-point tests typically involve a measurement or estimate of initial or average reservoir pressure and a flow rate and flowing bottomhole pressure measurement. Multipoint tests are used to establish gas well deliverability and absolute open flow potential.

Fluid a reservoir gas or liquid.

Fluid injection injection of gases or liquids into a reservoir to force oil toward and into producing wells.

Fluid saturation portion of the pore space occupied by a specific fluid; oil, gas, and water may exist.

Fold a fold occurs when one or a stack of originally flat and planar surfaces, such as sedimentary strata, are bent or curved as a result of permanent deformation; synsedimentary *folds* are formed due to slumping of sedimentary material before it is lithified; folds in rocks vary in size from microscopic crinkles to mountain-sized folds and occur singly as isolated folds and in extensive fold trains of different sizes, on a variety of scales.

Formation an interval of rock with distinguishable geologic characteristics.

Formation damage the reduction in permeability in reservoir rock due to the infiltration of drilling or treating fluids into the area adjacent to the wellbore.

Formation pressure the pressure at the bottom of a well when it is shut in at the wellhead.

Formation resistance factor (F) ratio between the resistance of rock saturated 100% with brine divided by the resistance of the saturating water.

Formation volume factor (B) the factor that relates the volume unit of the fluid in the reservoir with the surface volume. There are volume factors for oil, gas, in both phases, and for water. A sample may be directly measured, calculated, or obtained through empirical correlations.

Formation water salt water underlying gas and oil in the formation; water that is naturally present in the reservoir that is recovered during oil and gas production; also called *brine* (and may contain high mineral or salt content).

Fossil fuel resources a gaseous, liquid, or solid fuel material formed in the ground by chemical and physical changes (diagenesis, q.v.) in plant and animal residues over geological time; natural gas, petroleum, coal, and oil shale.

Fractional composition the composition of petroleum as determined by fractionation (separation) methods.

Fractional distillation the separation of the components of a liquid mixture by vaporizing and collecting the fractions, or cuts, which condense in different temperature ranges.

Fractional flow the ratio of the volumetric flow rate of one fluid phase to the total fluid volumetric flow rate within a volume of rock.

Fractional flow curve the relationship between the fractional flow of one fluid and its saturator during simultaneous flow of fluids through rock.

Fracture a natural or man-made crack in a reservoir rock.

Fracturing a method of breaking down a formation by pumping fluid at very high pressure; the objective is to increase production rates from a reservoir.

Free associated gas natural gas that overlies and is in contact with the crude oil of the reservoir. It may be gas cap.

Fuel gas gas used for field and plant operations. Substantial savings can be achieved to the operating cost of a project by avoiding the purchase of alternative supplies of gas or refined fuels such as diesel.

Fuller's earth a clay which has high adsorptive capacity for removing color from oils; attapulgus clay is a widely used fuller's earth.

Gallon measurement of volume in the petroleum industry (42 gallons = 1 barrel).

Gas cap a part of a hydrocarbon reservoir at the top that will produce only gas.

Gas cap gas natural gas which overlies and is in contact with crude oil in the reservoir; a subset of Associated gas.

Gas compressibility ratio (Z) the ratio between an actual gas volume and an ideal gas volume. This is a nondimensional amount that usually varies between 0.7 and 1.2.

Gas field a field containing natural gas but no crude oil.

Gas injection the process whereby separated associated gas is pumped back into a reservoir for conservation purposes or to maintain the reservoir pressure.

Gas lift artificial production system that is used to raise the well fluid by injecting gas down the well through tubing, or through the tubing-casing annulus.

Gas–oil ratio (GOR) ratio of reservoir gas production to oil production, measured at atmospheric pressure.

Gas-to-liquids (GTL) the conversion of natural gas to a liquid form so that it can be transported easily—typically the liquid is converted back to natural gas prior to consumption; alternatively the conversion of natural gas to higher molecular weight hydrocarbon liquids or alcohols.

Geological province a region of large dimensions characterized by similar geological and development histories.

Geostatistical methods a variety of mathematical techniques and processes dealing with the collection, analysis, interpretation, and presentation of masses of geological, geophysical, and engineering data to (mathematically) describe the variability and uncertainties within any reservoir unit or pool; specifically related here to resource and reserve estimates, including the definition of (all) well and reservoir parameters in one, two, and three dimensions and the resultant modeling and potential prediction of various aspects of performance. Examples of such processes include: Monte Carlo simulation, 2D gridding and modeling of the spatial variability of geological and petrophysical properties, simulated annealing, object-based simulation, multiple-point statistics, the use of (semi) variograms, and 3D stochastic modeling. New applications include fuzzy mathematics, fast flow simulation, well intervention, and lithology and fluid prediction.

Glycol dehydrator field equipment used to remove water from natural gas by using triethylene glycol or diethylene glycol.

Graben dip or depression formed by tectonic processes, limited by normal type faults.

Gravitational segregation reservoir driving mechanism in which the fluids tend to separate according to their specific gravities. For example, since oil is heavier than water it tends to move towards the lower part of the reservoir in a water-injection project.

Gravity see API gravity.

Gravity drainage the movement of oil in a reservoir that results from the force of gravity.

Gravity segregation partial separation of fluids in a reservoir caused by the gravity force acting on differences in density.

Gravity-stable displacement the displacement of oil from a reservoir by a fluid of a different density, where the density difference is utilized to prevent gravity segregation of the injected fluid.

Greenhouse gas intensity (GHG intensity) the average emission rate of a given greenhouse gas from a specific source; for example: greenhouse gases released per barrel of production.

Greenhouse gases (GHGs) a type of gas (such as carbon dioxide and nitrogen oxides) that contributes to the greenhouse effect by absorbing infrared radiation; these include carbon dioxide, methane, nitrous oxide, and water vapor; these gases occur through natural processes (such as ocean currents, cloud cover, volcanoes) and human activities (such as the burning of fossil fuels).

Handling efficiency shrinkage factor (HESF) this is a fraction of natural gas that is derived from considering self-consumption and the lack of capacity to handle such. It is obtained from the gas handling statistics of the final period in the area corresponding to the field being studied.

HCPV hydrocarbon pore volume.

Hearn method a method used in reservoir simulation for calculating a pseudo relative permeability curve that reflects reservoir stratification.

Heat value the amount of heat released per unit of mass, or per unit of volume, when a substance is completely burned. The heat power of solid and liquid fuels is expressed in calories per gram or in BTU per pound. For gases, this parameter is generally expressed in kilocalories per cubic meter or in BTU per cubic foot.

Heavy crude oil see Heavy oil.

Heavy ends the highest boiling portion of a petroleum fraction; see also Light ends.

Heavy oil petroleum having an API gravity of less than 20°.

Heavy petroleum see Heavy oil.

Heteroatom compounds chemical compounds which contain nitrogen and/or oxygen and/or sulfur and/or metals bound within their molecular structure(s).

Heterogeneity lack of uniformity in reservoir properties such as permeability.

Higgins-Leighton model stream tube computer model used to simulate waterflood.

Horizontal drilling drilling a well that deviates from the vertical and travels horizontally through a producing layer.

Horst a block of the earth's crust rising between two faults; the opposite of a graben.

Horizontal sweep efficiency the fraction of the flood pattern area that is effectively swept by the injected fluids.

Hot production the optimum production of heavy oils through use of enhanced thermal recovery methods.

Hot water process a method for separating bitumen from oil sand using hot water and caustic soda.

Huff-and-puff a cyclic enhanced oil recovery method in which steam or gas is injected into a production well; after a short shut-in period, oil and the injected fluid are produced through the same well.

Hydration the association of molecules of water with a substance usually as water of crystallization.

Hydraulic fracturing the opening of fractures in a reservoir by high-pressure, high-volume injection of liquids through an injection well; the process of pumping a fluid or gas down a well which causes the surrounding rocks to crack and allows natural gas or oil to be produced from tight formations; also known as fracking.

Hydrocarbon a compound containing only the elements hydrogen and carbon which may exist as a solid, a liquid or a gas; the word is also used in a catch-all sense for gas, condensate, and other fossil fuel-related materials.

Hydrocarbon index an amount of hydrocarbons contained in a reservoir per unit area.

Hydrocarbon compounds chemical compounds containing only carbon and hydrogen.

Hydrocarbon-producing resource a resource such as coal and oil shale (kerogen) which produces derived hydrocarbons by the application of conversion processes; the hydrocarbons so-produced are not naturally occurring materials.

Hydrocarbon reserves volume of hydrocarbons measured at atmospheric conditions that will be produced economically by using any of the existing production methods at the date of evaluation.

Hydrocarbon resource resources such as petroleum and natural gas which can produce naturally occurring hydrocarbons without the application of conversion processes.

Hydrocarbons chemical compounds fully constituted by hydrogen and carbon.

Hydrocracker an oil refining unit in which light or heavy gas oils or residue hydrocarbons are mixed with hydrogen under conditions of high temperature and pressure in the presence of a catalyst, yielding light oils.

Hydrocracking a refining process for converting middle distillates to high-octane gasoline, jet fuel, or high grade diesel through the introduction of a hydrogen catalyst under very high pressure.

Hydrodenitrogenation the removal of nitrogen—usually as ammonia or a derivative thereof—by hydrotreating.

Hydrodesulfurization the removal of sulfur—usually as hydrogen sulfide or a derivative thereof—by hydrotreating.

Hydrogen (H_2) the lightest of all gases, occurring chiefly in combination with oxygen in water; exists also in acids, bases, alcohols, petroleum, and other hydrocarbons.

Hydrogen addition an upgrading process in the presence of hydrogen, for example, hydrocracking; see Hydrogenation.

Hydrogenation the chemical addition of hydrogen to a material. In nondestructive hydrogenation, hydrogen is added to a molecule only if, and where, unsaturation with respect to hydrogen exists.

Hydrogen transfer the transfer of inherent hydrogen within the feedstock constituents and products during processing.

Hydroprocessing a term often equally applied to hydrotreating and to hydrocracking; also often collectively applied to both.

Hydrotreater a refining unit whereby processed material from the crude units are treated in the presence of catalysts and hydrogen, often to remove sulfur and other unwanted substances: often the critical unit for producing jet fuel and low-sulfur diesel fuel.

Hydrotreating the removal of heteroatomic (nitrogen, oxygen, and sulfur) species by treatment of a feedstock or product at relatively low temperatures in the presence of hydrogen.

Hydrovisbreaking a noncatalytic process, conducted under similar conditions to visbreaking, which involves treatment with hydrogen to reduce the viscosity of the feedstock and produce more stable products than is possible with visbreaking.

Hydropyrolysis a short residence time high temperature process using hydrogen.

Immiscible two or more fluids that do not have complete mutual solubility and coexist as separate phases.

Immiscible carbon dioxide displacement injection of carbon dioxide into an oil reservoir to effect oil displacement under conditions in which miscibility with reservoir oil is not obtained; see Carbon dioxide augmented waterflooding.

Immiscible displacement a displacement of oil by a fluid (gas or water) that is conducted under conditions so that interfaces exist between the driving fluid and the oil.

Improved recovery the extraction of additional petroleum, beyond primary recovery, from naturally occurring reservoirs by supplementing the natural forces in the reservoir; includes water flooding, secondary processes, tertiary processes, and any other means of supplementing natural reservoir recovery processes; also called Enhanced Recovery.

Impurities and plant liquefiables shrinkage factor (IPLSF) it is the fraction obtained by considering the nonhydrocarbon gas impurities (sulfur, carbon dioxide, nitrogen compounds, etc.) contained in the sour gas, in addition to shrinkage caused by the generation of liquids in gas processing plant.

Impurities shrinkage factor (ISF) it is the fraction that results from considering the nonhydrocarbon gas impurities (sulfur, carbon dioxide, nitrogen compounds, etc.) contained in the sour gas. It is obtained from the operation statistics of the last annual period of the gas processing complex (GPC) that processes the production of the field analyzed.

Incompatibility the *immiscibility* of petroleum products and also of different crude oils which is often reflected in the formation of a separate phase after mixing and/or storage.

Incremental ultimate recovery the difference between the quantity of oil that can be recovered by EOR methods and the quantity of oil that can be recovered by conventional recovery methods.

Infill drilling drilling additional wells within an established pattern; wells drilled between established producing wells on a lease in order to increase production from the reservoir.

Injection the forcing or pumping of substances into a porous and permeable subsurface rock formation. Examples of injected substances can include either gases or liquids.

Initial boiling point the recorded temperature when the first drop of liquid falls from the end of the condenser.

Initial vapor pressure the vapor pressure of a liquid of a specified temperature and 0% evaporated.

Injection profile the vertical flow rate distribution of fluid flowing from the wellbore into a reservoir.

Injection well a well in an oil field used for injecting fluids into a reservoir; a well used for injecting fluids (such as air, steam, water, natural gas, gas liquids, surfactants, alkali, and polymers) into an underground formation for the purpose of increasing recovery efficiency.

Injectivity the relative ease with which a fluid is injected into a porous rock.

In situ in its original place; in the reservoir; in position; in-situ recovery refers to various methods used to recover deeply buried bitumen deposits, including steam injection, solvent injection, and fireflood.

In-situ combustion an EOR process consisting of injecting air or oxygen-enriched air into a reservoir under conditions that favor burning part of the in-situ petroleum, advancing this burning zone, and recovering oil heated from a nearby producing well.

Instability the inability of a petroleum product to exist for periods of time without change to the product.

Integrity maintenance of a slug or bank at its preferred composition without too much dispersion or mixing.

Interface the thin surface area separating two immiscible fluids that are in contact with each other.

Interfacial film a thin layer of material at the interface between two fluids which differs in composition from the bulk fluids.

Interfacial tension the strength of the film separating two immiscible fluids, for example, oil and water or microemulsion and oil; measured in dynes (force) per centimeter or milli-dynes per centimeter; the amount of work required to create a new unit of surface area at the interface between two fluids.

Interfacial viscosity the viscosity of the interfacial film between two immiscible liquids.

Interference testing a type of pressure transient test in which pressure is measured over time in a closed-in well while nearby wells are produced; flow and communication between wells can sometimes be deduced from an interference test.

Interphase mass transfer the net transfer of chemical compounds between two or more phases.

Ions chemical substances possessing positive charges or negative charges in solution.

Isopach a line on a map designating points of equal formation thickness.

Kaolinite a clay mineral formed by hydrothermal activity at the time of rock formation or by chemical weathering of rock with high feldspar content; usually associated with intrusive granite rock with high feldspar content.

Kata-condensed aromatic compounds compounds based on linear condensed aromatic hydrocarbon systems, for example, anthracene and naphthacene (tetracene).

Kerogen a complex carbonaceous (organic) material that occurs in sedimentary rock and shale; generally insoluble in common organic solvents; produces hydrocarbons when subjected to a heat.

Kinematic viscosity the ratio of viscosity to density, both measured at the same temperature.

Known accumulation a term accumulation used to identify an individual body of moveable petroleum. The key requirement to consider an accumulation as known, and hence contain reserves or contingent resources, is that each accumulation/reservoir must have been penetrated by a well. In general, the well must have clearly demonstrated the existence of moveable petroleum in that reservoir by flow to surface or at least some recovery of a sample of petroleum from the well. However, where log and/or core data exist, this may suffice, provided there is a good analogy to a nearby and geologically comparable known accumulation.

Kriging a technique used in reservoir description for interpolation of reservoir parameters between wells based on random field theory.

Lease legal document giving an operator the right to drill for or produce oil or gas; also, the land on which a lease has been obtained.

Light crude oil liquid petroleum that has a low density and flows freely at room temperature.

Light ends the lower-boiling components of a mixture of hydrocarbons; see also Heavy ends, Light hydrocarbons.

Light hydrocarbons hydrocarbons with molecular weights less than that of heptane (C_7H_{16}).

Light oil the specific gravity of the oil is more than 27° API, but less than or equal to 38°.

Light petroleum petroleum having an API gravity greater than 20°.

Limestone calcium carbonate-rich sedimentary rocks in which oil or gas reservoirs are often found.

Limolite fine grain sedimentary rock that is transported by water; the granulometric sizes range from fine sand to clay.

Liquefied natural gas (LNG) oil field or naturally occurring gas, chiefly methane, liquefied for transportation.

Liquefied petroleum gas (LPG) low-boiling hydrocarbon material, gaseous at atmospheric temperature and pressure, held in the liquid state by pressure to facilitate storage, transport, and handling; commercial liquefied gas consists essentially of either propane or butane, or mixtures thereof.

Liquefied refinery gases (LRG) liquefied petroleum gases fractionated from refinery or still gases—through compression and/or refrigeration, they are retained in the liquid state; the reported categories are ethane/ethylene, propane/propylene, normal butane/butylene, and iso-butane/isobutylene; this category excludes gas from the distillation units (*still gas*).

Liquid chromatography a chromatographic technique that employs a liquid mobile phase.

Lithology the geological characteristics of the reservoir rock.

Lorenz coefficient a permeability heterogeneity factor.

Low-best-high estimates the range of uncertainty that reflects a reasonable range of estimated potentially recoverable volumes for an individual accumulation or a project. In the case of reserves, and where appropriate, this range of uncertainty can be reflected in estimates for proved reserves (1P), proved plus probable reserves (2P), and proved plus probable plus possible reserves (3P) scenarios. For other resource categories, the equivalent terms Low Estimate, Best Estimate, and High Estimate are recommended.

Lower-phase microemulsion a microemulsion phase containing a high concentration of water that, when viewed in a test tube, resides near the bottom with oil phase on top.

Lower-phase microemulsion a microemulsion phase containing a high concentration of water that, when viewed in a test tube, resides near the bottom with oil phase on top.

Macroscopic displacement efficiency a measure of the efficiency of the displacing fluid in contacting the oil-bearing parts of the reservoir.

Maltene that fraction of petroleum that is soluble in, for example, pentane or heptane; deasphaltened oil; also the term arbitrarily assigned to the pentane-soluble portion of petroleum that is relatively high boiling (>300°C, 760 mm) (see also Petrolenes).

Marsh an area of spongy waterlogged ground with large numbers of surface water pools. Marshes usually result from: (1) an impermeable underlying bedrock; (2) surface deposits of glacial boulder clay; (3) a basin-like topography from which natural drainage is poor; (4) very heavy rainfall in conjunction with a correspondingly low evaporation rate; (5) low-lying land, particularly at estuarine sites at or below sea level.

Marx-Langenheim model mathematical equations for calculating heat transfer in a hot water or steam flood.

Medium crude oil liquid petroleum with a density (API gravity typically 20–25) between that of light and heavy crude oil.

MEOR microbial-enhanced oil recovery.

Metagenesis the alteration of organic matter during the formation of petroleum that may involve temperatures above 200°C (390°F); see also Catagenesis and Diagenesis.

Metamorphic rocks a group of rocks resulting from the transformation that commonly takes place at great depths due to pressure and temperature. The original rocks may be sedimentary, igneous, or metamorphic.

Methane the principal constituent of natural gas; the simplest hydrocarbon molecule, containing one carbon atom and four hydrogen atoms.

Metric tonne equivalent to 1000 kg or 2204.61 pounds or 7.5 barrels.

Mica a complex aluminum silicate mineral that is transparent, tough, flexible, and elastic.

Micellar fluid (surfactant slug) an aqueous mixture of surfactants, cosurfactants, salts, and hydrocarbons. The term micellar is derived from the word micelle, which is a submicroscopic aggregate of surfactant molecules and associated fluid.

Micelle the structural entity by which asphaltene constituents are dispersed in petroleum.

Microcarbon residue the carbon residue determined using a themogravimetric method. See also Carbon residue.

Microcrystalline wax wax extracted from certain petroleum residua and having a finer and less apparent crystalline structure than paraffin wax.

Microemulsion a stable, finely dispersed mixture of oil, water, and chemicals (surfactants and alcohols).

Microemulsion or micellar/emulsion flooding an augmented waterflooding technique in which a surfactant system is injected in order to enhance oil displacement toward producing wells.

Microorganisms animals or plants of microscopic size, such as bacteria.

Microscopic displacement efficiency the efficiency with which an oil displacement process removes the oil from individual pores in the rock; a measure of the efficiency of the displacing fluid in mobilizing the residual oil once the fluid has come in contact with the oil.

Middle distillates medium-density refined petroleum products, including kerosene, stove oil, jet fuel, and light fuel oil.

Middle-phase microemulsion a microemulsion phase containing a high concentration of both oil and water that, when viewed in a test tube, resides in the middle with the oil phase above it and the water phase below it.

Midstream the processing, storage, and transportation sector of the petroleum industry; a term sometimes used to refer to those industry activities that fall between exploration and production (upstream) and refining and marketing (downstream); the term is most often applied to pipeline transportation of crude oil and natural gas.

Migration (primary) the movement of hydrocarbons (oil and natural gas) from mature, organic-rich source rocks to a point where the oil and gas can collect as droplets or as a continuous phase of liquid hydrocarbon.

Migration (secondary) the movement of the hydrocarbons as a single, continuous fluid phase through water-saturated rocks, fractures, or faults followed by accumulation of the oil and gas in sediments (traps, *q.v.*) from which further migration is prevented.

Mineral hydrocarbons petroleum hydrocarbons, considered *mineral* because they come from the earth rather than from plants or animals.

Mineral interest mineral interests in properties including (1) a fee ownership or lease, concession or other interest representing the right to extract oil, or gas subject to such terms as may be imposed by the conveyance of that interest; (2) royalty interests, production payments payable in oil or gas, and other nonoperating interests in properties operated by others; and (3) those agreements with foreign governments or authorities under which a reporting entity participates in the operation of the related properties or otherwise serves as producer of the underlying reserves (as opposed to being an independent purchaser, broker, dealer, or importer). Properties do not include other supply agreements or contracts that represent the right to purchase, rather than extract, oil and gas.

Mineral oil the older term for petroleum; the term was introduced in the 19th century as a means of differentiating petroleum (rock oil) from whale oil which, at the time, was the predominant illuminant for oil lamps.

Mineralogy a subject specializing in the scientific study of the chemistry, crystal structure, physical, and optical properties of minerals; other studies within mineralogy include the processes of mineral origin and formation, classification of minerals, their geographical distribution, as well as their utilization.

Minerals naturally occurring inorganic solids with well-defined crystalline structures; on occasion in some states and countries, petroleum, heavy oil, and tar sand bitumen fall under the definition of minerals.

Mineral seal oil a distillate fraction boiling between kerosene and gas oil.

Mineral wax yellow to dark brown, solid substances that occur naturally and are composed largely of paraffins; usually found associated with considerable mineral matter, as a filling in veins and fissures or as an interstitial material in porous rocks.

Mine tailings process water remaining after oil sands mining and stored in settling basins called tailings ponds.

Minimum miscibility pressure (MMP) see Miscibility.

Miscibility an equilibrium condition, achieved after mixing two or more fluids, which is characterized by the absence of interfaces between the fluids: (1) *First-contact miscibility:* miscibility in the usual sense, whereby two fluids can be mixed in all proportions without any interfaces forming. Example: At room temperature and pressure, ethyl alcohol and water are first-contact miscible. (2) *Multiple-contact miscibility (dynamic miscibility):* miscibility that is developed by repeated enrichment of one fluid phase with components from a second fluid phase with which it comes into contact. (3) *Minimum miscibility* pressure: the minimum pressure above which two fluids become miscible at a given temperature, or can become miscible, by dynamic processes.

Miscible flood an oil recovery process in which a fluid, capable of mixing completely with the oil it contacts, is injected into an oil reservoir to increase recovery.

Miscible flooding an oil recovery process in which a fluid, capable of mixing completely with the oil it contacts, is injected into an oil reservoir to increase recovery; see EOR process.

Miscible fluid displacement (miscible displacement) is an oil displacement process in which is an oil displacement process in which an alcohol, a refined hydrocarbon, a condensed petroleum gas, carbon dioxide, liquefied natural gas, or even exhaust gas is injected into an oil reservoir, at pressure levels such that the injected gas or fluid and reservoir oil are miscible; the process may include the concurrent, alternating, or subsequent injection of water.

Mmboe million Barrels Oil Equivalent.

Mobility a measure of the ease with which a fluid moves through reservoir rock; the ratio of rock permeability to apparent fluid viscosity.

Mobility buffer the bank that protects a chemical slug from water invasion and dilution and assures mobility control.

Mobility control ensuring that the mobility of the displacing fluid or bank is equal to or less than that of the displaced fluid or bank.

Mobility ratio ratio of mobility of an injection fluid to mobility of fluid being displaced and is a key parameter in determining the efficiency of the water/oil displacement process, with the recovery efficiency increasing as the water–oil-mobility ratio decreases; in a porous medium, the mobility ration of a fluid is defined as its endpoint relative permeability divided by its viscosity—hence, a fluid with a low viscosity (≤1 cP) has a high mobility unless its relative permeability is very low and, similarly, a low-API crude oil (≤20°API) has a high viscosity and a very low mobility unless it is heated to high temperatures; because the viscosity of water at reservoir temperatures generally is much lower than or, at best, equal to that of the crude oil in the reservoir, the water–oil-viscosity ratio is generally much greater than 1:1.

Modified alkaline flooding the addition of a cosurfactant and/or polymer to the alkaline flooding process.

Monte Carlo simulation a type of stochastic mathematical simulation which randomly and repeatedly samples input distributions (eg, reservoir properties) to generate a results distribution (eg, recoverable petroleum volumes).

Mud (drilling mud) fluid circulated down the drill pipe and up the annulus during drilling to remove cuttings, cool and lubricate the bit, and maintain desired pressure in the well.

Naft pre-Christian era (Greek) term for naphtha and sometime used as the name for crude oil or heavy oil.

Naphtha a petroleum product off of the distillation process (general boiling range: 30–200°C, 86–390°F) that is subsequently upgraded as a precursor to gasoline blend-stock.

Naphthenes cycloparaffins.

Native asphalt see Bitumen.

Natural gas a naturally occurring mixture of hydrocarbons existing in reservoirs in the gaseous phase or in solution in the oil, which remains in the gaseous phase under atmospheric conditions. It may contain some impurities or nonhydrocarbon substances (hydrogen sulfide, nitrogen, or carbon dioxide).

Natural gas liquids (NGL) the hydrocarbon liquids that condense during the processing of hydrocarbon gases that are produced from oil or gas reservoir; see also Natural gasoline.

Natural gas–liquids ratio the ratio of natural gas liquids to gas (in barrels/million cubic feet) in an oil field, calculated using measured natural gas liquids and gas volumes at stated conditions.

Natural gasoline a mixture of liquid hydrocarbons extracted from natural gas suitable for blending with refinery gasoline.

Net thickness (HN) the thickness resulting from subtracting the portions that have no possibilities of producing hydrocarbon from the total thickness.

Nonassociated gas the natural gas found in reservoirs that do not contain crude oil at the original pressure and temperature conditions.

Nonhydrocarbon gas gas that is nonhydrocarbon in nature; in the event that nonhydrocarbon gases are present, the reported volumes should reflect the condition of the gas at the point of sale. Correspondingly, the accounts will reflect the value of the gas product at the point of sale. Hence, if gas sold as produced includes a proportion of carbon dioxide, for example, the reserves and production should also include that carbon dioxide. In the case of the carbon dioxide being extracted before sale and the sales gas containing only hydrocarbon gases, the reserves and production should reflect only the hydrocarbon gases that will be sold.

Nonionic surfactant a surfactant molecule containing no ionic charge.

Nonproducing reserves reserves that are subcategorized as nonproducing and include shut-in and behind-pipe reserves. Shut-in reserves are expected to be recovered from (1) completion intervals which are open at the time of the estimate, but which have not started producing; (2) wells which were shut-in for market conditions or pipeline connections; or (3) wells not capable of production for mechanical reasons. Behind-pipe reserves are expected to be recovered from zones in existing wells, which will require additional completion work or future recompletion prior to the start of production.

Nonproved reserves volumes of hydrocarbons and associated substances, evaluated at atmospheric conditions, resulting from the extrapolation of the characteristics and parameters of the reservoir beyond the limits of reasonable certainty, or from assuming oil and gas forecasts with technical and economic scenarios other than those in operation or with a project in view.

Normal fault the result of the downward displacement of one of the blocks from the horizontal. The angle is generally between 25° and 60° and it is recognized by the absence of part of the stratigraphic column.

NO_x oxides of nitrogen.

Observation wells wells that are completed and equipped to measure reservoir conditions and/or sample reservoir fluids, rather than to inject or produce reservoir fluids.

Offset well location a potential drill location adjacent to an existing well. The offset distance may be governed by well spacing regulations. Proved volumes on the existing well are indicated by either conclusive formation test or production. For proved volumes to be assigned to an offset well location there must be conclusive, unambiguous technical data which supports the reasonable certainty of production of hydrocarbon volumes and sufficient legal acreage to economically justify the development without going below the shallower of the fluid contact or the lowest known hydrocarbon.

Oil the portion of petroleum that exists in the liquid phase in reservoirs and remains as such under original pressure and temperature conditions. Small amounts of nonhydrocarbon substances may be included. It has a viscosity of less than or equal to 10,000 cP at the original temperature of the reservoir, at atmospheric pressure and gas-free (stabilized): oil is commonly classified in terms of its specific gravity and it is expressed in degrees API.

Oil bank see Bank.

Oil breakthrough (time) the time at which the oil-water bank arrives at the producing well.

Oil equivalent (OE) total of crude oil, condensate, plant liquids, and dry gas equivalent to liquid.

Oil field a geographic area under which an oil reservoir lies.

Oil-in-place (OIP) an estimated measure of the total amount of oil contained in a reservoir, and, as such, a higher figure than the estimated recoverable reserves of oil.

Oil mining in the context of tar sands, a strip-mining process involving the removal and subsequent processing of tar sand from shallow reservoirs containing bitumen which is immobile at reservoir. For strip mining to be economic, the reservoir must be shallow and have high bitumen saturation. Strip mining is energy intensive, both in mining the tar sand and in its subsequent processing. Large amounts of waste material (gangue) are produced. Companies are usually required to landscape the area when mining is completed.

Oil originally in-place (OOIP) the quantity of petroleum existing in a reservoir before oil recovery operations begin.

Oil sand a naturally occurring mixture of bitumen, water, sand, and clay that is (in the Canadian context) found mainly in three areas of Alberta—a Athabasca, Peace River, and Cold Lake—a typical sample of oil sand might contain about 12% bitumen by weight; also found in other parts of the world but not generally developed commercially; see Tar sand.

Oil shale a fine-grained impervious sedimentary rock which contains an organic material called kerogen.

Once-through steam generator a vaporization of water unit in which water is pushed through tubes that are in contact with a heat source such as natural gas fired burners; this system does not incorporate any feedback loops such as an economizer that *heat recovery steam generators* (HRSGs) use for generating steam

OOIP see Oil originally in-place.

Operator the company or individual responsible for managing an exploration, development or production operation.

Optimum salinity the salinity at which a middle-phase microemulsion containing equal concentrations of oil and water results from the mixture of a micellar fluid (surfactant slug) with oil.

Organic sedimentary rocks rocks containing organic material such as residues of plant and animal remains/decay.

Organization of Petroleum Exporting Countries (OPEC) countries which have organized for the purpose of negotiating with oil companies on matters of oil production, prices, and future concession rights. Current members are Algeria, Angola,

Ecuador, Iran, Iraq, Kuwait, Libya, Nigeria, Qatar, Saudi Arabia, the United Arab Emirates, Venezuela, and The Neutral Zone (between Kuwait and Saudi Arabia).

Original gas volume in-place amount of gas that is estimated to exist initially in the reservoir and that is confined by geologic and fluid boundaries, which may be expressed at reservoir or atmospheric conditions.

Original oil volume in-place amount of petroleum that is estimated to exist initially in the reservoir and that is confined by geologic and fluid boundaries, which may be expressed at reservoir or atmospheric conditions.

Original pressure pressure prevailing in a reservoir that has never been produced. It is the pressure measured by a discovery well in a producing structure.

Original reserve volume of hydrocarbons at atmospheric conditions that are expected to be recovered economically by using the exploitation methods and systems applicable at a specific date. It is a fraction of the discovered and economic reserve that may be obtained at the end of the reservoir exploitation.

Overburden a layer of sand, gravel, and shale between the surface and the underlying tar sand (oil sand) and which must be removed before the tar sand can be mined; overburden underlies muskeg in many places in the Canadian tar sand area.

Override the gravity-induced flow of a lighter fluid in a reservoir above another heavier fluid.

Oxidation a process which can be used for the treatment of a variety of inorganic and organic substances.

Ozone ground-level ozone (O_3) is a colorless gas that forms just above the earth's surface.

1P reserve proved reserve.

2P reserves total of proved plus probable reserves.

3P reserves total of proved reserves plus probable reserves plus possible reserves.

Paraffin wax the colorless, translucent, highly crystalline material obtained from the light lubricating fractions of paraffin crude oils (wax distillates).

Partial upgrading upgrading heavy oil or bitumen to products that are suitable for pipeline transportation or for other forms of transportation without producing the final products that are ready for sales.

Particulate matter microscopic solid or liquid particles that remain suspended in the air for sometime.

Pattern the horizontal pattern of injection and producing wells selected for a secondary or enhanced recovery project.

Pattern life the length of time a flood pattern participates in oil recovery.

Pay zone rock in which oil and gas are found in exploitable quantities.

Penetration the intersection of a wellbore with a reservoir.

Peri-condensed aromatic compounds compounds based on angular condensed aromatic hydrocarbon systems, such as phenanthrene, chrysene, and picene.

Permeability rock property for permitting a fluid pass; a property that indicates whether or not a reservoir has producing characteristics.

Petrolenes the term applied to that part of the pentane-soluble or heptane-soluble material that is low boiling (<300°C, <570°F, 760 mm) and can be distilled without thermal decomposition (see also Maltenes).

Petroleum mixture of hydrocarbons composed of combinations of carbon and hydrogen atoms found in the porous spaces of rocks. Crude oil may contain other elements of a nonmetal origin, such as sulfur, oxygen, and nitrogen, in addition to trace metals as minor constituents. The compounds that form petroleum may be a gaseous, liquid or solid state, depending on their nature and the existing pressure and temperature conditions.

Petroleum-in-place the total quantity of petroleum that is estimated to exist originally in naturally occurring reservoirs. Oil-in-place, gas-in-place, and bitumen-in-place are defined in the same manner.

Petroleum refining a complex sequence of events that result in the production of a variety of products.

Phase a separate fluid that coexists with other fluids; gas, oil, water, and other stable fluids such as microemulsions are all called phases in EOR research.

Phase behavior the tendency of a fluid system to form phases as a result of changing temperature, pressure, or the bulk composition of the fluids or of individual fluid phases.

Phase diagram a graph of phase behavior. In chemical flooding a graph showing the relative volume of oil, brine, and sometimes one or more microemulsion phases. In carbon dioxide flooding, conditions for formation of various liquid, vapor, and solid phases.

Phase properties types of fluids, compositions, densities, viscosities, and relative amounts of oil, microemulsion, or solvent, and water formed when a micellar fluid (surfactant slug) or miscible solvent (eg, CO_2) is mixed with oil.

Phase separation the formation of a separate phase that is usually the prelude to coke formation during a thermal process; the formation of a separate phase as a result of the instability/incompatibility of petroleum and petroleum products.

Physical limit the limit of the reservoir defined by any geological structures (faults, unconformities, change of facies, crests and bases of formations, etc.), caused by contact between fluids or by the reduction, to critical porosity, of permeability limits, or the compound effect of these parameters.

Pilot project a project that is being executed in a small representative sector of a reservoir where tests performed are similar to those that will be implemented throughout the reservoir. The purpose is to gather information and/or obtain results that could be used to generalize an exploitation strategy in the oil field.

PINA analysis a method of analysis for paraffins, *iso*-paraffins, naphthenes, and aromatics.

Pinnacle reef a conical formation, higher than it is wide, usually composed of limestone, in which hydrocarbons might be trapped.

PIONA analysis a method of analysis for paraffins, *iso*-paraffins, olefins, naphthenes, and aromatics.

Pipestill a still in which heat is applied to the oil while being pumped through a coil or pipe arranged in a suitable firebox.

Pipestill gas the most volatile fraction that contains most of the gases that are generally dissolved in the crude. Also known as pipestill light ends.

Pipestill light ends see Pipestill gas.

Pitch the nonvolatile, brown to black, semisolid to solid viscous product from the destructive distillation of many bituminous or other organic materials, especially coal.

Plant liquefiables shrinkage factor (PLSF) the fraction arising from considering the liquefiables obtained in transportation to the processing complexes.

Plant liquids natural gas liquids recovered in gas processing complexes, mainly consisting of ethane, propane, and butane.

Plant liquids recovery factor (PLRF) the factor used to obtain the liquid portions recovered in the natural gas processing complex. It is obtained from the operation statistics of the last annual period of the gas processing complex that processes the production of the field analyzed.

Platform an offshore structure that is permanently fixed to the seabed.

Play a group of oil or gas fields that share geological similarities and where the reservoir and the trap control the distribution of oil and gas.

PNA analysis a method of analysis for paraffins, naphthenes, and aromatics.

Polar aromatics resins; the constituents of petroleum that are predominantly aromatic in character and contain polar (nitrogen, oxygen, and sulfur) functions in their molecular structure(s).

Polycyclic aromatic hydrocarbons (PAHs) polycyclic aromatic hydrocarbons are a suite of compounds comprised of two or more condensed aromatic rings. They are found in many petroleum mixtures, and they are predominantly introduced to the environment through natural and anthropogenic combustion processes.

Polymer in EOR, any very high molecular weight material that is added to water to increase viscosity for polymer flooding.

Polymer augmented waterflooding waterflooding in which organic polymers are injected with the water to improve horizontal and vertical sweep efficiency.

Polymer flood an enhanced oil recovery technique whereby a gel material called polymer is added to water to increase viscosity and further improve production in water flood operations.

Polynuclear aromatic compound an aromatic compound having two or more fused benzene rings, for example, naphthalene, phenanthrene.

PONA analysis a method of analysis for paraffins (P), olefins (O), naphthenes (N), and aromatics (A).

Pool a natural underground reservoir containing an accumulation of petroleum.

Pore diameter the average pore size of a solid material, such as a mineral or catalyst.

Pore space a small hole in reservoir rock that contains fluid or fluids; a four inch cube of reservoir rock may contain millions of interconnected pore spaces.

Pore volume total volume of all pores and fractures in a reservoir or part of a reservoir; also applied to catalyst samples.

Porosity the ratio between the pore volume existing in a rock and the total rock volume; a measure of rock's storage capacity; the percentage of rock volume available to contain water or other fluid.

Porphyrins organometallic constituents of petroleum that contain vanadium or nickel; the degradation products of chlorophyll that became included in the protopetroleum.

Possible reserves the volume of hydrocarbons where the analysis of geological and engineering data suggests that they are less likely to be commercially recoverable than probable reserves.

Potential reserves reserves based upon geological information about the types of sediments where such resources are likely to occur and they are considered to represent an educated guess.

Pour point the lowest temperature at which oil will pour or flow when it is chilled without disturbance under definite conditions.

Precipitation number the number of milliliters of precipitate formed when 10 mL of lubricating oil is mixed with 90 mL of petroleum naphtha of a definite quality and centrifuged under definitely prescribed conditions.

Preflush a conditioning slug injected into a reservoir as the first step of an EOR process.

Pressure cores cores cut into a special coring barrel that maintains reservoir pressure when brought to the surface; this prevents the loss of reservoir fluids that usually accompanies a drop in pressure from reservoir to atmospheric conditions.

Pressure gradient rate of change of pressure with distance.

Pressure maintenance augmenting the pressure (and energy) in a reservoir by injecting gas and/or water through one or more wells.

Pressure pulse test a technique for determining reservoir characteristics by injecting a sharp pulse of pressure in one well and detecting it surrounding wells.

Pressure transient testing measuring the effect of changes in pressure at one well on other well in a field.

Primary recovery the extraction of petroleum by only using the natural energy available in the reservoirs to displace fluids through the reservoir rock to the wells; the production of oil and gas from reservoirs using the natural energy available in the reservoirs and pumping techniques.

Primary tracer a chemical that, when inject into a test well, reacts with reservoir fluids form a detectable chemical compound.

Probabilistic estimate an estimate of the reserves derived from known geological engineering, and economic data that have been used to generate a range of estimates and their associated probabilities.

Probable reserves nonproven reserves where the analysis of geological and engineering data suggests that they are more likely to be commercially recoverable than not.

Produced water water naturally present in the reservoir that is recovered during oil and gas production; may include water from the reservoir, water that has been injected into the formation, and any chemicals added during the production/treatment process; also called *brine* (and may contain high mineral or salt content) or *formation water*; some produced water is quite fresh and may be used for livestock watering or irrigation (where allowed by law).

Producibility the rate at which oil or gas can produced from a reservoir through a wellbore.

Producing reserves reserves that are expected to be recovered from intervals which are open and producing at the time of the estimate. Improved recovery is considered producing only after the improved recovery project is in operation.

Producing well a well in an oil field used for removing fluids from a reservoir.

Propane (C_3H_8, $CH_3CH_2CH_3$) a straight-chain hydrocarbon; a colorless paraffinic gas extracted from natural gas or refinery gas streams.

Propylene (C_3H_6, $CH_3CH = CH_2$) an olefin hydrocarbon recovered from refinery processes or petrochemical processes.

Prorationing (pro-rationing) the Government allocation of demand among pools and wells; pipeline allocation of demand among shippers.

Prospect a potential accumulation that is sufficiently well defined to represent a viable drilling target.

Prospective resource the amount of hydrocarbons evaluated at a given date of accumulations not yet discovered, but which have been inferred, and which are estimated as recoverable.

Protopetroleum a generic term used to indicate the initial product formed changes have occurred to the precursors of petroleum.

Proved area plant projection of the known part of the reservoir corresponding to the proved volume.

Proved developed reserves the proved reserves that can be expected to be recovered through existing wells and facilities and by existing operating methods. Improved recovery reserves can be considered as Proved Developed Reserves only after an improved recovery project has been installed and favorable response has occurred or is expected with a reasonable degree of certainty. Developed reserves are expected to be recovered from existing wells, including reserves behind pipe. Improved recovery reserves are considered developed only after the necessary equipment has been installed, or when the costs to do so are relatively minor. Developed reserves may be subcategorized as producing or nonproducing.

Proved reserves (proven reserves) the volume of hydrocarbons or associated substances evaluated at atmospheric conditions, which by analysis of geological and engineering data, may be estimated with reasonable certainty to be commercially recoverable from a given date forward, from known reservoirs and under current economic conditions, operating methods and government regulations. Such volume consists of the developed proved reserve and the undeveloped proved reserve.

Proved undeveloped reserves reserves that are expected to be recovered from future wells and facilities, including future improved recovery projects which are anticipated with a high degree of certainty in reservoirs which have previously shown favorable response to improved recovery projects.

Proven field an oil and/or gas field whose physical extent and estimated reserves have been determined.

Pulse-echo ultrasonic borehole televiewer well-logging system wherein a pulsed, narrow acoustic beam scans the well as the tool is pulled up the borehole; the amplitude of the reflecting beam is displayed on a cathode-ray tube resulting in a pictorial representation of wellbore.

Pyrolysis exposure of a feedstock to high temperatures in an oxygen-poor environment.

Quench the sudden cooling of hot material discharging from a thermal reactor.

Quadrillion 1×10^{15}

Ramsbottom carbon residue see Carbon residue.

Raw materials minerals extracted from the earth prior to any refining or treating.

Raw natural gas natural gas as it is produced from the reservoir. It includes varying amounts of the heavier hydrocarbons which liquefy at atmospheric conditions, and water vapor; and may also contain sulfur compounds such as hydrogen sulfide, and other nonhydrocarbon gases such as carbon dioxide, nitrogen, or helium, but which, nevertheless, is exploitable for its hydrocarbon content. Raw natural gas is often not suitable for direct utilization by most types of consumers.

Reasonable certainty intended to express a high degree of confidence that the quantities will be recovered. If probabilistic methods are used, there should be at least a 90% probability that the volumes actually recovered will equal or exceed the estimate. The Securities and Exchange Commission considers the concept of reasonable certainty to imply that, as more technical data becomes available, a positive, or upward, revision is much more likely than a negative, or downward, revision.

Recovery efficiency a numeric expression of that portion of in-place quantities of petroleum estimated to be recoverable by specific processes or projects, most often represented as a percentage.

Recovery factor (RF) the ratio between the original volume of oil or gas, at atmospheric conditions, and the original reserves of the reservoir.

Reduced crude a residual product remaining after the removal, by distillation or other means, of an appreciable quantity of the more volatile components of crude oil.

Refinery a series of integrated unit processes by which petroleum can be converted to a slate of useful (salable) products.

Refinery gas a gas (or a gaseous mixture) produced as a result of refining operations.

Refining the processes by which petroleum is distilled and/or converted by application of a physical and chemical processes to form a variety of products are generated.

Reformulated fuels gasoline, diesel, or other fuels which have been modified to reflect environmental concerns, performance standards, government regulations, customer preferences, or new technologies.

Regression a geological term used to define the elevation of one part of the continent over sea level, as a result of the ascent of the continent or the lowering of the sea level.

Relative permeability the capacity of a fluid, such as water, gas or oil, to flow through a rock when it is saturated with two or more fluids.The value of the permeability of a saturated rock with two or more fluids is different to the permeability value of the same rock saturated with just one fluid.

Releases a rupture is an instantaneous tearing or fracturing of the pipeline material that immediately impairs the operation of the pipeline; leaks can be an opening, crack or hole in a pipeline causing some product to be released, but not immediately impairing the operation of the pipeline.

Remaining reserves volume of hydrocarbons measured at atmospheric conditions that are still to be commercially recoverable from a reservoir at a given date, using the applicable exploitation techniques. It is the difference between the original reserve and the cumulative hydrocarbon production at a given date.

Reserve replacement rate it indicates the amount of hydrocarbons replaced or incorporated by new discoveries compared with what has been produced in a given period. It is the coefficient that arises from dividing the new discoveries by production during the period of analysis and it is generally referred to in annual terms and is expressed as a percentage.

Reserve–production ratio the result of dividing the remaining reserve at a given date by the production in a period.This indicator assumes constant production, hydrocarbon prices, and extraction costs, without variation over time, in addition to the nonexistence of new discoveries in the future.

Reserve replacement rate a rate that indicates the amount of hydrocarbons replaced or incorporated by new discoveries compared with what has been produced in a given period. It is the coefficient that arises from dividing the new discoveries by production during the period of analysis. It is generally referred to in annual terms and is expressed as a percentage.

Reserves well-identified resources that can be profitably extracted and utilized with existing technology; an estimate of the amount of crude oil located in a specific region. Reserves must be accessible with current technology.

Reservoir the portion of the geological trap containing hydrocarbons that acts as a hydraulically interconnected system, and where the hydrocarbons are found at an elevated temperature and pressure occupying the porous spaces.

Reservoir-drive mechanisms the natural forces (also called *natural drives*) in the reservoir that displace crude oil out of the reservoir into the wellbore and up to surface. Reservoir-drive mechanisms include (1) gas drive, (2) solution gas drive, (3) water drive, (4) combination drive, and (5) gravity drainage. Water drive is the most efficient drive mechanism, followed by gas drive and gravity drainage.

Reservoir simulation analysis and prediction of reservoir performance with a computer model.

Residual oil the oil that remains in a reservoir after primary and secondary recovery; also the oil that remains in an energy-depleted reservoirs; similar in composition to the recovered oil but this is dependent upon the recovery processes so there might be some dissimilarities to the composition of the recovered oil and the residual oil cannot be recovered by conventional primary or secondary techniques; primary and secondary techniques can recover approximately 33–50% (v/v) of the original oil-in-place, the amount of residual oil can be considerable.

Residual resistance factor the reduction in permeability of rock to water caused by the adsorption of polymer.

Residuum (plural: residua) the nonvolatile residue from crude oil after distilling off all but the heaviest components; typically has a boiling range in excess of 510°C (950°F); used as feedstock to a cracking process or to produce road asphalt.

Resins (resin constituents) that portion of the maltene fraction that is adsorbed by a surface-active material such as clay or alumina; the fraction of deasphaltened oil that is insoluble in liquid propane but soluble in *n*-heptane.

Resistance factor a measure of resistance to flow of a polymer solution relative to the resistance to flow of water.

Resource total volume of hydrocarbons existing in subsurface rocks; also known as original in-situ volume.

Retention the loss of chemical components due to adsorption onto the rock's surface, precipitation, or to trapping within the reservoir.

Retention time the time it takes for an eluate to move through a chromatographic system and reach the detector; all things being equal, retention times are reproducible and can therefore be compared to a standard for analyte identification.

Reverse fault the result of compression forces where one of the blocks is displaced upwards from the horizontal.

Revision the reserve resulting from comparing the previous year's evaluation with the new one in which new geological, geophysical, operation, and reservoir performance information is considered, in addition to variations in hydrocarbon prices and extraction costs. It does not include well drilling.

Rock asphalt bitumen which occurs in formations that have a limiting ratio of bitumen-to-rock matrix.

Rock matrix the granular structure of a rock or porous medium.

Royalty the owner's share of production or revenues retained by government or freehold mineral rights holders. In natural gas operations, the royalty is usually based on a percentage of the total production.

Run-of-the-river reservoirs reservoirs with a large rate of liquid flow-through compared to the volume.

Saline groundwater (brackish water) deep groundwater that is high in dissolved salt and unsuitable for domestic or agricultural uses.

Salinity the concentration of salt in water.

Sand a course granular mineral mainly comprising quartz grains that is derived from the chemical and physical weathering of rocks rich in quartz, notably sandstone and granite.

Sand face the cylindrical wall of the wellbore through which the fluids must flow to or from the reservoir.

Sandstone a sedimentary rock formed by compaction and cementation of sand grains; can be classified according to the mineral composition of the sand and cement.

SARA analysis a method of fractionation by which petroleum is separated into saturates, aromatics, resins, and asphaltene fractions.

Saturates paraffins and cycloparaffins (naphthenes).

Saturation the ratio of the volume of a single fluid in the pores to pore volume, expressed as a percent and applied to water, oil, or gas separately; the sum of the saturations of each fluid in a pore volume is 100%.

Saturation pressure pressure at which the first gas bubble is formed, when it goes from the liquid phase to the two-phase region.

Saybolt Furol viscosity the time, in seconds (Saybolt Furol Seconds, SFS), for 60 mL of fluid to flow through a capillary tube in a Saybolt Furol viscometer at specified temperatures between 70°F and 210°F; the method is appropriate for high-viscosity oils such as transmission, gear, and heavy fuel oils.

Saybolt Universal viscosity the time, in seconds (Saybolt Universal Seconds, SUS), for 60 mL of fluid to flow through a capillary tube in a Saybolt Universal viscometer at a given temperature.

Screen factor a simple measure of the viscoelastic properties of polymer solutions.

Screening guide a list of reservoir rock and fluid properties critical to an EOR process.

Secondary recovery techniques used for the additional extraction of petroleum after primary recovery. This includes gas or water injection, partly to maintain reservoir pressure.

Secondary tracer the product of the chemical reaction between reservoir fluids and an injected primary tracer.

Section a term used in reference to a parcel of land—an area of land comprising approximately 1 $mile^2$ or 640 acres.

Sediment a naturally occurring material that is broken down by processes of weathering and erosion and is subsequently transported by the action of wind, water, or ice, and/or by the force of gravity acting on the particles; an insoluble solid formed as a result of the storage instability and/or the thermal instability of petroleum and petroleum products.

Sedimentary formed by or from deposits of sediments, especially from sand grains or silts transported from their source and deposited in water, as sandstone and shale; or from calcareous remains of organisms, as limestone.

Sedimentary basin a geographical area in which much of the rock is sedimentary (as opposed to igneous or metamorphic) and therefore likely to contain hydrocarbons.

Sedimentary rocks rocks formed by the accumulation of sediments; there are three basic types of sedimentary rocks: (1) clastic sedimentary rocks such as breccia, conglomerate, sandstone, siltstone, and shale are formed from mechanical weathering debris; (2) chemical sedimentary rocks, such as rock salt, iron ore, chert, flint, some dolomites, and some limestones, form when dissolved materials precipitate from solution; and (3) organic sedimentary rocks such as coal, some dolomites, and some limestones, form from the accumulation of plant or animal debris.

Sedimentary strata typically consist of mixtures of clay, silt, sand, organic matter, and various minerals; formed by or from deposits of sediments, especially from sand grains or silts transported from their source and deposited in water, such as sandstone and shale; or from calcareous remains of organisms, such as limestone.

Seismic section seismic profile that uses the reflection of seismic waves to determine the geological subsurface.

Seismic studies studies done to gather and record patterns of induced shock wave reflections from underground layers of rock, which are used to create detailed models of the underlying geological structure.

Separation the process of separating liquid and gas hydrocarbons and water; typically accomplished in a pressure vessel at the surface, but newer technologies allow separation to occur in the wellbore under certain conditions.

Shale fine-grained impermeable rock formed from clay.

Shear mechanical deformation or distortion, or partial destruction of a polymer molecule as it flows at a high rate.

Shear rate a measure of the rate of deformation of a liquid under mechanical stress.

Shear-thinning the characteristic of a fluid whose viscosity decreases as the shear rate increases.

Shutdown a production hiatus during which the platform ceases to produce while essential maintenance work is undertaken.

Shut-in reserves reserves that are expected to be recovered from (1) completion intervals which are open at the time of the estimate, but which have not started producing, (2) wells which were shut-in for market conditions or pipeline connections, or (3) wells not capable of production for mechanical reasons.

Slim tube testing laboratory procedure for the determination of minimum miscibility pressure using long, small-diameter, sand-packed, oil-saturated, stainless steel tube.

Slime a name used for petroleum in ancient texts.

Sludge a semisolid to solid product which results from the storage instability and/or the thermal instability of petroleum and petroleum products.

Slug a quantity of fluid injected into a reservoir during enhanced oil recovery.

Solubility parameter a measure of the solvent power and polarity of a solvent.

Solution gas natural gas that is found with crude oil in underground reservoirs; at the surface, the gas expands and comes out of the solution.

Solvent deasphalting a refining process for removing asphalt compounds from petroleum fractions, such as atmospheric or vacuum residua; a process for removing asphaltic and resinous materials from reduced crude oils, lubricating-oil stocks, gas oils, or middle distillates through the extraction or precipitant action of low-molecular-weight hydrocarbon solvents; see also Propane deasphalting.

Solvent gas an injected gaseous fluid that becomes miscible with oil under reservoir conditions and improves oil displacement.

Solvent naphtha a refined naphtha of restricted boiling range used as a solvent; also called petroleum naphtha; petroleum spirits.

Sonic log a well log based on the time required for sound to travel through rock, useful in determining porosity.

Sour crude oil crude oil containing an abnormally large amount of sulfur compounds; see also Sweet crude oil.

Sour gas natural gas at the wellhead may contain hydrogen sulfide (H_2S), a toxic compound; natural gas that contains more than 1% of H_2S is called sour gas.

Spacing the optimum distance between hydrocarbon-producing wells in a field or reservoir.

Specific gravity an intensive property of the matter that is related to the mass of a substance and its volume through the coefficient between these two quantities. It is expressed in grams per cubic centimeter or in pounds per gallon.

Spills spills include accidental release of crude oil, produced water or other hydrocarbon products from well sites, batteries or storage tanks; these spills can affect land, vegetation, water bodies, and groundwater.

Spud-in the operation of drilling the first part of a new well.

Stabilization the removal of volatile constituents from a higher-boiling fraction or product (stripping); the production of a product which, to all intents and purposes, does not undergo any further reaction when exposed to the air.

Stakeholders industry activities often affect surrounding areas and populations; people with an interest in these activities are considered stakeholders and may be landowners, municipalities, Aboriginal (native) communities, recreational land users, other industries, environmental groups, governments, and regulators.

Standard conditions the reference amounts for pressure and temperature. In the English system, it is 14.73 pounds/inch^2 for the pressure and 60°F for temperature.

Steam-assisted gravity drainage (SAGD) a recovery technique for extraction of heavy oil or bitumen that involves drilling a pair of horizontal wells one above the other; one well is used for steam injection and the other for production.

Steam cracking a conversion process in which the feedstock is treated with superheated steam.

Steam distillation distillation in which vaporization of the volatile constituents is effected at a lower temperature by introduction of steam (open steam) directly into the charge.

Steam drive injection (steam injection) EOR process in which steam is continuously injected into one set of wells (injection wells) or other injection source to effect oil displacement toward and production from a second set of wells (production wells); steam stimulation of production wells is *direct steam stimulation* whereas steam drive by steam injection to increase production from other wells is *indirect steam stimulation.*

Steam injection an improved recovery technique in which steam is injected into a reservoir to reduce the viscosity of the crude oil.

Steam stimulation injection of steam into a well and the subsequent production of oil from the same well.

Stiles method a simple approximate method for calculating oil recovery by waterflood that assumes separate layers (stratified reservoirs) for the permeability distribution.

Stimulation the process of acidifying or fracturing carried out to expand existing ducts or to create new ones in the source rock formation.

Strata layers including the solid iron-rich inner core, molten outer core, mantle, and crust of the earth.

Stratigraphy the part of geology that studies the origin, composition, distribution, and succession of rock strata.

Stripper well a well that produces (strips from the reservoir) oil or gas.

Stripping the removal of volatile constituents from a mixture to leave a nonvolatile residue; a means of separating volatile components from less volatile components ones in a liquid mixture by the partitioning of the more volatile materials to a gas phase of air or steam (stabilization).

Structural nose a term used in structural geology to define a geometric form protruding from a main body.

Sucker rod pumping system a method of artificial lift in which a subsurface pump located at or near the bottom of the well and connected to a string of sucker rods is used to lift the well fluid to the surface.

Sulfur a yellow mineral extracted from petroleum for making fertilizers, pharmaceuticals, and other products.

Sulfur dioxide a major component of a group of airborne contaminants termed *acidifying emissions*.

Sulfur recovery sour gas is processed at recovery plants to extract sulfur for sale to fertilizer manufacturers and other industries.

Supercritical fluid an extraction method where the extraction fluid is present at a pressure and temperature above its critical point.

Super-light oil the specific gravity is more than 38° API.

Surface-active material a chemical compound, molecule, or aggregate of molecules with physical properties that cause it to adsorb at the interface between *two* immiscible liquids, resulting in a reduction of interfacial tension or the formation of a microemulsion.

Surfactant a type of chemical, characterized as one that reduces interfacial resistance to mixing between oil and water or changes the degree to which water wets reservoir rock.

Suspended well a well that has been capped off temporarily.

Sweep efficiency the ratio of the pore volume of reservoir rock contacted by injected fluids to the total pore volume of reservoir rock in the project area. (See also Horizontal sweep efficiency and Vertical sweep efficiency.)

Sweet crude oil crude oil containing little sulfur; see also Sour crude oil.

Sweetening plant industrial plant used to treat gaseous mixtures and light petroleum fractions in order to eliminate undesirable or corrosive sulfur compounds to improve their color, odor, and stability.

Sweet gas natural gas containing little or no hydrogen sulfide.

Sweet oil petroleum containing little or no hydrogen sulfide.

Swelling increase in the volume of crude oil caused by absorption of EOR fluids, especially carbon dioxide. Also increase in volume of clays when exposed to brine.

Swept zone the volume of rock that is effectively swept by injected fluids.

Synthetic crude oil (syncrude) a hydrocarbon product produced by the conversion of coal, oil shale, or tar sand bitumen that resembles conventional crude oil; can be refined in a petroleum refinery; in the case of tar sand bitumen, a manufactured crude oil comprised of naphtha, distillate, and gas oil-boiling range material which can range from high-quality, light sweet crude (without any residuum) to heavy, sour blends.

Tar the volatile, brown to black, oily, viscous product from the destructive distillation of many bituminous or other organic materials, especially coal; a name used for petroleum in ancient texts.

Tar sand a deposit of sand saturated with bitumen: tar sands have been historically been referred to as tar, pitch, or asphalt; in Canada, the term *oil sand* is used most often to describe the naturally occurring bitumen deposits; see Bituminous sand.

Technical reserves accumulative production derived from a production forecast in which economic criteria are not applied.

Tertiary recovery the third major phase of crude oil recovery that involves using more sophisticated techniques, such as steam flooding or injection of chemicals, to increase recovery.

THAI process (Toe-Heel Air Injection) a method of in situ bitumen recovery in which air is continuously injected into a bitumen deposit to propagate a wave of combustion, which is intended to push recoverable and partially upgraded oil towards a production well or production portion of a well; in the process, air is injected at the end (*toe*) of a horizontal well and the wave travels along the horizontal well length to the other end (*heel*) of the horizontal well.

Thermal coke the carbonaceous residue formed as a result of a noncatalytic thermal process; the Conradson carbon residue; the Ramsbottom carbon residue.

Thermal process any recovery or refining process which utilizes heat, without (in the case of a refining process) the aid of a catalyst.

Thermal recovery see EOR process.

Thief zone any geologic stratum not intended to receive injected fluids in which significant amounts of injected fluids are lost; fluids may reach the thief zone due to an improper completion or a faulty cement job.

Tight gas gas with very low flow rates and found in sedimentary layers of rock that are cemented together so tightly that it greatly hinders the extraction process; extracting tight gas from the formation usually requires enhanced technology such as *hydraulic fracturing* where fluid is pumped into the ground to make it more permeable.

Time-lapse logging the repeated use of calibrated well logs to quantitatively observe changes in measurable reservoir properties over time.

Topped crude petroleum that has had volatile constituents removed up to a certain temperature, for example, 250°C+ (480°F+) topped crude; not always the same as a residuum.

Total thickness (*h*) thickness from the top of the formation of interest down to a vertical boundary determined by a water level or by a change of formation.

Tracer test a technique for determining fluid flow paths in a reservoir by adding small quantities of easily detected material (often radioactive) to the flowing fluid, and monitoring their appearance at production wells. Also used in cyclic injection to appraise oil saturation.

Transgression a geological term used to define the immersion of one part of the continent under sea level, as a result of a descent of the continent or an elevation of the sea level.

Transmissibility (transmissivity) an index of producibility of a reservoir or zone, the product of permeability and layer thickness.

Transport liquefiables shrinkage factor (TLSF) the fraction obtained by considering the liquefiables obtained in transportation to the processing complexes.

Trap geological geometry that permits the concentration of hydrocarbons; a sediment in which oil and gas accumulate from which further migration is prevented.

Triaxial borehole seismic survey a technique for detecting the orientation of hydraulically induced fractures, wherein a tool holding three mutually seismic detectors is clamped in the borehole during fracturing; fracture orientation is deduced through analysis of the detected microseismic perpendicular events that are generated by the fracturing process.

Trillion 1×10^{12}

Ultimate analysis elemental composition.

Ultimate recovery the cumulative quantity of oil that will be recovered when revenues from further production no longer justify the costs of the additional production.

Uncertainty the range of possible outcomes in an estimate.

Unconformity a surface of erosion that separates younger strata from older rocks.

Undeveloped proved area plant projection of the extension drained by the future producing wells of a producing reservoir and located within the undeveloped proved reserve.

Undeveloped proved reserves volume of hydrocarbons that is expected to be recovered through wells without current facilities for production or transportation and future wells. This category may include the estimated reserve of enhanced recovery projects, with pilot testing, or with the recovery mechanism proposed in operation that has been predicted with a high degree of certainty in reservoirs that benefit from this kind of exploitation.

Undeveloped reserves reserves that are expected to be recovered: (1) from new wells on undrilled acreage; (2) from deepening existing wells to a different reservoir; or (3) where a relatively large expenditure is required to (a) recomplete an existing well or (b) install production or transportation facilities for primary or improved recovery projects.

Undiscovered recoverable resources those resources estimated to be recoverable from accumulations believed to exist based on geological and geophysical evidence but not yet verified by drilling, testing, or production.

Undiscovered resource volume of hydrocarbons with uncertainty, but whose existence is inferred in geological basins through favorable factors resulting from the geological, geophysical, and geochemical interpretation. They are known as prospective resources when considered commercially recoverable.

Universal viscosity see Saybolt Universal viscosity.

Unproved reserves reserves that are based on geologic and/or engineering data similar to that used in estimates of proved reserves; but technical, contractual, economic, or regulatory uncertainties preclude such reserves being classified as proved. Unproved reserves may be further classified as probable reserves and possible reserves. Unproved reserves may be estimated assuming future economic conditions different from those prevailing at the time of the estimate. The effect of possible future improvements in economic conditions and technological developments can be expressed by allocating appropriate quantities of reserves to the probable and possible classifications.

Unrecoverable resources the portion of discovered or undiscovered petroleum-initially-in-place quantities not currently considered to be recoverable. A portion of these quantities may become recoverable in the future as commercial circumstances change, technological developments occur, or addition data are acquired.

Upgrading the process of processes for converting heavy oil or tar sand bitumen into synthetic crude oil.

Upgrading, partial upgrading heavy oil or bitumen to products that are suitable for pipeline transportation or for other forms of transportation without producing the final products that are ready for sales.

Upper-phase microemulsion a microemulsion phase containing a high concentration of oil that, when viewed in a test tube, resides on top of a water phase.

Upstream the companies that explore for, develop and produce petroleum resources are known as the upstream sector of the petroleum industry.

Vacuum distillation unit a distillation column that operates in a vacuum (where operating pressure is less than atmospheric pressure) and which allows better separation of higher-boiling feedstocks; the lower-boiling are removed from the top of the column, whereas the less valuable, very heavy hydrocarbons, such as resin constituents and asphaltene constituents, are removed from the bottom of the column.

Vacuum residuum the nonvolatile residuum obtained by distillation of a crude oil under atmospheric pressure and which boils above 510°C (950°F).

VAPEX process a gravity drainage process in which solvent vapors are dissolved into bitumen to reduce its viscosity and allow it to be pumped to the surface via a traditional well.

Vertical seismic profiling a method of conducting seismic surveys in the borehole for detailed subsurface information.

Vertical sweep efficiency the fraction of the layers or vertically distributed zones of a reservoir that are effectively contacted by displacing fluids.

VGC (viscosity-gravity constant) an index of the chemical composition of crude oil defined by the general relation between specific gravity, s_g, at 60°F and Saybolt Universal viscosity, SUV, at 100°F:

$$a = 10s_g - 1.0752\log(\text{SUV} - 38)/10s_g - \log(\text{SUV} - 38)$$

The constant, a, is low for the paraffin crude oils and high for the naphthenic crude oils.

VI (Viscosity index) an arbitrary scale used to show the magnitude of viscosity changes in lubricating oils with changes in temperature.

Visbreaking a process for reducing the viscosity of heavy feedstocks by controlled thermal decomposition.

Viscosity a measure of the ability of a liquid to flow or a measure of its resistance to flow; the force required to move a plane surface of area 1 m^2 over another parallel plane surface 1 m away at a rate of 1 m/seconds when both surfaces are immersed in the fluid.

Viscous fingering a condition whereby the interface of two fluids, such as oil and water, bypasses sections of a reservoir as it moves along, creating an uneven, or fingered, profile; fingering is a relatively common condition in reservoirs with water-injection wells and the result is an inefficient sweeping action that can bypass significant volumes of recoverable oil and, in severe cases, an early breakthrough of water into adjacent production wellbores.

Volatile organic compounds (VOCs) gases and vapors, such as benzene, released by petroleum refineries, petrochemical plants, plastics manufacturing and the distribution and use of gasoline;VOCs include carcinogens and chemicals that react with sunlight and nitrogen oxides to form ground-level ozone, a component of smog.

Volumetric sweep the fraction of the total reservoir volume within a flood pattern that is effectively contacted by injected fluids.

VSP vertical seismic profiling, a method of conducting seismic surveys in the borehole for detailed subsurface information.

Vugs large holes which occur in limestone formations.

Vugular porosity porosity due to the presence of vugs in limestone formation.

WAG process injection of alternating slugs of water and gas into a reservoir through an injection well.

Waterflood injection of water to displace oil from a reservoir (usually a secondary recovery process).

Waterflood mobility ratio mobility ratio of water displacing oil during waterflooding. (See *also* Mobility ratio.)

Waterflood residual the waterflood residual oil saturation; the saturation of oil remaining after waterflooding in those regions of the reservoir that have been thoroughly contacted by water.

Watson characterization factor see Characterization factor.

Wax distillate a neutral distillate containing a high percentage of crystallizable paraffin wax, obtained on the distillation of paraffin or mixed-base crude, and on reducing neutral lubricating stocks.

Wax fractionation a continuous process for producing waxes of low oil content from wax concentrates; see also MEK deoiling.

Wax manufacturing a process for producing oil-free waxes.

Wax a solid or semisolid material consisting of a mixture of hydrocarbons obtained or derived from petroleum fractions, or through a Fischer-Tropsch type process, in which the straight-chained paraffin series predominates.

Weathered crude oil crude oil which, due to natural causes during storage and handling, has lost an appreciable quantity of its more volatile components; also indicates uptake of oxygen.

Well abandonment the final activity in the operation of a well when it is permanently closed under safety and environment preservation conditions; the permanent plugging of a dry hole or of a well that no longer produces petroleum or is no longer capable of producing petroleum profitably. Several steps are involved in the abandonment of a well: permission for abandonment and procedural requirements are secured from official agencies; the casing is removed and salvaged if possible; and one or more cement plugs and/or mud are placed in the borehole to prevent migration of fluids between the different formations penetrated by the borehole.

Wellbore the hole in the earth comprising a well.

Well completion the complete outfitting of an oil well for either oil production or fluid injection; also the technique used to control fluid communication with the reservoir.

Wellhead that portion of an oil well above the surface of the ground.

Well logs the information concerning subsurface formations obtained by means of electric, acoustic and radioactive tools inserted in the wells. The log also includes information about drilling and the analysis of mud and cuts, cores and formation tests.

Wells-to-wheels the full product life cycle is considered—from production (wells) to the use of the fuel in a vehicle (wheels); wells-to-wheels analysis can be used to assess total life cycle greenhouse gas emissions from production to combustion of different crude oils.

Wet gas a mixture of hydrocarbons obtained from processing natural gas from which nonhydrocarbon impurities or compounds have been eliminated, and whose content of components that are heavier than methane is such that it can be commercially processed.

Wettability the relative degree to which a fluid will spread on (or coat) a solid surface in the presence of other immiscible fluids; the tendency for a solid to prefer one fluid over another which is a function of the chemical composition of both the fluids and the rock.

Wettability number a measure of the degree to which a reservoir rock is water-wet or oil-wet, based on capillary pressure curves.

Wettability reversal the reversal of the preferred fluid wettability of a rock, for example, from water-wet to oil-wet, or vice versa.

Wet gas natural gas containing significant amounts of liquefiable hydrocarbons.

Wildcat a well drilled in an area where no oil or gas production exists; not an irate feline!

Wobbe Index (or Wobbe Number) the calorific value of a gas divided by the specific gravity.

Workover operations on a producing well to restore or increase production; a workover may be performed to stimulate the well, remove sand or wax from the wellbore, to mechanically repair the well, or for other reasons.

WTI West Texas Intermediate crude oil—a type of crude oil commonly used as a price benchmark.

Xylenes ($CH_3C_6H_4CH_3$) colorless liquid of the aromatic group of hydrocarbons made the catalytic reforming of certain naphthenic petroleum fractions; used as high-octane motor and aviation gasoline blending agents, solvents, chemical intermediates; isomers are (alphabetically) meta-xylene (m-xylene, 1,3-dimethyl benzene), ortho-xylene (o-xylene, 1,2-dimethyl benzene), and para-xylene (p-xylene, 1,4-dimethyl benzene)—the numbers refer to the relative positions of the methyl (CH_3) groups on the benzene ring.

Zeolite a crystalline aluminosilicate used as a catalyst and having a particular chemical and physical structure.

Zone designation a description of the stratigraphic interval for a particular set of mineral rights.

INDEX

Note: Page numbers followed by "*f*" and "*t*" refer to figures and tables, respectively.

A

AA spectrometry. *See* Atomic absorption (AA) spectrometry
AAPG. *See* American Association of Petroleum Geologists (AAPG)
Abel closed-cup apparatus, 213
Abiogenic origin, 52
Abiogenic theory, 51
Adsorption chromatography, 236
Acid fracturing, 315
Acid number of reservoir fluids, 205–209
Acidity rate, 32–33
Acidizing, 264, 315
Aerial remote sensing, 59
Aerobic thermophiles, 327
AET. *See* Atmospheric equivalent temperature (AET)
After recovery basis, 230
After-the-fact methods, 139
Aggregation effect, 240
Aggregation factor, 240
Aging, 140–141
Air consumption, 440
Air-blowing of asphaltene constituents, 160
Alberta tar sand, 54
Alkaline flooding, 279–280, 305, 372–373
Alkanes, 326
Alkaline-surfactant polymer EOR method, 372–373
Allogenic minerals, 90
Alsace, France, 387
American Association of Petroleum Geologists (AAPG), 72–73
American petroleum, 5
American Petroleum Institute (API) gravity, 4, 7*t*–10*t*, 13, 28–29, 30*t*–31*t*, 36*t*, 76, 125, 186, 199, 201, 290–291
American Society for Testing and Materials (ASTM)
  ASTM D1160, 212, 214–215, 218
  ASTM D2007, 236
  ASTM D2224, 238
  ASTM D2502, 238
  ASTM D2503. *See* Vapor phase osmometry (VPO)
  ASTM D2878, 238
  ASTM D2887. *See* Simulated distillation
  ASTM D2892. *See* True boiling point (TBP) distillation method
  ASTM D4124, 236
  ASTM D5236. *See* Vacuum pot still method
  ASTM D5296. *See* Size exclusion chromatography
  ASTM D86, 214, 218–219
Anaerobic production, 383
Anaerobic thermophiles, 327
Anglo-Persian Oil Company, 58
Anticline trap, 106
Appraisal stage, 68
Aquaconversion technology, 466, 467*f*, 480–481
Aquifer water, 362
Aragonite ($CaCO_3$), 85, 95
Arkose, 91
Aromatic hydrocarbons, 326. *See also* Polynuclear aromatic hydrocarbons
  aerobic and anaerobic thermophiles, 327
  anaerobic consortia, 327
  microbial degradation of alkyl tetralin derivatives, 327
  thermophilic aerobic bacteria, 326–327
Aromatic(s), 14–15
  constituents, 138*t*
Artificial lift methods, 258–259
Artificial water drives, 270, 361
Ashing, 227
Asphalt, 3–4, 14
Asphaltene, 14–15
  pyrolysis, 158
  separation, 164, 230–235

Asphaltene constituents, 135, 136*f*, 138*t*, 142–143, 148–150, 334–335
air-blowing, 160
deposition, 150–151
effect, 149
irreversible flocculation, 164
mechanism of deposition, 168
order of deposition, 205*f*
oxidation, 160
separation, 153*f*
thermal decomposition, 158–159
linear path for, 233*f*
multireaction sequence for, 233*f*
Asphaltene deposition, 148–149. *See also* Wax deposition
composition, 152–155
fractions, 151–152
heavy oil separation, 150*f*
maltenes, 149–150
molecular weight, 155–157
properties and behavior of petroleum, 151
reactions, 157–161
other reactions, 160–161
oxidation, 160
thermal reactions, 158–159
resin constituents, 149
solubility parameter, 161–164
Asphaltene fraction, 152, 153*f*, 158, 233–234, 335
from different crude oils, 154*f*
by HPLC, 167*f*
molecular weights, 155–156
Associated natural gas, 262
ASTM D1160 method, 215
Athabasca bitumen, 43
Athabasca oil sands, 389
Athabasca tar sands, 389
Atmospheric equivalent temperature (AET), 214
Atmospheric pressure, 214
Atomic absorption (AA) spectrometry, 226
Authigenic minerals, 90–91
Auto-ignition, 468
Average molecular weight, 237–238
Average parameters, 234–235
Average properties, 235

B

Backflow, 299, 436, 442
Bacterial flooding process. *See* Microbial flooding: process
Bacterial growth, 335
Bacterial metabolites, 336
Bakken crude oil, 23–24
Baku, Azerbaijan, 5, 50–51, 57
Baumé scale, 200
Beam pumps. *See* Horsehead pumps
Bed forms, 99–100
Bedding, 98–99
Before recovery basis, 230
Benz[*a*]anthracene, 329–330, 330*f*
Benzene, toluene, ethyl benzene, and xylene (BTEX), 326–327
Benzo[*a*]pyrene, 328, 328*f*
Best estimate, 75
Biochemical
chert, 93
limestone, 93
sedimentation, 87
sediments, 93–94
Biodegradation, 323
Biodesulfurization, 484–486
Biogenic origin, 52–53
Alberta tar sand, 54
bitumen formation, 54
catagenesis process, 54
diagenesis process, 54
lower mobility of heavy oil, 53–54
protopetroleum, 53
transformation of sedimentary material, 53
Biogenic theory, 51, 56
Biopolymers, 343, 381–382
Bioremediation, 323–324
Biosurfactants, 343
Bitumen, 3–4, 34, 39–40. *See also* Tar sand bitumen
formation, 54
recovery, 391–393
Bituminous rock, 37. *See also* Oil shale; Sapropel
Bituminous sand, 36–37
Black oil, 28, 69
models, 128

Blind techniques, 115
Blowout preventer, 64, 260, 395–396
Boiling point elevation methods, 241
Boiling ranges (cut points), 214
Borehole stability, 410
Bottom-hole pressure, 261–263
Bottom hole sampling. *See* Subsurface sampling
Bottom water zone, 399, 442, 445
Bottom-water influx, 261–263
Branched alkanes, 326
Breccia, 91
Bridle, 267
Brine, 132–133
British Petroleum, 58
BTEX. *See* Benzene, toluene, ethyl benzene, and xylene (BTEX)
Bucket-wheel reclaimers, 389–390
Bulk composition, 228–242
Bulk fractions, 150*f*, 151, 180*f*, 181, 230
Bulk rock volume, 103–104
Bulk thermal processes, 464
  aquaconversion process, 467*f*
  CAPRI process, 470
  chemical reactions, 469
  forward combustion, 468
  hydrogen, 470
  hydrovisbreaking in situ, 465
  ISC, 468
  mobilization of heavy oil, 464
  modified in situ extraction processes, 465
  ORMAT process, 466
  in situ upgrading technique, 469
  Tervahl T process, 467
  THAI process (Toe-to-Heel Air Injection process), 470
  visbreaking, 466–467
Burma Oil Company, 58
Butane, liquid, 199–200, 232–233
Butane deasphalting, 58

C

Canadian tar sand deposits, 132
C7+ fraction of reservoir fluid, 187
CAGD. *See* Combustion-assisted gravity drainage (CAGD)
Calcite, 85, 95
Capillary number, 254–255
  effect, 312–313
CAPRI Process (Catalytic Upgrading Process In-Situ), 446, 470
Captain Field in North Sea, 363
Carbenes, 150*f*, 153–154, 180*f*, 209
Carboids, 150*f*, 153–154, 180*f*, 209
Carbon
  content, 197–198
  rejection process, 455
  residue, 210
Carbon dioxide (CO_2), 268, 284–285, 373–374
  augmented waterflooding, 304
  CO_2-miscible flooding, 289
  CO_2-sand fracturing, 318–319
  cyclic CO_2stimulation, 375–377
  flooding, 373–375
  miscible displacement, 268
  for miscible fluid displacement processes, 286–287, 289–290
Carbonate rocks, 85
Carbonated crude, solution gas–oil ratio for, 290
Carbonates, 55–56, 89–90
Carboxylic acid function, 208*t*
Casing head gasoline. *See* Natural gasoline
Catagenesis, 52–54
Catalysts, 53, 164, 224
  metal, 209
  zeolite, 225
Catalytic cracking, 21–22, 219, 224, 475
Catalytic hydrovisbreaking process, 480
Catalytic visbreaking process, 466
Caustic flooding. *See* Alkaline flooding
Cavities, 106
Cementation, 95
Centipoises (cp), 405
Centistokes, 201, 293
Centrifugation, 212
CEOR. *See* Chemically enhanced oil recovery (CEOR)
Chemical
  chert, 92–93
  flood recovery methods, 304
  sedimentation, 87
  sediments, 92–93

Chemical processes, 61, 278
alkaline flooding, 279–280
chemical flooding, 279
micellar-polymer flooding, 282–283
polymer flooding, 281–282
Chemical reactions, 469
Chemically enhanced oil recovery (CEOR), 369
Chert, 86
Choice of solvent, 464
CHOPS process. *See* Cold heavy oil production with sand (CHOPS) process
Christmas Tree, 259–260, 259*f*
Chrysene, 329–330, 330*f*
Classes of fluids, 124–131
Clastic sedimentation, 87
Clastic sediments, 91–92
Clearwater formation, 429–430
Cleveland Open-Cup method, 212–213
Cloud point, 222
Coal, 6, 93–94
Coefficients
of expansion, 200
of viscosity, 201–202
COFCAW. *See* Combination of forward combustion and waterflooding (COFCAW)
Coke, 158, 407
coke-like residue, 434
in situ upgrading, 471
Coking
delayed, 466
effect, 134
fluid coking, 393–394, 474
technologies, 481
Cold finger testing technique, 223
Cold flow, 354–355
Cold heavy oil production with sand (CHOPS) process, 190, 355–356, 365–368
Cold production, 355–357, 364–365
high recovery factor, 368
sand production, 365
surface footprint for CHOPS wells, 367
Cold-water process, 393
Combination of forward combustion and waterflooding (COFCAW), 299, 435, 442–443
Combustion
displacement process, 443, 447–448
Combustion-assisted gravity drainage (CAGD), 458–459
Complex displacement mechanisms, 354–355
Compositional models, 128
Condensate reservoirs, 260–261
Condensation front, 419–420
Condenser temperature, 217
Condensing vapor, 295, 421–422
Conditioning, 193, 230
Conglomerate, 91
Coning processes, 428
Connate water, 131–132, 181
Conradson method, 210
Contact time, 234
Contingent resources, 73–74
Control, 114
Conventional crude oil, 6–7, 42, 65
light, 79–80
Conventional light oil, 76
Conventional oil production methods, 62
Conventional petroleum, 3, 13, 28. *See also* Petroleum
asphalt, 14
cyclic microbial method, 19–20
distillation, 21–22
geologic techniques, 15
hydrocarbons, 14–15
MEOR, 18–19
metal-containing constituents, 13
microbial products and contribution to EOR, 20*t*
petroleum map, 13
petroleum underground, 14
plugging approaches, 20–21
primary recovery, 15
protopetroleum, 13–14
recovery using chemicals or detergents, 18*f*
reservoir energy, 17
reservoirs, 14
residual oil, 21
secondary recovery, 17
steam flooding, 17, 18*f*
stimulation of oil production, 21
sulfate reducers, 19

target for EOR processes, 21
tertiary recovery, 17
Conventional waterflooding, 306–307
Conversion, 21–22
refinery, 479–480
factors, 202
Corrosion, 263
cp. *See* Centipoises (cp)
Cracked oil, 446
Critical gas saturation point, 359
Cross-bedding, 98–99
Cross-polarization microscopy technique, 224
Crude oil, 148–149, 198. *See also* Heavy oil
API gravity and sulfur content of crude oils, 7*t*–10*t*
challenges associated with production, 144
changing parameters, 45*f*
composition, 145–146
elemental analyses, 185
increasing demand for, 6
initial production of, 253
primary recovery methods, 256–257
quality, 7–12, 219
reservoirs, 124
subfractionation of various fractions, 167*f*
from tight formations, 22–23
Bakken crude oil, 23–24
existence of large quantities of crude oil, 23
paraffin dispersants, 24
properties, 24
shale plays, 23
tight oil play, 24
variations in composition and properties, 184–185
Crystal growth stage, 142
Crystallization, 142
CSS process. *See* Cyclic steam stimulation (CSS) process
Cyclic carbon dioxide stimulation, 375–377
Cyclic microbial method, 19–20
Cyclic oil recovery, 469
Cyclic steam injection, 178, 297–298, 304, 410, 414, 418, 422, 447
cyclic steam stimulation, 422–423
downhole fluid-flow measurements, 425
drawback, 426
gravity override, 424–425
in heavy-oil reservoirs, 424
steam drive technology, 425
Cyclic steam stimulation (CSS) process, 316
Cycloalkanes, 326

D

Data acquisition, 186–189
Dean-Stark extraction method, 133–134
Deasphalting, 50–51, 234, 455
in situ, 486
solvent, 476–477
Decalin, 327–328, 328*f*
Dehydration, 128
Density of reservoir fluids, 198–200
Deoxygenation tower, 363
Deposit formation. *See* Instability
Deposition process, 96–97
Deposits, 102–104, 116
Detergent flooding. *See* Micellar-polymer flooding
Detergents, 303*f*
DHI. *See* Direct hydrocarbon indicators (DHI)
Diagenesis, 52–53
process, 54
sediments, 94
Diatomaceous earth. *See* Diatomite
Diatomite, 93
Differential thermal expansion, 316, 429
Dilated zones, 365
Dilution, 457
Direct hydrocarbon indicators (DHI), 50–51
Direct line-drive pattern, 272–273
Directional drilling, 65, 258*f*
Discovered petroleum-initially-in-place, 73
Dispersed gas injection, 269
Displacement efficiency, 273–274, 309–310
Dissolved gas drive. *See* Solution gas drive
Distillation, 21–22, 296–297, 455
loss, 217
method, 214
temperatures, 216–217
Dolomite ($CaCO_3$ $MgCO_3$), 85
Dolostone, 92

Downhole
drill motor concept, 63–64
fluid analyzers, 195
heating, 412–413
separation, 114
steam generator, 443
Downhole sampling. *See* Subsurface sampling
Draglines, 389–390
Drake, Colonel Edwin, 50–51
Drill bit, 86
Drilling, 15, 57, 63
Drive mechanism, 103, 117
Dropping point, 220
Dry bitumen basis, 392
Dry forward combustion. *See* Forward combustion
Dry point, 216–217
Dry sludge, 138
Dry underground combustion, 433. *See also* Fire flooding; In situ combustion (ISC) processes
Dynamic process, 285–286, 289
Dynamic wax deposition, 166

E

Ebullioscopic constant, 241
Ebullioscopic methods. *See* Boiling point elevation methods
EIA. *See* Energy Information Administration (EIA)
Electric downhole heating systems, 412–413
Electrical heating, 412–413, 460–461, 471, 473
Electrical submersible pumps (ESPs), 361
Electrostatic precipitation, 38–39
Elemental analysis of reservoir fluids, 191–197
Emulsions, 32
End-point, 216–217
Energy Information Administration (EIA), 80
Enhanced oil recovery (EOR) methods, 17–19, 42, 60–61, 104, 129, 178, 182, 268, 275, 315, 369. *See also* Primary recovery methods; Secondary recovery methods
alkaline flooding, 372–373
chemical processes, 278–283
chemical-enhanced recovery processes, 370*f*
CO_2 flooding, 373–375
cyclic carbon dioxide stimulation, 375–377
fluid phase process, 278, 283–285
fluid-flow simulations, 277–278
IOR, 275–276
MEOR process, 381–386
Micellar-polymer flooding, 379
miscible fluid displacement processes, 285–292
miscible-enhanced recovery processes, 371*f*
nitrogen flooding, 377
oil production methods, 276*f*
pilot test, 278
polymer flooding, 278–279, 377–379
process definition by US Tax Code, 303–308
project design, 277
screening criteria, 308–309
solvent-based processes, 379–381
tertiary oil recovery, 275–276
thermal process, 278, 292–303
Enhanced recovery, 256–257
Enriched gas, 269
Entrapment, 55
EOR methods. *See* Enhanced oil recovery (EOR) methods
Equation of state (EOS), 130–131
Equivalent oil saturation, 441
ESPs. *See* Electrical submersible pumps (ESPs)
Estimated ultimate recovery (EUR), 74
Euphrates valley, 387
Evaluation of reservoir fluids, 179, 181. *See also* Reservoir fluids
asphaltene separation, 230–235
bulk composition, 228–242
chemical and physical properties, 189–209
acid number, 205–209
density, 198–200
elemental analysis, 191–197
sample preparation, 191–197

sampling, 191–197
solubility, 203–205
specific gravity, 198–200
viscosity, 200–203
data acquisition and QA/QC, 186–189
fluid assay, 183–186
fluid evaluation, 181–183
fractionation, 235–237
metals content, 224–228
molecular weight, 237–242
recommended inspection data, 184*t*
thermal properties, 209–210
carbon residue, 210
heat of combustion, 211
liquefaction, 220–222
solidification, 220–222
specific heat, 210–211
volatility, 211–220
wax analysis, 222–224
wax appearance temperature, 222–224
Evaporites, 92, 133
Existent dry sludge, 138
Exploration, 317–318
Extra heavy oil, 29–30, 34, 36, 38
Extracellular lactase activity, 336
Extraction, 234

F

Faja (Orinoco belt), 413
FAST. *See* Fracture-assisted steam technology (FAST)
Fault trap, 106
Favorable mobility ratio, 254–255
FCC. *See* Fluid catalytic cracking (FCC)
Federal Energy Administration (FEA), 405
Feedstock, 13, 151
heavy feedstock, 24–25, 151–152, 184*t*
refinery, 28, 49
Feldspar minerals, 89–91
Fermentation products, 18–19, 383
Fines migration, 365
Fire flooding, 432, 468
Fire point, 212–213
Fischer-Tropsch (FT) synthesis, 458
Five-spot pattern, 271–272
Flame front control. *See* Process control
Flash point, 213
of petroleum, 212
Flint, 93
Flocculation, 156–157
Flow assurance, 178
Flow loops, 224
Flue gas desulfurization, 38–39
Fluid catalytic cracking (FCC), 461
Fluid phase process, 61, 278, 283–285
Fluid-transmitting capacity, 106
Fluid(s), 125, 264–265, 453–454
assay, 183–186
classes of, 124–131
distribution in reservoir rock, 126
evaluation, 181–183
fluid-flow simulations, 277–278
fluid-holding capacity, 106
hydrocarbon, 128
phase process, 278, 283–285
production, 309–310
Fluidity, 201. *See also* Viscosity
Fluoranthene, 329–330, 329*f*
Flute casts, 100
Foam fracturing, 314–315, 318–319
Foamy flow, 366–367
Foamy oil, 24–27, 129–130
Formation(s), 86, 98
Forward combustion, 435, 437, 468. *See also* Reverse combustion
air consumption, 440
characteristics, 438
fuel, 440
hydrocarbon products, 437
in situ combustion project, 439–440
wet combustion method, 438
Forward dry combustion process, 298–299
Fossil energy resource, 40, 41*f*
Fossil fuels, 49
resource, 70–71
Fossil water. *See* Connate water
Four-spot pattern, 271–272
4D seismic, 67
Fracking. *See* Hydraulic fracturing
Fractionating distillation, 217–218
Fractionation, 152–153, 235–237
Fracture-assisted steam technology (FAST), 430–431
Fracture gradient, 428–429
France, 58–59
Free-radical process, 455–456

Freezing point depression, 240
FT synthesis. *See* Fischer-Tropsch (FT) synthesis
Fuel
availability, 439–440
consumption, 439–440
content, 439–440

G

G as cap drive, 15, 16*f*
Garciaella petrolearia TERIG02, 338
Gas
cap, 64–65
drive, 261, 261*f*
condensate, 23–24
flood, 267
recovery methods, 304
injection, 119
lift, 62, 259, 284–285, 374
gas oil conversion plants, 479–480
Gas–oil and oil–water contacts, 127
Gas–oil ratio (GOR), 27, 187, 356
Gel permeation chromatography. *See* Size exclusion chromatography
Geochemistry, 59
Geologic techniques, 15
Geological mapping, 51, 58
Geothermal gradient, 14
GHI drill bit. *See* Grit-hot-pressed insert (GHI) drill bit
GOR. *See* Gas–oil ratio (GOR)
Graded bedding, 99
Grain supported rock, 96
Gravity
drive, 263
override, 424–425
segregation, 269, 373
separation, 416–417
Gravity-driven processes, 428
Greek fire, 5
Grit-hot-pressed insert (GHI) drill bit, 86
Gusher, 64
Gypsum, 95

H

Halite, 95
Hansen solubility parameters, 162*t*
Hard formations, 86
Heat efficiency, 112–113
Heat of combustion, 211
Heat recovery techniques, 467
Heaviness. *See* Viscosity
Heavy asphaltic fractions, 181
Heavy crude oil. *See* Heavy oil
Heavy metals, 225
Heavy oil, 3–4, 28, 49–50, 80–81, 127, 129, 177, 183, 199, 405. *See also* Crude oil
acidity rate, 32–33
API gravity and sulfur content of selected heavy oils and bitumen, 30*t*–31*t*
asphaltene separation, 230–235
and bitumen upgrading, 432
black oil, 28
bulk composition, 228–242
changing parameters, 45*f*
chemical and physical properties, 189–209
acid number, 205–209
density, 198–200
elemental analysis, 191–197
sample preparation, 191–197
sampling, 191–197
solubility, 203–205
specific gravity, 198–200
viscosity, 200–203
constituents, 461, 473
elemental analyses, 185
emulsions, 32
exploration, 57
aerial remote sensing, 59
anticlinal formation, 58–59
conventional crude oil, 65
conventional oil production methods, 62
development alternatives, 60
directional drilling, 65
downhole drill motor concept, 63–64
drilling technology, 63
EOR processes, 61
gas cap, 64–65
geology, 68
gusher, 64
mapping of gravity and magnetic anomalies, 66–67
mud pulse telemetry, 63

for oil and gas, 60
other tools, 59
proppants, 64
recovering reserves, 60–61
remote sensing of onshore areas, 67
secondary oil recovery, 62
sedimentary basins, 66
seismic acquisition and processing, 67
seismic reflection method, 63
steam flood, 61–62
tight formations, 66
extra heavy oil, 29–30
flash point, 213–214
fractionation, 235–237
high molecular weight fraction of, 4
large resources, 28–29
metals content, 224–228
molecular weight, 237–242
occurrence and distribution, 54–57
origin of, 51–54
abiogenic origin, 52
biogenic origin, 52–54
production, 353
PVT data interpretation and modeling, 188
quality of, 219–220
reservoirs, 254
resources, 109–110
role, 81–82
sampling program, 187
separation, 150*f*, 180*f*
specialized recovery and refining processes, 28
strategies for recovering, 179
thermal properties, 209–210
carbon residue, 210
heat of combustion, 211
liquefaction, 220–222
solidification, 220–222
specific heat, 210–211
volatility, 211–220
wax analysis, 222–224
wax appearance temperature, 222–224
understanding properties and behavior, 179–181
variation
in composition and properties, 184–185
of viscosity with temperature, 33*f*
viscosity, API gravity, and density, 29, 32*f*
viscous nature, 3
Heavy Oil Belt, 359, 367–368
Hematite, 95
n-Heptane, 233–234
Heteroatomic species, 40
Heteroatom(s), 224
constituents, 138*t*
Heterogeneity, 24–27
reservoirs, 107
crude oil and heavy oil reservoirs, 107
effects of, 107
heavy oil, 109–110
hydrophilic reservoirs, 108
input and output data measurements, 111
oil wet, 108
properties of oil, 110
reservoir lithology, 110
reservoir rock, 107–108
wettability, 108–109
High acid crudes, 24–27
High performance liquid chromatography (HPLC), 152–153, 179, 237
High-permeability channels, 358–359, 366
Homogenous front, 273
Horizontal injectors, 363
Horizontal production wells, 357, 399
Horizontal wells, 29–30, 178, 258–259, 300, 303–304, 316
Horsehead pumps, 62, 266, 266*f*
Hot water
drive, 417
extraction process, 392
flooding, 119, 296, 408, 416, 419
process, 391
Hot-fluid injection processes, 413
effective mobility ratio, 415
gravity separation, 416–417
hot-water drive, 417
hot-water flood, 416
huff and puff process, 414
parameters, 417*f*
seawater, 417
surface lines, 415–416
thermal recovery processes, 418
HPLC. *See* High performance liquid chromatography (HPLC)

Hubbert theory, 81–82
Huff "n" puff process. *See* Cyclic steam injection
Hunt, Thomas, 57–58
Hydrate compounds, 263
Hydraulic fracturing, 12, 264, 313–319
Hydrocarbon(s), 14–15, 376
degrading bacteria and fungi, 325
detection, 50–51
fluid, 128
Hansen solubility parameters, 162*t*
hydrocarbon-based gas injection, 268–269
hydrocarbon-producing resources, 40
resource, 40
Hydrocracking, 219–220, 458, 479, 482–483
Hydrocyclones, 114
Hydrodesulfurization, 485
Hydrogen
addition process, 455
content, 198
deficiency index, 206–209
incorporation, 480–481
Hydrogen sulfide (H_2S), 143–144
Hydrophilic reservoirs, 108
Hydroskimming refinery, 479–480
Hydrostatic pressure, 262–263
Hydrotransport, 389–390
Hydrotreating, 457
Hydrovisbreaking, 455, 466
in situ upgrading by, 465*f*

I

ICAP method, 227–228
ICAP spectrometry. *See* Inductively coupled argon plasma (ICAP) spectrometry
IGI. *See* Inert gas injection (IGI)
Imbricate bedding, 99
Immiscible carbon dioxide
displacement, 304
flooding, 285
Immiscible nonhydrocarbon gas displacement, 304
Imported Refiner Acquisition Cost, 80
Improved oil recovery (IOR) techniques, 275–276
Impurities, 453–454
carbon black, 453–454
hydrocarbons, 453–454
hydrogen sulfide, 453–454
sour gas, 453–454
In situ bacterial fermentation, 21
mechanisms, 21
In situ combustion (ISC) processes, 120, 298–300, 304, 431–432, 436, 458–459
air injection, 434
chemistry of process, 433
factors, 436
forward combustion, 435
forward combustion, 437–440
to reservoirs, 432
reverse combustion, 440–443
South Belridge work, 434
THAI process (Toe-Heel Air Injection process), 443–446
In situ deasphalting, 476
In situ processes, 220, 387, 390–391, 410–411
In situ recovery processes, 443, 460, 464, 484
In situ steam processes, 431
In situ thermal process, 426
In situ upgrading, 458, 476. *See also* Partial upgrading at surface
advantages for refinery operators, 463
approaches for heating reservoir, 460–461
bulk thermal processes, 464–470
combustion technology, 462–463
fuel, 459
future, 470–471
MEOR, 477–478
mild thermal cracking, 473
partial combustion, 474–475
solvent deasphalting, 476–477
steam distillation, 471–473
by hydrovisbreaking, 465*f*
primary recovery operations, 460
red mud process, 461
in situ conversion, 462
solvent-based processes, 463–464
upgrading of heavy oil, 460
Incompatibility, 136–140, 163

Incubation period, 384
Inductively coupled argon plasma (ICAP) spectrometry, 226–227
Induration, 97
Inert gas injection (IGI), 381, 399
  in Canada, 399
  gas coning, 399
  high recovery ratios, 399
  interfaces, 399
  phase densities, 399
  residual oil saturation value, 399
  top-down process, 399
  water coning, 399
Inferred reserves, 71
Initial boiling point, 216–217
Injected stream, 290
Injection fluids, 271, 281, 285–286, 362, 417
  acquifer water, 362
  produced water, 362
  river water, 362
  seawater, 362
Injection water, 363
  deoxygenation tower, 363
Injectivity, 112, 188, 297
Input gas types
  enriched gas, 269
  liquefied petroleum gas (LPG), 269
  wet casinghead gas, 269
Inspection assays, 186
Instability, 136–137
  properties related to, 138*t*
Interface boundary conditions, 108–109
Interfacial tension, 343
Interfacial tension reduction, 279–280
Interference, 157
Interstitial water, 126–127
Inverted seven-spot pattern, 272
IOR techniques. *See* Improved oil recovery (IOR) techniques
IP 170. *See* Abel closed-cup apparatus
Iron, 225
ISC processes. *See* In situ combustion (ISC) processes

J

Jasper, 93

K

Kerogen, 37–38, 40
Kinematic viscosity, 201–202
Knuckle joint, 65
KOH. *See* Potassium hydroxide (KOH)

L

Lateral stress, 429. *See also* fracture gradient
Layering, 98–99
Lift gas, 258–259
Light conventional crude oil, 79–80
Light crude oil, 49
Light naphtha, 23–24
Light transmittance technique, 223–224
Liquefaction, 220–222
Liquefied petroleum gas (LPG), 269
Lithification, 94–95
Lithology, 58–59, 106, 270
Lloydminster, Alberta, 367
Logan, William, 57–58
Logging-while-drilling, 359
Logging-while-drilling technology, 360
Long horizontal wells, 29–30, 178, 353, 380, 398
Long-term waterflooding, 182, 286
Low pressure gradient, 428–429

M

Macroscopic displacement efficiency, 256, 311–312
Magnetic and gravity surveys, 62–63
Maltenes, 149–150
Matrix supported rock, 96
Maximum temperature. *See* End-point
MEA. *See* Mono-ethanolamine (MEA)
Measurement of the produced fluids, 425
  cross-well seismic measurements, 425
  downhole fluid-flow measurements, 425
  fiber-optic, 425
  high-resolution imaging, 425
  nuclear spectroscopy logs, 425
  surface seismic measurements, 425
  time-lapse, cross-well electrical imaging, 425
  wire line technology, 425
Measurement-while-drilling, 63, 360
Medium formations, 86

Melting point test, 220
MEOR. *See* Microbial enhanced oil recovery (MEOR)
Metagenesis, 52–53
Metals, 155
content, 224–228
Micellar solution, 305
Micellar-polymer flooding, 282–283, 305–306, 379
micellar slug, 305
Micelles, 305
Microbes, 337
Microbial alteration of petroleum constituents, 325
alkanes, 326
aromatic hydrocarbons, 326–327
polynuclear aromatic hydrocarbons, 327–331
process parameters, 338–341
processes types, 341, 344
bacterial flooding process, 345
general principles, 341–343
mechanisms, 343–344
Microbial Huff'n' Puff, 344
residua, microbial degradation of, 331–338
Microbial degradation, 327
of alkyl tetralin derivatives, 327
of naphthalene process, 329
of petroleum, 324–325
rates, 340–341
Microbial enhanced oil recovery (MEOR), 17–19, 21, 323, 381–382, 477–478. *See also* Enhanced oil recovery (EOR) methods
advantage, 385–386
biodegradation, 323
bioremediation, 323–324
chemical characteristics of contaminants, 324
enhanced oil recovery processes, 385
hydrocarbon degrading bacteria and fungi, 325
microbial alteration of petroleum constituents, 325
alkanes, 326
aromatic hydrocarbons, 326–327
polynuclear aromatic hydrocarbons, 327–331
process parameters, 338–341
residua, microbial degradation of, 331–338
microbial degradation of petroleum, 324–325
for microbial metabolism, 385
microbial products and contribution, 383*t*
oil recovery, 384
organic acids, 383
processes types, 341, 344
bacterial flooding process, 345
general principles, 341–343
mechanisms, 343–344
microbial Huff'n' Puff, 344
types, 382*t*
Microbial flooding, 19–20
oil recovery by, 384
process, 61, 345
Microbial Huff'n' Puff process, 344
Microbial technology, 382
Microcarbon residue, 210. *See also* Thermogravimetric method
Microemulsion, 305
Microemulsion flooding. *See* Micellar-polymer flooding
Micropaleontology, 59
Microscopic displacement efficiency, 287–288, 310–313
capillary number effect, 312–313
macroscopic displacement efficiency, 311–312
Microwave technology, 449, 478
Migration, 55–56
Mild thermal cracking, 473
Milled tooth (MT) bit, 86
Mineralogical maturity, 90–91
Minerals, 90
Mine spoils, 390
Minimum miscibility pressure (MMP), 268
Miscible carbon dioxide enhanced oil recovery, 375–377
Miscible displacement. *See* Miscible fluid displacement processes

Miscible fluid displacement processes, 61, 182, 285, 304
affecting parameters, 288–289
using carbon dioxide, 286–287, 289–290
differences in density and viscosity, 291–292
using hydrocarbon gases and condensates, 290
microscopic displacement efficiency, 287–288
microscopic observations, 287
multiple-contact process, 285–286
using nitrogen and flue gases, 290–291
procedures for, 287
recovery process parameters, 288*t*
reservoir operating pressure, 289
single-contact miscible process, 285–286
solution gas–oil ratio, 290
solvent displacing oil, 286
Miscible fluid flooding, 274
Miscible process. *See* Fluid phase process
Mixing step, 392
MMP. *See* Minimum miscibility pressure (MMP)
Mobility, 311–312
control, 306
ratio, 254–255, 270–271
Modern open-hole wireline formation testers, 195, 229–230
Modified combustion, 299
Modified in situ extraction, 300–303, 300*f*, 316
Molecular diffusion, 145
Molecular dispersion, 144–145
Molecular polarity, 153–154
Molecular topology. *See* Ring linkage patterns
Molecular weight, 153–157, 237–242
Mono-ethanolamine (MEA), 144
MT bit. *See* Milled tooth (MT) bit
Mud cracks, 100
Mud pulse telemetry, 63
Mudrocks, 85, 92
Mudstone, 92
Multilateral wells, 62
Multiple-contact process. *See* Dynamic process

N

N-Solv process, 381, 477
Naphtha, 14–15
Naphthenates, 209
Naphthenes, 14–15
Naphthenic acids, 25, 206–209
alternate structural types, 208*t*
compounds classes in California crude oil, 208*t*
representative structures, 207*t*
Native asphalt. *See* Bitumen
Natural asphalt lake, 336
Natural diamond drill bit, 86
Natural energy of reservoir. *See* Primary recovery methods
Natural gas, 181, 262
Natural gas gasoline. *See* Natural gasoline
Natural gasoline, 262
Natural recovery. *See* Primary recovery methods
Natural state, 405
Net pay thickness, 429. *See also* Zone thickness limit
Net rock volume, 103–104
Net-to-gross ratio, 103–104
Nickel, 155
porphyrins, 225
Nine-spot pattern, 272
Nitrogen
content, 198
disposition, 154
flooding, 377
nitrogen-carbon dioxide flooding, 377
Nitrogen oxides (NO_x), 224
Nonaqueous phase liquid, 368
Nonasphaltene constituents, 138*t*
Nonbiogenic chert, 93
Noncondensable gases, 381, 415, 426–428, 477
Noncondensable hydrocarbons, 400
Nonhydrocarbon constituents of petroleum, 14–15
Nonindurated rock, 97
Nonmining methods, 396–397. *See also* Tar sand mining
bitumen recovery by, 397–398
heavy oil production, 398

Nonmining methods (*Continued*)
horizontal wells, 398–399
hybrid approaches, 400
IGI, 399
nonmining recovery of bitumen, 397
oil–water interface, 399
upgrading of heavy oil and bitumen, 400
Nonsorted sediment, 99
Nonthermal methods of recovery. *See also* Thermal methods of recovery
cold production, 355–356
enhanced recovery methods, 369–386
heavy oil
recovery, 354–355
resource, 353
nonmining methods, 396–400
oil production and recovery, 356
primary recovery methods, 356–360
recovery processes for heavy oil and tar sand bitumen, 355*t*
secondary recovery methods, 360–369
steam injection, 353–354
tar sand mining, 386–396
water flooding of heavy oil, 355
Number average molecular weight, 239, 241
Numerical modeling, 128
Nutrients, 382
effect, 338–339
N-Solv process, 381, 431, 477
heated solvent vapor, 431

O

Ocean drilling, 65–66
Ocean rigs, 65–66
Oil Creek, Pennsylvania, 50–51, 57
Oil exploration, 50–51
anticlinal theory, 57–58
geological mapping, 51
plate tectonics, 66, 87–89
seep identification, 57–58
seismic acquisition, 59
site specificity, 63
tools, 63
Oil field, 6
province, 14
sedimentary basin, 14
Oil fingerprinting, 123–124
Oil foam, 366
Oil in place, 70, 117–118, 129, 275
Oil industry, 25, 76, 148–149, 398
Oil mining, 386–387
bucket-wheel reclaimers, 389–390
fracture system, 388–389
production rate, 388–389
reservoir properties, 388
truck and shovel method, 389–390
Oil pricing, 77
heavy oil role, 81–82
history, 78–79
petroleum reserves and resources, 77–78
strategies, 79–81
tar sand bitumen, 81–82
Oil recovery
capillary number, 254–255
EOR methods, 275–309
fluids production, 309–310
hydraulic fracturing, 313–319
methods for, 257*f*
microscopic displacement efficiency, 310–313
capillary number effect, 312–313
macroscopic displacement efficiency, 311–312
mobility ratio, 254–255
primary recovery methods, 256–264
secondary recovery methods, 264–275
viscous fingering, 254–255
Oil sand. *See* Bituminous sand
Oil shale, 37–38, 40
shale oil, 37–38
Oil(s)
displacement tests, 283
field, 14
fraction, 233
mining, 386–387
oil–rock interactions, 181
production intensification, 315
shale, 37–38
swelling, 285
wet, 108–109
sands, 38
Oil reserves, 69–70
inferred reserves, 71

possible reserves, 71
potential reserves, 71
probable reserves, 71
proven reserves, 71
undiscovered reserve, 71
Oil trap. *See* Reservoir(s)
Oil viscosity, 21, 293, 296–297
Oil-bearing transition zone, 375
Oil-water interface, 399
Oleophilic sieve process, 394
OOIP. *See* Original oil in place (OOIP)
OPEC. *See* Organization of Petroleum Exporting Countries (OPEC)
Open-flash point, 213
Opportunity crudes, 24–27
Organic
acids, 383
fluids, 181
residues, 37
sedimentation, 87
sediments, 93–94
Organization of Petroleum Exporting Countries (OPEC), 78
Organo-metallic constituents, 225
Original oil in place (OOIP), 357, 427–428
Origin of carbon fuels, 51. *See also* Abiogenic theory; Biogenic theory
Orimulsion fuel, 38–39
Orinoco belt, 413
ORMAT process, 466
Osmometer, 238
Oxidation, 160
Oxygen, 155
content, 198
Oxygen scavenging agent, 363

P

Packer, 425
Packing, 96–97
Paraffin dispersants, 24
Paraffins, 14–15, 159
Paraffins, *iso*-paraffins, naphthenes, and aromatics (PINA), 237
Paraffins, *iso*-paraffins, olefins, naphthenes, and aromatics (PIONA), 237
Paraffins, naphthenes, and aromatics (PNA), 237
Paraffins, olefins, naphthenes, and aromatics (PONA), 237
Partial combustion, 474
laboratory experiments, 475
in situ heavy oil upgrading, 475
in situ recovery, 474–475
Partial upgrading at surface, 478–479. *See also* In situ upgrading
aspects of upgrading, 479
aquaconversion technology, 480–481
coking technologies, 481
hydrogen incorporation, 480–481
partial upgrading process, 483
refinery, 479–480
sequential operation, 484
in situ recovery processes, 484
traditional heavy feedstock processing, 482–483
visbreaking, 481
biodesulfurization, 484–486
PDC drill bit. *See* Polycrystalline diamond compact (PDC) drill bit
Pechelbronn oil field, 387
Pensky-Marten apparatus, 212–213
n-Pentane, 149–150, 233–234
Pentane-precipitated asphaltenes, 235–236
Peptized state, 232
Percent evaporated, 217
Permeability, 55–56, 311–312
Petrolenes. *See* Maltenes
Petroleum, 3, 12, 405
exploration, 57–68
flash point of, 212
genesis, 52–53
history, 4–5
American petroleum, 5
API gravity and sulfur content of crude oils, 7*t*–10*t*
coal, 6
crude oil quality, 7–12
hydraulic fracturing, 12
increasing demand for crude oil, 6
simplified differentiation, 11*t*
tar sand bitumen, 6–7
US petroleum industry, 5
use of petroleum in Middle East, 5
viscous fluid, 12

Petroleum (*Continued*)
map, 13
members of petroleum family, 22
crude oil from tight formations, 22–24
foamy oil, 24–27
high acid crudes, 24–27
opportunity crudes, 24–27
occurrence and distribution, 54
biogenic theory, 56
drilling, 57
petroleum deposits, 55
reservoir, 55–56
trapping mechanism, 56
origin of, 51–54
abiogenic origin, 52
biogenic origin, 52–54
refining, 479
reservoirs, 258–259
Phase behavior tests, 283
Physical composition. *See* Bulk composition
Physical property data, 190–191
PINA. *See* Paraffins, *iso*-paraffins, naphthenes, and aromatics (PINA)
PIONA. *See* Paraffins, *iso*-paraffins, olefins, naphthenes, and aromatics (PIONA)
Piping tubes, 365
Piston-like displacement, 406
Pitch, 36–37
Plate tectonics, 66, 87–89
PNA. *See* Paraffins, naphthenes, and aromatics (PNA)
Poise, 201. *See also* Viscosity
Polyacrylamides, 281–282
Polycrystalline diamond compact (PDC) drill bit, 86
Polycrystalline diamond drill bit, 63–64
Polymer augmented waterflooding. *See* Polymer flooding
Polymer flooding, 119, 278–279, 281–282, 306–307, 377–379
Polymer solutions, 307
Polymers
biologically produced, 281–282
polymer technology, 369
synthetically produced, 306–307
Polynuclear aromatic hydrocarbons, 327–328, 331
bacterial degradation of pyrene, 330–331
chemical properties, 328
degradation, 329
fluoranthene, 330
four-ring, 329–330
microbial degradation
of naphthalene process, 329
petroleum-related constituents, 331
size and angularity, 329
Polysaccharides, 17, 281–282, 306–307, 336
PONA. *See* Paraffins, olefins, naphthenes, and aromatics (PONA)
Poorly-sorted sediment, 101–102, 102*f*
Pore scale, 182, 286, 368–369
Porosity, 96–97
Porous medium. *See* Reservoir rocks
Possible reserves, 71
Potassium hydroxide (KOH), 280
Potential dry sludge, 138
Potential reserves, 71
Potentially recoverable resources, 73
Pour point, 42–43, 44*f*, 220, 221*f*, 222
PPT. *See* Pressure pulse technology (PPT)
Precipitation, 234
Preheating, 442
Pressure, 143, 147
coring system, 63
maintenance, 253
Pressure pulse technology (PPT), 368–369
Pressure–volume–temperature (PVT), 130, 179–181
data interpretation and modeling for heavy oil, 188
Pressure-driven instabilities, 428
Pressure-driven thermal processes, 429
Primary conversion step, 480
Primary oil production, 260
Primary oil recovery, 259–260
Primary recovery. *See* Primary recovery methods
Primary recovery methods, 15, 60–61, 104, 256–259, 356. *See also* Enhanced oil recovery (EOR) methods; Secondary recovery methods
artificial lift methods, 258–259
associated natural gas, 262
Christmas Tree, 259–260, 259*f*

cold production, 357
oil reservoir, 358–359
conventional oil production, 264
corrosion, 263
directional drilling, 258*f*
drilling, 360
gas cap drive, 261, 261*f*
gravity drive, 263
heavy oil
production, 359
reservoirs, 358
hydrate compounds, 263
initial reservoir pressure, 257–258
macroscopic displacement efficiency, 256
parameters contributing site specificity, 258*f*
petroleum reservoirs, 258–259
recovery efficiency, 256
recovery factors, 357
retrograde condensate pools, 262
solution gas drive, 261, 261*f*
water drive, 262, 262*f*
Primary upgrading, 463, 477. *See also* Visbreaking
Probable reserves, 71
Process control, 439–440
Process efficiency, 441–442
Process parameters, 338
microbial degradation rates, 340–341
nutrients effect, 338–339
temperature effect, 339–340
Produced water, 362
Production method, 26, 183
enhanced, 110, 454
primary, 368
secondary, 454
tertiary, 110
Propane asphalt, 232–233
Propane deasphalting, 232–233
Proppants, 64, 260, 315
Protopetroleum, 13–14, 53
Proved plus probable plus possible reserves (3P), 74–75
Proved plus probable reserves (2P), 74–75
Proved reserves (1P), 74–75
Proven reserves, 71
Province. *See* Sedimentary basin
Pulsing, 368, 392
Pump jack, 266–267
Pumps
beam, 62
electrical submersible, 62, 266, 361
horsehead, 62, 266, 266*f*
nodding donkey, 266
pump jack, 266
rotary progressive cavity pumps, 366
sucker rod, 266
PVT. *See* Pressure–volume–temperature (PVT)
Pyrene, 329–331, 330*f*
Pyrite (FeS_2), 433
Pyrolysis, 461

Q

Quality assurance and quality control (QA/QC method), 182–183, 186–189
Quartz, 89–91
Quartz arenite, 91

R

$R(CH_2)_nCOOH$. *See* Naphthenic acids
Radio frequency (RF), 460–461
Raindrop marks, 100
Ramsbottom method, 210
Range of uncertainty, 74–75
Raoult's law, 240
Ratio of Injectors, 273*t*
Ratio of solvent to feedstock, 234
RCOO–. *See* Naphthenates
Reactions, 157–161
other reactions, 160–161
oxidation, 160
thermal reactions, 158–159
Reconnaissance techniques, 62–63
Recovery, 217
efficiency, 256
factor, 70
Recovering reserves, 60–61
Recrystallization, 95
Red beds, 100–101
Red mud process, 461
"Reduced footprint" developments, 398
Reduced pressure, 214
Refinery, 479–480

Refinery process. *See* Distillation
Reflection seismic surveys, 63
Remediation strategies, 164
  asphaltene deposition, 166–168
  wax deposition, 164–166
Reserves, 68
  best estimate, 75
  black oil, 69
  contingent resources, 73–74
  conventional crude oil, 75
  discovered petroleum-initially-in-place, 73
  growth, 70
  inferred reserves, 71
  oil reserves, 69–70
  potentially recoverable resources, 73
  probable reserves, 71
  proven reserves, 71
  quantities, 68–69
  range of uncertainty, 74–75
  real numbers, 75
    conventional light oil, 76
    regional distribution of recoverable heavy oil and natural bitumen, 77*t*
    total resources of heavy oil in known accumulations, 76
    USGS, 76
  reclassification, 71–72
  recovery factor, 70
  resource, 70–71
    estimation representation, 72*f*
  Society for Petroleum Engineers, 72–73
  total petroleum-initially-in-place, 73
  undiscovered petroleum-initially-in-place, 74
Reservoir fluids, 123–124, 130. *See also* Evaluation of reservoir fluids
  characterization, 183, 228
  classes of fluids, 124–131
    zones in reservoir, 126*f*
  data, 123–124
  properties, 123, 182–183
  PVT
    characterization, 182–183
    properties, 182
  solids deposition during recovery, 135–168
    asphaltene deposition, 148–164
    remediation strategies, 164–168
    wax deposition, 140–148
  water, 131–135
Reservoir productivity, 117
  production, 117–118
    techniques, 118–120
Reservoir(s), 14, 55–56, 102–104, 418
  brines, 342
  deposition, 104–105
  energy, 17
  evaluation, 111–116
    other considerations, 114–115
    reservoir thickness, 112–113
    well spacing, 113–114
  heterogeneity, 107
    crude oil and heavy oil reservoirs, 107
    effects of, 107
    heavy oil, 109–110
    hydrophilic reservoirs, 108
    input and output data measurements, 111
    oil wet, 108
    properties of oil, 110
    reservoir lithology, 110
    reservoir rock, 107–108
    wettability, 108–109
  matrix, 445
  properties, 388
  rock, 181
  rocks, 105
  structural types, 106
  temperature, 221, 221*f*
  thickness, 112–113
  uniformities, 273
Reservoir rocks, 105
Residua
  asphaltene constituents, 334–337
  asphaltenic fraction, 335
  bacterial consortium, 335–336
  bacterial metabolites, 336
  biodegradation, 334
  Garciaella petrolearia TERIG02, 338
  limiting factors, 334
  microbes, 337
  microbial degradation, 331–332
  natural asphalt lake, 336
  polycyclic aromatics and hydrocarbons, 337

properties, 332, 333t
resin constituents, 332–334
Residual material, 277–278
Residual oil, 21, 275
saturation, 255
zone, 375
Residue, 217
Residuum, 331–332
Resin, 149–150
constituents, 149, 232–233, 332–334
fractions, 152
oxidation, 160
Resource, 70–71
Retrograde condensate pools, 262
Reverse combustion, 298–299, 435, 440–441. *See also* Forward combustion
dry combustion, 443
primary recovery operations, 443
process efficiency, 441–442
to reservoirs, 441
variation of combustion process, 442–443
Reverse graded bedding, 99
RF. *See* Radio frequency (RF)
Rhythmic layering, 98
Ring linkage patterns, 328
Ripple marks, 99–100
River water, 362
Rock asphalt, 34
Rock Color, 100–101
Rock formation, 15
acidizing, 264
fracturing, 264
improving formation, 264
Roller cone bit, 86
Rounding, 102

S

SAGD process. *See* Steam-assisted-gravity-drainage (SAGD) process
Salt dome trap, 106
Sample preparation of reservoir fluids, 191–197
Sampling of reservoir fluids, 191–197
Sand filters, 363
Sand production, 365
Sand-reduction process, 393–394
Sandstone, 14
Sandstone formations, 312
Sandstone reservoirs, 36–37
Sandstone rocks, 85, 319
Sapropel, 38
Saturates, aromatics, resins, and asphaltenes (SARA) analysis, 146, 237, 470
separation, 179
Satellite and high-altitude imagery, 62–63
Saybolt furol viscosity, 202
Saybolt universal viscosity, 202
Sea water, 362
Secondary oil recovery methods, 62, 266–267
Secondary recovery. *See* Secondary recovery methods
Secondary recovery methods, 17, 60–61, 104, 256–257, 264–265, 360. *See also* Enhanced oil recovery (EOR) methods; Primary recovery methods
carbon dioxide miscible displacement, 268
cold production, 364–368
direct line-drive pattern, 272–273
dispersed gas injection, 269
displacement efficiency, 273–274
five-spot pattern, 271–272
four-spot pattern, 271–272
gas injection, 364
horsehead pump, 266, 266f
hydrocarbon-based gas injection, 268–269
inverted seven-spot pattern, 272
miscible fluid flooding, 274
MMP, 268
mobility ratio, 270–271
nine-spot pattern, 272
PPT, 368–369
ratio of injectors, 273t
reservoir uniformities, 273
secondary oil recovery methods, 266–267
seven-spot pattern, 272
staggered line-drive pattern, 272–273
sweep efficiency, 267
vaporization, 269
WAG process, 274–275
water, 265
waterflooding, 267, 270, 274, 361–364
well patterns, 271f

Secondary wells, 62
Sediment formation. *See* Instability
Sediment(s), 85, 87, 196
  bedding, 98
  characteristics of, 101*t*
  classification, 101–102
  diagenesis, 94
  formation, 87
  geologic timescale, 88*t*
  layering and bedding, 98–99
  lithification, 94–95
  markers, 99–101
  maturity, 102
  mineralogy, 89–91
  occurrence, 87–89
  packing, 96–97
  sedimentary facies, 99, 100*t*
  stratification, 98
  structure, 97
  syncline-anticline petroleum-natural gas reservoir, 89*f*
  texture, 95–96
  transportation, 97–98
Sedimentary basin, 14, 66
Sedimentary facies, 99, 100*t*
Sedimentary rocks, 85, 131–132
  biochemical and organic sediments, 93–94
  chemical sediments and, 92–93
  clastic sediments and, 91–92
Sedimentation, 87
Sedimentology, 85–86
Seismic
  reflection method, 63
  acquisition method, 67
  refraction method, 58–59
  surveys, 59–60
Selective catalytic reduction, 38–39
Selective plugging, 343, 385
Separation class, 230
Separation processes, 455
Seven-spot pattern, 272
Shale, 92
  oil, 37–38
  play, 22–23
  shale streaks, 316
  shale zones, 62
Shale thermal fracturing, 429
Shear dispersion, 144–145
Shear rate, 32
Sheared zones, 365
Shell, 93, 363, 442
Short displacement mechanism, 445
Short path distillation, 216
Silliman, Benjamin, 5
Siltstone, 14, 92
Simulated distillation, 218–219
Sinai Peninsula, 387
Single-component solid model, 149
Single-contact miscible process, 285–286
Size exclusion chromatography, 242
Sludge formation. *See* Instability
Slurry fracture injection, 366–367
Soak period, 376, 384
Society for Petroleum Engineers (SPE), 72–73
Society of Petroleum Evaluation Engineers (SPEE), 72–73
Sodium chloride (NaCl), 132–133
Soft formations, 86
Softening point test, 220
Sole marks, 100
Solidification, 220–222
Solids deposition during recovery, 135
  asphaltene constituents, 136*f*
  asphaltene deposition, 148–164
  dry sludge, 138
  incompatibility, 136–140
  instability, 136–137
  phase relationships during feedstock cracking, 136*f*
  phase separation, 136
  remediation strategies, 164–168
  stability, 137–138
  thermal oxidative stability, 137–138
  thermal stability, 137–138
  thermodynamic equilibrium, 135–136
  wax deposition, 140–148
Solubility of reservoir fluids, 203–205
Solubility parameter, 161–164
  of asphaltene fractions, 204, 205*f*
  of petroleum fractions, 204*f*
Soluble oil, 305
Solution gas drive, 15, 16*f*, 261, 261*f*, 359, 376
  reservoir, 358–359

Solvent
deasphalting, 464, 476–477
displacing oil, 286
extraction process, 394
solvent-based heavy oil recovery process, 476
solvent-based processes, 379–381, 463–464
solvent-heavy oil system, 476
Sour gas, 453–454
South Belridge project, 432, 434
Specific gravity of reservoir fluids, 198–200
Specific heat, 210–211
SPEE. *See* Society of Petroleum Evaluation Engineers (SPEE)
Spherical agglomeration process, 394
Spindletop, 57–58
Spinning band method, 215
Spontaneous emulsification, 279–280
State of combustion, 432
Stability, 137–138
Staggered line-drive pattern, 272–273
Standard conditions, 426–427
Standard test methods, 222
Steam chamber, 427–428
Steam cracking, 461
Steam distillation, 211–212, 301–302, 471–473
Steam drive injection, 119–120, 294–297, 303, 308, 353–354, 421–422, 426–427. *See also* Cyclic steam injection
Steam flooding, 17, 18*f*, 303*f*, 424
Steam injection. *See* Steam drive injection
Steam soak process, 297
Steam stimulation processes, 419
Steam-and combustion-drive processes, 408
Steam-assisted-gravity-drainage (SAGD) process, 178, 316–317, 354–355, 410, 427, 458–459
benefits, 427
cost of heat, 429
FAST-SAGD operation, 430
gravity-driven processes, 428
heat losses, 429
hybrid variants for thermal recovery processes, 431*t*
steam chamber, 427–428
VAPEX process, 431
Steam-based processes, 418
condensation front, 420
cyclic steam injection, 422–426
displacement mechanisms, 419
hot fluid, 421
SAGD technology, 427–431
steam drive, 426–427
steam drive injection, 421–422
steam stimulation processes, 419
steam-stripping process, 419–420
tar-sand formation, 421
Steam-based thermal-recovery processes, 413
Steam-distilled material
aromatics, 420
naphthenic constituents, 420
Steam-saturated zone. *See* Steam chamber
Steam-stripping process, 419
Stoke, 201
Stratification, 98
Stratigraphic trap, 106
Stratum, 98
Structural basin, 89
Structural traps, 55
Subsurface sampling, 194–195, 197, 229–230
Sucker rod, 258–259
pump, 266
Sulfate reducers, 19
Sulfate-reducing bacteria, 327
Sulfur, 155
content, 198
Sulfur oxides (SO_x), 224
Sulfuric acid method, 236
Suncor Energy, 389–390
Suncor process, 392
Surface
mining, 389
production facility, 128
sampling, 194–195, 229–230
Surfactant flooding, 304–305
Surfactant-polymer flooding. *See* Microemulsion flooding
Sweep efficiency, 267, 312
Swollen micelles, 305
Syncline-anticline petroleum-natural gas reservoir, 89*f*

Synclines, 89
Syncrude Canada Limited, 389
Synthetic crude oil, 42

T

Tailored pulse fracturing, 318–319
TAN. *See* Total acid number (TAN)
Tar sand, 33–34, 37
  direct heating, 394–395
  feed, 392
  formation, 421
Tar sand bitumen, 6–7, 33–36, 42–43, 81–82, 177
  asphaltene separation, 230–235
  bituminous rock and sand, 37
  bulk composition, 228–242
  chemical and physical properties, 189–209
    acid number, 205–209
    density, 198–200
    elemental analysis, 191–197
    sample preparation, 191–197
    sampling, 191–197
    solubility, 203–205
    specific gravity, 198–200
    viscosity, 200–203
  degree of porosity, 34
  deposits, 34
  elemental analyses, 185
  expression, 36–37
  extra heavy oil, 36, 38
  flash point, 213–214
  fractionation, 235–237
  metals content, 224–228
  molecular weight, 237–242
  native asphalt, 34
  oil shale, 37–38
  Orimulsion fuel, 38–39
  properties and composition, 38
  sapropel, 38
  specific gravity, API gravity, and viscosity of bitumen samples, 36*t*
  strategies for recovering, 179
  thermal properties, 209–210
    carbon residue, 210
    heat of combustion, 211
    liquefaction, 220–222
    solidification, 220–222
    specific heat, 210–211
    volatility, 211–220
    wax analysis, 222–224
    wax appearance temperature, 222–224
  understanding properties and behavior, 179–181
  variations in composition and properties, 184–185
Tar sand mining, 386. *See also* Nonmining methods
  bitumen mining recovery, 393
  bitumen recovery, 391–393
  cold-water process, 393
  direct heating of tar sand, 394–395
  excavation of shafts and tunnels, 396
  improved mining, 395
  mining operations, 389
    oil mining method of recovery, 390–391
    Suncor Energy, 389–390
    surface mining, 389
    synthetic crude oil, 391
  oil mining, 386–387
  recovery efficiency, 395–396
  reservoir properties, 388
  spherical agglomeration process, 394
TBP distillation method. *See* True boiling point (TBP) distillation method
TCI bit. *See* Tungsten carbide insert (TCI) bit
Temperature, 146–147
  difference, 239
  effect, 339–340
Tensile stress, 429
TEOR methods. *See* Thermal-enhanced oil recovery (TEOR) methods
Terminology, 3
  of convenience, 228
Tertiary oil recovery, 275–276
Tertiary recovery. *See* Enhanced oil recovery (EOR) methods
Tervahl H process, 467
Tervahl T process, 467
Tetralin, 327–328, 328*f*
Texture, 95–96
THAI process. *See* Toe-to-heel air injection (THAI) process
Thermal cracking processes, 465–466
Thermal decomposition, 410

Thermal drives, 408, 410–411, 415
Thermal expansion, 296–297
of oil, 301–302, 416–417
Thermal methods of recovery. *See also* Nonthermal methods of recovery
comparison of recovery methods, 446–449
geomechanics, 410
heavy oil, 405
methods for oil recovery, 409*f*
oil saturation, 406
production method, 409–410
in situ combustion processes, 431–446
in situ processes, 410–411
steam-based processes, 418–431
thermal decomposition, 410
thermal methods, 407
thermal recovery processes, 408
thermal stimulation, 411–418
volumetric sweep efficiency, 406–407
Thermal oxidative stability, 137–138
Thermal process, 61, 278, 292
cyclic steam injection, 297–298
large heat losses, 293–294
modified in situ extraction, 300–303, 300*f*
poor injectivity, 294
in situ combustion, 298–300
steam drive injection, 294–297
technical problems, 293
thermal EOR processes, 293
viscosity, 292–293
Thermal reactions, 158–159
Thermal recovery methods, 302–304, 408
Thermal stability, 137–138
Thermal stimulation, 411–412
downhole heating, 412–413
hot-fluid injection processes, 413–418
wellbore heating, 412
Thermal-enhanced oil recovery (TEOR) methods, 353
Thermally-enhanced oil recovery methods, 308
Thermogravimetric method, 210
Thermophilic aerobic bacteria, 326–327
Thief zone, 428
Tight light oil, 23–24
Tight oil, 23–24
Tight reservoirs, 314
Tight shale oil, 23–24
Toe-to-heel air injection (THAI) process, 443–444, 444*f*, 461–462
advantages, 446
benefit, 445–446
CAPRI process, 446
configuration, 444*f*
in reservoirs, 445
vertical air injection well, 444
Toluene, 134, 143, 239
Top-down process, 399
Topping refinery, 479–480
Total acid number (TAN), 32–33, 205–206, 209
Total petroleum-initially-in-place, 73
Total recovery, 217
Trapping mechanism, 56
Travertine, 92
Trinidad Asphalt Lake, 5
Trinidad Pitch Lake. *See* Trinidad Asphalt Lake
Truck and shovel method, 389–390
True boiling point (TBP) distillation method, 214
Tungsten carbide insert (TCI) bit, 86
Tubing, 267, 294
Turn the corner, 398
Tweaking, 177

U

Underground mining, 390
Undiscovered petroleum-initially-in-place, 73–74
Undiscovered reserves, 71
Unfavorable mobility ratio, 254–255
United States Federal Energy Administration, 405
United States Geological Survey (USGS), 76
Unlimited dilution, 239
Upgrading during recovery. *See also* In situ upgrading
heavy crude oil and bitumen, 453
oil recovery, 454–455
partial upgrading at surface, 478–486
reservoir rock, 454
technologies for refinery, 455
thermal methods, 456
thermal processes, 457

Upgrading in situ. *See* In situ upgrading
US Tax Code, process definition by, 303–308

V

Vacuum gas oil, 332
Vacuum pot still method, 215
Vacuum residuum, 332
Vanadium porphyrins, 225
Vapor phase osmometry (VPO), 238
Vapor pressure, 214
Vapor pressure osmometry (VPO), 237–238
Vapor-assisted petroleum extraction (VAPEX), 189–190, 354–355, 380, 458–459, 476
  dilution process, 381
  with SAGD, 190, 380
Vaporization, 269
Varves, 98
Venango Sands, Pennsylvania, 58–59
Venezuela, 6
  Heavy Oil Belt (FAJA), 359
  Orinoco belt, 38, 413
  Venzuelan heavy oil deposits, 460–461
Visbreaking, 407, 410, 466–467, 481. *See also* Hydrovisbreaking
Viscometry technique, 223
Viscosity, 178. *See also* Saybolt furol viscosity; Saybolt universal viscosity
  capillary types, 201
  coefficient of viscosity, 201–202
  instruments used, 201–202
  effect of temperature on, 202
  term *extremely viscous*, 43–44
  of reservoir fluids, 200–203
Viscosity breaking. *See* Visbreaking
Viscosity scale, 39–40
Viscosity reduction. *See* Visbreaking
Viscous crude oil. *See* Heavy oil
Viscous fingering, 254–255
Volatility, 211–220
Volatilize, 225
Volume fraction of solvent, 162
Volumetric sweep efficiency, 406–407
VPO. *See* Vapor phase osmometry (VPO); Vapor pressure osmometry (VPO)

W

Wacke sandstone, 91
WAG process. *See* Waterflood-alternating-gas (WAG) injection process
Walking beam, 258–259, 267
WAT. *See* Wax appearance temperature (WAT)
Water, 131–135, 196, 265
  drive system, 15, 16*f*, 262, 262*f*
  injection pumps, 363
Waterflooding, 267, 270, 274, 361
  augmented
    carbon dioxide, 364
    polymer-augmented, 363–364
  deoxygenation tower, 363
  location of injectors and producers, 270
  long-term, 182, 286
  mobility of oil, 270
  patterns
    direct line-drive pattern, 272–273
    five-spot pattern, 271–272
    four-spot pattern, 271–272
    inverted nine-spot pattern, 272
    inverted seven-spot pattern, 272
    nine-spot pattern, 272
    seven-spot pattern, 272
    staggered line-drive pattern, 272–273
  produced water, 362
  reservoir uniformities, 273
  steam injection, 363–364
Waterflood-alternating-gas (WAG) injection process, 274–275
Water–rock interactions, 132
Wax analysis, 222–224
Wax appearance temperature (WAT), 141–142, 147, 165, 222–224
Wax deposition, 140. *See also* Asphaltene deposition
  aging, 140–141
  challenges associated with production crude oil, 144
  crystal growth stage, 142
  crystallization, 142
  factors leading to, 144–148
    crude oil composition, 145–146
    other factors, 147–148
    pressure, 147

temperature, 146–147
pressure, 143
WAT, 141–142
Weathering, 91
Well conditioning. *See* Conditioning
Well corrosion, 25
Well indurated rocks, 97
Well maintenance, 196–197
Well patterns, 113, 271*f*
Well pressure, declining, 257–258
Well spacing, 113–114
sand continuity, 113
Well-sorted sediment, 101–102, 102*f*
Well, workover, 263, 398
Wellbore heating, 412
Wells, multilateral, 62, 110, 360
Wells, secondary, 62
West Texas Intermediate (WTI), 79
Wet and partially quenched combustion. *See* Combination of forward combustion and waterflooding (COFCAW)
Wet casinghead gas, 269
Wet combustion, 438, 448
addition of water, 438–439
benefits, 439
cracking process, 439
dry combustion *vs.*, 443, 448
gravity-dominated operations, 443
in low-permeability reservoirs, 439
in thin reservoirs, 438
at different water/air ratios, 439
Wet diluted basis, 392
Wettability
alteration, 342–343
interface boundary conditions, 108–109
of reservoir rocks, 310, 343–344
Whipstock, 65
Williams, James Miller, 5
World Petroleum Council (WPC), 72–73
Wormholes, 27, 358–359, 365–367
WPC. *See* World Petroleum Council (WPC)
WTI. *See* West Texas Intermediate (WTI)

X

X-ray fluorescence (XRF) spectrometry, 226

Z

Zeolite catalysts, 225
Zone thickness limit, 429. *See also* Net pay thickness
Zones
bottom water zone, 399, 442
burned zone of COFCAW, 435
combustion zone, 433–434, 440–441, 475
dilated zones, 365
residual oil zone (ROZ), 375
shale zones, 62
sheared zones, 365
steam chamber zone, 428–429
steam-saturated zone, 427–428
thermal cracking zone, 468–469

FSC
www.fsc.org
MIX
Papier | Fördert gute Waldnutzung
FSC® C083411

Kolibri 360 GmbH
Ferdinand-Jühlke-Straße 7
99095 Erfurt, Deutschland
produktsicherheit@kolibri360.de